Nonhuman Primates
and Medical Research

Contributors

GEOFFREY H. BOURNE

G. M. CHERKOVICH

DONALD N. FARRER

AMBHAN D. FELSENFELD

OSCAR FELSENFELD

PEDRO GALINDO

ROBERT C. GOOD

G. A. GRESHAM

J. H. GROENEWALD

S. S. KALTER

MICHALE KEELING

ROBERT E. KUNTZ

B. A. LAPIN

ROBERT C. MACDONELL, JR.

GILBERT W. MEIER

RICHARD METZGAR

JOSEPH H. PATTERSON

LAWRENCE R. PINNEO

KEITH REEMTSMA

DUANE M. RUMBAUGH

HIDEO SAKATA

H. F. SEIGLER

CLARKE STOUT

J. J. W. VAN ZYL

S. L. WASHBURN

FRANCIS A. YOUNG

MARTIN D. YOUNG

GERALD T. ZWIREN

NONHUMAN PRIMATES
AND MEDICAL RESEARCH

Edited by Geoffrey H. Bourne

Yerkes Regional Primate Research Center
Emory University
Atlanta, Georgia

ACADEMIC PRESS   New York and London   1973

ACADEMIC PRESS, INC.
111 Fifth Avenue, New York, New York 10003

United Kingdom Edition published by
ACADEMIC PRESS, INC. (LONDON) LTD.
24/28 Oval Road, London NW1

LIBRARY OF CONGRESS CATALOG CARD NUMBER: 72-9421

PRINTED IN THE UNITED STATES OF AMERICA

Contents

List of Contributors

Numbers in parentheses indicate the pages on which the authors' contributions begin.

GEOFFREY H. BOURNE (487), Yerkes Regional Primate Research Center, Emory University, Atlanta, Georgia

G. M. CHERKOVICH (307), Institute of Experimental Pathology and Therapy of the U.S.S.R. Academy of Medical Science, Sukhumi, Abkhasia, U.S.S.R.

DONALD N. FARRER (407), Radiology Division, School of Aerospace Medicine, Brookes Air Force Base, Texas

AMBHAN D. FELSENFELD (25), Department of Virology, Tulane University, Delta Primate Research Center, Covington, Louisiana

OSCAR FELSENFELD (281), Tulane University, Delta Primate Research Center, Covington, Louisiana

PEDRO GALINDO (1), Gorgas Memorial Laboratory, Panama, Republic of Panama

ROBERT C. GOOD (39), Hazleton Laboratories, Vienna, Virginia

G. A. GRESHAM (225), Department of Morbid Anatomy, John Bonnet Clinical Laboratories, Addenbrookes Hospital, Cambridge, England

J. H. GROENEWALD (269), Department of Surgery, Faculty of Medicine, University of Stellenbosch and Karl Bremer Hospital, Bellville, South Africa

S. S. KALTER (61), Division of Microbiology and Infectious Diseases, Southwestern Foundation for Research and Education, San Antonio, Texas

MICHALE KEELING (257), Yerkes Regional Primate Research Center, Emory University, Atlanta, Georgia

ROBERT E. KUNTZ (167), Division of Infectious Diseases, Southwestern Foundation for Research and Education, San Antonio, Texas

B. A. LAPIN (213, 307), Institute of Experimental Pathology, and Therapy of U.S.S.R., Academy of Medical Science, Sukhumi, Abkhasia, U.S.S.R.

ROBERT C. MacDONELL, JR. (257), Department of Pediatrics, Emory University School of Medicine, The Henrietta Egleston Hospital for Children, Atlanta, Georgia

GILBERT W. MEIER (431), Department of Psychology, George Peabody College, Nashville, Tennessee

RICHARD METZGAR (257), Yerkes Regional Primate Research Center, Emory University, Atlanta, Georgia, and Department of Microbiology, School of Medicine, Duke University, Durham, North Carolina

JOSEPH H. PATTERSON (257), Department of Pediatrics, Emory University School of Medicine, The Henrietta Egleston Hospital for Children, Atlanta, Georgia

LAWRENCE R. PINNEO (329), Neurophysiology Program, Stanford Research Institute, Menlo Park, California

KEITH REEMTSMA (203), Columbia-Presbyterian Medical Center, New York, New York

DUANE M. RUMBAUGH* (415), Department of Psychology, Georgia State University, and Yerkes Regional Primate Research Center, Atlanta, Georgia

HIDEO SAKATA† (381), Department of Neurophysiology, Tokyo Metropolitan Institute for Neurosciences, Tokyo, Japan

H. F. SEIGLER (257), Yerkes Regional Primate Research Center, Emory

* Present address: Department of Psychology, Georgia State University, Atlanta, Georgia.
† Present address: Department of Physiology, The Johns Hopkins University School of Medicine, Baltimore, Maryland.

University, Atlanta, Georgia, and Department of Surgery, School of Medicine, Duke University, Durham, North Carolina

CLARKE STOUT (249), Department of Pathology and Medicine, University of Oklahoma Medical Center, Oklahoma City, Oklahoma

J. J. W. VAN ZYL (269), Department of Surgery, Faculty of Medicine, University of Stellenbosch and Karl Bremer Hospital, Bellville, South Africa

S. L. WASHBURN (467), Department of Anthropology, University of California, Berkeley, California

FRANCIS A. YOUNG (353), Primate Research Center, Washington State University, Pullman, Washington

MARTIN D. YOUNG (17), Gorgas Memorial Laboratory, Panama, Republic of Panama

GERALD T. ZWIREN (257), Department of Pediatrics, Emory University School of Medicine, The Henrietta Egleston Hospital for Children, Atlanta, Georgia

Preface

This book merely covers a fraction of the contributions made by non-human primates to biomedical research. Each issue of *Current Primate References,* published weekly by the Primate Information Center of the University of Washington Primate Research Center, lists seventy or eighty references to nonhuman primate research done throughout the world; an impressive total of three to four thousand references a year, and *Current Primate References* does not claim a hundred percent coverage. These references cover an immense range of subjects—viruses, cancer, cardiovascular physiology and diseases, parasitology, microbiology, immunology, organ and tissue grafts, endocrinology, neurology and neurophysiology, radiation, behavior—to name just a few.

The part being played by nonhuman primates in biomedical research has thus become massive, and there is no doubt that this impetus stems from the establishment of the Primate Research Center program of the National Institutes of Health. This program has not only stimulated non-human primate research in this country but has spurred interest in the use of these animals all over the world, including the Soviet Union. The Soviets have had a Primate Center at Sukhumi on the Black Sea since 1928 when Dr. Yerkes was engaged in building up his collection of chimpanzees and two years before he established the Primate Laboratories at Orange Park. More recently, however, the growing interest in and appreciation of the importance of nonhuman primates in biomedical research have led the Soviets to construct additional Centers.

The United States leads the rest of the world in publications on non-human primate research. A series of twenty-two issues of recent *Current Primate References* lists a total of 1755 papers of which 1067 or 60.6%

were published by American authors. Over a twelve-month period, the Yerkes Center alone produced nearly one hundred publications, including twelve books; this is about two publications a week. Over the period covered by the twenty-two issues mentioned above, the Primate Centers alone produced over 6% of the world's publications on nonhuman primates.

The buildup of biomedical knowledge from research with nonhuman primates has already made important practical contributions to human health and welfare, and its promise for the future is impressive. This book attempts to bring to the biomedical research community a portion of that research. It helps to identify some of the areas in which research with nonhuman primates has made or is making an important contribution. Among the subjects covered is the use of nonhuman primates in the study of infectious disease: bacterial, viral, and parasitological; in degenerative diseases, especially of the cardiovascular system and in cancer; in neurology, neurological disease, and sensory physiology; and their direct use in the therapy of hepatic coma. The significance of research with monkeys and apes in the area of psychology with special attention to mental retardation is also among the subjects treated in this book. What primate research has contributed to our knowledge of human evolution is described by Professor S. L. Washburn in a very interesting chapter. I have described some of the contributions to nonhuman primate research made by the NIH Primate Center program.

I am indebted to all the distinguished contributors for their hard work and patience, and hope the end product will be ample reward for their labors. As always, the staff of Academic Press has done a fine job, and I express my grateful thanks to them.

Geoffrey H. Bourne

CHAPTER 1

Monkeys and Yellow Fever

PEDRO GALINDO

I. The Rhesus Monkey and Early History of Yellow Fever Research

Yellow fever occurs in Africa and the Western Hemisphere tropics. It has also caused great epidemics in the temperate zone of the New World. It was one of the world's great plagues for several centuries.

It is now known that there are two different types of this disease. The virus causing the disease is the same but the hosts and insect vectors usually are different. Urban yellow fever occurs in city dwellers and is transmitted by the domesticated mosquito, *Aedes aegypti.* Jungle yellow fever has monkeys as its usual important hosts and is transmitted by jungle-breeding, usually arboreal, mosquitoes. It has been only recently, after the great epidemics in the cities abated, that the emerging picture of jungle yellow fever began to take shape. However, monkeys played an important part in the early development of knowledge about yellow fever,

1

even in the urban type, and are more importantly involved now in the more common jungle type.

Reed (1902) in Cuba produced yellow fever in a human volunteer by injecting filtered serum from another infected human. Thus for the first time, a filtrable virus was proven to be the cause of a specific human disease.

Reed and his commission in Cuba also proved that the virus was transmitted by a mosquito then called *Stegomyia fasciata* (now known as the *Aedes aegypti*) and that the disease could be transmitted by the mosquito only under certain conditions. These findings indicated the necessity for the control or eradication of the insect vector, which led to the eradication of yellow fever from Cuba and many other populated areas of the Western Hemisphere. Associated with the control of the malaria-bearing mosquitoes, belonging to a different genus, these measures eradicated yellow fever and almost eradicated malaria from the Canal Zone in Panama making the construction of the Canal possible.

After it was known that yellow fever virus grows in man and is transmitted by mosquitoes, it then became important to find experimental animal hosts to study the disease in more detail. Great efforts were displayed by scientists of the Rockefeller Foundation in Africa during the third decade of this century to unravel some of the mysteries involved in the epidemiology of this disease. In 1927 members of the West African Yellow Fever Commission, produced the disease in an Indian monkey, *Macaca mulatta* by injecting it with the blood of an ailing West African named Asibi. This first transmission of the yellow fever virus to a non-human animal opened up entire new possibilities for research (Stokes *et al.*, 1928). The strain of yellow fever virus isolated from this rhesus monkey became known as the Asibi strain, after its human donor, and has been one of the most widely used strains in yellow fever research. It was also the strain which eventually gave rise to development of the vaccine. In the same year it was shown that yellow fever could be transmitted from monkey to monkey by *Aedes aegypti*.

Laboratory-acquired infections of yellow fever sometimes resulted fatally. The world lost some of its most brilliant scientists in its battle against yellow fever. The names of Lazear, Stokes, Noguchi, Young, Lewis, Cross, and Hayne, among others, stand out in the history of medicine as true heroic soldiers, who, above and beyond the call of duty, gave their lives trying to conquer one of the worst scourges of mankind. Some form of immunization was urgently needed. Dr. Sawyer and his associates of the International Health Division, Rockefeller Foundation, observed that monkeys inoculated with highly virulent strains of yellow fever virus 4–6 hr following an injection of yellow fever immune serum possessed a solid permanent immunity after the passive immunity had disappeared. By this time Theiler had successfully adapted the Asibi virus to mice and had produced a new strain with all the properties of the yellow fever virus but completely harmless to monkeys. This mouse-adapted strain became known as 17D. Dr. Sawyer, using the 17D strain, devised a vaccine consisting of a 10% suspension of infected mouse-brain tissue in fresh human immune sera. This material, when used with supplementary immune sera, gave solid immunity in monkeys without the development of symptoms. After thorough testing in monkeys, 10 persons were vaccinated in May and June 1931. This was the first vaccine against yellow fever. After its improvement and introduction into public health services, no further cases of the disease occurred among workers involved in yellow fever research. The mass vaccination of people throughout the tropics helped in averting development of the devastating urban epidemics which swept the western world from the sixteenth through the nineteenth century.

After the brilliant discoveries in yellow fever by Finlay, Reed, Lazear, Carroll and Agramonte in Cuba and the successful control work against the disease by Gorgas and Chagas in Cuba, Panama, and Brazil, the limelight of yellow fever research had passed from America to Africa. During the fourth decade of this century the stage for some of the most dramatic episodes in the history of yellow fever shifted back from the Dark Continent to the New World.

This does not mean that yellow fever research in Africa came to a standstill. During the last 30 years, through the continued efforts of the West African Yellow Fever Commission of the Rockefeller Foundation and of the East African Virus Research Institute at Entebbe, Uganda, many important advances in our knowledge of the epidemiology of yellow fever have been made. This work has contributed, in particular, to a clear understanding of the basic ecologic differences that exist between the African and the American manifestations of the disease.

II. The Role of the New World Monkeys in Yellow Fever Research

A. HISTORICAL

Yellow fever was a well-known pathologic entity in the Western Hemisphere during the seventeenth century (Carter, 1931), but it was not until early in the twentieth century when its viral etiology and the mosquito-borne nature of the disease became firmly established (Reed, 1902). The discoveries of the U. S. Army Yellow Fever Commission in Cuba led to development of the "Gorgas doctrine," which postulated that yellow fever could not exist in the absence of the highly domestic mosquito, *Aedes aegypti,* then known as *Stegomyia fasciata,* and that adequate control of this insect would invariably lead to eradication of the disease.

Franco *et al.* (1911) advanced the first hints that human cases of yellow fever could be contracted in uninhabited jungle areas, when they reported on an epidemic outbreak which occurred in Muzo, Colombia in 1907. They concluded that the cases they studied had contracted the disease during daylight hours while working in the forest and that the vector had been *Stegomyia calopus* (= *Aedes aegypti*). This last statement invalidated their finding for many years, because the mosquito they reported as a forest vector was known to be strictly a house mosquito in the Western Hemisphere. More recent investigations indicate that Dr. Franco and associates probably mistook a forest mosquito common in the area, *Aedes dominicii,* for *A. aegypti.*

One of the first positive reports of the existence of a jungle form of yellow fever, with possible involvement of monkeys as natural hosts of the virus, was that of Balfour (1914). Instructed by Dr. Patrick Manson to look for possible reservoirs of yellow fever among the lower animals of the Western Hemisphere, this scientist picked up stories in Trinidad that old residents could always tell when there was going to be an epidemic of yellow fever because, prior to its appearance, red howler monkeys were found dead and dying in the "high woods." Dr. Balfour confirmed these stories from reliable sources and was able to trace back some human cases to infection in the jungle. Balfour's observations in Trinidad were confirmed many years later by Downs (1955).

The epidemiologic term "jungle yellow fever" was coined by Soper *et al.* (1933) in reporting an outbreak of yellow fever in the Valle do Chanaan, Brazil, among woodcutters and agricultural workers who acquired the disease while working in or near the forest during the daytime and in the complete absence of the mosquito *Aedes aegypti.* Monkeys were not specifically mentioned in this report as possible sources of virus. However, the existence of natural vertebrate hosts other than man was implied by the authors who stated: ". . . man may not be an essential factor in the continuance of endemicity nor in the spread of the virus."

The discovery of jungle yellow fever in Valle do Chanaan, was followed by numerous studies of similar outbreaks in other areas of Brazil and of Colombia. In 1934, monkeys were found for the first time to be natural hosts of the virus, when five white-faced monkeys, captured in the jungles of Brazil, were found to have naturally acquired immunity against the yellow fever virus (Levi Castillo, 1948). The first isolations of the virus from the blood of naturally infected monkeys were reported by Laemmert and Castro-Ferreira (1945), who isolated it from several marmosets captured in the jungle in the vicinity of Ilheus, Brazil.

Up until 1948 outbreaks of jungle yellow fever had occurred in South America. In November and December 1948, "like a bolt from the blue," yellow fever reappeared in Panama in the jungle form, killing five farmers who contracted the disease along the edge of the Pacora woods, scarcely 15 miles east of Panama City (Elton, 1952). The last autochthonous case of yellow fever, prior to 1948, was diagnosed in Panama during the year 1905.

The yellow fever virus moved further east in 1949 reaching the shores of the Panama Canal and killing additional persons there. Early in 1950 the virus managed to cross the Canal, appearing on the western shores of this natural and artificial barrier, which normally blocks passage of yellow fever from South to Central America. From here a typical wave of jungle yellow fever was to start on a long, slow, but relentless journey north that was to reach the border area between Guatemala and Mexico in 1956. This wave left in its wake hundreds of human deaths and utter destruction of the howler monkey population, whose cadavers left the woods reeking with the odor of dead animal flesh (Trapido and Galindo, 1956).

Despite the sequelae of death and devastation that this wave of jungle yellow fever left behind, many important points on the ecology of the disease were elucidated during its passage through Central America.

In June 1951 clinical diagnosis of a human case was made by the late Dr. Gustav Engler at the Almirante Hospital. A serum sample from this patient was sent to Dr. Enid de Rodaniche of Gorgas Memorial Laboratory, who isolated from it a strain of yellow fever virus. This became the first time that this virus had been isolated from any source in Panama, where thousands of people had fallen victims of the disease since the seventeenth century.

Vargas-Mendez and Elton (1953) collected 35 dead monkeys in forests of different areas of Costa Rica during a reported yellow fever outbreak and made a histopathologic diagnosis of the disease in 24 of them, including 14 howlers, 9 red spider monkeys, and 1 squirrel monkey. This became the first time that monkeys were definitely shown to be involved in the natural transmission of jungle yellow fever in Middle America.

Dr. Jorge Boshell, then with the Pan-American Sanitary Bureau, collected several dead howler monkeys in Guatemala and sent five formalinized and 1 glycerinated liver specimens to Gorgas Memorial Laboratory. All five formalinized livers were histopathologically diagnosed as yellow fever and the virus was isolated from the glycerinated specimen. This became the first yellow fever virus isolate obtained from a naturally infected monkey in Middle America (Johnson and Farnsworth, 1956).

Rodaniche and Galindo (1957) isolated yellow fever virus from the mosquitoes *Sabethes chloropterus, Haemagogus mesodentatus,* and *H. equinus* collected in Guatemala by Dr. Jorge Boshell, and Galindo *et al.* (1956) demonstrated experimentally the ability of these mosquitoes to transmit yellow fever virus from monkey to monkey. This work represented the first proof that Central American species of *Haemagogus* were involved in the natural transmission of yellow fever and the first time that a sabethine mosquito was shown to be capable of transmitting the virus through its bite.

Ever since the 1948 outbreak in Panama, scientists of Gorgas Memorial Laboratory have kept up a surveillance for yellow fever activity in eastern Panama. In 1956, while the crest of the wave that gained momentum on the western shores of the canal in 1950 was still killing howlers along the Usumacinta river valley, between Guatemala and Mexico, a new series of events began to take place in Panama. In September 1956 a strain of yellow fever virus was isolated at Gorgas Memorial Laboratory from a pool of *Haemagogus lucifer.* These mosquitoes were collected in the canopy of the forest near Mandinga, on the Atlantic coast of the isthmus, about 25 miles NE of the Pacora woods, theater of the 1948 outbreak. This was followed in October 1956 by the appearance of human cases at the old Pacora site of the 1948 epidemic. Yellow fever virus was also isolated from mosquitoes captured in the canopy of the forest where the human cases had supposedly contracted the disease. The species of mosquitoes involved were: *Haemagogus lucifer, H. equinus, H. spegazzinii falco, Sabethes chloropterus,* and *Anopheles neivai* (Rodaniche *et al.,* 1957). These isolations constituted the first time that yellow fever virus was isolated from any mosquito collected in Panama.

The sequence of events during this outbreak paralleled the timetable of the 1948 epidemic. Following the Pacora flare-up, yellow fever appeared in 1957 on the eastern shores of the Panama Canal, where a human case was diagnosed through isolation of yellow fever virus from its blood and where diurnal arboreal mosquitoes captured in the forest also yielded virus.

In contrast to what occurred during the previous outbreak, when early in 1950 yellow fever crossed to the western shores of the canal, in 1957 the virus failed to cross and faded out of the picture on the eastern shores of the waterway. Trapido and Galindo (1956), who conducted long-term studies on the vector populations in this critical area, have offered an explanation for the usual failure of the yellow fever virus to cross into Central America. These authors believe that conditions around the canal are marginal for sustaining yellow fever transmission, not only because human activity has greatly reduced the acreage of forests, but, more important still, because the type of forest covering this area is not conducive to production of high densities of vectors at the usual time that yellow fever virus arrives at the eastern shores of the canal. Therefore, it is only in years of unusually heavy rainfall during this time that conditions become favorable to permit passage of virus into western Panama and Central America.

Galindo and Srihongse (1967), as part of the team of Gorgas Memorial Laboratory scientists who have kept up a long and tedious vigil in the wide expanses of forests of eastern Panama, reported yellow fever activity among the monkey population in the extreme southeastern corner of Darien province during 1965. These same investigators, working in the northern part of the province during 1965 and 1966, failed to find any signs of yellow fever activity in the spiders and howlers of that area and conjectured that an unusually dry year, following the 1965 outbreak, could have disrupted virus transmission by drastically reducing the vector population, thus preventing the formation of one of the typical waves that periodically move westward toward the Panama Canal.

The question as to the origin of the yellow fever waves that from time to time reach the eastern shores of the Panama Canal remains unanswered. Gorgas Memorial Laboratory investigators, working in Darien province since 1958, have failed to find signs of enzootism in this area of eastern Panama, which led Galindo and Rodaniche (1964) to reach the conclusion that yellow fever was neither endemic nor enzootic in eastern Panama, and that the periodic incursions of the virus into this area are probably epizootic extensions of some enzootic center in South America. Kerr (1967) has questioned the validity of this theory with the argument that the negative evidence gathered to date is not sufficient to justify the conclusion arrived at by the above-mentioned authors. Regardless of which side of the question one takes, there is no doubt that somewhere east and south of the Panama Canal the yellow fever virus lurks, hidden in silent transmission cycles, occasionally spilling over into the monkey population and giving rise to the devastating waves of the disease, so well known in the folklore of some areas of tropical America.

B. EXPERIMENTAL WORK WITH MONKEYS AND YELLOW FEVER

The classification of New World monkeys is still in a chaotic state and different modern authors disagree extensively as to the names and numbers of families, genera, and species.

For practical purposes we may divide them into six groups, namely, marmosets and tamarins, squirrel monkeys, night monkeys, howlers, spider monkeys, and white-faced monkeys. Susceptibility and tolerance to infection with the yellow fever virus varies greatly among these groups and even among species of each group. It is, however, difficult to present an accurate comparative picture of the response to the virus of the six groups mentioned above, because of the wide range of susceptibility and tolerance that exists among individuals of one species, as well as between different strains of the virus. In general, all New World monkeys appear to be susceptible to infection with the yellow fever virus and they all seem to circulate enough concentration of virus in the blood to be capable of infecting the mosquito vectors (Bugher, 1951). We may thus conclude that monkeys play a major role in sustaining the natural transmission chain of jungle yellow fever in the American tropics. Comparative estimates of susceptibility and tolerance to yellow fever infection in six groups of New World monkeys are presented in Table I. It must be borne in mind that these are gross estimates, subject to limitations imposed by the variabilities noted previously and based, in part, on inconclusive experimental work. The table will be useful, however, in interpreting basic points on the ecology of jungle yellow fever in Middle America. Following is a presentation of the salient points of laboratory research with yellow fever virus and each of the groups of monkeys mentioned above.

1. *Howlers*

These monkeys belong in a single genus, *Alouatta,* as generally accepted by all modern mammalogists. Speciation in the group is still not well understood and there is general disagreement as to the names that should be applied to the different populations of howlers that extend from Mexico south to Argentina and Bolivia.

Little experimentation has been carried out on the susceptibility of howlers to the yellow fever virus, because of the difficulties experienced

TABLE I

COMPARATIVE SUSCEPTIBILITY AND TOLERANCE TO YELLOW FEVER
VIRUS IN NEW WORLD MONKEYS

Monkeys	Susceptibility to infection	Morbidity	Mortality
Howlers	Very high	Very high	Very high
Marmosets	Very high	Very high	Very high
Squirrel monkeys	Very high	Very high	Very high
Night monkeys	Very high	High	High
Spider monkeys	High	Moderate-to-low	Moderate-to-low
White-faced monkeys	Probably moderate	Low-to-negative	Low-to-negative

everywhere in keeping them alive under laboratory conditions. Davis (1931) infected a single specimen of the red howler by mosquito bite and was able to recover virus by blood transfusion to a rhesus monkey and by feeding *Aedes aegypti* mosquitoes on its blood. Laemmert and Kumm (1950), working with the howler *A. caraya,* were successful in carrying on several cyclic transmissions from monkey to mosquito to monkey, using two forest mosquitoes of Brazil, namely, *Aedes leucocelaenus* and *A. scapularis.* Results of large-scale neutralization tests for yellow fever immunity among natural populations of howlers of Mexico, Central America, Panama, Trinidad, and South America indicate that howlers frequently become infected in nature with the virus.

2. *Marmosets and Tamarins*

These primates belong to some six genera of the family Calithricidae and extend from Panama to southern Brazil. A great deal of experimental work has been carried out with marmosets and yellow fever and all species have been shown to be extremely susceptible to infection, with development of severe illness and high mortality. In some regions of South America they have been shown to be important as links in the transmission chain of the virus. However, the only species known from Panama does not appear to play an important role in yellow fever transmission, as it seldom comes in contact with the vectors, because it is mainly an inhabitant of second-growth forests which do not sustain high population densities of the natural vectors of the disease.

3. *Squirrel Monkeys*

There is but a single recognized species, namely, *Saimiri sciurius,* known from Venezuela, the Guianas, the Amazon area of Brazil, Colombia, the Pacific coast of western Panama, and adjacent areas in Costa Rica. This monkey is extremely susceptible to infection and capable of sustaining yellow fever transmission through many monkey–mosquito–monkey passages. They usually develop severe symptoms of the disease and often die of it. Because of its restricted distribution, it can only be considered as locally important in the natural transmission of yellow fever. In Panama and Costa Rica it may have played a very important role in an outbreak which became localized in its area of distribution within these countries.

4. *Night Monkeys*

A single species is recognized today, *Aotus trivirgatus,* which extends from eastern Panama south to the Orinoco and Amazon basins. The species is nocturnal, sleeping during the day in well-protected tree holes. It is extremely susceptible to infection with yellow fever, circulates high concentrations of virus in the blood, being capable of infecting vector mosquitoes. It is affected by the virus with severe illness which is frequently fatal. Because of its restricted distribution and its habits, which keep it away from the diurnal vectors, the species does not seem to be very important in sustaining transmission cycles of yellow fever in nature.

5. *Spider Monkeys*

These monkeys belong to the single genus *Ateles*. There are several species which extend from Mexico south to Brazil. Experimental work carried on by Davis (1930a) with *A. ater* (= *A. paniscus*) and by Galindo *et al.* (1956) with *A. fusciceps,* the black spider monkeys of Brazil and Panama, respectively, has shown that they are highly susceptible to infection and are capable of infecting natural vectors. *Ateles fusciceps,* the black spider of eastern Panama, tolerates yellow fever infections well, developing light-to-moderate symptoms but seldom dying of the disease.

6. *White-faced Monkeys*

All the white-faced or capuchin monkeys belong to the genus *Cebus,* which has been divided into many species, extending its range from Honduras south to Brazil and Peru. Experimental work by Aragao (1928) and Davis (1930b, 1931), with several strains of yellow fever, has demonstrated that capuchin monkeys are susceptible to the virus and are capable of infecting mosquito vectors. However, infections are usually asymptomatic and very seldom fatal.

C. FIELD OBSERVATIONS ON MONKEYS AND YELLOW FEVER

The massive experimental data accumulated on the susceptibility and tolerance of New World monkeys to the yellow fever virus, summarized above, have been amply confirmed by field observations.

The susceptibility to yellow fever infection and high morbidity and mortality in howlers, demonstrated in the few laboratory experiments carried out with them, has been amply corroborated in the field by the incredible mortality of these monkeys observed during jungle yellow fever outbreaks in Central America and Trinidad. It has been demonstrated in these areas that howlers are extremely sensitive indicators of the presence of yellow fever virus. When howlers cease to be heard in the forest, or when groups of them are found dead of natural causes, there is a good likelihood that human cases of yellow fever will soon begin to appear. Stories of the "silence of approaching death" are commonly picked up by scientists from country folk of Trinidad, Guatemala, Nicaragua, and parts of Brazil, who live in intimate contact with the forest. They tell of the

Yellow Scourge sweeping through the countryside soon after the booming calls of the howlers, that always fill the morning air as dawn breaks over the tropical rain forest, are suddenly replaced by an ominous silence, which they consider a presage of imminent tragedy. The vivid accounts given by Dr. Jorge Boshell of his trips through the Central American jungles during yellow fever outbreaks, when he observed literally hundreds of howler monkeys falling out of trees like rotten fruit, constitute an unforgettable experience for those of us fortunate to have heard them firsthand.

Field observations on spider monkeys and yellow fever do not fall in the same pattern as that described for howlers. There are two species of spider monkeys in Central America which appear to have played important roles in maintaining the transmission chain of jungle yellow fever in the forest: the black spider of eastern Panama and the red spider of Western Panama and Central America. When human cases of yellow fever appeared near Almirante, in extreme northwestern Panama during 1951, they were preceded by unusual mortality of red spiders in the mountains to the southwest of Almirante, reported by an engineer in charge of a road-surveying crew. It is interesting to note that the first yellow fever case diagnosed in Almirante, which resulted in death was that of a member of this crew. The story of how this man died of yellow fever, after being vaccinated a few days too late, is one example of the many tragic incidents that occurred as Yellow Jack moved relentlessly north through the Central American forests.

One day in April 1951, the engineer and two of his chainmen were walking across the continental divide from Chiriquí to Bocas del Toro and came to rest at noon by a stream, under the branches of a huge *almendro* tree whose fruit is a favorite food of spider monkeys. He noted many "blue mosquitoes" biting them at the edge of the clearing in the forest made by the stream and also was puzzled by finding, under the same *almendro* tree, about a dozen fresh cadavers of red spider monkeys which appeared to have died of natural causes. Next day he arrived at Almirante and immediately took a plane bound for Panama City, where he related his findings to Gorgas Memorial Laboratory scientists. Upon being told that the monkeys had probably died of yellow fever and that he and his men should be immediately vaccinated, he returned to Almirante the following day. It took him all of 24 hr to round up his men and take them to the Almirante hospital where they received the yellow fever vaccine in the morning. The crew walked back a few hours later to their jungle camp located some 12 km from Almirante. That evening one of the chainmen, who had accompanied the engineer across the mountains, came down with chills and fever and by the next day he was jaundiced and had developed the dreaded "coffee-colored vomit." He was carried to the

Almirante hospital where he died a few days later. A piece of his liver removed at autopsy showed the typical pathologic lesions of yellow fever.

As this epidemic wave moved through Costa Rica and Nicaragua, Vargas-Mendez and Elton (1952) and later Boshell (1952) picked up a number of dead red spiders in the forest whose cause of death was histopathologically diagnosed as yellow fever. During a visit to Chontales province in Nicaragua by Boshell, Trapido, and Galindo (Trapido and Galindo, 1956), immediately after the epidemic wave of yellow fever had passed through the area, observations were made on the populations of howlers, red spiders, and white-faced monkeys which occur in the area. Natives reported finding huge numbers of howlers, some red spiders, but no white-faced monkeys dead in the forest. The scientists neither saw nor heard howlers, but ran across small groups of red spiders and large bands of white-faced monkeys. These observations fit the general tolerance picture among these monkeys given in Table I. It is clear from these accounts that red spider monkeys frequently die of yellow fever infection during outbreaks. However, in general, the species appears to be more tolerant of the virus than the howler, as a good number of them survive the infection and develop permanent immunity against the disease.

The black spider monkey of Panama does not react to yellow fever in the same manner as its kin from Central America. Laboratory experiments carried out at Gorgas Memorial Laboratory have demonstrated that black spiders seldom perish as the result of the disease. Observations carried out in the forests of eastern Panama, after yellow fever has swept through them, also show the great tolerance of the black spider to the virus, as populations of this primate do not appear to be affected at all.

D. Monkeys in the Detection of Jungle Yellow Fever

Gorgas Memorial Laboratory has been interested in studying the ecology of yellow fever in eastern Panama, as a means to elucidate the origin of the epidemic waves which periodically move from east to west across the isthmus. Through these studies, it has been possible to detect two such waves that made their appearance after the rediscovery of yellow fever in Panama in 1948. The existence of one was demonstrated in the Pacora area in 1956, and the second one was discoverel in Darien in 1965.

Detecting activity of jungle yellow fever in the virgin forests of eastern Panama is a difficult task at best. There are no human inhabitants in these forests and the few Indians and Negroes that venture into them are immune to yellow fever, so that attempts to diagnose the presence of the disease by picking up human cases, through viscerotomy or other means, do not offer good possibilities of success. The finding of dead howlers in these forests is far less feasible than in other areas, because the frequent waves of yellow fever that pass through keep the growth of howler populations under restraint. For this reason, during outbreaks in these sparsely inhabited wide expanses of forest, there have been no reports of dead howlers. However, in studying populations of this species of monkey before and after yellow fever episodes, it has been noted that individual groups of this highly gregarious animal are drastically reduced in numbers. Since howlers are easy to locate because of the noisy calls of males, a few weeks spent in a particular forest may serve to estimate the average size of howler groups and determine whether they are normal or greatly reduced in numbers. In this way, useful information may be obtained as to possible recent yellow fever activity.

Black spider monkey populations are not affected during yellow fever outbreaks, so population densities offer no information as to possible recent activity of yellow fever in an area. It is known, however, that black spiders are quite susceptible to infection with the yellow fever virus and that they develop permanent immunity to the virus soon after infection. By bleeding and testing a sample of the black spider population of a particular forest, and determining the approximate age of each specimen bled, it becomes possible to arrive at a reasonable conclusion regarding recent jungle yellow fever activity. In areas of eastern Panama that have just been visited by jungle yellow fever, a high percentage of the spider monkey population will be found to have immune antibodies in the blood, with many positive samples among young specimens. In areas not recently affected by a yellow fever wave, the rate of individuals with antibodies will be much lower, with no positives among the younger age groups. While white-faced monkeys do not seem to be affected by yellow fever, they are not as good indicators of jungle yellow fever activity as the black spiders, since they appear to be somewhat more refractory to infection in eastern Panama, as shown by the significantly lower antibody rates found in the white-faced monkeys.

Another successful method of detecting the presence of yellow fever virus is through the use of sentinel monkeys. There are two ways by which monkeys may be used as sentinels. One is by exposing them in the canopy of the forest continuously and checking them daily for development of symptoms. Another is by direct inoculation into them of the blood of wild-caught monkeys. Rhesus monkeys were widely used at first as sentinels because of their high susceptibility to the virus, but in later years they have been replaced by native American species, such as spider monkeys, which are more gentle and easier to handle in the field.

These methods of field observations, blood-testing of the wild primate populations and exposure of sentinel monkeys for detecting waves of jungle yellow fever in eastern Panama, are the methods of choice utilized by Gorgas Memorial Laboratory in yellow fever surveillance activities being developed in Darien province. They are extremely useful, as they yield information regarding movements of yellow fever waves from east to west, much before arrival of the virus at densely populated centers in the general vicinity of the Panama Canal.

III. Monkeys and Yellow Fever in the Future

In this chapter we have attempted to present a concise summary of the role that the monkey has played in yellow fever research. This disease, which has plagued the world for many centuries, will continue to haunt the tropics as long as wide expanses of forests remain standing. The monkeys, in their role as amplifiers of the yellow fever virus, will also continue to be a threat to the health of human beings dedicated to the conquest of this disease of the tropics. However, through the efforts of countless men of science, which we have attempted to summarize above, these same monkeys have now become useful tools in the hands of scientists to help detect approaching danger, when silent yellow fever waves slowly move through uninhabited forests and creep up on areas densely populated by susceptible human beings. These warning signals offered by monkeys, will allow public health authorities to initiate intense vaccination campaigns before arrival of the wave, thus saving many human lives that would otherwise have fallen victims to the Black Vomit.

In summary, monkeys were important in the experiments that were concerned with the following: the discovery of the virus of yellow fever; the first transmission of the virus to a nonhuman animal; the production of the first vaccine; and the elucidation of the differences between urban and jungle types of the disease. Monkeys are the major reservoirs of jungle yellow fever. Also, they are now being used in surveillance for yellow fever in the Panama jungles, to warn public health authorities of the approaching danger to densely populated centers around the Panama Canal.

Yellow fever is only one of some hundreds of viruses. As with yellow fever, many of these viruses are known to be, and others will be shown to be, common pathogens of man and monkey. The important part played by the monkeys in the yellow fever work presages their possible greater role in studying other viral agents in the future.

REFERENCES

Aragao, H. de B. (1928). *Mem. Inst. Oswaldo Cruz, Suppl.* 2, 35.
Balfour, A. (1914). *Lancet* 1, 1176.

Boshell, J. (1952). Personal communication.

Bugher, J. C. (1951). *In* "Yellow Fever" (G. K. Strode, ed.), p. 200. McGraw-Hill, New York.

Carter, H. R. (1931). "Yellow Fever: An Epidemiological and Historical Study of It's Place of Origin." Williams & Wilkins, Baltimore, Maryland.

Davis, N. C. (1930a). *J. Exp. Med.* **51**, 703.

Davis, N. C. (1930b). *Amer. J. Hyg.* **11**, 321.

Davis, N. C. (1931). *Amer. J. Trop. Med.* **11**, 113.

Downs, W. G. (1955). *Pan Amer. Sanit. Bur. Sci. Publ.* **19**, 584.

Elton, N. W. (1952). *Amer. J. Trop. Med. Hyg.* **1**, 436.

Franco, R., Martínez-Santamaría, J., and Torre-Villa, G. (1911). *Acad. Nac. Med. Ses. Cient. Centenario, Bogota* **1**, 169.

Galindo, P., and Rodaniche, E. (1964). *Amer. J. Trop. Med. Hyg.* **13**, 844.

Galindo, P., and Srihongse, S. (1967). *Bull. WHO* **36**, 151.

Galindo, P., Rodaniche, E., and Trapido, H. (1956). *Amer. J. Trop. Med. Hyg.* **5**, 1022.

Johnson, C. M., and Farnsworth, S. F. (1956). *Bol. Of. Sanit. Panamer.* **41**, 182.

Kerr, J. A. (1967). Personal communication.

Laemmert, H. W., Jr., and Castro-Ferreira, L. de (1945). *Amer. J. Trop. Med.* **25**, 231.

Laemmert, H. W., Jr., and Kumm, H. W. (1950). *Amer. J. Trop. Med.* **30**, 723.

Levi-Castillo, R. (1948). *Rev. Kuba Med. Trop.* **4**, 37.

Reed, W. (1902). *J. Hyg.* **2**, 101.

Rodaniche, E., and Galindo, P. (1957). *Amer. J. Trop. Med. Hyg.* **6**, 232.

Rodaniche, E., Galindo, P., and Johnson, C. M. (1957). *Amer. J. Trop. Med. Hyg.* **6**, 681.

Sóper, F. L., Penna, H. A., Cardoso, E., Serafim, J., Jr., Frobisher, M., Jr., and Pinheiro, J. (1933). *Amer. J. Hyg.* **18**, 555.

Stokes, A., Bauer, J. H., and Hudson, N. P. (1928). *J. Amer. Med. Ass.* **90**, 253–254.

Trapido, H., and Galindo, P. (1956). *Exp. Parasitol.* **5**, 285.

Vargas-Mendez, O., and Elton, N. W. (1953). *Amer. J. Trop. Med. Hyg.* **2**, 850.

CHAPTER 2

Monkeys and Malaria

MARTIN D. YOUNG

Malaria is one of the world's most devastating diseases. Its intermittent fevers have been recognized since the beginning of recorded history. Because the parasites which cause malaria are very small it was not possible to detect these pathogens until after the invention of the microscope. In fact the first description of the parasite was just over 90 years ago (Laveran, 1880). Four species of malaria in man are recognized now, the first three being described during the late 1800's and the fourth and last in 1922, viz., *Plasmodium vivax, P. falciparum, P. malariae,* and *P. ovale.*

About five years after the finding of the human malaria parasites, birds were found also to be infected with malaria. But it was not until 1898 that malaria was seen in monkeys (Koch, 1898; Kossel, 1899). It seems odd that bird malarias were found before the monkey malarias and that there was a lapse of 18 years from the discovery of the human parasite until they were found in these lower primates, animals which are recognized to have many similarities to man.

One of the earliest known attempts to transmit human malaria from man to nonhuman primates was by Koch (1900). He injected malarious blood apparently containing *P. vivax* and *P. falciparum* into orangutans and gibbons. The negative results led that famous scientist to conclude that manlike apes are not susceptible to human malaria, but that it should not be taken for granted that other animals farther from man could not harbor the human malaria parasites. Although we now know that he was partially wrong in his conclusion that human malaria will not grow in the apes, he was correct in the prediction that they might grow in the lower primates farther removed from man. Chimpanzees and gibbons have been shown experimentally to be receptive to the human malarias, if splenectomized. At least six of the monkeys of the Western Hemisphere can be infected with one or more of the human malaria species and in some cases the infections can go back and forth between humans and non-human primates by inoculation with infected blood or with sporozoites from the mosquito (Young, 1970).

Benefits from the study of simian malarias can fall into two categories, i.e. (1) the immediately practical when used for the treatment of diseases or as a testing model for the development of new drugs; (2) the production of knowledge which relates to the disease condition in man and other primates.

The first and most dramatic practical benefit was the use of simian malaria for the therapy of central nervous system syphilis in man. Wagner von Juaregg (1922), a psychiatrist in Vienna, noted that, although neuro-syphilitic patients usually died within a few years after the diagnosis was made, some of those with fevers lived considerably longer, and, in fact, some recovered. After experimenting with various pathogens causing fevers, he determined that infections with human malarias could be indeed a treatment for neurosyphilis, a late stage of syphilis in man which causes insanity. This was the first and, for many years, the only known specific treatment for a mental disease. After this became recognized, the use of the therapeutic malaria became widely established. *Plasmodium vivax* was the most widely used species, but as it would not infect negroes well, *P. malariae* was used for members of this race.

In 1932, Knowles and Das Gupta in India found that *P. knowlesi* from the macaque monkey could infect man. This monkey malaria then was used in several countries for the treatment of neurosyphilis. It appeared to have certain advantages over the human malarias as it produced a shorter and milder course of infection, and apparently there was less danger of its being transmitted by the local anopheline mosquitoes. However, some of these same characteristics were considered by some to be disadvantages. Often the course of the infection was not long enough to produce the 10 to 20 paroxysmal bouts considered optimum for the best results, the parasitemias often were low, and it showed some of the lower infectivity to Negroes that characterizes human *P. vivax* also.

Another practical application of the study of simian malarias is their use in nonhuman primate hosts for the testing of candidate compounds for antimalarial activity. This has been widely used during the past 25 years and has been useful in detecting compounds that are effective in humans against human malaria. For these studies much use has been made of the macaque monkeys usually infected with *P. cynomolgi* or other malarias similar to *P. vivax* of man.

The other class of important benefits is the scientific knowledge resulting from the study of simian malarias in simians. The early facts concerning the biology of malaria tended to result from the study of the disease in man. After the discovery of the parasite in the red blood cells, it was determined that there were two stages of these pathogens, one causing the fever (the asexual forms), and the other forms (the sexual forms or gametocytes) harmless to man but producing infection in mosquitoes. It was then found that these mosquito forms produce a third type of parasite, the sporozoites, which, upon injection into man, would cause the infection. As there was a lag between the injection of the sporozoites into man and the appearance of parasites in the red blood cells, it became obvious that there had to be still other types of parasites living in other tissues.

In 1948, it was dramatically shown that this hidden cycle was in the parenchymal cells of the liver. Shortt and Garnham (1948) using *P. cynomolgi* in the rhesus monkey demonstrated such a developmental stage in the liver. *Plasmodium cynomolgi* closely resembles *P. vivax* of man. This led to experiments with the human *P. vivax* and it was found that this parasite grows in the liver cells of man just as does the *P. cynomolgi* in the monkeys. Subsequently, the liver stages for all of the four species of human malaria were discovered. These hidden cycles in the liver gave the first logical explanation for the missing stages between the bite of the infected mosquito and the appearance of the bloodstream parasites. It was shown further that these liver stages could persist for months in some cases, and years in others, after the first primary attack. An exacerbation of these forms released into the bloodstream would reinvade the red blood cells producing repeated separate attacks known as relapses.

Thus a great deal of knowledge was produced by the study of human malaria in humans, monkey malaria in humans, and monkey malarias in monkeys. As a result many chemical compounds were tested for antimalarial activity, toxicity, pharmacologic activity, and other biologic parameters. Out of these investigations came better malarial drugs, some with remarkable advantages in the treatment of malaria. Coupled with this progress, was even more definitive progress in the control of malaria-bearing mosquitoes by insecticides in preventing the mosquito transmission of malaria. The demonstrated usefulness of the latter, together with the complementary benefits of drugs which reduced the supply of malaria parasites available to the mosquitoes led to the concept in 1955 that malaria eradication was a feasible concept and should be undertaken as an international, even global, goal. Much progress was made in eradicating malaria in some areas and greatly reducing its prevalence in others. Millions of people were spared the malignant effects of malaria and there was a great improvement of the health and of the economy in many of the malarious areas of the world.

However, adverse conditions began to appear. Most serious was the appearance of resistance to the insecticides by the malaria vectors. This was then followed by the appearance of resistance of the malaria parasites to practically all of the recently developed drugs, which had had such good characteristics except for this weakness. However, some of the 4-aminoquinoline compounds were still excellent drugs for the treatment of clinical malaria and some of the 8-aminoquinolines would attack the liver stages of the parasite and thus reduce relapses. Confidence in these two types of compounds masked the difficulties of resistance appearing in the other drugs.

This confidence was greatly shaken when, in 1961 (Moore and Lanier, 1961; Young and Moore, 1961), it was shown that the worst of the malaria parasites, *P. falciparum,* could in some cases become resistant to the 4-aminoquinoline drugs which had for many years been the best agents against clinical malaria. Although the areas of resistance to these drugs were circumscribed, one being in upper South America and the other in Southeast Asia, nevertheless the presence of malaria-resistant parasites in the latter area was of major concern because of the war existing there and the large numbers of troops exposed to and contracting *P. falciparum* malaria, which often was resistant to many standard drugs.

Renewed investigations began to find better drugs for the prevention, treatment, and cure of the disease, hopefully for drugs which would not have the weakness of failing to cure all strains and types of malaria because of resistance. For the necessary testing of thousands of compounds to find better therapeutic agents, a large-scale program was put into effect using various models, including avian malaria in chicks, rat malaria in rodents, monkey malaria in nonhuman primates, and finally malaria in man. Although the lower test systems were useful, the final testing of the compounds had to be done in man, or hopefully, if possible, in nonhuman primates. Large monkeys were used extensively, especially the rhesus, but these were costly in acquisition and husbandry. The need for a better model was indicated. One greatly desired model was a small monkey in which human malaria could be grown.

Taliaferro and Taliaferro (1934) had tried to grow human malaria in Panama monkeys. Although the parasites of *P. falciparum* would persist for a few days, the shortness of the parasitemia and the failure to maintain the infections by serial transfers made this model impractical. Surprisingly few attempts had been made previous to this time or even after this time to find a small monkey in which human malaria would grow. In 1965, workers at the Gorgas Memorial Laboratory in Panama renewed the attempts to grow human malaria in monkeys. Within one year it was shown that human *P. vivax* malaria could be grown in the small *Aotus* (night or owl) monkey, would infect mosquitoes, could be passed back to man by mosquito bites, and could be successfully maintained in these monkeys by serial passage by the injection of infected blood (Young *et al.*, 1966). This led to the use of this model by other investigators and as a result much valuable information is being produced in many diverse lines.

Further testing by the scientists at the Gorgas Memorial Laboratory showed that of the seven species of monkeys in Panama, six of them could be infected with one or more of the different species of human malaria.

Of immediate and urgent use of this human malaria–monkey host model is the possible employment for the final or almost final testing of chemical compounds for their use as antimalarial drugs. Hopefully it may be possible to substitute this system partially for the use of human volunteers. The use of human volunteers is restricted in scope, probably becoming more so, and is very expensive. Obviously their use does not have the experimental capabilities of a laboratory animal model.

However, before these monkey models can be used to their fullest capacity, a great deal of information must be produced. We need to know much more about the biology of the human malaria in the monkeys, especially on the points of important similarities and dissimilarities to the disease of human malaria in humans. Mosquito transmission of the disease resulting in the liver stages is a necessity to determine the effects of the drugs upon the persistent parasite stages in the liver which cause relapses. Therefore, the ability of various mosquitoes to become infected with and to transmit these human malarias between monkeys must be learned. This should lead to the development of techniques for the mass production of infected mosquitoes for this experimental work.

In addition to the obvious benefits of using this model for drug-testing, other information not readily available by the study of malaria induced in human volunteers can be produced. These would include the examination of internal tissue to follow the progress of the disease, experimental procedures which would attempt to change the response of the host to the disease, such as splenectomy, removal of the thymus, the use of immunosuppressant drugs, the effects of stress, and other procedures difficult or impossible to do in man. Such results could create knowledge in many lines such as immunology, pathogenesis, hematology, biology of the parasite, etc.

Other necessary research areas have to do with the potential insect vectors. There is only one known natural vector of monkey malaria at present in the Western Hemisphere. To test mosquitoes adequately they must be colonized. This procedure itself is difficult and laborious. Only a few of the known mosquitoes of the world have been colonized. It is likely that vectors of monkey malaria in nature would be better vectors of human malaria in monkeys than are the presently known vectors of human malaria to humans. However, only one or two proven vectors of monkey malaria in nature are known, much less colonized. Once colonized, additional studies of human malaria in monkey hosts can be done.

The study of monkey malaria in monkeys has produced helpful information on the epidemiology of human malaria. Anopheline mosquitoes are the accepted vectors of mammalian malarias. In areas where human malaria is present it has been assumed that the naturally infected anopheline mosquitoes were the vectors. In areas where humans and monkeys live close together little is known about the vectors of the monkey malarias. Recently, workers in Malaysia (Warren and Wharton, 1963) have found that some of the mosquitoes thought to be vectors of human malarias were actually transmitters of monkey malarias and probably were not responsible for the human malarias as had been supposed previously. Such knowledge is helpful in developing control measures against human malaria.

Also important is the question of whether monkey malarias can be a zoonosis, i.e., will they infect man in nature? If so, what would be the impact on malaria eradication programs and would the monkeys serve as a reservoir of human malaria and a focus for reinfection after the human malaria was eradicated. It has been difficult to answer these questions because so few vectors of monkey malarias are known nor has the potential of the infection going from monkey to man been assessed adequately.

In view of our present knowledge, it seems improbable that the transmission of monkey malaria to man would occur very often, if at all, under ordinary conditions. But again our knowledge is so scant that more information needs to be obtained before the question can be answered satisfactorily. And in obtaining the answers to these questions, information would be obtained on the various mosquito vectors which would have immediate application to the epidemiology of human malaria.

Some scientists now believe that some of the malarias of man and monkey are the same species, viz., *P. malariae* of man is the same as *P. brasilianum* of the Western Hemisphere monkeys. Obviously this is an important point to elucidate.

There are many important problems which can, should, and probably will be investigated in this new monkey host–human parasite model. However, the major difficulties now have to do with the supply of the experimental animals, both quantitatively and qualitatively.

Because of the growing use of small monkeys for biomedical investigations, many thousands are taken from the jungles for export to research centers. Fears are being raised in some quarters that this will lead to a depletion and perhaps disappearance of certain species. However, the use of monkeys for biomedical research does not seem to be the most important factor, if indeed it is a factor, in the possible depletion of a species in the Western Hemisphere. It seems apparent that the numbers of monkeys from Central America are decreasing but this began long before there was any demand for these species for research. One cause for the decrease appears to be due to the destruction of the forests in which the monkeys live. Also, epizootics of disease, such as yellow fever, at times has almost wiped out populations of certain monkeys, such as the howler (*Alouatta*). The possibility that other epizootic diseases can create similar devastating effects indicates an area of needed investigations.

The second, and very important factor, in considering the supply of monkeys is the quality of the research animal. When captured in the jungle, these animals have been exposed to an unknown, but obviously very large, number of infectious agents and trauma. This would result in a varied and equally unknown immunologic profile. It is logical that some of this feral preconditioning may affect adversely the tests undertaken with these specimens in the laboratory.

There is inadequate knowledge at present for the breeding of these animals under controlled conditions. Of the Western Hemisphere monkeys, only a few can be bred in captivity. For some of much importance at present, no breeding is possible. In fact, so little is known of the husbandry that it is difficult to maintain some of the animals in captivity for very long. The lack of husbandry knowledge and of ability to breed in captivity indicate areas in which tremendous effort is required.

With the recent burst of information showing that these small monkeys are good models for the study of human diseases, it appears patent that there is at present a need, and that the need will rapidly increase in the future, for the establishment of breeding colonies to supply some of the research needs. In addition to leading to a higher quality of experimental subjects, this would perhaps aid in the preservation of the species, should the numbers taken for biomedical research show to have any importance in this area.

REFERENCES

Knowles, R., and Das Gupta, B. M. (1932). *Indian Med. Gaz.* **67**, 213–268.
Koch, R. (1898). *Zentralbl. Bakteriol., Parasitenk-Infektionskr., Abt. 1* **24**, 200–204.
Koch, R. (1900). *Deut. Med. Wochenschr.* **26**, 88–90.

Kossel, H. (1899). *Z. Hyg. Infektionskr.* **32**, 25–32.
Laveran, A. (1880). *Bull. Acad. Med., Paris* **9**, 1235–1236.
Moore, D. V., and Lanier, J. E. (1961). *Amer. J. Trop. Med. Hyg.* **10**, 5–9.
Shortt, H. E., and Garnham, P. C. C. (1948). *Brit. Med. J.* **1**, 1225–1228.
Taliaferro, W. H., and Taliaferro, L. G. (1934). *Amer. J. Hyg.* **19**, 318–334.
Wagner von Jauregg, J. (1922). *J. Nerv. Ment. Dis.* **55**, 369–375.
Warren, McW., and Wharton, R. H. (1963). *J. Parasitol.* **49**, 892–904.
Young, M. D. (1970). *Lab. Anim. Care* **20**, 361–367.
Young, M. D., and Moore, D. V. (1961). *Amer. J. Trop. Med. Hyg.* **10**, 317–320.
Young, M. D., Porter, J. A., and Johnson, C. M. (1966). *Science* **153**, 1006–1007.

CHAPTER 3

Cell Cultures

AMBHAN D. FELSENFELD

The etymology of the expression "tissue culture" is one of the most interesting of terms used and abused in the biological sciences. We shall disregard here the historical aspects of "culture" as applied to the maintenance and growth of unicellular and multicellular microorganisms as well as individual cells, tissues, and organs under conditions usually characterized by artificiality of the external factors, composing the so-called culture media. However, the origin of the word "tissue," rather recently derived from the French *tissé,* meaning "woven," and accepted for scientific English usage, probably by Grew in 1682, deserves attention. It would appear that "cell culture" is a more precise term for such growth procedures, when one considers the historical development of this science and the actual experimental process.

When the historical development of cell cultures is examined by analyz-

25

ing the data compiled by Studnicka (1926), Eagle (1958), White (1963), Grundmann (1964), and Paul (1965), Grew's "little boxes" of compartmentalized "tissue" may have been preceded by the term "cells" used by Hooke in 1667. The actual cell theory, a subject of sometimes heated debate reaching far into the present century, was promulgated by Schleiden and Schwann in 1838–1839. Virchow's famous *omnis cellula a cellula,* originated some decades later, means that all cells are derived from antecedent cells. This excluded the idea of the so-called "spontaneous generation," but the latter misconception haunted biology until Pasteur and others furnished definitive and irrefutable proof that it is not tenable. Pasteur was principally interested in chemistry and microbiology, whereas Virchow's field was anatomy and pathology. Nevertheless, we are still burdened with many unanswered questions concerning generation and growth.

One of these problems grew out of the observation that the cell is not merely a structural unit, but also a physiologic laboratory, often with a highly specialized field of operations. The animal body consists of an orderly society of such cells, be they solitary, grouped, patterned in tissues, or arranged in organs or what used to be called "histions" or "tissue units." All divergent cellular components of the body originated from one single totipotent cell, the fertilized ovum. The specialized cell is derived from the dividing fertilized ovum. Is the cell division leading to loss-variants of totipotent cells to be regarded as a sign and expression of regression? Do the variations in chromosomal numbers and patterns that often accompany the development of specialized cells denote irreversible functional recession? How should the polyploidy of specialized cells (e.g., in the adrenals) and the actively growing diploid cells (as seen, e.g., in the regenerating liver) be correlated? What is the relationship of such cells to their neighbors, to other cells of the body, and to the internal and external environment? Can a unipotent cell revert to omnipotency when its external environment undergoes modifications? Today, with cell-culture methods progressing rapidly and with long strides, one may hope that answers to some of these problems may be forthcoming in the near future, including those in which the internal environment plays a role. The importance of this factor has been emphasized by Claude Bernard but its exploration is still lagging.

In view of these problems, the cell culture is principally a model for the study of the anatomy and physiology of normal and pathologic cells *in vitro,* and only secondarily a tool in microbiology, especially in virology. Undoubtedly the need for better and more efficient cell cultures for virologic work contributed greatly to the efforts to develop novel approaches and more efficient cell-culture procedures.

A summary of the salient points of the history of cell culture shows that probably Roux, in 1885, was the first to culture chick embryo medullary plates in saline solution. Arnold is usually credited with being the first to grow leukocytes (of the frog) in 1887. Lyunggren, in 1898, inoculated skin into ascitic fluid. Jolly in 1908 employed serum as a culture medium. Further information on the history and development of single-cell cultures has been summarized by Sanford *et al.* (1948).

Animal cells require specific nutrients into which they must be transferred without being damaged when the culture is initiated. As a rule, the cells adhere to a support (e.g., wall of the culture vessel). If the experiment is successful, the cells show metabolic activity and its physical prerequisites, including phagocytosis or pinocytosis. Their movement is revealed especially when observed by time-lapse cinematography. A further sign of a growing culture is that the cells multiply.

Harrison, in 1907--1910, used clotted lymph as a support for the culture of frog neuroblasts. Burrows, in 1910, showed that plasma may be used for this purpose. It appears also that Burrows introduced the term "tissue culture."

Carrel's work in 1910–1913 led to the introduction of embryonic extracts in cell-culture procedures. He developed a culture flask that was named after him and replaced the previously used hanging-drop culture method. Carrel's work demonstrated the potential immortality of the cell, i.e., that a cell can be propagated indefinitely when conditions are feasible for its multiplication and for the survival of further cell generations.

At the same time Warren and Margaret Lewis attempted to carry out a complete analysis of the nutrients necessary for the maintenance and propagation of cells *in vitro*. Their efforts did not bear the expected fruits because biochemistry was in its infancy at the time of their studies. Decades later, Paul (1962) was able to present a more adequate picture of the mechanism of metabolic control in cultured mammalian cells.

It appears that both the Embden-Meyerhof and Krebs cycles of carbohydrate metabolism are followed by cells in culture. Lactic acid may accumulate, especially in N_2 atmosphere. The utilization of O_2 may be correlated to growth and may vary with glycolysis and respiration. Cells derived from adult animals require higher oxygen tension. Proteins are formed from amino acids, of which about 12 are designated as essential, and the rest of them may be formed by transamination. Some cells lack proteolytic enzymes and must be furnished with amino acids. Lipids appear to be produced readily from their basic building stones, e.g., from acetate. Their synthesis seems to take place *de novo* from excreted or extruded chemical compounds. Nucleosides are preferred by most cells for the production of nucleic acids but these can be manufactured also from simple compounds, such as formate, glycine, and hydrocarbonate, as well as from intermediates of the nucleotide synthesis.

I. Cell-Culture Media and Methods

In addition to basic nutrients, a buffer is necessary to maintain the pH of the cellular environment. Hanks' balanced salt solution, Eagle's solution (based on Hanks' fluid), Krebs' solution, and triscitrate are among the most frequently used. One has to recall that these buffers are based on Locke's, Tyrode's and Ringer's salt solutions that have been used in physiology and therapy for a long time. White (1946) demonstrated that metabolites alone are not sufficient for cell multiplication, whereas metabolite mixture alone may support the survival of many cells. The addition of some serum is required for rapid growth. Waymouth (1956) replaced the formerly used peptone with protein. The publications of Morgan et al. (1950), Morton et al. (1956), and Parker (1961) contain numerous additional data on these questions. The commonly recommended salt solutions are No. 199 and M150 of Morgan et al. (1950), Parker 858 (Healy et al., 1955), Earle's NCTC 109 (Evans et al., 1956), Eagle's (1955), and Hanks' (Hanks and Wallace, 1949). Puck (1958) also developed a medium that, like Hanks', has a low sodium bicarbonate content and assists in the maintenance of the pH. Melnick's lactalbumin hydrolysate medium (Melnick, 1955) is widely used. While these solutions are available commercially, those who have to make their own will find Eagle's the simplest combination of essential ingredients.

When protein is added to the culture fluid, fetal serum and lactalbumin are often used as the source of protein because of their lack of antibodies.

It has been mentioned that the original hanging-drop cell-culture method was replaced by the use of the Carrel flasks. These are considered handy by many workers but are difficult to clean. Large-scale cell cultures in flasks of larger size was introduced by Maitland and Maitland (1928) and again by Enders et al. (1949). Today either glass or plastic bottles, spinner flasks for continuous cultures, and even apparatuses suitable for aeration and shake cultures are available. Leighton tubes for histologic and radioisotope studies are considered routine supply items.

Virologists also use substitutes for the originally employed plasma clot to enhance adherence, such as agar gel or cellophane sheets. Trypsin and Versene are frequently employed to separate the cells when indicated.

Cloning of cells is carried out by the methods of Puck or Dulbecco. The development of clones is indispensable for the study of morphogenesis, metabolism, interrelationship between cells and cell lines, and pathologic phenomena.

There is no agreement concerning the best method of cleaning glassware used for cell-culture work. The reader may find details of this and other laboratory procedures in the textbooks of Paul (1965), White (1963), and other authors.

II. Cells Suitable for Cultivation

Great hopes have accompanied the first attempts to culture fertilized ova. It appeared, however, that the eggs of viviparous animals are quite refractory to attempts to grow them in media known to date. Lewis and Wright (1935) observed a few divisions of the fertilized ova of rhesus monkeys in culture. To the knowledge of this writer, no success has been reported in culturing fertilized nonhuman primate ova beyond the stage of the blastula. Other primate cells follow the rules valid for mammalian tissue cells in general.

When adult tissue is explanted into a cell culture, there is a lag phase before growth begins. Embryonic tissue, however, does not show this phase and cell division is initiated shortly after inoculation.

Most cells are constantly replaced in the animal after they have been used up. Such replacement is vigorous in highly active organs, as in the intestines, as well as on exposed surfaces, as of the skin. Heart and neuron cells are not replaced. When a tissue is explanted, there is a tendency of fibroblasts and histiocytes to outgrow other elements. Heart, adult nerve, and liver parenchyma cells of mammals, principally of nonhuman primates, are very difficult to culture. In virology, cells derived from the kidneys, lungs, and testes of nonhuman primates are frequently used, as well as primary cultures of embryonic cells. The latter are not readily available in many instances, because of the scarcity of primate embryos accessible to the virologist antemortem. Established nonhuman primate cell lines, both fetal and from young animals, however, can be purchased from cell-culture supply houses.

Cell lines derived from tumors have been widely studied. The best known of them are HeLa cells. The literature consulted for this contribution did not reveal references to cell lines derived from malignant growths of monkeys and apes. Transmissible tumors abound in rodents but in nonhuman primates only those induced by the Yaba virus have been studied in adult grivet monkey cells (Niven *et al.,* 1961). There is a wide-open field for students of tumor-derived cell lines of nonhuman primate origin. It seems to be important to investigate also explants of pathologic growth of nonhuman primates. The results may be gratifying not only for the value of the observations per se but also for possible conclusions concerning similar or identical tumors found in the human primate.

A further step in the study of nonhuman primate cells suitable for cultivation may be in the direction that borders with organ culture as well as with transplantation. Neither is the subject of this chapter. One may recall, however, the studies of ossification patterns in mice in relation to vitamin A (Fell and Mellanby, 1956) that deserves further elaboration and examination for suitability in studies of primate tissues. Another example is the work of Gaillard (1957) who followed the principles worked out for the thyroid by Stone *et al.* (1934). Gaillard cultured parathyroid glands in the sera of prospective recipients through several passages, then implanted them into parathyroidectomized individuals. There was no rejection of the transplant. The implanted tissue assumed parathormone production. Perhaps this procedure could be applied also to other endocrine glands, taking into account the results of recent results achieved with the aid of transplantation-facilitating factors, including immunosuppressive drugs and antilymphocytic serum.

III. Interaction between Cell Cultures and Parasites

We include under the heading of parasites microorganisms classified as bacteria, yeasts and molds, viruses and rickettsiales, and parasites *sensu stricto,* which themselves, or with their toxic products, cause pathologic changes in primates, including man.

The changes observed in tissue culture may be none after primary inoculation but may appear after serial passages. This applies particularly to some viral agents.

A. BACTERIA AND THEIR TOXINS

Perhaps Suzuki (1918) was the first to demonstrate the action of diphtheria toxin on growing tissues. Placido Sousa and Evans (1957) applied cell-culture methods to diphtheria-toxin evaluation. They were followed by Strauss and Hendee (1959) and Kato and Pappenheimer (1960). It appears that severe metabolic disturbances accompany the incorporation of this toxin into cell cultures.

Vicari *et al.* (1960) established the influence of dysentery bacilli on KB cell cultures. This method is widely used today with various cell lines as a proof of toxigenicity of Enterobacteriaceae, including enteropathogenic *Escherichia coli.* The method is yet to be standardized.

Felton *et al.* (1954) studied pertussis bacteria by the cell-culture method.

Staphylotoxin became one of the agents most frequently titrated in cell cultures after its cytoplasm-vacuolizing effect was reported by Felton and Pomerat (1962). Our group (Felsenfeld and Felsenfeld, 1966) has been primarily concerned with the cytopathic effect of the alpha toxin of staphylococci and with its evaluation in growing cells.

Cholera vibrio toxin was studied in cell cultures by Read (1965) and by our group (Felsenfeld *et al.,* 1966). We found that its effect can be demonstrated with ease in primary embryonic cell lines derived from primates.

Acid-fast organisms have been extensively studied and the giant-cell formation induced by them is described in standard textbooks. Nevertheless the studies of Hanks (1941) with explanted lepromatous tissue, in which human leprosy bacilli appeared approximately in the same numbers in the daughter cells as in the respective mother cells, deserve special mention as an example of an applied quantitative cytologic study concerned with the interaction of two genetically distant cells.

B. RICKETTSIALES AND VIRUSES

The cell-culture method is offering the virologist an excellent means for the isolation of numerous rickettsial and viral agents, as well as a growth medium for the production of some vaccines.

According to Schaechter *et al.* (1957) rickettsiae actively and dynamically penetrate cells in cultures. Their localization in the cytoplasm and/or in the nucleus of the cultured cell may not be the same as in infected primates *in vivo.*

Virus–cell interactions at the macromolecular level were reviewed by Bernhard (1964) with special regard to the cytopathic effect (CPE) of viruses on cultured cells.

Apparently Steinhardt *et al.* (1913) were the first to grow vaccinia virus on the explanted cornea, then Li and Rivers (1930) cultured vaccinia and other pox viruses successfully. Pox viruses appear to be taken up by pinocytosis rather than phagocytosis.

Herpes viruses appear to influence primarily the nuclei of the cells in culture. This effect depends on the particular herpes strain. Fusion, nuclear migration, metabolic changes affecting DNA, and aberrant mitosis follow (Roizman, 1962). Chromosomal breaks, deletions, and translocation also have been seen. The general impression made by herpes-infected cell lines is not dissimilar to that observed during malignant transformation of primate tissues *in vivo* if accompanied by giant-cell formation.

Adenoviruses include some tumor-inducing viruses. Adenoviruses usually cause the separation of the cell culture from the glass and an increase in nucleic acid formation (Nosik and Klisenko, 1963).

Papova viruses also include some that may cause tumors. Of special interest is the vacuolizing virus of the vervet monkey (*Cercopithecus aethiops*) (Melnick, 1962), called also SV20. It has been grown in patas monkey cell cultures (G. D. Hsiung and Gaylord, 1961). Among the characteristic changes evoked by this virus are protoplasmic vacuoles and nuclear inclusion bodies. Its growth is slow (Easton, 1964).

Other viruses of special interest for the cytologist are that of the mouse parotid tumor which grows very slowly in monkey tissue cells, and "monkey virus" SV40 which causes transformation of human cell lines (Schein and Enders, 1962). The growth of the cells is accelerated, syncytium formation may take place, and chromosomal aberrations, such as monosomy and aneuploidy, have been observed. The nuclei of the cells increase in volume, then fragmentation takes place (Pruniéras *et al.,* 1964).

Myxoviruses, which include the agents of influenza, parainfluenza, measles, and mumps, affect in different degrees and ways cells in cultures. Giant cells, spindle cells, and antigen, located either in the cytoplasm or in the nucleus, may be observed, according to the viral strain.

Picornaviruses, among which that of poliomyelitis belongs, were reviewed with regard to their CPE by Klöne (1958). Their biochemistry was analyzed by Darnell and Eagle (1960). It seems that loss of nucleoli and chromatin and the formation of lipid-containing paranuclear masses may be observed in suitable cell lines after infection with these viruses.

Apparently Haagen and Theiler (1932) were the first to culture an arbovirus (that of yellow fever) in growing cells. Of the other arboviruses, Hotta (1959) cultured dengue virus in rhesus monkey cells. Buckley (1959) and Henderson and Taylor (1960) extended the use of tissue-culture methods to the study of other arboviruses.

Echoviruses were divided into two groups by G. D. Hsiung and Melnick (1957) according to their different behavior in patas and in rhesus monkey cell cultures.

The CPE effect of viruses may depend on the species from which the virus was isolated and from which the cell culture was derived. For instance round, large cell clusters are observed after inoculation of some adenoviruses into cells cultured from organs of the same animal species from which the virus was grown. This phenomenon, with special regard to primate species, requires further investigation. The vacuoles produced by SV40 virus may be observed in cultured cells of a limited number of primate species, whereas intranuclear changes are seen in cells derived from man as well as from all hitherto tested species of nonhuman primates.

The "why's" of this limitation are not yet known. Neither is it clear why pox, herpes, varicella zoster, measles, and myxoviruses (including mumps, parainfluenza, and the respiratory syncytial virus, but not related viruses), cause giant-cell formation in cell cultures.

C. PARASITES

The relationship of protozoa and cell cultures was reviewed by Pipkin and Jensen (1956, 1960).

According to Belding (1965), Meyer and Zuckerman, in 1946, established protozoa in splenic cultures, and Hawking in 1948 was the first to grow leishmanias and plasmodia in cell cultures.

Trypanosoma (*Schizotrypanosoma*) *cruzi* appears to grow well in fibroblasts, histiocytes and other macrophages, epithelial cells, and HeLa cells (Bock *et al.,* 1959; Neva *et al.,* 1961). Trypanosomes causing African sleeping sickness are, however, very difficult to grow in cell cultures (Warren and Borsos, 1959).

Toxoplasma appears to multiply easily in monkey testis and kidney cells (Chernin and Weller, 1957) and in embryonic tissue cell lines. Fibroblast cultures are preferred, but other cells will also permit the multiplication of this organism (Pulvertaft *et al.,* 1954; Balducci and Tyrrell, 1956), whereas the related *Pneumocystes carinii* is difficult to establish in cell culture (Moser, 1959).

IV. Antibody Formation and Pharmacologic Studies

Nossal (1966) pointed out the difficulties encountered in attempts to study antibody formation in cell cultures. Apparently only blast cells multiply in such cultures when isolated from lymphoid-plasma cell organs. They are easily outgrown by other cells. The transformation of small lymphocytes into immunologically active blast cells in the sense of Gowans (1962) is still under experimental study.

Immunologic as well as pharmacologic investigations require standardization of the cell cultures. Numerous methods have been suggested to avoid cell clumping. We employ determinations in blood-cell counting chambers and the optical density method elaborated originally for nonhuman primate cells by Youngner (1954). The study of mitotic activity is indispensable. Metabolic studies, such as glucose utilization, radioactive-labeled amino acid(s), and precursors of DNA and RNA are of great value. Cell morphology, differentiation, and migration also require close observation. Only under stable conditions can cytotoxicity, including the action of antimetabolites, be studied, as well as the influence of hormones. It appears, however, that organ cultures rather than cell cultures are being used in pharmacology today.

V. Radiobiology

Trowell (1966) stated that experiments with tissue cultures in radio-biology contributed little to that branch of science but they gave the cytologist an excellent tool, permitting the exploration of cell damage.

Radiation injury of cells does not manifest itself with initial lesions that can be observed under the optical microscope immediately after ir-radiation. The susceptibility of cells to radiation damage is about the same *in vitro* as it is *in vivo,* namely, young cells are more sensitive than mature cells. The effects of radiation reveal themselves by mitotic retardation in the metaphase, chromosomal abnormalities, immediate (nonmitotic) or late (mitotic) cell death, and repression of the ability of total inability of single cells to develop into cell sheets. Giant-cell formation is not unusual.

The greatest advantage of radiation is that its quality and quantity can be controlled with ease. X-Rays and gamma, hard beta, and alpha emissions are, therefore, excellent tools in cell-culture studies in well-equipped laboratories which strictly adhere to the rules and regulations aimed at the elimination of radiation hazards.

Soft beta radiation emitters (^{14}C, ^{3}H) are frequently used in metabolic studies, often by autoradiography, which requires special skill, but the results appear to be gratifying and informative.

VI. Discussion

This chapter has not been devised to present all information available on cell cultures, nor was it written with the intention to enumerate all publications concerning the study of such cultures in general, and cell growth derived from nonhuman primates in particular. Reference has been made to standard and easily available texts for those who wish to become more intimately acquainted with the pertinent problems of *in vitro* cell-culture methods. Emphasis was placed on the first revelant publication dealing with each of the problems involved, as well as on reviews that are a few years old but have withstood the test of time and are authoritative.

Nonhuman primates have been used for the study of many important facets of biomedicine, such as atherosclerosis, infectious diseases, and nutritional disorders, to mention only a few. Because of the close relationship of nonhuman primates to man, they should be of great value in many scientific approaches when a feasible experimental animal model is being sought. Owing to progress in surgical techniques, it is no longer required that a monkey be killed when his kidney is needed to establish a fibroblast cell culture. Neither is he destined to become useless in the reproductive animal colony when testicular tissue is wanted. A properly performed biopsy, carried out by a competent surgeon, under anesthesia and strict aseptic conditions, will provide the so-called tissue-culture laboratory (*recte:* cell-culture laboratory) with sufficient material for a primary culture.

It is regrettable that the monkey kidney cell culture that has been a standard for polio vaccine production may have to be replaced by some other cell line because slow-growing and human tissue-transforming SV viruses may be present in it. It is possible that a concentrated effort to develop cell cultures free from transforming viruses as well as raising animal colonies devoid of such infective agents should be made. Primate centers with carefully selected and checked animals, producing a sufficient number of "clean" offspring, could fulfill this mission.

G. D. Hsiung and Atoynatan (1966) investigated the incidence of naturally acquired virus infections in monkeys and surveyed the pertinent literature related to several agents which were isolated from cell cultures derived from nonhuman primates. Using rhesus (*Macaca mulatta*), vervet (*Cercopithecus aethiops*), and patas (*Erythrocebus patas*) tissues, these authors isolated principally myxovirus SV5, foamy agent, SV40, and measles virus, alone or in mixtures, from animals recently acquired by a primate colony. Most of these agents disappeared from the animals and specific serologic tests became negative in 5–6 weeks when cross-infection was prevented by proper quarantine measures. However, the SV40 virus proved to be more persistent. Therefore it is possible to avoid viral contamination of the primate colony by preventive measures.

More intensive studies of antiviral agents in nonhuman primates also may help to achieve this goal. The number of known "slow" viruses is increasing from maedi to kuru. The much-dreaded SV40 virus also may be classified as a slow virus because of its delayed growth in tissue cultures. While apparently of no consequence to the nonhuman primate, this and related transforming viruses should be followed up in every primate colony to facilitate the use of nonhuman primates in cytologic and virologic studies.

G. D. Hsiung and Atoynatan (1966) demonstrated the value of quarantine of newly arriving animals in the primate colony, and the need of longitudinal virologic studies in them, not only with the aid of tissue sampling, but also by serologic means. This task is not beyond the capability of well-equipped and well-manned primate centers. The primary requirements—strict quarantine, proper isolation, and virologic studies—can be carried out at least on those animals from which cell cultures for basic research, viral or pharmacological studies, and cell lines for vaccine production will be derived.

Many human lives could not have been saved nor permanent invalidism prevented if vaccines prepared from "monkey cells" had not been used. Moreover, adult and embryonic nonhuman primate cells served in the primary isolation of a considerable number of viral agents. Their use in diagnostic investigations should not be abandoned but, on the contrary, it should be extended because many physiologic and anatomic characteristics of these cells have not yet been fully investigated.

Radiobiological experiments also demonstrated the similarity of primate cells. Perhaps a more intensive biochemical study of cultured monkey and ape cells may shed more light on problems still unresolved in human cytoarchitecture and cytochemistry.

Closely allied are problems of drug-testing. Screening, with the aid of cell cultures, titration of toxins and antitoxins, and other pharmacologic and toxicologic tasks can be performed in cell cultures before expensive animal experiments are undertaken. Finally, we must hope that the development of organ-culture methods will be applied to a greater extent in the studies of primates. The results achieved on them, principally on apes closely related to man, should be helpful in understanding human medical problems related to cells, tissues, and organs.

REFERENCES

Balducci, D., and Tyrrell, D. (1956). *Brit. J. Exp. Pathol.* 37, 168.

Belding, D. L. (1965). "Textbook of Parasitology," 3rd ed. Appleton, New York.

Bernhard, E. (1964). *Cell. Injury, Ciba Found. Symp., 1963* p. 209.

Bock, M., Kollert, W., and Gönnert, R. (1959). *Z. Tropenmed. Parasitol.* 10, 284.

Buckley, S. M. (1959). *Ann. N. Y. Acad. Sci.* 81, 172.

Chernin, E., and Weller, T. H. (1957). *J. Parasitol.* 43, 33.

Darnell, J. E., Jr., and Eagle, H. (1960). *Advan. Virus Res.* 7, 1.

Eagle, H. (1955). *Science* 122, 501.

Eagle, H. (1958). *Bact. Rev.* 22, 217.

Easton, J. M. (1964). *J. Immunol.* 93, 716.

Enders, J. F., Weller, T. H., and Robbins, F. C. (1949). *Science* 109, 85.

Evans, V. J., Bryant, J. C., Fioramonti, M. C., McQuilkin, W. T., Sanford, K. K., and Earle, W. R. (1956). *Cancer Res.* 16, 77.

Fell, H. B., and Mellanby, E. (1956). *J. Physiol.* 133, 89.

Felsenfeld, O., and Felsenfeld, A. D. (1966). *Proc. Soc. Exp. Biol. Med.* 122, 442.

Felsenfeld, O., Felsenfeld, A. D., Greer, W. E., and Hill, C. W. (1966). *J. Infec. Dis.* 116, 329.

Felton, H. M., and Pomerat, C. M. (1962). *Exp. Cell Res.* 27, 280.

Felton, H. M., Gagero, A., and Pomerat, C. M. (1954). *Tex. Rep. Biol. Med.* **12**, 960.

Gaillard, J. P. (1957). *Colloq. Int. Cont. Nat. Rech. Sci.* **78**, 157.

Gowans, J. L. (1962). *Ann. N. Y. Acad. Sci.* **99**, 432.

Grundmann, E. (1964). "Allgemeine Cytologie." Thieme, Stuttgart.

Haagen, E. M., and Theiler, M. (1932). *Proc. Soc. Exp. Biol. Med.* **29**, 435.

Hanks, J. H. (1941). *Int. J. Lepr.* **9**, 275.

Hanks, J. H., and Wallace, R. E. (1949). *Proc. Soc. Exp. Biol. Med.* **71**, 196.

Healy, G. M., Fisher, D. C., and Parker, R. C. (1955). *Proc. Soc. Exp. Biol. Med.* **89**, 71.

Henderson, J. R., and Taylor, R. M. (1960). *J. Immunol.* **84**, 590.

Hotta, S. (1959). *Acta Trop.* **16**, 108.

Hsiung, G. D., and Atoynatan, T. (1966). *Amer. J. Epidemiol.* **83**, 38.

Hsiung, G. D., and Gaylord, Jr., W. H. (1961). *J. Exp. Med.* **114**, 974.

Hsiung, G. D., and Melnick, J. L. (1957). *J. Immunol.* **78**, 128.

Kato, I., and Pappenheimer, Jr., A. M. (1960). *J. Exp. Med.* **112**, 329.

Klöne, W. (1958). *In* "Handbuch für Virusforschung" (C. Hallauer and K. F. Meyer, eds.), 2nd ed., p. 203. Springer-Verlag, Berlin and New York.

Lewis, W. H., and Wright, E. S. (1935). *Contrib. Embryol. Carnegie Inst.* **25**, 161.

Li, C. R., and Rivers, T. M. (1930). *J. Exp. Med.* **52**, 465.

Maitland, H. B., and Maitland, M. C. (1928). *Lancet* **2**, 596.

Melnick, J. L. (1955). *Ann. N. Y. Acad. Sci.* **61**, 754.

Melnick, J. L. (1962). *Science* **135**, 1128.

Morgan, J. F., Morton, H. J., and Parker, R. C. (1950). *Proc. Soc. Exp. Biol. Med.* **73**, 1.

Morton, H. J., Pasieka, H. J., and Morgan, J. F. (1956). *J. Biophys. Biochem. Cytol.* **2**, 589.

Moser, L. (1959). *Zentralbl. Bakteriol., Parasitenk., Infektionskr. Hyg., Abt. I: Orig.* **174**, 457.

Neva, T. A., Malone, M. F., and Myers, R. R. (1961). *Amer. J. Trop. Med.* **10**, 140.

Niven, J. S. F., Armstrong, G. A., Andrewes, C. H., Pereira, H. G., and Valentine, R. C. (1961). *J. Pathol. Bacteriol.* **81**, 1.

Nosik, N. N., and Klisenko, G. A. (1963). *Acta Virol.* **7**, 42.

Nossal, G. J. V. (1966). *In* "Cells and Tissues in Culture" (E. N. Willmer, ed.), Vol. 3, p. 317. Academic Press, New York.

Parker, R. C. (1961). "Methods of Tissue Culture," 3rd ed. Harper (Hoeber), New York.

Paul, J. (1962). *In* "New Developments in Tissue Culture" (J. W. Green, ed.), p. 211. Rutgers Univ. Press, New Brunswick, New Jersey.

Paul, J. (1965). "Cell and Tissue Culture," 3rd ed. Livingstone, Edinburgh.

Pipkin, A. G., and Jensen, D. V. (1956). *Exp. Parasitol.* **7**, 491.

Pipkin, A. G., and Jensen, D. V. (1960). *Exp. Parasitol.* **9**, 167.

Placido Sousa, G., and Evans, D. G. (1957). *Brit. J. Exp. Pathol.* **38**, 644.

Pruniéras, M., Chardounet, Y., and Schier, R. (1964). *Ann. Inst. Pasteur, Paris* **106**, 1.

Puck, T. T. (1958). *Proc. Nat. Acad. Sci. Wash.* **44**, 772.

Pulvertaft, R. J. V., Valentine, J. C., and Lane, W. F. (1954). *Parasitology* **44**, 478.

Read, J. K. (1965). *Proc. Cholera Res. Symp., 1965* p. 151.

Roizman, B. (1962). *Proc. Nat. Acad. Sci. U. S.* **48**, 228.

Sanford, K. K., Earle, W. R., and Likely, G. D. (1948). *J. Nat. Cancer Inst.* 9, 229.

Schaechter, M., Bozeman, F. M., and Smadel, J. E. (1957). *Virology* 3, 160.

Schein, H. M., and Enders, J. F. (1962). *Proc. Nat. Acad. Sci. U. S.* 48, 1164.

Steinhardt, E., Israeli, C., and Lambert, R. A. (1913). *J. Infec. Dis.* 13, 294.

Stone, H. B., Owings, G. C., and Gey, G. O. (1934). *Amer. J. Surg.* 24, 386.

Strauss, N., and Hendee, E. D. (1959). *J. Exp. Med.* 109, 145.

Studnicka, F. K. (1926). "Die Organisation der lebendigen Masse." Springer-Verlag, Berlin and New York.

Suzuki, Y. (1918). *J. Immunol.* 3, 238.

Trowell, O. A. (1966). *In* "Cells and Tissues in Culture" (E. N. Willmer, ed.), Vol. 3, p. 63. Academic Press, New York.

Vicari, G., Olitzki, A. L., and Olitzki, Z. (1960). *Brit. J. Exp. Pathol.* 41, 179.

Warren, L. G., and Borsos, T. J. (1959). *J. Immunol.* 82, 585.

Waymouth, V. (1956). *J. Nat. Cancer Inst.* 17, 315.

White, P. R. (1946). *Growth* 10, 231.

White, P. R. (1963). "The Cultivation of Animal and Plant Cells," 2nd ed. Ronald, New York.

Youngner, J. S. (1954). *Proc. Soc. Exp. Biol. Med.* 85, 202.

CHAPTER 4

Tuberculosis and Bacterial Infection

ROBERT C. GOOD

I. Introduction

Tuberculosis and bacterial diseases have been recognized for many years as major factors in illness and death of caged primates maintained in zoological exhibits or laboratory colonies (Haberman and Williams, 1957; Dolowy et al., 1958; Lund and Petersen, 1959; Ruch, 1959; Greenstein et al., 1965; Hunt et al., 1968; Good et al., 1969; Kourany and Porter, 1969; Good and May, 1971). Even though free-ranging nonhuman primates may live near human habitations, the infections common in humans have not been detected in animals immediately after capture. In their native environment weak or dead animals are usually not found since

39

they are rapidly disposed of by scavengers. However, once the animals are maintained in compounds, clinical signs of disease can develop. The infections probably result from undetected disease, carriers, or contacts with human carriers; but the stresses imposed by shipping, caging, and diet establish increased susceptibility of the host (Fiennes, 1970).

Bacterial infections of the respiratory and gastrointestinal tracts may run a rapidly fatal course. Tuberculosis, on the other hand, progresses more slowly, but if undetected and untreated it is just as deadly. Interest in control of these diseases in primate colonies initiated research which led investigators to recognize the similarities of the natural infections in the nonhuman and human primate hosts. With this relationship established it was logical to proceed with induction of disease in animals in order to develop standardized conditions for study. Encouraged by the success of these studies, other investigators attempted to infect monkeys with bacteria which were responsible for human diseases even though the pathogens had not been isolated from monkeys. Their combined efforts have resulted in unique animal models of human diseases.

Studies of bacterial diseases have relied heavily on experimental infections in mice, guinea pigs, and rabbits. To turn to more expensive and laborious studies in primate species must therefore be justified by the total biologic evaluation of the disease obtainable in these subjects. The obvious advantages of using naturally susceptible *Macaca mulatta* for studies of shigellosis rather than an artificial system in guinea pigs are readily apparent. Studies of induced tuberculosis in nonhuman primates are far more definitive in providing information for treatment and prevention of human disease than studies in rodent systems. However, the latter do provide a primary screen for many investigations, a role which the primate model should not usurp.

Specific human bacterial diseases which have been reliably replicated and studied in nonhuman primates are discussed in succeeding sections.

II. Tuberculosis

A. Natural Susceptibility

Most species of Old World and New World primates are subject to infections with *Mycobacterium tuberculosis*. The rhesus monkey (*M. mulatta*) is highly susceptible to both human and bovine tubercle bacilli, and infections spread rapidly among grouped animals. The disease is principally pulmonary in naturally infected monkeys (Habel, 1947; Francis, 1956; L. H. Schmidt, 1956a, Dolowy *et al.,* 1958), although lesions may occur in liver, spleen, kidneys, and mesenteric lymph nodes. In 1923 Fox described the disease in a group of 192 primates and indicated that 45% were probably "pulmonary-aerogenic" in origin, while 55% were "intestinal-lymphatic." Since *M. bovis* was isolated and identified from four of the infected primates, it is possible that many of the infections represented ingestion of contaminated food.

Susceptibility to infection with *M. tuberculosis* is not consistent throughout all primate subjects. Kennard and associates (1939) believed that the mangabey was more susceptible than the rhesus monkey and that lemurs, galagos, and marmosets were less susceptible. Haberman and Williams (1957) found 16% of 615 rhesus monkeys with macroscopic disease at autopsy, while 93 cynomolgus monkeys (*M. fascicularis*) housed in same quarters were free of disease and only 3 gave positive tuberculin reactions. Bywater *et al.* (1962) found 1.6% of 2301 rhesus monkeys held in a laboratory colony for 7–9 weeks to be tuberculous at autopsy, whereas none of 5328 vervet monkeys (*Cercopithecus aethiops*) handled in a like manner exhibited macroscopic lesions.

Infections in the larger apes have been reviewed by Ruch (1959), and rare cases have been reported in *Saimiri sciureus* (Dunn, (1968), *Aotus trivirgatus* (Snyder *et al.,* 1970), *M. speciosa* (Wolf *et al.,* 1967), *M. nemestrina* (Sedgewick *et al.,* 1970), *M. radiata* and *M. fascicularis* (Good and May, 1971), *C. aethiops* (Stones, 1969), *Erythrocebus patas* and *M. fascicularis* (Tribe, 1969), and *Papio cynocephalus* and *P. anubis* (Heywood and Hague, 1969).

B. DISEASE MODEL

Rhesus monkeys have been the most widely used nonhuman primate for laboratory studies, even though they are highly susceptible to natural infection with *M. tuberculosis* and *M. bovis*. Tuberculosis was observed in one colony (L. H. Schmidt, 1956a) to resemble the fulminating disease which occurs in some segments of the human population. The dissemination of disease from a single tuberculin-hypersensitive animal with radiologically detectable lesions to all of 9 or 10 cagemates occurred within the space of 15 months. Within 6 months of the first positive tuberculin reaction 75% of the animals were dead. However, roentgenographic evidence of pulmonary tuberculosis could not necessarily be obtained at the same time; yet, when pulmonary lesions were observed, death invariably followed within 2–4 months. From these preliminary studies of "naturally acquired" tuberculosis, a background of data was obtained which could be compared to studies of induced tuberculosis in a large group of rhesus monkeys.

Since natural infection was not acquired by all animals in a reasonably short period of time, experiments were protracted and out of phase. Through a series of systematic studies L. H. Schmidt *et al.* (1955) found that pulmonary tuberculosis could be induced in rhesus monkeys by the intratracheal instillation of 10–1000 tubercle bacilli. In these instances tuberculin hypersensitivity developed in 2–4 weeks, pulmonary infiltrate was detected on roentgenograms in 4–6 weeks, and death occurred in 6–12 weeks. In all instances large areas of caseating pneumonia developed, leading to necrosis and massive cavitation. Although the events could be modified somewhat by the strain of *M. tuberculosis* used, the result was eventually the same. The basic, reproducible disease has been used for almost 20 years by Schmidt and his colleagues to investigate a number of parameters concerned with the biology of tuberculosis. These are discussed in more detail in later sections.

The studies discussed in succeeding sections were carried out in young juvenile rhesus monkeys which weighed 2–5 kg. The disease is similar to the fulminating infection encountered in the human white infant and young black adult. The disease that develops in older animals weighing in excess of 7 kg is equivalent to the chronic cavitary infection encountered in the white adult. However, the availability of younger animals and the validity of results obtained with monkeys in this age group have promoted utilization of the fulminating disease for therapeutic and prophylactic studies. In interpretation of these reports it is important to note that the characteristics of induced disease were similar to the naturally acquired disease originally described by L. H. Schmidt (1956a).

C. CHEMOTHERAPY STUDIES

1. *Isoniazid and Streptomycin*

The preliminary studies described by L. H. Schmidt (1956a) were done shortly after the efficacy of isoniazid was recognized for the treatment of tuberculosis. His experiments, aimed at defining the utility of pulmonary tuberculosis in rhesus monkeys for evaluating therapeutic compounds, were centered on the use of the newly discovered agent. A group of 127 tuberculin-positive rhesus monkeys with evidence of moderate to far-advanced pulmonary tuberculosis and a group of 51 tuberculin-negative animals were given a total dose of 5 mg isoniazid/kg body weight/day in the diet. After 2 months, radiologic evidence of progressive involvement was found in about 20% of the animals, so the total daily dosage of isoniazid was increased to 20 mg/kg. On the higher dose, marked clearing of pulmonary infiltrate occurred within a 3-month period. Prolonged treatment was accompanied by cavity closure and continued reduction in pulmonary infiltrate. Histopathologically, untreated disease produced caseous coagulation type of necrosis in the lungs. The centers of these lesions softened and were expelled, resulting in rough-walled cavities. Proliferative caseous lesions were surrounded by rapidly spreading pneumonic areas or areas of granulomatous inflammation which contained large numbers of tubercle bacilli. Pulmonary disease was accompanied by enlargement and caseation of tracheobronchial lymph nodes which occasionally caused partial obstruction of the bronchi and trachea. However, rhesus monkeys treated with isoniazid for 12 months were remarkably clear, with 50% showing no gross evidence of residual disease. Examination of sections disclosed areas of interstitial fibrosis, chronic emphysema, and cystic bronchiectasis. The remaining animals exhibited residua of disease which varied from small caseous nodules in the tracheobronchial lymph nodes or lungs, to open cavities, one of which was showing evidence of progression. Tubercle bacilli sensitive to easily obtainable blood levels of the drug were isolated from residual lesions. Progression of the disease, then, was consistent with that observed in human tuberculosis (Rich, 1951).

L. H. Schmidt et al. (1955) inoculated monkeys intratracheally with 0.5 ml of a 10^{-4} dilution of a 7-day culture (presumably 10^3 viable units) of M. tuberculosis (strain 5159). Within 3 weeks evidence of disease was obtained by tuberculin hypersensitivity and detection of pulmonary lesions by roentgenography in all but 1 of the 120 monkeys in the study. Lesions detected on roentgenograms were limited to one lobe or were observed in as many as five lobes. After grading the disease level of each animal they were assigned to one of seven groups to assure equal distribution in each. Groups of 18 to 19 animals were treated with streptomycin (20 or 80 mg/kg body weight) or isoniazid (5, 20, or 80 mg/kg body weight) by incorporation of drug into the diet for the duration of the 6-month study.

Six untreated monkeys included in the studies died with massive pulmonary disease within 96 days of infection. Only two treated animals died of tuberculous infection, both on the low dose of isoniazid. The parameters of radiologic clearance of pulmonary lesions, gross examination at autopsy, histopathologic evaluation, and recovery of tubercle bacilli from residual lesions confirmed the therapeutic benefits of the drugs and the variations to be expected with dose.

In the two experiments described above drugs were administered in the diet. In a debilitated, anorexic animal subject to the social stratification that exists in gang cages, dosing via voluntary ingestion would be irregular and fall below the estimated dosage based on calculated food intake. The lack of absolute dosage clouds the results of these studies but does not detract from their overall worth of providing a tool for detailed study of the therapy of tuberculosis. The dose of drugs used in succeeding studies was standardized by intramuscular injection or oral intubation.

2. *Other Potential Antituberculous Drugs*

Many parameters of pulmonary tuberculosis in the rhesus monkey may be studied concurrently. In addition to survival data, serial chest roentgenographs follow the course of disease; histopathologic studies provide a basis for extent of tissue destruction or repair; microbiologic studies provide data on the sterilization of lesions and emergence of resistance; pharmacologic studies establish the correlation of blood level to therapeutic response, a correlation that is not based on the administered dose alone; and observations of the clinical course identify toxic manifestations. These parameters characterize chemotherapeutic agents so that regimens composed of single drugs or combinations may be devised for treatment of human disease.

A study in monkeys on the therapeutic activities of cycloserine administered alone or in combinations with streptomycin and isoniazid was reported by L. H. Schmidt (1956b). The conclusions were that combinations of cycloserine and isoniazid were beneficial, whereas combinations of cycloserine and streptomycin offered little improvement in response. These results are of interest because cycloserine had been found effective *in vitro* but of little value in treatment of induced tuberculosis in rodents. Conzelman (1955, 1956) and Conzelman and Jones (1956) reported similarities of absorption and excretion of this drug in rhesus monkeys and human subjects as opposed to the pattern observed in rodent species, which would account for the differences in therapeutic response noted in rodent and primate species.

The studies on cycloserine indicate one area where results in rodents did not accurately predict therapeutic activity in humans, whereas studies in monkeys did. A slightly different pattern exists in the case of the therapeutic activities of thiocarbanidin. Youmans *et al.* (1958) and Steenken *et al.* (1958) presented evidence that this compound possessed antituberculous activity both *in vitro* and *in vivo* in rodents. However, studies by L. H. Schmidt and colleagues (1959) showed that the drug had little effect on the course of tuberculosis in rhesus monkeys when administered alone and did not exert beneficial effects when administered with suboptimal doses of isoniazid. The limited value of thiocarbanidin was confirmed in the human studies reported by Larkin (1959). A similar series of events surround studies of the therapeutic effects of 4,4-diisoamyloxythiocarbanilide (Isoxyl). Although reported active in mouse tuberculosis (Crowle *et al.,* 1963; Freerksen and Rosenfeld, 1963), Good *et al.* (1964)

presented data which indicated that the compound had very little activity against induced tuberculosis in the rhesus monkey. The poor effects of therapy were ascribed to poor absorption of orally administered drug and a high degree of binding of absorbed drug to plasma proteins. Subsequent studies in human subjects have confirmed the conclusions of the monkey study, but emotional judgments arising from preliminary studies in rodents and humans have only slowly been put to rest.

The therapeutic and toxic relationships of ethambutol were defined in experiments using induced tuberculosis in rhesus monkeys. The compound was initially found to affect favorably the course of induced tuberculosis in mice (Thomas *et al.*, 1961), but extrapolation of dosage schedules to human subjects resulted in a toxic reaction manifested by progressive blindness (Carr and Henkind, 1962). A series of experiments conducted in simians by L. H. Schmidt and associates (1962, 1963; L. H. Schmidt, 1966) confirmed that ethambutol exerted a significant therapeutic effect on pulmonary tuberculosis either alone or in combination with isoniazid. More importantly, the blood level required to bring about the therapeutic effect could be identified and met in human subjects without toxic symptoms developing. Concurrent studies of I. G. Schmidt (1966) and I. G. Schmidt and Schmidt (1966) defined the neurotoxicity of ethambutol in the monkey. The clarification of blood levels, therapeutic effects, and toxicity in the monkey was largely responsible for the acceptance of ethambutol as a companion drug to be used in conjunction with isoniazid in the treatment of primary tuberculosis.

The utility of induced tuberculosis in the rhesus monkeys for the development of therapeutic agents has been proven in the studies discussed in the preceding section. The simian model has accurately predicted the response of human tuberculosis and provided a means of defining various parameters of that response which are impractical in rodent or human disease. Although this model has been used in only one laboratory since it was first developed, the reproducibility of the disease and ease of study recommend it for wider acceptance.

D. STUDIES OF HYPERSENSITIVITY AND IMMUNE REACTIONS

In a study of 128 rhesus monkeys with various histories of contact with tuberculosis, Good *et al.* (1961) were able to show a dissociation between delayed hypersensitivity and immunity to induced disease. In these studies the hypersensitive response had waned in many animals vaccinated with BCG or treated with isoniazid for a prolonged period after induced infection. Upon challenge with 45 viable units of a streptomycin-resistant variant of the 5159 strain of *M. tuberculosis,* disease developed in vaccinated or previously exposed groups at the same rate independent of the presence of tuberculin allergy. The length of the anergic period was not a significant factor in anticipating the immune state.

Good (1968) summarized 10 years experience in the use of the rhesus monkey for immunologic studies. The basic reasoning for the use of the model was expounded through detailed discussions of experimental procedures and parameters of evaluation. Rhesus monkeys vaccinated with the attenuated BCG strain of *M. bovis* attained maximum resistance to challenge infection in 8–12 weeks. Therefore, in direct comparisons of experimental agents to the standard BCG vaccine, monkeys were challenged intratracheally 8–12 weeks postvaccination. The route of vaccine administration was varied according to the needs of the experimental plan, and the resulting hypersensitivity was quantitated by injecting graded dilutions of Koch's Old Tuberculin into the abdominal skin. As in the therapeutic model, extent of infection or protection was based on roentgenographic evidence of pulmonary disease, estimation of involvement of lungs and tracheobronchial lymph nodes by examination of gross specimens, and microscopic evaluation of lesions. Since preliminary evaluations of the vaccines had been made in rodents, concurrent tests were set up in guinea pigs to provide a direct comparison. Using the described procedures, comparisons of the protection afforded by tubercle bacilli killed by ultraviolet irradiation, isoniazid-resistant BCG, and a cell wall preparation were made to one or more standard BCG preparations. In all cases the experimental vaccines did not offer the degree of protection afforded by BCG. However, as an outgrowth of these experiments, routes of administering BCG were studied (Good and Schmidt, 1968). Intravenous vaccination with BCG was found to provide the maximum protection against challenge infection; intratracheal vaccination afforded more protection than subcutaneous vaccination.

Barclay and associates (1970) felt the intratracheal route did not simulate natural infections and proceeded to modify the rhesus monkey model by using inhalation of 12 to 49 infectious units of the H37Rv strain from an aerosolized suspension for challenge. The resulting disease was found primarily in either the right or left lower lobe, a finding which is consistent with studies of monkeys infected by the intratracheal route. However, the erratic results of tuberculin skin hypersensitivity measurements in the palpebral fold and in abdominal skin are very unusual. Nevertheless, these studies with small challenge doses confirmed the protection afforded by intravenous administration of BCG.

From the foregoing discussion it is evident that tuberculin hypersensitivity is of value in following the course of disease in rhesus monkeys. However, tuberculin hypersensitivity is not an accurate indicator of tuberculous infection in all nonhuman primate species. Recently Good (1970) observed that tuberculin reactions in naturally infected cynomolgus monkeys (*M. fascicularis*) fluctuated from month to month yet each of the animals had tuberculous lesions containing viable bacilli at autopsy. A tentative conclusion from observations in this relatively resistant species is that tuberculin hypersensitivity is detectable during periods when there is multiplication of organisms in lesions; when the infection is quiescent, the allergen is no longer able to elicit a response. This reaction may be of value for clarification of immunologic responses to tubercle bacilli.

Immunologic studies in rhesus monkeys permit definitive correlations of certain events with the time after infection, status of disease, and evaluations of protection afforded by vaccination procedures. Minden and Farr (1969) used antisera from vaccinated monkeys in an attempt to identify substances for primary binding tests to detect humoral antibodies which would distinguish between strains of mycobacteria. Chaparas *et al.* (1970) have reported the utility of the lymphocyte transformation test for detection of tuberculosis in monkeys. The results of the tests were compared to tuberculin-hypersensitivity reactions and appearance of lesions at autopsy. This study indicates an area of further study which will have direct bearing on detection of tuberculosis in both human and nonhuman primates. However, these authors found that five of nine tuberculin-positive rhesus monkeys did not have gross lesions of tuberculosis at necropsy. Similar statements reporting false tuberculin reactions have also been made by Tribe (1969). However, properly administered Old Tuberculin or PPD will elicit a delayed hypersensitive reaction which can be read by an experienced observer. Anergic animals in terminal stages of disease are easily detected by their marked debilitation. Weak reactions are more accurately observed when the abdominal site is used, chiefly because the monkey must be removed from the cage for examination. Tuberculin-positive monkeys have gross evidence of disease, which may appear as a slightly enlarged tracheobronchial lymph node or a small area of induration in a carefully examined lung. Positive evidence of disease can be obtained by histopathologic study, culture of the lesions, or inoculation of lesion homogenates into guinea pigs. Recent studies by Fife *et al.* (1970) have indicated that the soluble antigen fluorescent antibody test may provide a more objective tool for detection of tuberculosis in nonhuman primates.

E. STUDIES WITH ATYPICAL AND ISONIAZID-RESISTANT MYCOBACTERIA

A discussion of mycobacterial infections in nonhuman primates would not be complete without inclusion of atypical mycobacteria. Atypical mycobacteria were isolated from infections in a pigtailed macaque (*M. nemis-*

trina) (Sedgwick *et al.*, 1970) and rhesus monkeys (Karassova *et al.*, 1965). However, attempts to experimentally induce progressive infections in rhesus monkeys with two atypical mycobacteria (presumably *M. kansasii*) were not successful (L. H. Schmidt *et al.*, 1957). A spectrum of primate susceptibility to atypical mycobacteria has yet to be determined, but isolated infections indicate that one of the primate species may provide a ready model for detailed study.

Isoniazid-resistant strains of tubercle bacilli produce self-limiting pulmonary infections following massive inocula (10^7 viable organisms) (L. H. Schmidt *et al.*, 1958). The residual lesions found 12 weeks after inoculation often contain isoniazid-sensitive bacilli. Study of the isoniazid-resistant bacilli showed that the isoniazid-sensitive mutant could be selected *in vitro* (Good, 1965). Comparative pathogenicity studies in rhesus monkeys and guinea pigs indicated that the recovered strain was not necessarily identical to the original isoniazid-sensitive strain isolated from patients prior to isoniazid therapy (Good, 1967). These findings are significant to the treatment of human tuberculosis caused by isoniazid-resistant bacilli.

III. Bacterial Diseases

A. NATURAL INFECTION

Even though enteric pathogens have not been isolated from nonhuman primates in their natural habitat (Takasaka *et al.*, 1964; Carpenter and Cooke, 1965; Mysorekar *et al.*, 1966; Agarwal and Chakravarti, 1969), diarrhea and dysentery present major health problems in newly imported and colonized simians. The history of enteric infections has been reviewed by Ruch (1959) and current reports of the problem in large colonies have been presented by Takasaka *et al.* (1964), Good *et al.* (1969), and Kourany and Porter (1969). Whereas *Shigella* species appeared to pose the primary health hazard to Old World primates, *Salmonella* species were the significant isolates from New World species. The owl monkey (*A. trivirgatus*) presents an interesting situation in that *Shigella sonnei* is often found in the bloodstream, even in cases where the pathogen cannot be isolated from rectal swabs (Good, 1968).

Natural susceptibility of nonhuman primates to infection with *Shigella* species is a characteristic shared only with man. The selection of nonhuman primates for systematic studies is, then, a logical choice to study the pathologic processes involved in bacterial dysentery.

The principal pathogens causing respiratory diseases in a large colony of primates were found to be *Klebsiella pneumoniae, Diplococcus pneumoniae, Pasturella multocida,* and *Bordetella bronchiseptica* (Good and

May, 1971). Significant numbers of monkeys also had respiratory infections due to *Staphylococcus aureus, Escherichia coli,* and α and β-streptococci. Although contributing to disease, the role of the latter organisms as primary or secondary pathogens was uncertain. However, natural susceptibility to klebsiella and diplococcal infections makes nonhuman primates a desirable choice for studies of these infections which are important in human health.

B. Enteric Infections

1. *Salmonellosis*

Weinberg (1960) first reported the successful induction of experimental typhoid fever in *"Macacus cynocephalus."* Edsall *et al.* (1960) found that chimpanzees developed a disease syndrome similar to that seen in man after oral infection with approximately 10^9 *Salmonella typhi* cells. Cvjetanovic *et al.* (1970) used the susceptibility of the chimpanzee to study the effects of vaccination with a living streptomycin-dependent strain of *S. typhi.* Four doses of the vaccine strain were given orally with 3-day intervals between doses for a total inoculation of about 2.6×10^{11} bacilli per animal. Ten days after the final dose the animals were challenged orally with 2.6×10^{10} virulent *S. typhi* cells. Based on bacteriologic, serologic, and clinical evaluations the vaccine appeared to offer some degree of protection against challenge infection. The authors felt that the vaccine strain may be useful for the protection of man, but further experiments to determine safety and efficacy are required prior to human inoculation.

Dack *et al.* (1929) found that typical signs of salmonella enteritis in humans could be induced in rhesus monkeys by the administration of 95 to 816 billion organisms by stomach tube. Both *S. aertrycke* and *S. enteritidis* produced clinical disease manifested by weight loss, a slight increase in temperature, leukocytosis, and principally the development of diarrhea the day following infection. Diarrhea continued for 2–3 days and cleared spontaneously with apparent complete recovery. The causative organisms could be isolated from stools collected before signs of disease occurred and during the period of diarrhea, but were usually not found after formed stools were passed. The slight rise in antibody titer did not provide immunity to infection since the monkeys were again susceptible 3 weeks after the initial challenge.

Kent *et al.* (1966), in a study designed to describe the lesions which develop in *Salmonella* gastroenteritis, fed rhesus monkeys 3×10^{10} *S. typhipurium* cells. The animals were serially sacrificed for examination of the gastrointestinal tract. Gastroenteritis developed within 24 hr with the most severe lesions occurring in the colon. Using the fluorescent antibody and cultural techniques, organisms were found in the ileum and colon up to 4 days after challenge. The infecting organisms invaded colonic epithelial cells and intestinal lamina propria, and penetration appeared to be necessary for the production of lesions. Seven days after infection few salmonellae appeared in the colonic mucosa and organisms could not be isolated from the contents of the ileum or colon. Owing to the many similarities of clinical signs and pathologic lesions these studies appear to be applicable to the disease in man. A later study by the same group (Kent *et al.,* 1967a) described the results of virulence tests of strains of *S. typhimurium* and *S. anatum* in guinea pigs and monkeys. Assessments of mortality data, organ cultures, lesions, and the number of organisms in fluorescent antibody-treated sections indicated strain differences, but individual strains had approximately the same virulence properties in the two animal species.

Induced salmonellosis in primates has offered valuable insights into the understanding of the disease in man. The continued use of this model will provide better understanding of the human disease and aid in the preparation of a more efficient vaccine against typhoid.

2. *Shigellosis*

A study of the pathogenic mechanisms, pathology, and immunologic factors of bacilliary dysentery in monkeys has been reported in a series of papers from the Department of Applied Immunology of the Walter Reed Army Institute of Pathology. The first of the series (LaBrec *et al.,* 1964) reported the comparative results of experiments using mice, guinea pigs, and monkeys to study the virulence of two colonial types of *Shigella flexneri* 2a. Clinical infections in guinea pigs depended on a prechallenge, 4-day period of starvation and injection of opium at the time of oral challenge. Virulence tests in mice were performed by intraperitoneal injection of the test strains. Monkeys culturally shown to be free of enteric pathogens were challenged by stomach tube with 5×10^{10} organisms suspended in 20 ml of infusion broth. In these experiments the parent *S. flexneri* strain in mice caused fatal infections (peritonitis); in guinea pigs, produced a fatal infection characterized by ulcerative lesions; in 3 of 10 monkeys, produced classical dysentery with typical lesions in the colon and histopathologic alterations in the ileum. Using fluorescent antibody techniques it was found that bacilli had penetrated to the intestinal lamina propria in infected guinea pigs and monkeys, even though the mucosal epithelium remained intact. Comparative observations with the mutant, avirulent strain indicated that penetration of the epithelial cells was a necessary characteristic of virulent bacilli. This observation was confirmed when only the virulent mutant was found to infect and multiply in HeLa cells and to invade the epithelial cells of guinea pig cornea.

Induced infections resulting from oral inoculation of an *E. coli–S. flexneri* hybrid into guinea pigs and monkeys were reported by Formal *et al.* (1965a). The hybrid strain penetrated the mucosal epithelium of the intestine and produced a transient inflammatory reaction which cleared within 4 days. Formal *et al.* (1965b) tested the ability of the avirulent mutant and hybrid to protect against infection with virulent homologous and heterologous strains. Five oral doses (each containing 5×10^{10} organisms) of the avirulent mutant were required in monkeys to increase resistance to experimental challenge, but a single dose of the hybrid strain was effective. Using heterologous challenge some evidence of specificity was found, but response was often difficult to evaluate since clinical disease could not be routinely induced (between 26 and 80% of nonvaccinated monkeys showed clinical signs following challenge). Inconsistent increases in serum antibody and the lack of detectable coproantibody following vaccination seem to rule out these protective mechanisms as the specific factors in the induced resistance. However, examinations utilizing fluorescent antibody and histologic procedures showed that the hybrid vaccine protected the animals from tissue damage by preventing the penetration of virulent bacilli into the intestinal mucosa (Formal *et al.,* 1966a). The utility of pool vaccines was shown (Formal *et al.,* 1966b) when two doses, consisting of 2×10^7 cells each of *S. flexneri* 1b, *S. flexneri* 2a, *S. flexneri* 3, and *S. sonnei* I hybrid strains, gave significant protection against challenge infection with virulent homologous types. However, administration of such large numbers of bacilli resulted in a transient diarrhea which usually occurred after the first vaccine dose. Good and Walker (1967) confirmed the protection afforded by the *S. flexneri* 2a hybrid by challenge of 48 vaccinated and 48 control animals. In these studies 10 vaccinated and 36 control monkeys showed clinical signs of disease which persisted for 2 or more days. Serum titers remained stationary or increased up to eightfold following vaccination but did not further increase following challenge. Serum antibody titers in nonvaccinated animals increased up to 32-fold following challenge. Wynne *et al.* (1970) administered a *S. flexneri* 4b hybrid to monkeys but have not tested the protection afforded against challenge with this serologic type which is most prevalent in natural monkey infections.

In Japan, Honjo *et al.* (1964, 1967), Ogawa *et al.* (1964, 1967), and Takasaka *et al.* (1968) studied experimentally induced shigellosis in the cynomolgus monkey (*M. fascicularis*) and reported establishment of clinical infection, gross and histopathologic lesions, and duration of infections which were similar to those in rhesus monkeys. Honjo *et al.* (1967) found that repeated (two to four) challenges with *S. flexneri* 2a at 2- to 4-month intervals caused the development of resistance to subsequent challenge, but if challenge were delayed for 7 months, the resistance had disappeared. The transient protection was not associated with marked rises in serum antibody titer; therefore, humoral antibodies appear to be as unrelated to protection against shigellosis in cynomolgus monkeys as in rhesus monkeys. The large numbers of bacilli required to infect a host which is assumed to naturally develop disease following exposure to a few bacilli concerned Takasaka *et al.* (1968), who injected graded numbers of organisms directly into the cecal lumen of 20 cynomolgus monkeys. Whereas 10^9 to 10^{10} bacilli were required to produce clinical disease in 50–60% of monkeys challenged by the oral route, 88% of monkeys injected intracecally with 10^9, 50% of monkeys injected with 10^7, and 16% of animals injected with 10^5 virulent bacilli developed dysentery.

The progressive events which occur following induced shigellosis in rhesus monkeys have been described by Kent *et al.* (1967b), who found that gastric mucosa showed shallow ulcerations with degeneration of epithelial cells, polymorphonuclear leukocyte infiltration, and exudation with surrounding edema and hemorrhage within 2 days. The lesions progressed after 3–4 days extending deeper into the mucosa; but the lesions terminated at the esophageal and duodenal junctions. By 7–8 days healing processes were evident. Two days after infection, examination of the large intestine showed degeneration of epithelial cells, polymorphonuclear infiltration of the lamina propria, and the presence of an exudate in the lumen. By 7–8 days the colonic lesions were either mildly inflammed and healing, or were extensively ulcerated and abscessed. *Shigella flexneri* was isolated from the gastric, ileal, and colonic contents of monkeys sacrificed 3–8 days after challenge. Electron microscopy revealed that shigellae were present predominantly in the epithelial cells near the lumen of the colon (Takeuchi *et al.,* 1968). Invasion of these cells caused abnormal responses, but the invading bacilli appeared capable of multiplication and lateral spread to adjacent cells.

Immunoglobulin formation following experimental infection of patas monkeys (*Erythrocebus patas*) with a weakly pathogenic strain of *S. flexneri* 2 was followed by Felsenfeld *et al.* (1968a). The immunoalphaglobulin (IgA) antibody titers in tissues were high 3 days following infection and remained high when retested 10–11 days after infection. Immunomacroglobulin (IgM) antibodies were also elevated in all tissue but did not exceed IgA titers in the intestines. Immunogammaglobulin (IgG) antibody titers were elevated 3 days after infection and continued to rise through 11 days but did not reach the levels of IgA and IgM.

The above studies, carried out in three nonhuman primate species, have provided information on the etiologic agent and host responses to shigellosis which have greatly increased the understanding of the human infection. Further studies in naturally susceptible monkey hosts may lead to the development of methods to successfully control this disease.

3. *Cholera*

Although nonhuman primate species infected with *Vibrio cholerae* have not been found in their natural environments, Felsenfeld and associates (1966b), noting that immunoelectrophoretic patterns in sera of nonhuman primates resembled those in man, utilized *Cercopithecus aethiops* to examine immunologic responses following administration of cholera vaccines. Since *V. cholerae* does not invade the tissues, the disease results from the liberation of toxins into the lumen of the gastrointestinal tract. Immunologic protection, then, is dependent upon release of both toxin-neutralizing antibodies and vibriocidal antibodies. Following a regime which included subcutaneous and intramuscular vaccination with lipopolysaccharide antigen prepared from an El Tor vibrio, psychological stress was found to retard or inhibit the formation of serum agglutinins and coproantibodies, vibriocidal activity, and toxin-neutralizing capability of serum. However, when the monkeys adjusted to stress, as indicated by serum levels of cortisol, significant differences in immunologic response disappeared.

Studies by Felsenfeld *et al.* (1966a, 1967) and Felsenfeld (1966) were carried out to examine the immunoglobulin response to vaccines. Using three vaccine preparations (phenolized preparation containing 8×10^8 organisms/ml, lipopolysaccharide extracted from El Tor vibrios, and whole-body cholera toxin) the immunoglobulin response was found to resemble the pattern seen in brucellosis. Precipitins in peripheral blood were found in all three forms of immunoglobulin. Antibody-forming cells in local lymph glands, stimulated by vaccine injection, produced immunoglobulins two to five times more rapidly than circulating cells, and 85% of the manufactured IgM and 60% of the IgG appeared in the circulating blood. Antitoxic activity was found primarily in IgG and IgA, but the delivery of these active agents to the site of the toxin-producing microbial infection in the intestine is necessary for immunologic control of disease. When phenolized cells were injected, 15 μg of IgG were manufactured per hour by 10^8 splenic cells, about 20 μg when lipopolysaccharide was used, but approximately 35 μg after an oral dose of 0.1 LD_{50} unit of toxin. Also, manufacture of immunoglobulins by mesenteric lymph nodes and antibody-forming cells in the intestinal wall was significantly greater following oral administration of toxin than following injection of vaccine. The antitoxin produced in the intestinal wall, spleen, and mesenteric lymph nodes was transferred to the duodenum and jejunum. The results of these studies indicate that vaccines against cholera should be administered via the oral route to stimulate antibody-forming cells which can deliver immune globulins to the gastrointestinal tract.

Greer and Felsenfeld (1966) utilized *E. patas* to show that the injection of lipopolysaccharide antigen was more effective for the stimulation of antibody formation than phenolized cells. Furthermore, the response following injection of lipopolysaccharide was not dose-related between 1 and 10 mg. The increased quantity of antibody in the gastrointestinal tract following oral vaccine was reported again in 1968 by Felsenfeld and Greer. In this study, which utilized gelada baboons (*Theropithecus gelada*), the vibriocidal activity of the mesenteric nodes and the intestinal lymphatic tissue occurred more quickly and was more intense following oral administration of antigen (supernatant fluid collected after disruption of *V. cholerae* by ultrasonic treatment) than following intramuscular injection.

Additional experiments (Felsenfeld *et al.,* 1968b, 1969) utilized the chimpanzee to follow early antibody formation and to study the avidity of immunoglobulins for cholera toxin. In the latter study relatively large amounts of toxin were bound by IgG and secretory IgA, and avidity of these two immunoglobulins did not change during the 14 days following antigen administration.

C. MISCELLANEOUS BACTERIAL DISEASES

The intravenous injection of bacterial toxins or whole organisms into *M. mulatta* has been used by investigators such as Vaughn *et al.* (1968), Staab *et al.* (1969), Elsberry *et al.* (1969), Guenter and Hinshaw (1970), and Rhoda and Beisel (1970) to clarify the pathophysiologic effects of endotoxin shock. Since the patterns of host response in the monkey are similar or identical to those in man, the induced toxemias can be used to study individual organ response as well as total host response. The results of these studies are directly applicable to the early diagnosis and specific therapy of patients with endotoxin shock.

The pathophysiologic effects of bacterial exotoxins have also been studied in nonhuman primates. Herrero *et al.* (1967) reported the intravenous LD_{50} of type A partially purified botulinum toxin to be 40 mouse units (MU)/kg in the rhesus monkey and 66 MU/kg in the squirrel monkey. The intragastric LD_{50} in the rhesus monkey was 30,000 MU/kg. Clinical signs of intoxication followed the same sequence that occurs in the human. The induction of a toxemia that mimics the succession of events in the human disease provides an animal model which is large enough to study physiologic parameters and to investigate responses to the cellular level.

A description of the pathogenesis of anthrax induced by the inhalation of 10^5 to 10^6 spores of *Bacillus anthrasis* by *M. mulatta* has been presented by Berdjis *et al.* (1962). The gross and microscopic pathology of the ensuing infection was followed by sacrifice of animals daily. Significantly, dissemination of bacilli occurred without establishment of localized pulmonary lesions. Fish and Lincoln (1968) utilized rhesus monkeys and guinea pigs to study *in vivo*-produced anthrax toxin. The advantages of using the monkey for these experiments were that sufficient specimens were available from individual animals to determine the course of bacteremia, toxemia, and the immunologic response.

Zimmerman and associates (1970) briefly reviewed experimentally induced streptococcal infections in simian species and further studied the immunologic response to infection. Chimpanzees inoculated by the intranasal instillation of 2.7×10^7 chains developed a rather mild disease characterized by pharyngitis and tonsillitis which appeared along with a slight elevation in temperature and leukocytosis 2–4 days following inoculation. (See also Krushak *et al.,* 1970.) Antibodies were detectable 14–21 days after infection; however, chimpanzees did not produce antibodies to the group A streptococcal polysaccharide antigen. Treatment of the infection within 12 days of inoculation was found to block the formation of type specific antibodies in 6 of 11 chimpanzees, suggesting that the treatment of carriers would prevent the development of protective mechanisms.

Recent studies by Finegold *et al.* (1968) investigated the hematologic changes and tissue responses of rhesus monkeys with induced pneumonic plague. In addition to the typical plague pneumonia and septicemia that developed following inhalation of 3×10^4 cells of *Pasteurella pestis,* fibrin thrombi regularly occluded glomerular capillaries and fibrin deposits were observed in splenic sinusoids. These findings suggested to the authors that the release of endoxin may be one mechanism of the pathogenesis of plague. Walker (1968) investigated changes in physiologic parameters following intraperitoneal administration of plague endotoxin (lipopolysaccharide) to mice, guinea pigs, and monkeys. In *M. mulatta* depletion of liver glycogen and an increase in plasma glucose were inferred, but normal variations in these levels prevented clear-cut conclusions from being drawn. However, an increase in blood urea nitrogen was observed 15 hr following toxin injection. Pathologic changes noted, particularly in the capillaries of kidneys and lungs, were consistent with the findings of Finegold *et al.*

Chen and Meyer (1965) reviewed the comparative susceptibilities of nonhuman primate species to infection with *P. pestis.* Due to the consistent disease reported in the langur (*Semnopithecus entellus,* currently *Presbytis entellus*), this species was selected to determine if reproducible infection could be developed. Thirteen monkeys were infected by subcutaneous injection of 1×10^2 to 1.25×10^5 viable cells. As a result, 10 of the 13 infected monkeys developed temperatures in excess of 40°C, 7 had septicemia, and 6 died; however, these reactions were not dose-related. Even though humoral antibodies did not rise in all animals, five surviving

langurs were challenged by the intratracheal instillation of 1160 viable
P. pestis cells 3.5 months after the primary subcutaneous challenge to determine acquired immunity. With the exception of one animal which had
initially been challenged with 102 viable bacilli and had shown no humoral
antibody response, the survivors had acquired protective immunity. Even
though langurs may exhibit individual levels of resistance to subcutaneous
infection with *P. pestis,* the authors stated they were more consistently
susceptible than rhesus or cynomolgus monkeys. Pathologic manifestations
were similar to findings in human plague and in agreement with those reported by Finegold *et al.* and Walker in rhesus monkeys. Chen and Meyer
further stated that injection of acetone-killed plague bacilli protected four
langurs from challenge with 1.25×10^5 *P. pestis* cells for 8 months. It is
anticipated that further studies with this species will result in a model
which can be utilized for reliable tests of vaccine efficiency and potency.

Saslaw and Carlisle (1969) have reviewed their use of nonhuman primates to study serious infections which occur in the human population.
Infections with streptococci (group A), staphylococci, and diplococci
(type 3) could not be established via aerosol challenge but protracted,
sometimes fatal, disease could be induced by intravenous inoculation of
10^9 to 10^{11} viable cells. These models have been used for the *in vivo*
evaluation of various established and experimental therapeutic agents. The
advantages of the simian model were that definitive, in-depth studies of
clinical and laboratory parameters could be performed for evaluation of
new chemotherapeutic agents. This value was shown in a comparative
study (Saslaw and Carlisle, 1970) of the antistreptococcal activities of
cephalexin, nafcillin, and three isoxazolyl penicillins. The results of the
study suggest that orally administered cephalexin is an effective agent in
the treatment of β-hemolytic streptococcal infections.

IV. Summary

The etiologic agents of the principal bacterial diseases responsible for
morbidity and mortality in the human population have been recognized
for many years. Yet, pathogenic mechanisms, specific protective responses
of the host, and general host–parasite relationships are still only poorly
understood. The discovery of antibiotics and their subsequent widespread
use significantly reduced overall interest in these infectious diseases, but
it is now obvious that antimicrobial agents are not a panacea.

Until the advent of the antibiotic era, investigations of infectious diseases were limited to *in vivo* studies in rodents. Unfortunately, induced
disease in rodents does not simulate naturally occurring infections in the

human host. In the last 20 years, reproducible models of human disease have been established in nonhuman primate species, many of which are naturally susceptible to the infectious agents. These models of human disease permit studies of many parameters of infection at one time, but principally progression of infection in relation to time, immunologic response, and the effects of therapy or prophylactic procedures on the course of disease.

Induced disease in the nonhuman primate is not intended as a primary screen of therapeutic or prophylactic agents, but should be an essential step prior to preclinical trials in the systematic investigation of new drugs or vaccines. In this way relationships of dose or serum levels of therapeutic agents to response can be determined; toxic manifestations can be noted; immunologic response and length of protection can be investigated; and related measurements of host–parasite interaction can be followed.

The massive inocula required to induce progressive infections in nonhuman primates reflect the health status at the beginning of the experiment. Studies of these nonspecific defense mechanisms should lead to a better definition of natural resistance as well as providing a means for determining effects of pollutants or other agents on host susceptibility to infection.

The specific disease models discussed in preceding sections indicate the value of studies in nonhuman primates to control of infectious diseases in the human population. Further studies are indicated with these models; but models of other diseases should be established, particularly those of respiratory diseases. With the ongoing work and new areas of research yet to be developed, we can anticipate developing the knowledge which will be the means of eradicating many of the world's major infectious diseases.

REFERENCES

Agarwal, K. C., and Chakravarti, R. N. (1969). *J. Ass. Physicians, India* 17, 409.
Barclay, W. R., Anacker, R. L., Brehmer, W., Leif, W., and Ribi, E. (1970). *Infect. Immunity* 2, 574.
Berdjis, C. C., Gleiser, C. A., Hartman, H. A., Kuehne, R. W., and Gochenour, W. S. (1962). *Brit. J. Exp. Pathol.* 43, 515.
Bywater, J. E. C., Hartley, E. G., Rutty, D. A., and Ackerley, E. T. (1962). *Vet. Rec.* 74, 1414.
Carpenter, K. P., and Cooke, E. R. N. (1965). *J. Comp. Pathol.* 75, 201.
Carr, R. E., and Henkind, P. (1962). *Arch. Ophthalmol.* 67, 566.
Chaparas, S. D., Hedrick, S. R., Clark, R. G., and Garman, R. (1970). *Amer. J. Vet. Res.* 31, 1437.
Chen, T. H., and Meyer, K. F. (1965). *J. Infec. Dis.* 115, 456.
Conzelman, G. M., Jr. (1955). *Antibiot. Chemother. (Washington, D. C.)* 5, 444.
Conzelman, G. M., Jr. (1956). *Amer. Rev. Tuberc. Pulm. Dis.* 74, 739.
Conzelman, G. M., Jr., and Jones, R. K. (1956). *Amer. Rev. Tuberc. Pulm. Dis.* 74, 802.

58   ROBERT C. GOOD

Crowle, A. J., Mitchell, R. J., and Petty, T. L. (1963). *Amer. Rev. Resp. Dis.* **88**, 716.

Cvjetanovic, B., Mel, D. M., and Felsenfeld, O. (1970). *Bull. WHO* **42**, 499.

Dack, G. M., Jordan, E. O., and Wood, W. L. (1929). *J. Prev. Med.* **3**, 153.

Dolowy, W. C., Frank, M. H., Cox, G. E., and Hesse, A. L. (1958). *Amer. J. Vet. Res.* **19**, 225.

Dunn, F. L. (1968). *In* "The Squirrel Monkey" (L. A. Rosenblum and R. W. Cooper, eds.), pp. 31–68. Academic Press, New York.

Edsall, G., Gaines, S., Landy, M., Tigertt, W. D., Sprinz, H., Trapani, R. J., Mandel, A. D., and Benenson, A. S. (1960). *J. Exp. Med.* **112**, 143.

Elsberry, D. D., Rhoda, D. A., and Beisel, W. R. (1969). *J. Appl. Physiol.* **27**, 164.

Felsenfeld, O. (1966). *Bull. Pathol.* **7**, 303.

Felsenfeld, O., and Greer, W. E. (1968). *Immunology* **14**, 319.

Felsenfeld, O., Felsenfeld, A. D., Greer, W. E., and Hill, C. W. (1966a). *J. Infec. Dis.* **116**, 329.

Felsenfeld, O., Hill, C. W., and Greer, W. E. (1966b). *Trans. Roy. Soc. Trop. Med. Hyg.* **60**, 514.

Felsenfeld, O., Greer, W. E., and Felsenfeld, A. D. (1967). *Nature (London)* **213**, 1249.

Felsenfeld, O., Greer, W. E., and Jiricka, Z. (1968a). *Lab. Invest.* **19**, 146.

Felsenfeld, O., Greer, W. E., Kirtley, B., and Jiricka, Z. (1968b). *Trans. Roy. Soc. Trop. Med. Hyg.* **62**, 278.

Felsenfeld, O., Felsenfeld, A. D., and Greer, W. E. (1969). *Trans. Roy. Soc. Trop. Med. Hyg.* **63**, 846.

Fiennes, R. N. T-W- (1970). *In* "Infections and Immunosuppression in Subhuman Primates" (H. Balner and W. I. B. Beverdige, eds.), pp. 149–154. Munksgaard, Copenhagen.

Fife, E. H., Jr., Kruse, R. H., Toussaint, A. J., and Staab, E. V. (1970). *Lab. Anim. Care* **20**, 969.

Finegold, M. J., Petery, J. J., Berendt, R. F., and Adams, H. R. (1968). *Amer. J. Pathol.* **53**, 99.

Fish, D. C., and Lincoln, R. E. (1968). *J. Bacteriol.* **95**, 919.

Formal, S. B., LaBrec, E. H., Kent, T. H., and Falkow, S. (1965a). *J. Bacteriol.* **89**, 1374.

Formal, S. B., LaBrec, E. H., Palmer, A., and Falkow, S. (1965b). *J. Bacteriol.* **90**, 63.

Formal, S. B., Kent, T. H., Austin, S., and LaBrec, E. H. (1966a). *J. Bacteriol.* **91**, 2368.

Formal, S. B., Kent, T. H., May, H. C., Palmer, A., Falkow, S., and LaBrec, E. H. (1966b). *J. Bacteriol.* **92**, 17.

Fox, H. (1923). "Disease in Captive Wild Mammals and Birds." Lippincott, Philadelphia, Pennsylvania.

Francis, J. (1956). *J. Comp. Pathol.* **66**, 123.

Freerksen, E., and Rosenfeld, M. (1963). *Beitr. Klin. Spezifischen Tuberk. Forsch.* **127**, 386.

Good, R. C. (1967). Unpublished data.

Good, R. C. (1968). Unpublished data.

Good, R. C. (1970). Unpublished data.

Good, R. C. (1965). *Trans. 24th VA, Armed Forces Res. Conf. Pulm. Dis.* p. 46.

Good, R. C. (1968). *Ann. N. Y. Acad. Sci.* **154**, 200.

Good, R. C., and May, B. D. (1971). *Infect. Immunity* **3**, 87.

Good, R. C., and Schmidt, L. H. (1968). Unpublished data.

Good, R. C., and Walker, R. V. (1967). Unpublished data.

Good, R. C., Hoffmann, R. A., and Schmidt, L. H. (1961). *Trans. 20th VA, Armed Forces Res. Conf. Pulm. Dis.* p. 246.

Good, R. C., Hoffmann, R. A., Smith, C. C., and Schmidt, L. H. (1964). *Trans. 23rd VA, Armed Forces Res. Conf. Pulm. Dis.* p. 10.

Good, R. C., May, B. D., and Kawatomari, T. (1969). *J. Bacteriol.* **97**, 1048.

Greenstein, E. T., Doty, R. W., and Lowy, K. (1965). *Lab. Anim. Care* **15**, 74.

Greer, W. E., and Felsenfeld, O. (1966). *Ann. Trop. Med. Parasitol.* **60**, 417.

Guenter, C. A., and Hinshaw, L. B. (1970). *Proc. Soc. Exp. Biol. Med.* **134**, 780.

Habel, K. (1947). *Amer. Rev. Tuberc.* **55**, 77.

Haberman, R. T., and Williams, F. P., Jr. (1957). *Amer. J. Vet. Res.* **18**, 419.

Herrero, B. A., Ecklund, A. E., Streett, C. S., Ford, D. F., and King, J. K. (1967). *Exp. Mol. Pathol.* **6**, 84.

Heywood, R., and Hague, P. H. (1969). *Lab. Anim. Handb.* **4**, 43–50.

Honjo, S., Takasaka, M., Fujiwara, T., Nakagawa, M., Andoo, K., Ogawa, H., Takahashi, R., and Imaizumi, K. (1964). *Jap. J. Med. Sci. Biol.* **17**, 307.

Honjo, S., Takasaka, M., Fujiwara, T., Kaneko, M., Imaizumi, K., Ogawa, H., Mise, K., Nakamura, A., and Nakaya, R. (1967). *Jap. J. Med. Sci. Biol.* **20**, 341.

Hunt, D. E., Pittillo, R. F., Deneau, G. A., Schabel, F. M., Jr., and Mellett, L. B. (1968). *Lab. Anim. Care* **18**, 182.

Karassova, V., Weissfeiler, J., and Krasznay, E. (1965). *Acta Microbiol. Acad. Sci. Hung.* **12**, 275.

Kennard, M. A., Schroeder, C. R., Trask, J. D., and Paul, J. R. (1939). *Science* **89**, 442.

Kent, T. H., Formal, S. B., and LaBrec, E. H. (1966). *Arch. Pathol.* **82**, 272.

Kent, T. H., Formal, S. B., and LaBrec, E. H. (1967a). *Arch. Pathol.* **84**, 300.

Kent, T. H., Formal, S. B., LaBrec, E. H., Sprinz, H., and Maenza, R. M. (1967b). *Amer. J. Pathol.* **51**, 259.

Kourany, M., and Porter, J. A., Jr. (1969). *Lab. Anim. Care* **19**, 336.

Krushak, D. H., Zimmerman, R. A., and Murphy, B. L. (1970). *J. Amer. Vet. Med. Ass.* **157**, 742.

LaBrec, E. H., Schneider, H., Magnani, T. J., and Formal, S. B. (1964). *J. Bacteriol.* **88**, 1503.

Larkin, J. C., Jr. (1959). *Trans. 18th VA, Armed Forces Conf. Chemother. Tuberc.* p. 324.

Lund, E., and Petersen, K. B. (1959). *Acta Pathol. Microbiol. Scand.* **45**, 309.

Minden, P., and Farr, R. S. (1969). *J. Exp. Med.* **130**, 931.

Mysorekar, N. R., Chakravarti, R. N., Chawla, L. S., and Chhuttani, P. N. (1966). *J. Ass. Physicians, India* **14**, 583.

Ogawa, H., Takahashi, R., Honjo, S., Takasaka, M., Fujiwara, T., Ando, K., Nakagawa, M., Muto, T., and Imaizumi, K. (1964). *Jap. J. Med. Sci. Biol.* **17**, 321.

Ogawa, H., Nakamura, A., Nakaya, R., Mise, K., Honjo, S., Takasaka, M., Fujiwara, T., and Imaizumi, K. (1967). *Jap. J. Med. Sci. Biol.* **20**, 315.

Rhoda, D. A., and Beisel, W. R. (1970). *Amer. J. Vet. Res.* **31**, 1846.

Rich, A. R. (1951). "The Pathogenesis of Tuberculosis." Thomas, Springfield, Illinois.

Ruch, T. C. (1959). "Diseases of Laboratory Primates." Saunders, Philadelphia, Pennsylvania.

Saslaw, S., and Carlisle, H. N. (1969). *Ann. N. Y. Acad. Sci.* **162**, 568.

Saslaw, S., and Carlisle, H. N. (1970). *Amer. J. Med. Sci.* **259**, 383.

Schmidt, I. G. (1966). *Ann. N. Y. Acad. Sci.* **135**, 759.

Schmidt, I. G., and Schmidt, L. H. (1966). *J. Neuropathol. Exp. Neurol.* **25**, 40.

Schmidt, L. H. (1956a). *Amer. Rev. Tuberc. Pulm. Dis.* **74**, Part II, 138.

Schmidt, L. H. (1956b). *Trans. 15th VA, Armed Forces Conf. Chemother. Tuberc.* p. 353.

Schmidt, L. H. (1966). *Ann. N. Y. Acad. Sci.* **135**, 747.

Schmidt, L. H., Hoffmann, R., and Jolly, P. N. (1955). *Trans. 14th VA, Armed Forces Conf. Chemother. Tuberc.* p. 226.

Schmidt, L. H., Hoffmann, R., and Steenken, W., Jr. (1957). *Amer. Rev. Tuberc. Pulm. Dis.* **75**, 169.

Schmidt, L. H., Grover, A. A., Hoffmann, R., Rehm, J., and Sullivan, R. (1958). *Trans. 17th VA, Armed Forces Conf. Chemother. Tuberc.* p. 264.

Schmidt, L. H., Grover, A. A., and Hoffmann, R. (1959). *Trans. 18th VA, Armed Forces Conf. Chemother. Tuberc.* p. 312.

Schmidt, L. H., Lang, J., Good, R. C., and Hoffmann, R. (1962). *Trans. 21st VA, Armed Forces Res. Conf. Pulm. Dis.* p. 355.

Schmidt, L. H., Good, R. C., Mack, H. P., Zeek-Minning, P., and Schmidt, I. G. (1963). *Trans. 22nd VA, Armed Forces Res. Conf. Pulm. Dis.* p. 262.

Sedgwick, C., Parcher, J., and Durham, R. (1970). *J. Amer. Vet. Med. Ass.* **157**, 724.

Snyder, S., Peace, T., Soave, O., and Lund, J. (1970). *J. Amer. Vet. Med. Ass.* **157**, 712.

Staab, E. V., Niederhuber, J., Rhoda, D. A., Faulkner, C. S., II, and Beisel, W. R. (1969). *Appl. Microbiol.* **17**, 394.

Steenken, W., Jr., Montalbine, V., Smith, M. M., and Wolinsky, E. (1958). *Trans. 17th VA, Armed Forces Conf. Chemother. Tuberc.* p. 368.

Stones, P. B. (1969). *Lab. Anim. Handb.* **4**, 11–17.

Takasaka, M., Honjo, S., Fujiwara, T., Hagiwara, T., Ogawa, H., and Imaizumi, K. (1964). *Jap. J. Med. Sci. Biol.* **17**, 259.

Takasaka, M., Honjo, S., Fujiwara, T., Imaizumi, K., Ogawa, H., Nakaya, R., and Nakamura, A. (1968). *Jap. J. Med. Sci. Biol.* **21**, 275.

Takeuchi, A., Formal, S. B., and Sprinz, H. (1968). *Amer. J. Pathol.* **52**, 503.

Thomas, J. P., Baughn, C. O., Wilkinson, R. G., and Shepherd, R. G. (1961). *Amer. Rev. Resp. Dis.* **83**, 891.

Tribe, G. W. (1969). *Lab. Anim. Handb.* **4**, 19–23.

Vaughn, D. L., Gunter, C. A., and Stookey, J. L. (1968). *Surg. Gynecol. Obstet.* **126**, 1309.

Walker, R. V. (1968). *J. Infec. Dis.* **118**, 188.

Weinberg, M. (1906). *C. R. Soc. Biol.* **61**, 648.

Wolf, R. H., Bullock, B. C., and Clarkson, T. B. (1967). *J. Amer. Vet. Med. Ass.* **151**, 914.

Wynne, E. S., Henrikson, D. M., and Daye, G. T., Jr. (1970). *Appl. Microbiol.* **19**, 731.

Youmans, G. P., Youmans, A. S., and Doub, L. (1958). *Trans. 17th VA, Armed Forces Conf. Chemother. Tuberc.* p. 365.

Zimmerman, R. A., Krushak, D. H., Wilson, E., and Douglas, J. D. (1970). *J. Infec. Dis.* **122**, 280.

CHAPTER 5

Virus Research

S. S. KALTER

I. Introduction

Use of nonhuman primates in biomedical research is not a recent development. Examination of the literature concerned with experimental medicine provides evidence for the use of one or another species of monkey or ape from early recorded time. One need only quickly examine the reports of a variety of conferences (Whitelock, 1958, 1960; Vagtborg, 1965, 1967, 1968; R. N. T-W Fiennes, 1966; Beveridge, 1969a,b; Goldsmith and Moor-Jankowski, 1969, 1972; Hofer, 1969a,b; Balner and Beveridge, 1970) to realize that these animals are now firmly entrenched among the accepted laboratory animals, and, perhaps more importantly, are replacing the mouse, guinea pig, rabbit, rat, and others as the desired animals for certain definitive experiments. Surprisingly, however, in spite of the early realization that there was a certain inherent danger associated with the use of monkeys in the laboratory (Sabin and Wright, 1934), it was not until the rhesus (*Macaca mulatta*) monkey kidney cell was used for cultivation of human viruses *in vitro* (Enders, 1952) that more extensive employment of monkeys and apes for virus research was undertaken and a greater curiosity regarding their disease potential developed.

Significantly, it is this usage as well as man's encroachment upon arboreal areas and development of the pet trade that has become so extensive that certain species of simians are now practically extinct and many other species are endangered. Some others are so costly that this fortunately places a limitation on their availability to man's desires.

Difficulties arising in procurement of monkeys and especially apes has suggested the need for the development of breeding programs. Such programs, if conceived in sufficient time and numbers, may be of value in supplying the needed numbers of animals. Laboratory-bred animals may also have the added attraction of being "cleaner" than their counterparts that have been brought from capture in the "wild" to the laboratory through a multitude of unhygienic and contaminating experiences (Kalter, 1971b; Kalter and Heberling, 1971b; Kalter *et al.,* 1966a; World Health Organization, 1971).

This chapter is concerned with providing information on nonhuman primates used in virus research.

II. Early Virus Research—Prior to 1950

As one would expect, Pasteur was among the pioneers in the experimental use of monkeys in the laboratory. He and his collaborators (Pasteur *et al.,* 1884a,b) demonstrated that rabies virus would lose its virulence for dogs by passage in monkeys. Halberstaedter and von Prowazek (1907) demonstrated the susceptibility of apes to trachoma* and Thomas (1907) that same year reported the occurrence of yellow fever in the chimpanzee. Shortly thereafter, Landsteiner and Popper (1909) demonstrated the susceptibility of monkeys (*Papio hamadryas* and *Macaca mulatta*) to experimentally transmitted poliomyelitis. Triturated spinal cord obtained from human cases produced paralysis in these

---

* Now considered as member of a separate family and not true viruses, the psittacosis–lymphogranuloma venereum (Bedsoniae) group of agents (Chlamydozoaceae) will be included here for historical purposes.

animals within 6 days after inoculation. Animals that died or were sacrificed showed at necropsy cord lesions similar to those observed in man. Flexner and Lewis (1910a), were similarly successful in infecting monkeys (rhesus) by intracerebral, subcutaneous, intraperitoneal, intravenous, intrathecal, and intraneural inoculation of human poliomyelitis patient cord material. Cord suspensions obtained from infected monkeys produced disease when passaged serially in other monkeys. Filtrates of infected cords as well as nasopharyngeal washings were likewise capable of infecting monkeys. Similar studies were also reported by Thomsen (1912).

In a series of experiments, J. F. Anderson and Goldberger (1911a,b,c) and Goldberger and Anderson (1911) were able to demonstrate the susceptibility of monkeys to measles. As a result, it was shown that the blood and buccal and nasal secretions of measles patients contained the virus. These findings were rapidly confirmed by Nicolle and Conseil (1911) and Lucas and Prizer (1912). More detailed studies regarding the susceptibility of monkeys to measles virus were reported also by Nicolle and Conseil (1920). As a result of monkey inoculations, M. H. Gordon (1914) considered mumps to be due to a filterable agent. Kraus and Kantor (1917) also demonstrated the susceptibility of monkeys to poliomyelitis. Blake and Trask (1921a) inoculated rhesus monkeys intratracheally with filtered and unfiltered nasopharyngeal washings from measles patients obtained 6 days prior to, and 22 hr following, appearance of rash. The disease in monkeys was described as closely resembling human measles. The incubation period was 6–10 days and was followed by characteristic symptoms—listlessness, leukopenia, catarrhal conjunctivitis, rash on labial mucosa, and a discrete red maculo-papular skin rash. Disease was transmitted from monkey to monkey by intratracheal inoculation of skin and buccal mucosa suspensions or by intravenous inoculation of blood obtained 7–13 days after infection. The symptomatology and pathology of measles infections of monkeys were described separately (Blake and Trask, 1921b).

One of the earliest literature references to the wild monkey as a possible reservoir for yellow fever virus was that of Balfour (1914). He reported that the natives of Trinidad could always tell when yellow fever would occur because of the increased number of dead howler monkeys noted prior to the epidemic.

Blaxall (1923) suggested that the monkey was the most susceptible animal to variola virus. Five days after intracutaneous inoculation of human pustular material, papules, which in turn developed into typical pocks by the 10th day, were observed on the monkeys. These findings were confirmed by the studies of Bleyer (1922) and M. H. Gordon (1925). Meanwhile Bras (1922) reported the occurrence of smallpox in an orangutan (*Pongo pygmaeus*) at the Djakarta Zoo. Attempts to recover the virus of

chickenpox (herpes zoster) resulted in mixed findings. R. Cole and Kuttner (1925) inoculated rhesus monkeys with patient material by a variety of pathways (intracerebral, intracutaneous, intraneural, and corneal
routes) and vervets by the testicular route with negative results; Rivers
(1926), however, inoculated human chickenpox material intratesticularly
and reported the subsequent presence of acidophilic intranuclear inclusion
bodies. Another herpesvirus (herpes simplex), was found by Teisser
et al. (1925) to produce vesiculopustules and very small papules after
cutaneous inoculation of *Callithrix*.

During this decade there was an upsurge in interest regarding yellow
fever. The classic studies of Reed and co-workers on the role of mosquitoes in the spread of this disease was well established. The 1928 Rio de
Janeiro epidemic introduced the sylvatic concept of yellow fever. Thomas
(1907) had suggested the possibility that chimpanzees might act as a
reservoir for the virus of yellow fever. Then Stokes *et al.* (1928) clearly
demonstrated that rhesus monkeys were susceptible to yellow fever by
inoculation of virulent material or by infected *Aedes aegypti* mosquitoes,
thus establishing the role monkeys may play in the epidemiology of this
disease. Hindle (1929a), the following year, demonstrated that monkey
livers from animals infected with yellow fever maintained their virulence
for about 3 months past infection. An attempt was made to produce a
vaccine by using phenolized or formalized infected monkey liver and
spleen.

There was also during the late 1920's a renewed interest in using monkeys for studies on poliomyelitis. Aycock and Kagan (1927) demonstrated that monkeys may be actively immunized against poliomyelitis by
repeated intracutaneous inoculation with live virus. This was substantiated by the studies of Kling *et al.* (1929). Meanwhile Amoss (1928)
described poliomyelitis in monkeys and noted the similarity to the human
disease.

Other viruses were also studied in monkeys during this period. Although
rabies as a natural infection of monkeys is apparently quite rare, Levaditi
et al. (1926) reported on the susceptibility of a number of simians
(*M. sinicus, M. cynomolgus, M. callithrix, Cercocebus fuliginosus, P. mor-
mom,* and *Troglodytes niger*) to rabies infection. Mouquet (1926) indicated that chimpanzees were susceptible to influenza virus with development of clinical symptoms. Hindle (1929b) reviewed the relationship of
monkeys to human virus diseases.

The next 10 years again saw some moderate involvement of nonhuman
primates in virus research. Most activity was associated with studies attempting to understand poliomyelitis. Aycock and Kramer (1930) demonstrated the presence of poliovirus-neutralizing antibodies by testing sera in

rhesus monkeys. The neutralizing properties of normal rhesus and cebus monkey sera toward the virus of poliomyelitis were described by Jungeblut and Engle (1932). Paul and Trask (1932) observed that poliomyelitis virus was also able to produce abortive forms of disease in monkeys. *Macaca cynomolgus* monkeys, after feeding with poliomyelitis virus, developed intravascular lesions according to Saddington (1932). These findings were confirmed by Kling *et al.* (1934), Vignec *et al.* (1939), Lepine and Sedallian (1939), and Burnet and his collaborators (1939a). Strain differences among the polioviruses (not completely recognized until several years later) was suspected by the studies of Flexner (1937) as a result of reinfection experiments in monkeys. It was not until the 1930's that investigators began to realize, however, that poliomyelitis might be a complex disease as a result of studies with simians. Mackay and Schroeder (1935) reported that the spider monkey (*Ateles ater*) was resistant to infection with the virus of poliomyelitis. Similar findings were reported with another South American monkey (marmosets) by Grossman and Kramer (1936). Adaptation of poliovirus, Lansing strain (now recognized as type 2), to cotton rats (*Sigmodon hispidus hispidus*) was accomplished by C. Armstrong (1939), but only after serial passage through monkeys. Meanwhile, Kolmer and Rule (1934) were considering the possibility of a poliomyelitis vaccine and attempted to vaccinate monkeys against poliovirus infection by preparing a 4% suspension of infected monkey cord in 1% sodium ricinoleate. An outbreak of paralytic disease in monkeys was described in zoo animals by Goldman (1935).

During the 1930's there was also a certain amount of studies involving monkeys in an attempt to understand the arthropod-borne viruses (arboviruses). A variety of Old World monkeys was used by Simmons *et al.* (1931) to experimentally study dengue. Another virus disease, louping ill, was experimentally examined in monkeys by Hurst (1931). Similar investigations with this virus were conducted by Elford and Galloway (1933) in which it was demonstrated that infection could result from intranasal as well as by intracerebral inoculations. Experimentally Rift Valley fever was found by Findlay (1932) to be induced in monkeys by a variety of routes of inoculation—cutaneous, subcutaneous, or intranasal. Yellow fever in monkeys was again examined by F. L. Soper *et al.* (1934) who demonstrated that silent yellow fever would be detected by routine postmortem removal of liver tissue from rapid yellow fever deaths. It was then shown by Findlay and Clarke (1935) that an encephalitis would be produced in monkeys by the intranasal instillation of a neurotropic strain of virus. It was also demonstrated by Findlay *et al.* (1936) that yellow fever antibodies existed in the sera of various African animals including monkeys. It will be recalled that one of the earliest virologic infections (yellow fever) of the chimpanzee was reported by Thomas (1907); this question was again raised by Pettit and Aguessy (1932). Findlay and MacCallum (1939) fed several strains of yellow virus to monkeys by means of a gastric tube producing a viremia and serum-neutralizing antibodies. These investigators also demonstrated that yellow fever virus was rapidly inactivated by the gastric juices of monkeys and could not be detected in feces.

Human cases of an encephalitis occurring in St. Louis (St. Louis encephalitis, SLE) were shown to be of viral etiology by Muckenfuss and co-workers (1933). Simultaneous inoculation of rhesus monkeys, intra-cerebral and intraperitoneal, with human brain material produced in these animals in 8–14 days an increased temperature, drowsiness, intention tremors, muscular weakness, and incoordination of the extremities. Upon sacrifice of the animals it was found that the histopathologic changes observed in sections of the central nervous system were similar to those observed in man. It was noted that only 40% of the inoculated animals responded with a detectable infection. C. Armstrong and Lillie (1934), in their studies with a fatal human case of SLE, isolated a virus that produced a lymphocytic choriomeningitis (LCM) in monkeys. Meanwhile, in Japan, it was reported that Japanese encephalitis virus (Japanese B) produced a nonsuppurative meningoencephalitis upon intracerebral inoculation of monkeys (Endo et al., 1960a,b). As a follow-up of their previous studies, both C. Armstrong et al. (1936) and Lillie (1936) independently studied LCM virus in monkeys and noted the extensive lymphocytic infiltration throughout the animals' bodies.

A number of other virus diseases were also studied in nonhuman primates. For example, Johnson and Goodpasture (1935) found that mumps virus, after several monkey passages, was still infectious for man. In other experiments these same investigators (Johnson and Goodpasture, 1936) reported that monkeys recovered from mumps were actively immune. Passive immunity using human serum did not protect the animals against intraperitoneal infection. It was then reported by Bloch (1937) that mumps virus produced in monkeys a parotitis with focal acinar necrosis with edema, hemorrhage, and a lymphocyte infiltration as secondary reactions. Dochez et al. (1930) reported on the transmission of the common cold to anthropoid apes by means of a filterable agent. Fowl plague virus was found to produce only a mild disease in rhesus monkeys according to Findlay and Mackenzie (1937a). Intratesticular inoculation of monkeys with variola virus resulted in an orchitis with rash on the scrotum. Teissier et al. (1931) also noted that passage from monkey to monkey could be easily accomplished with alastrim virus and that similar lesions could be produced by inoculating monkeys with pus from human cases.

Hurst (1936) described pseudorabies (infectious bulbar paralysis, mad itch) infections in rhesus monkeys. Chickenpox had previously been shown by Rivers (1926) to be able to proliferate in monkey testicular tissue. Eckstein (1933) also investigated this disease in monkeys describing the clinical and experimental results as they pertain to chickenpox encephalitis.

Two very important events also occurred during this decade. One was the observation by Cowdry and Scott (1935) on the occurrence of nuclear inclusions in the salivary glands of cebus (*Cebus fatuellus*) monkeys. This observation is now recognized as indicating the presence of a cytomegalovirus. The second observation was that of Sabin and Wright (1934) and by Sabin (1949) on the occurrence in monkeys of a herpesvirus that was highly lethal for man. Perhaps of greater significance than the actual recognition of the effect of this virus, was the indication that the individual who died had been bitten by a "normal" rhesus monkey. Burnet and collaborators (1939b) demonstrated the antigenic relationship of the "B virus"* to other herpesviruses.

Investigations also continued into the pathogenesis of the psittacosis–lymphogranuloma group of agents as seen in monkeys. Bedson and Western (1930) demonstrated that intracerebral inoculations of monkeys with psittacosis agents produced a meningoencephalitis. Intranasal or intratracheal instillation resulted in a pneumonia similar to that described for man. Similar findings were also reported by M. H. Gordon (1930), Rivers and Berry (1931a–d), and Rivers et al. (1931). Francis and Magill (1938) recorded the isolation from ferrets of the "meningopneumonitis virus" which was pathogenic for monkeys by subcutaneous or intraperitoneal inoculation, but not by intranasal inoculation.

The trachoma story is interesting as it illustrates a number of problems inherent to animal experimentation. Noguchi (1927, 1928), in a series of experiments, reported that pure cultures of a gram-negative bacillus, *Bacterium granulosis,* when inoculated beneath the conjunctiva and on the scarified conjunctiva of rhesus monkeys, produced in approximately 2–4 weeks a granular conjunctivitis. The lesion progressed slowly for a number of months until the inoculated conjunctiva appeared similar to that seen in a human trachomatous conjunctiva. Cicatrization, characteristic of the human disease, was reported to have occurred in one monkey. Histologically the human and monkey disease was similar. Monkey-to-monkey transmission was also reported. These findings were confirmed in a series of studies by Olitsky (1930) and Olitsky et al. (1931a,b,c). However, a number of other investigators were unable to isolate this bacterium from a large series of human cases. Continued investigations by Wilson (1928),

---

* Named for the patient dying of the disease.

Olitsky and Tyler (1933), and Olitsky *et al.* (1933) found that the confusion regarding trachoma was based upon the failure of all investigators to similarly define trachoma as a clinical entity and the spontaneous development in monkeys of a follicular conjunctivitis which simulates in many respects the human disease. Fairbrother and Hurst (1932) described the occurrence of spontaneous diseases in monkeys.

The 1940's were in many respects similar to the previous 10–20 years with regard to the use of monkeys in virus research. Abnormal findings in monkeys as seen at necropsy was described by Kennard (1941). Poliovirus studies in monkeys continued to be one of the major virologic programs. Again it was pointed out by Burnet (1940), Aycock (1940), and Faber *et al.* (1944) that the cynomolgus monkey was more susceptible than the rhesus monkey to poliovirus infection following oral feeding or swabbing of the nasopharyngeal area. Experimental poliomyelitis in the African green monkey (*Cercopithecus aethiops sabaeus*) following administrations of poliovirus by oral and other routes was reported by Trask and Paul (1941). Chimpanzees could be infected by feeding of poliovirus even after bilateral section of the olfactory tract as indicated by Howe and Bodian (1941). Howe and Bodian (1940) also described the portals of entry of poliovirus in the chimpanzee, how the virus penetrated the gastrointestinal tract of this animal (Howe and Bodian, 1942), and accidental poliomyelitis in the chimpanzee (Howe and Bodian, 1944). A few years later, these investigators (Howe and Bodian, 1945) found that poliovirus hyperimmune serum administered to chimpanzees prior to infection had no effect. Aisenberg and Grubb (1943), however, were able to produce poliomyelitis in rhesus monkeys by instillation of infectious materials into the pulp canal of three anterior teeth. The cavities were sealed to prevent leakage of the virus. Paralysis occurred in 8 days. Encephalomyelitis virus of mice ("mouse poliomyelitis") was found to be nonpathogenic for monkeys by Theiler and Gard (1940a,b). Encephalomyocarditis in apes was ascribed by Helwig and Schmidt (1945) as the result of infection with a filtrable virus.

Continued interest in use of monkeys for the study of arboviruses was evidenced by the investigations of Shortt *et al.* (1940) with the demonstration that rhesus monkeys would develop mild phlebotomus fever following inoculation of patient blood. Cultivated virus produced an inapparent infection. Simmons (1940) reported that monkeys from nonendemic areas may be induced to develop a very mild form of dengue. Two African viruses were studied in monkeys: Smithburn *et al.* (1940) produced an encephalitis in monkeys following the intracerebral inoculation of West Nile virus. Degenerative changes in the Purkinje cells were similar to those seen in louping ill. Another virus, this one isolated in the

Semliki forest of Western Uganda, killed rhesus monkeys following intra-cerebral inoculation according to Smithburn and Haddow (1949). The lesions produced in these monkeys were similar to those described for the equine encephalomyelitis viruses. It was suggested that the red-tailed (*Cercopithecus nictitans*) monkey may be one of the natural reservoirs for this virus. The relationship of baboons to yellow fever was discussed by Haddow *et al.* (1947). Development of vaccines for yellow fever and their evaluation in animals led to the use of monkeys. According to a number of studies more encephalitis was produced in man after use of a more neurovirulent strain of 17D yellow fever vaccine as evaluated in monkeys than the parent 17D strain (Fox *et al.*, 1942; Fox and Penna, 1943). Similar results were also reported for the French neurotropic strain of vaccine (Yellow Fever Vaccination, W.H.O. Mongr. No. 30, Geneva, 1956). Paul *et al.* (1948) were able to demonstrate the susceptibility of the chimpanzee to phlebotomus fever and dengue.

Measles virus carried through 20 embryonate egg passages was still capable of producing illness in monkeys, according to Rake and Shaffer (1940). Continued investigations by Shaffer *et al.* (1941) with measles virus in monkeys reported the successful transmission of the disease using blood drawn from measles patients. The incubation period for these animals varied from 6 to 16 days. According to these investigators the disease in monkeys was milder than that produced in children—no fever, cough, or coryza. Swan and Mawson (1943) reported that mumps could be successfully transmitted to *M. mulatta, M. irus, M. nemestrinus,* and *M. maurus.* Typical parotitis with glandular enlargement and facial edema resulted from the direct inoculation of virus via Stensen's duct. Serial passage in monkeys by parotid gland suspension occurred when the glands were removed 4–7 days after inoculation. It was also reported that monkeys would be infected with cerebrospinal fluid obtained from patients with mumps encephalitis.

Bland (1944a) attempted to resolve the question regarding the etiology of trachoma and noted that the grivet and vervet monkeys were susceptible to spontaneous folliculosis but could be infected with trachoma. Differentation of the two diseases was considered difficult, if not impossible. In another study (Bland, 1944b) it was reported that baboons (*P. hama-dryas*) also develop spontaneous folliculosis. It was then suggested by Bland (1944c) that the rhesus monkey was the most reliable simian for experimental trachoma studies; however, only 50% of the animals were susceptible.

In 1949, a report was published by Enders *et al.* describing the successful growth of the Lansing strain of poliomyelitis virus in culture of various human embryonic tissue. This was the beginning of a new era as regards the use of cell cultures for virus research. Without detracting from the impact of this report (these investigators received the Nobel Prize for their efforts) it should be pointed out that cell culture had been used for virus research for many years previous to this study. For example, Huang (1943) was able to use cultures of minced 9-day-old chick embryos for titrating and neutralization of western equine encephalomyelitis virus.

III. Current Virus Research in Monkeys

It is obvious that monkeys had been used in the virus laboratory prior to 1950; the numerous investigators mentioned above attest to this activity. Nonetheless, the current explosive demand for monkeys or their tissues was a direct consequence of the finding that poliovirus could be cultivated in proliferating tissue cells maintained in culture (Enders *et al.*, 1949; Robbins and Enders, 1950). As a result, many of the large number of simians now employed in biologic studies are used in the general area of virology. A large segment of these animals was used for the preparation of cell cultures for virus vaccine purposes; many are, however, used as experimental animals. According to Goodwin (1970) a total of 951 projects using primates was funded during 1967. Of these, the rhesus monkey (*M. mulatta*) was the most commonly used species. Approximately 15 genera of primates were indicated as currently used for research purposes in the United States and it is estimated that at least 100,000 monkeys and apes are imported annually; during the years 1954 to 1965 nearly 2 million nonhuman primates were imported into this country (Hughes, 1967). Table I lists the various monkeys and apes that have been reported as used in the laboratory these last 20 years.

As a point of interest, primarily because it emphasizes the lack of study in many instances with other simians, the majority of virus isolates have been derived from two genera: *Macaca* (*M. mulatta* and *M. cynomolgus*) and *Cercopithecus*. An ancillary problem has developed from finding these same viruses in other genera monkeys and apes. Are these viruses normal to all these monkeys and apes or are they the result of contamination? This problem may also apply to human and other animal viruses recovered from simian tissues. Simply stated, the following question may be asked: if a virus is isolated from a monkey or ape, is that virus a nonhuman primate virus? As information is developed on several of the so-called simian viruses, it is highly probable that a certain number will eventually be excluded from this classification.

In addition to the enhanced numbers of simians used as a result of vaccine development and general virologic research, increased employment

TABLE I

GENUS AND SPECIES NONHUMAN PRIMATES EMPLOYED IN THE VIRUS LABORATORY

Genus	Species	Common name
Old world		
Gorilla	*G. gorilla*	Gorilla
Pan	*P. troglodytes*	Chimpanzee
	P. paniscus	Chimpanzee (pigmy)
Pongo	*P. pygmaeus*	Orangutan
Hylobates	*H. lar*	Gibbon
Papio	*P. cynocephalus* group: includes *P. anubis, P. papio, P. ursinus* (Chacma)	Baboon
	P. hamadryas	Baboon (hamadryas)
Theropithecus	*T. gelada*	Baboon (gelada)
Cercopithecus	*C. aethiops*	Grivet *Aethiops* group
	C. sabaeus	Green other groups-*Mono,*
	C. pygerythrus	Vervet *Nicititans,* etc.
	C. talapoin	Talapoin
Presbytis	*P. entellus*	Langur
	P. cristatus	Langur
Erythrocebus	*E. patas patas*	Patas
Macaca	*M. mulatta*	Rhesus
	M. fascicularis (preferred to *M. irus, M. cynomolgus,* and *M. philippinensis*)	Cynomolgus
	M. radiata	Bonnet
	M. nemestrina	Pigtailed macaque
	M. cyclopis	Formosan rock macaque
	M. speciosa	Stumptailed macaque
	M. fuscata	Japanese macaque
New world		
Saimiri	*S. sciureus*	Squirrel
Aotus	*A. trivirgatus*	Owl
Alouatta	*A. belzebul*	Howler
Ateles	*A. paniscus*	Spider
Cebus	*C. capucinus*	Capuchin
Lagothrix	*L. lagothricha*	Woolly
Pithecia	*P. pithecia*	Sakis
Callithrix		Marmoset
Cebuella	*C. pygmaea*	Marmoset (pigmy)
Saguinus	*S. tamarin*	Marmoset (tamarin)
Leontideus	*L. rosalia*	Marmoset (golden lion tamarin)

of these animals resulted in the discovery of approximately 70 indigenous agents now described as simian virus (SV) or simian agents (SA). These viruses are of obvious concern to investigators (Kalter and Heberling, 1969). In addition to their nuisance value as contaminants in cultures of monkey cells, where they cause confusion in interpretation of results, they also pose as a potential threat to the health of the animal colony and to laboratory personnel. Enders and Peebles (1954) and Rustigian and co-workers (1955) were the first to describe the occurrence of "virus-like" agents in uninoculated monkey kidney culture that were referred to as "foamy virus." Shortly thereafter, Hull and his collaborators in a series of reports (1956, 1958; Hull and Minner, 1957) described the isolation of a large number of viruses from monkey tissue and excreta. Confirmation of these findings as well as description of isolates from monkeys other than *Macaca* sp. followed rapidly (Malherbe and Harwin, 1957; Hsiung and Melnick, 1958a; Hoffert *et al.*, 1958; Heberling and Cheever, 1960, 1966; Sweet and Hilleman, 1960; Fuentes-Marins *et al.*, 1963). The original articles and a number of reviews (Hull *et al.*, 1956, 1958; Hull and Minner, 1957; Malherbe and Harwin, 1957; Hoffert *et al.*, 1958; Heberling and Cheever, 1960) are suggested to those desirous of more detailed background to these early findings.

In evaluating these data, bear in mind that the majority of results reported are obtained from animals maintained in captivity for varying periods of time and with unknown and presumably frequent contacts with many and diverse animal species. These contacts with other species of simians markedly influence the results obtained, but are *not* generally recognized nor considered as relevant.

A. ADENOVIRUSES

Adenoviruses are one of the virus groups more frequently encountered as a result of working with nonhuman primates. Approximately 17 adenoviruses antigenically distinct from those originally recovered from other animal species have been described (Table II). As a group, these simian adenoviruses share the biologic characteristics common to all adenoviruses (Pereira *et al.*, 1963; Hull, 1968, 1969a,b; Hull and Minner, 1957; Hull *et al.*, 1956, 1958; Hsiung, 1968, 1969; Kalter, 1969a,b, 1971b; Heberling, 1972). Some question has been raised regarding the presence in all simian adenoviruses of a CF antigen common to the other adenoviruses, a distinction also shared by the avian adenoviruses. Adenoviruses, but not all types, have been recovered from respiratory and intestinal tracts of all simian species examined. Certain of the adenovirus types were isolated more readily from one or another of these locations—SV1, SV11, SV15,

TABLE II

SIMIAN ADENOVIRUSES

Virus	Originally isolated from
SV1, 11, 15, 17, 20, 23, 25, 30, 31, 32, 33, 34, 36, 37, 38	Rhesus, cynomolgus
SA7, 17, 18	African green
V340, AA153	Baboon
C1, PAN5, 6, 7, CV33	Chimpanzee
SqM-1	Squirrel monkey

SV17, SV23, and SV32 demonstrated a predilection for the respiratory tract, the others were encountered with greater regularity in the gastrointestinal tract. A listing and description of adenovirus isolates as well as their biologic activities have been reported in detail (Hull, 1968, 1969a,b; Hull and Minner, 1957; Hull et al., 1956, 1958; Hsiung, 1968, 1969; Kalter, 1969a,b, 1971b; Archetti and Bocciarelli, 1963; Burnett and Harrington, 1968; Fong et al., 1968a,b; Hillis and Goodman, 1969; Rapoza, 1967; Soike et al., 1969a,b).

Adenoviruses are recovered with great frequency and ease from primates; however, the role of adenoviruses in the production of disease in these animals apparently is not marked. Without question, adenoviruses (at least certain types) have a disease-producing capability as evidenced by the production in man, nonhuman primates, and other animals of respiratory disease, conjunctivitis, and possibly other clinical entities by at least some strains. Tyrrell et al. (1960) reported on the properties of SV17, which had been previously isolated by Goffe (unpublished data), and now recovered from nasal and conjunctival secretions of patas monkeys having a disease resembling pharyngoconjunctival fever of man. Adenoviruses SV17 and SV32 were reported by Bullock (1965) as associated with episodes of respiratory illness in captive monkeys. Interestingly, these same viruses, as well as SV15, were recovered from normal animals in the colony. Hull and his associates (1956, 1958; Hull and Minner, 1957; Hull, 1968) reported that SV32 was the predominating organism in about one-half of the nonbacterial conjunctivitis infections occurring in rhesus monkeys. Experimental verification of these observations was made by Heath et al. (1966) by inoculating vervet monkeys intranasally with SV17. Landon and Bennett (1969) recovered SV15 from rhesus monkeys with a mild, nonfebrile conjunctivitis. These investigators were able also to produce the disease in two rhesus monkeys by the conjunctival application of this isolate.

In an attempt to explore the use of monkeys as a laboratory model for studying adenovirus infections, Pedreira *et al.* (1968) inoculated *Macaca mulatta* with human adenovirus type 4. Their evidence was inconclusive, but infection and virus replication was obtained. It was felt, however, that this system would not serve as a useful model for testing adenovirus vaccines. In the Soviet Union, Gravrilov and his co-workers (1963) similarly studied adenovirus infections of the rhesus monkey. In an attempt to ascertain whether or not human adenoviruses (type 12), known to be oncogenic for newborn hamsters, were similarly oncogenic for newborn nonhuman primates, Kalter *et al.* (1966b,c) inoculated newborn baboons as well as baboons *in utero*. Tumors were not observed in any of these animals. Similar findings were reported by Cotes (1966) in rhesus monkeys with adenovirus type 12. However, thymectomized immature baboons developed a wasting syndrome or "runt" disease (Kalter *et al.,* 1966b). The association of a new adenovirus isolate, V340 (Table II), with a fatal pneumoenteritis of African green monkeys was reported by Kim *et al.* (1967). An outbreak in newborn baboons caused by this same agent was reported shortly thereafter by Eugster *et al.* (1969). Of 63 isolates recovered, 60 were identified as V340 and the remaining 3 viruses were neutralized by SA7 antiserum. Kohler and his collaborators (Kohler and Apodaca, 1968; Kohler *et al.,* 1968, 1969) have been interested in the possible association of adenoviruses with hepatitis, using the marmoset (*Saguinus oedipus*) as the experimental host. During the course of these experiments (Lange *et al.,* 1969), increased evidence of intestinal invaginations were observed in the experimentally infected animals. Noninfected animals were found only rarely (2 out of 450) to invaginate.

As will be discussed in greater detail below, there has been extensive interest in the relationship of simians, especially chimpanzees to the problem of hepatitis. In the course of such experiments, Hillis and his collaborators (1968) reported on the isolation of a number of adenoviruses serologically related to specific human adenoviruses, from chimpanzees in the study. In a subsequent study, Hillis *et al.* (1969) reported on a "new" simian adenovirus related to human adenovirus type 2 from a chimpanzee with "viral hepatitis." Final interpretation of these findings must be held in abeyance especially when we consider that adenoviruses are among the most frequently recovered agent from simian tissues. B. L. Murphy *et al.* (1970) reported on a method of assaying chimpanzee liver for evaluating hepatitis candidate agents that multiply in liver tissue. In their studies, these investigators found no differences between prototype adenovirus type 5 and newly isolated type 5 virus in their predilection for, and multiplication in, chimpanzee liver tissue.

Discovery of small viruses associated with adenoviruses (adeno-associated satellite viruses: AAV) derived from humans as well as simians—SV11, SV15, and SV34 (Atchison *et al.,* 1965; Mayor *et al.,* 1965; Hull *et al.,* 1965; Archetti *et al.,* 1966)—suggests that infection by adenoviruses may be mixed, as the relationship between the adenoviruses and AAV is extremely intimate. It also is possible that these viruses may be more extensively distributed among simians (and other animals) than originally recognized. Data are sorely needed to evaluate the status of the satellite viruses. These agents are small, defective, DNA, polyhedral, viruses requiring adenoviruses as helpers and are grouped with the parvoviruses. Rapoza and Atchinson (1967) reported that seven simian adenoviruses were found by electron microscopy to contain AAV-like particles (SV15, SV17, SV20, SV23, SV31, SV32, and SV36). CF antibody to AAV-1 were found in rhesus, but not in grivet (0/12), vervet (0/5), and patas (0/3) sera. An HI antibody study by Mayor and Ito (1967) suggests that AAV-4 is of simian rather than human origin.

Serologic surveys of numerous monkey and ape sera for evidence of infections by adenoviruses have been conducted in a number of laboratories. Table III lists those monkeys and apes found to have antibody to this virus group or individual types within the group. Frequently, in doing serologic surveys for antibody to adenoviruses as well as other viruses (particularly certain of the picornaviruses), an animal may be shedding viruses *without* serologic evidence of its presence. More recently, additional primate sera were assayed for SN antibody to V340, SA7, SV15, and SV23. This was done in order to ascertain the geographic distribution of viruses as well as, perhaps, some species susceptibilities. Confirmation of the geographic distribution of viruses was readily detected, but insufficient numbers of species were examined for determination of species susceptibilities. Aulisio *et al.* (1964) previously described neutralizing antibody to SV20 in human sera.

Because of the importance of adenoviruses in the development of tumors in hamsters, brief mention will be made of these studies. In addition to the human oncogenic strains, Hull *et al.* (1965) described several of the simian adenoviruses to be similarly tumorogenic. Included herein are SV1, SV11, SV20, SV23, SV25, SV33, SV34, SV37, SV38, and SA7. Gilden *et al.* (1968) have shown that these viruses may be subdivided by their CF reactions into three distinct serologic subgroups in a similar fashion to their human adenovirus counterparts (cross-reactions among tumor and infected cell T antigens). Riggs and his collaborators (1968) were able to demonstrate cross-reactions between oncogenic simian adenovirus T antigen with serum derived from hamsters with tumors induced by human types 7 or 13. Other aspects of simian virus oncogenicity may be found in the studies of Altstein *et al.* (1968) and Slifkin *et al.* (1969).

TABLE III

MONKEYS AND APES WITH ANTIBODY TO ADENOVIRUS

Simian source	CF group reaction	Viruses[a]	Test type	Reference
Baboon	+		CF	Kalter et al. (1964a)
Rhesus		+ Type 2; neg. 1, 3–7; + SV32, SV33	HI, SN	Bhatt et al. (1966a,b)
Bonnet		+ SV32, SV33		
Langur		+ SV32		
Rhesus				Hoffert et al. (1958)
Cynomolgus				
Gorilla	−	Neg. type 12, SV1, SV15, SV39, V340		Kalter et al. (1967a)
Chimpanzee	+	+ Type 12, SV1, SV15, SV39, V340		
Orangutan	+	+ SV39, V340; neg. type 12, SV1, SV15		
Giboon	+	Neg. type 12, SV1, SV15, SV39, V340		
Baboon	+	+ SV1, SV15, SV39, V340; neg. type 12		
Baboon (Africa)	+	+ V340, neg. type 12, SV1, SV15, SV39		
Vervet	+	+ SV1, SV15, V340; neg. type 12, SV39		
Rhesus	+	+ Type 12, SV1, SV15, SV39, V340		
Patas	N.D.	Neg. type 12, SV1, SV39, V340; SV15 N.D.		
Rhesus		+ SV20, SA7	SN	Shah and Morrison (1969)
Baboon (Africa)	−		CF	Kalter et al. (1968d)
Marmoset	−			Deinhardt et al. (1967b)
Baboon (Africa)		+ V340, SA7; neg. SV15, SV23		Kalter and Heberling (1970)
Baboon (U. S.)		+ V340, SA7, SV15, SV23		
African green		+ V340, SA7; neg. SV15, SV23		
Patas		+ V340, SA7; neg. SV15, SV23		
Rhesus		+ V340, SA7, SV15, SV23		

[a] SV39 and SV23 are now considered as identical.

B. Arboviruses

Considerable attention has been given to the use of monkeys and apes for the study of arbovirus infections either as model systems or for sentinel purposes. Very little, however, has been done to provide data on the prevalence of these viruses in simians in nature. Inasmuch as these animals reside in close proximity with the main vectors of most arboviruses, this lack of study is of concern to a number of investigators. With approximately 200+ arboviruses now identified, reconsideration of this area of research would be highly desirable. Early studies, as indicated above were concerned with yellow fever. Within the last 10 years, a limited number of investigations have been initiated to include other arboviruses.

Much of the early data with yellow fever virus have been reviewed (Bugher, 1951). It is now recognized that most nonhuman primates, including New World and Old World monkeys, are susceptible to yellow fever virus and are extensively involved in the epidemiology of this disease especially as pertains to jungle yellow fever. For example, Gaitan (1938) as well as Sanchez (1939) reported that monkey deaths preceded yellow fever and that this was common knowledge among the natives. Sanchez (1939) also reported that the 1901 yellow fever epidemic in Guatamala annihilated the monkey population. Collias and Southwick (1952) described concurrent decreases in marmosets and howler monkeys that occurred with the approach of yellow fever in Panama. An important epidemiologic observation was the finding that white-faced (*Cebus* sp.) monkeys outnumbered howler monkeys (*Alouatta palliata aequatorialis*) on Colorado Island in Lake Gatun (Panama). It was suggested that the howlers fed higher up in the trees than did the white-faced monkeys and were therefore closer to the mosquito fauna of the forest canopy.

Simians are extremely variable in their response to infection with the virus of yellow fever; reactions range from mild viremia to rapid deaths (Rosen, 1958a,b). More recently Galindo and Srihongse (1967) have shown that the level of viremia and the resistance to infection of Panamanian monkeys vary among the different genera. Spider and capuchin monkeys were highly resistant, with the howler and marmoset monkeys most susceptible.

On the assumption that yellow fever would continue northward through Central America to Mexico, Elton (1952) described the waves of disease as it progressed into Nicaragua from Honduras. A procedure involving human and monkey cases was presented in order to trace disease waves. In a follow-up of this study, Elton (1953) described a method to block the progress of jungle yellow fever. It is interesting to note that part of the project involves: (1) "extermination of all forest animals in barrier zone by systematic hunting and trapping . . . ," (2) "installation of non-immune sentinel monkeys, preferably *Macacus rhesus,* in wide mesh cages or free on swivel chains in tree tops in the barrier zone," and (3) establishment of sentinel zones . . . to detect the approach of the epizootic phase of the wave. These sentinel zones should be set up where monkeys are known to be abundant. Vargas-Mendez and Elton (1953) as well as H. C. Clark (1952) surveyed primate livers for evidence of yellow fever infection and as a cause of death. These studies not only defined the areas of infection but served as a reminder to the human population to insure vaccination. The relationship of monkeys to sylvan yellow fever is important inasmuch as their relationship to the epidemiology of the disease was not clearly defined until relatively recently. The early observation of Stokes and colleagues (1928) concerning monkeys and yellow fever virus dissemination has been mentioned. Skepticism continued for many years regarding the lethal effect of this virus on wild monkey populations. Historical evidence and numerous observations suggested that monkey populations were involved in sylvan outbreaks of yellow fever, though very little data were available to substantiate these observations and they were more often presumptive and not based upon pathologic or virologic studies. Laemmert and de Castro Ferreira (1945) were able to demonstrate liver lesions in marmosets pathologically similar to those described by Torres (1928) and experimentally produced in rhesus monkeys. A number of investigators persistently supported the viewpoint that arboreal primates were highly susceptible to yellow fever virus. The susceptibility of the howler monkey to yellow fever virus was also reported by Laemmert and Kumm (1950).

Waves of yellow fever have been observed among humans and monkeys in Central America (Trapido and Galindo, 1956; Rodaniche and Galindo, 1957; Galindo and Rodaniche, 1964; Galindo *et al.,* 1966) that decimated the population of howler monkeys (*Alouatta* sp). The last known outbreak was reported in 1956. Evidence for jungle yellow fever was reported by Galindo and Srihongse (1967) in Panama and again demonstrated how monkeys may serve as an indicator of yellow fever activity. Monkeys present in Panama represent the following genera: *Ateles* (spider), *Alouatta* (howler), *Cebus* (capuchin), *Saguinus* (marmoset), and *Aotes* (night monkey). The spider and capuchin monkeys showed the greatest resistance to yellow fever infection. Yellow fever in spider monkeys results in viremia with minimal symptoms. The capuchin appears to be more resistant than the spider monkey, rarely showing any symptomatology. The other three groups (howler, marmoset, and night monkeys) are highly susceptible, becoming sick, often fatally. Night monkeys, by virtue of living in tree holes, generally escape the mosquito vector, so cannot be used as a point of reference. These fatalities, along with serologic tests, offer a pragmatic mechanism for determining yellow fever activity in an area.

Existence of yellow fever in naturally infected howler monkeys (*A. seniculus insulanus*) was reported by C. R. Anderson and Downs in Trinidad, B.W.I. (1955). Groot (1962) reported that rhesus monkeys inoculated with the 17D yellow fever vaccine first produced HI antibody followed by CF antibody. These antibodies were followed by the appearance of HI antibody to other group B arboviruses, Ilheus, St. Louis, and dengue 2. Heterologous titers were lower than those obtained with the yellow fever antigen. Similar findings were observed in human yellow fever virus infections by Theiler and Casals (1958).

In Africa, Taylor *et al.* (1955) found by neutralization tests that baboons (*Papio* sp.) and grivet monkeys (*Cercopithecus aethiops*) had a high prevalence of yellow fever antibody, whereas the galago (*G. senegalensis*) did not appear to be involved in the yellow fever cycle. Haddow (1952) had previously shown that galagos (*G. crassicaudatus lasiotis*) in Kenya had yellow fever antibody but none of nine baboons and less than 2% of assorted other monkeys had this antibody. The epidemiology of yellow fever as it involves the bushbaby (*Galago* sp.) was reported in greater detail by Haddow and Ellice (1964). These findings may suggest species differences to yellow fever among various African primates and prosimians. By using rhesus monkeys, as well as mice, McNamara (1953) demonstrated an antigenic relationship between Uganda S and yellow fever viruses. Bearcroft (1957) had previously demonstrated that Zika virus could modify the hepatic lesions of yellow fever in rhesus monkeys thereby prolonging their survival. Epidemiologic considerations as relates to a number of laboratory animals and the susceptibility of the mangabey (*Cercocebus galeritus agilis*) and colobus monkeys to yellow fever have been described (Williams *et al.,* 1965; Woodall, 1969; Woodall *et al.,* 1969).

Haddow (1968) has reviewed the natural history of yellow fever in Africa, particularly as relates to the wild primate–mosquito (*Aedes africanus*) cycle in Uganda. Extensive surveys of primate sera reveals extensive immunity to this virus but overt disease may be rarely observed. More recently, Henderson *et al.* (1970) explored the immunologic relationships among five group B arboviruses (West Nile, Wesselsbron, Banzi, Uganda S, and Zika) and yellow fever in rhesus and *C. aethiops* monkeys. Primary inoculation resulted in viremia and varying febrile reactions. The immunologic responses obtained substantiated previously reported suggestions that the cross-protection between group B viruses has an influence upon the epidemiology of yellow fever in Africa. An epidemiologic aspect of yellow fever of concern to us is the potential danger of importing non-human primates carrying yellow fever virus. As mentioned previously, studies on sera of baboons showing only 3 of 10 presumably vaccinated animals with antibody (Kalter *et al.,* 1964a) prompted further investigation. Additional examination of other primate sera for antibody to yellow fever virus (Kalter *et al.,* 1967a) substantiated this observation. This lack of antibody is important as it was known that many of the animals tested had been vaccinated prior to their arrival in the United States because they originated in endemic yellow fever areas. Consequently, in order to determine the adequacy of vaccination, additional studies were initiated to test for the presence of yellow fever-neutralizing antibody in serum specimens from African baboons before and after vaccination (Kalter and Jeffries-Klitch, 1969). The results indicated that the vaccine was efficacious but emphasized the need for care in its handling and administration. Improper employment and handling resulted in decreased vaccine potency. F. L. Soper and Smith (1938) made an important observation, the full extent of which was not appreciated until some years later (see Section III,I,1). It was noted that 20–30% of the individuals receiving yellow fever vaccine prepared with rhesus monkey serum as a stabilizer developed jaundice.

Arboviruses other than yellow fever were also studied in nonhuman primates. Russian spring–summer and Japanese encephalitis virus infection and immunologic patterns in monkeys were described by Morris *et al.* (1955). Smorodintsev (1958) reported that tick-borne spring–summer encephalitis (TE) virus (also a group B virus) produced an encephalitis in monkeys following intracerebral inoculation. Variable results, however, were obtained as a result of strain differences, Central European strain less severe than the others. Libikova and Albrecht (1959) described the results of intranasal inoculation of *M. mulatta* monkeys with TE virus recovered in Slovakia. The sensitivity of rhesus and cynomologus monkeys to this virus was described by Benda *et al.* (1960b), who also demonstrated the susceptibility of these monkeys to airborne infection by this virus (1962). In a series of studies with this tick-borne encephalitis virus, Mayer (1966) described a strain that lost its virulence for nervous tissue of rhesus monkeys. This was followed by observations concerning the virulence of selected cloned virus for rhesus monkeys by intracerebral inoculation (Mayer and Rajcani, 1967) and after intranasal inoculation (Mayer and Rajcani, 1968). Similar TE virus studies in monkeys, especially as concerns its spread following inoculation, were performed by Simon *et al.* (1967a,b). Nathanson and his collaborators (1965, 1966, 1967, 1968) developed methods for studying the neurovirulence of group B arboviruses by evaluating the appearance of the lesion produced in the central nervous system after intracerebral inoculations. Significantly, it was found that there was a marked variation in response of monkeys, necessitating use of sufficient numbers of animals. Thus a mechanism for testing and evaluating vaccine candidates of low neurovirulence was offered. For example, Thind and Price (1966a,b) reported on an attenuated strain of Langat virus which, after passage in chick embryo, mice, monkeys, and human tissue culture, was very stable in its neurotropic potential. This virus could be then evaluated by using the suggested neurovirulence scale.

Spider monkeys were found by W. H. Price *et al.* (1961, 1963a,b) to be protected from challenge with all members of the TE group of viruses following vaccination with attenuated live Langat virus vaccine. These findings were confirmed by O'Reilly *et al.* (1965). Other investigators in addition to those indicated above, have found the Russian spring–summer complex of viruses to have a pronounced variability in production of encephalitis in monkeys after intracerebral inoculation (Ilyenko and Pokroskaya, 1959; W. H. Price *et al.* 1963a,b; Pond *et al.,* 1953; Work and Trapido, 1957; Smorodintsev, 1958). In another series of studies, Nathanson and his collaborators (Nathanson and Harrington, 1966, 1967; Nathanson *et al.,* 1969a) investigated the pathogenesis of Langat virus in spider monkeys following subcutaneous or intravenous inoculation. It was demonstrated that inapparent infection with viremia and antibody production occurred. Virus-feeding produced infection when large doses (10^{11} LD_{50}) were given but virus could not be recovered from feces. Intracerebral and intraspinal inoculation of rhesus monkeys with the JIR strain of TE virus produced histologic and encephalitic clinical symptoms according to Slonim *et al.* (1966). These investigators also reported that inapparent infection may be produced by the subcutaneous inoculations of monkeys with this virus. Viremia and serologic (CF) responses were noted.

Kyasanur Forest disease (KFD) was first recognized in India in 1957 when an epizootic occurred in monkeys (*Macaca radiata* and *Presbytis entellus*) and humans. This virus is also a member of the tick-borne complex (Work and Trapido, 1957). Bhatt *et al.* (1966a,b,c) described the isolation of KFD virus from various tissue of man and monkey (*M. radiata* and *P. entellus*). Webb and collaborators (Webb and Laxmana, 1961; Webb and Chatterjea, 1962; Webb and Burston, 1966; Webb, 1969) described the clinical picture in man and monkeys with particular reference to the disease as it applies to the central nervous system. The occurrence of a nonsuppurative encephalitis of monkeys in the field as a result of this disease was noted by Iyer *et al.* (1960). Shah (1965) was able to demonstrate the transmission of KFD virus by experimental infection of lactating macaque monkeys. Experimental exposure of monkeys to ticks in the Kyasanur Forest area was described by Boshell and Rajagopalan (1968). Kemerovo virus, another of the tick-borne arboviruses was studied in rhesus monkeys by Libikova *et al.* (1970). Intracerebral inoculation produced a mild meningitis or meningoencephalitis with eventual regression of symptoms. Neutralizing-antibody development started approximately 1 week after inoculation. Virus persisted in the brain for relatively long periods of time, a finding previously reported by Gresikova *et al.* (1966), and occurred also in the rhesus monkey after inoculation with Tribec virus.

Causey *et al.* (1961) and Shope and his co-workers (1961) were able to isolate new members of the group C arbovirus by inoculation of *Cebus apella* monkeys used as sentinel animals in the rain forests of Brazil. Disease in these animals was extremely mild. Infection of rhesus monkeys with group C arboviruses was reported by W. P. Allen *et al.* (1967). Overt illness in these animals was limited to fever seen only in a few animals. Cross-immunity was demonstrated in recovered monkeys to relate to heterotypic viruses of the group. Chimpanzees were reported as susceptible to Tahyna virus (another California arbovirus) by Simkova and Bardos (1969). Bardos *et al.* (1969) had previously found that rhesus monkeys were insusceptible to aerosol infection with this virus. Attempts to infect chimpanzees by the aerosol route with Tahyna virus were also unsuccessful (Simkova and Danes, 1968; Simkova and Bardos, 1969). Tahyna virus infection of chimpanzees after exposure to infected mosquitoes was reported by Simkova and Danielova (1969).

Other arbovirus studies in monkeys and apes involve the isolation of Eastern equine encephalomyelitis (EE) virus from naturally infected cynomolgus monkeys (Livesay, 1949). The distribution of EE lesions in the central nervous system of man and rhesus monkeys was reported by Nathanson and co-workers (1969b). Gerloff and Larson (1959) demonstrated the susceptibility of rhesus monkeys to experimental infection with Colorado tick fever virus. A new virus was isolated by Smithburn and Haddow (1944) in the Semliki Forest of Western Uganda that was fatal for rhesus monkeys on intracerebral inoculation. These authors suggested that the red-tailed monkey (*C. nictitans*) may be the natural reservoir. Rhesus, as well as African green, monkeys are also susceptible to Bunyamwera and Germiston viruses (Smithburn *et al.*, 1946).

A new virus (Manzanilla) isolated from howler monkeys in Trinidad was described by C. R. Anderson *et al.* (1960). In Lumbo, Mozambigue, Kokernot *et al.* (1962) described the recovery of a new arbovirus (Lumbo) belonging to the California group of virus from mosquitoes. African green and rhesus monkeys were shown to be susceptible. Monkeys could be protected from Japanese encephalitis virus infections by a vaccine grown in hamster diploid cell cultures (G. C. Y. Lee *et al.*, 1967). Verlinde (1968) demonstrated the susceptibility of cynomolgus monkeys to a number of different viruses: Mayaro and Mucambo (group A), Oriboca and Restan (group C), and an unidentified virus originating in South America (Surinam) and named kwatta. Cynomolgus monkeys could be protected against St. Louis encephalitis and West Nile arbovirus by previous dengue virus infection according to Sather and Hammon (1970). Semenov and Lapin (1967) reviewed the status of monkeys captured in the jungle of northern Liberia with regard to arbovirus infections. By means of experimental infection of monkeys with Bunyamwera and Germiston viruses, Schwartz and Allen (1970) demonstrated that the HI, CF, and SN tests were more reliable than clinical disease as an indicator of infection.

Hemagglutination-inhibition tests in this laboratory on (Kalter *et al.,* 1964a) baboon (*P. cynocephalus*) sera for antibodies to the viruses of EE New Jersey, WE Fleming, Chikungunya chick L, louping ill DXLIV, Sindbis Ar 1055, Semliki Ri-1, Calif H & R, WE 38873, Cache Valley-like A9171B, Bunyamwera RI-1, Guaroa J-C2, West Nile AR248, yellow fever asibi, Jap B G8924, Langat TP21, SLE F1a P-15, MVE11A, Marituba Be An 15, and Dengue II Tr 1751 resulted in one yellow fever-positive. CF testing of these sera indicated two sera with antibody for EE. Shah and Southwick (1965) surveyed free-living rhesus monkeys as well as captive monkeys maintained in Lagos, Nigeria and Poona, India. Included in this group, in addition to the rhesus monkeys, were *E. patas, C. tantalus, C. mona, C. nictitans, M. leucophaeus, C. erythrogaster, C. torquatus, M. radiata,* and *P. entellus.* Sera, when tested for antibodies to dengue 1, Chikungunya, and JBE were essentially negative. Marmoset (cotton-top and white-lipped tamarins) sera tested in the CF test by Deinhardt *et al.,* (1967b) against WE and EE antigens were also negative. Harrison *et al.* (1967b) examined sera from U. S.-born and wild-born chimpanzees (*Pan troglodytes*), gorilla (*Gorilla gorilla*), rhesus (*M. mulatta*), African greens (*Cercopithecus* sp.), orangutan (*Pongo pygmaeus*), and baboon (*Papio* sp.) for antibody to Chikungunya and other serologically related viruses (Semliki Forest, Onyong-nyong, Mayora). They found that 26% of the sera had neutralizing antibody to one or more of these antigens, and viral isolations on these animals were negative. It is highly probable that these reactions indicate cross-reactions with infectious agents common to the U. S. Similar findings have been reported by other investigators (McIntosh *et al.,* 1964; Osterrieth and Deleplanque-Biegeois, 1961; Osterrieth *et al.,* 1960). Binn *et al.* (1967) demonstrated that Asian monkeys (*M. mulatta*) were susceptible to Chikungunya and Mayaro virus but not to Onyong-nyong. High levels of antibody were, however, produced in all the animals. A vaccine prepared in green monkey kidney cells against Chikungunya virus was tested by Harrison *et al.* (1967a) in rhesus monkeys with good development of neutralizing antibodies (CF and HI responses were low). *Erythrocebus patas, C. mona,* and *C. aethiops* in the Lagos area of Nigeria were examined by Boorman and Draper (1968) for neutralizing antibodies to Pongola, Bunyamwera, and Chikungunya viruses. Antibodies to Chikungunya virus were present in all three species. Two of the *C. mona* monkeys had antibody to Pongola virus. All three viruses when used to challenge rhesus and *C. mona* monkeys yielded variable results. Pongola virus required two inoculations to produce antibody in the rhesus monkey, but took only one injection to produce antibody in *C. mona*. The Bunyamwera virus killed the rhesus monkeys in 16 days; both species, however, developed viremia and the *C. mona* produced antibody. Chikungunya virus infection resulted in viremia in both species with subsequent development of antibody.

Monkeys have been used in the preparation of arbovirus antisera by Smithburn (1952, 1954). Both type- and group-specific SN antisera usable for serologic differentiation of arboviruses may be obtained. Responses, however, were not always high nor always type-specific (H. W. Lee and Scherer, 1961). Scherer and Muira (1965) described successful production of polyvalent group B arbovirus antisera in *M. irus*. These sera had a broad spectrum of heterologous group B complement-fixing reactions as a result of the inoculation sequence used in preparation of the antisera. In attempting to characterize California encephalitis (CE), Hammon *et al.* (1952) reported that *all* inoculated monkeys did not develop antibody, even though clinical symptoms were noticed in the inoculated animals. Rhesus monkeys were used by Behbehani *et al.* (1967) for the production of monotypic antisera against CE, Sicilian sandfly fever, and Turlock arboviruses. Specificity was seen in HI tests but the CF test reacted non-specifically. Disease in the inoculated animals was relatively minor, generally without evidence of CNS disturbances. Russell and his collaborators (1968) used cynomolgus monkeys for preparation of antisera to viruses isolated during a dengue hemorrhagic fever outbreak in Thailand. Juvenile *C. aethiops* inoculated with unadapted type 1 dengue virus by Sweet *et al.* (1969) did not develop evidence of clinical illness. Serologic evidence of infection was noted with higher homotypic responses than heterotypic responses to the type 2 strain. Chimpanzees (Paul *et al.,* 1948), Old World monkeys (Simmons *et al.,* 1931), and New World monkeys (Rosen, 1958a,b) have all been reported as susceptible to dengue. Whitehead *et al.* (1970) showed that the gibbon (*Hylobates lar*) was uniformly susceptible to primary infection with any of the four dengue viruses. Viremia and subsequent antibody development was reported.

The current large-scale employment of immunosuppressive procedures suggest comparative studies in nonhuman primates with emphasis placed on the potential dangers involved. The effect of cortisone on tick-borne

encephalitis in cynomolgus monkeys (Benda *et al.,* 1960a) and Vene-
zuelan equine encephalomyelitis (Gleiser *et al.,* 1961) in rhesus monkeys
has been described. Nathanson and Cole (1970) were able to produce a
fatal encephalitis in spider monkeys (*Ateles* sp.) following intracerebral
inoculation and use of cyclophosphamide. Monkeys inoculated with virus
alone developed antibody; those on cyclophosphamide did not demonstrate
detectable responses. Similar studies by Zlotnik *et al.* (1970) indicated
that cyclophosphamide treatment of patas monkeys resulted in a fatal infection with louping ill virus. Untreated animals developed clinical illness
from which they recovered. Untreated rhesus monkeys, following inoculation with WE virus evidenced no clinical response. Treatment, however,
with cyclophosphamide infection resulted in a fatal encephalitis. Louping
ill, VE, and WE usually produced an acute inflammatory CNS lesion.
After cyclophosphamide the inflammatory reaction was replaced by a degenerative process with both neuronal neurosis and spongy degeneration
observed. The antibody response of rhesus monkeys after whole-body
X-irradiation to group A arboviruses was reported by S. L. Reynolds
et al. (1968). For a complete listing of arboviruses, their biologic characteristics and susceptible hosts (including simians), see the "Catalogue
of Arthopod-Borne Viruses of the World" (Taylor, 1967).

C. HERPESVIRUSES

Of the various viruses recovered from simian tissues, the herpesvirus
group has achieved the greatest prominence. Sabin and Wright (1934)
reported the development of an acute fatal ascending myelitis in man following a monkey (*M. mulatta*) bite. Greater details regarding the pathogenic potential of this B* virus were reported later that year in a series of
studies by Sabin (1934a,b) and Sabin and Hurst, 1935). A second report
concerning human infection was also reported by Sabin (1949). Intravenous inoculation of monkeys with this agent was followed by a generalized disease with rash on the face, buccal muocsa, and conjunctiva.
Attempts to transmit the disease to monkeys and chimpanzee were unsuccessful (Haber, 1935; Vienchange, 1935). It will be recalled that Blanc
and Caminopetros (1921) reported that rhesus monkeys were resistant to
infection with herpes simplex virus. The lack of relationship of antibody
to susceptibility was shown by Burnet *et al.* (1939b) and van Rooyen and
Rhodes (1948). Burnet *et al.* (1939b) showed antibody development
with evidence of clinical disease.

Breen *et al.* (1958) reported a human case of monkey-bite encephalo-

---

* Named after the patient, Dr. W. B.

myelitis with recovery. In retrospect, review of all human cases of B virus infection suggests careful interpretation of such reports. One or two similar reports are also now known. Burnet *et al.* (1939b) noted the similarity between B virus, herpes simplex, and pseudorabies virus. It is highly possible that *H. hominis* and *H. simiae* have been at one time the same virus. Subsequent passage through different host species may then have been responsible for the differences in pathogenicity now observed. *Herpesvirus simiae* in monkeys is important primarily as a hazard to man and possibly other monkeys and animals rather than to its natural host. Like other herpesvirus (as well as other viruses) in the natural host, inapparent or latent infection, rather than overt disease, is frequently the usual clinical course. *Herpesvirus simiae* appears to be primarily a mild or inapparent disease of *M. mulatta,* although it has been recovered from other species. All herpesviruses produce intranuclear inclusion bodies (Cowdry and Scott, 1935), invade the CNS by the peripheral nerves, and produce deaths in animals other than their natural hosts. Pseudorabies in rhesus monkeys was studied by Hurst (1936). Nicolau and co-workers (1938) reported neuroinfection of chimpanzees by pseudorabies (Aujeszky's disease). Melnick and Banker (1954) also were able to recover B virus from the CNS of rhesus monkeys. Recovery of this virus from tissue cultures prepared from rhesus and cynomolgus monkeys was reported by W. Wood and Shimada (1954). Endo *et al.* (1960a,b) in studies on B virus in Japan reported the presence of antibody to this virus in *M. fuscata, M. cyclopis,* and *M. irus.* It would appear that *H. simiae,* while primarily a disease of rhesus monkeys, may be indigenous to all macaques. Kirschstein (quoted by Hull, 1968) demonstrated that African simians (African green and patas) could be infected with B virus.

B virus infections of man and monkeys have been the subject of numerous reports. These have included case reports involving man and monkeys (including occurrence of "natural" disease in cynomolgus monkeys), diagnosis, distribution in nature, serum surveys for existence of antibody in "normal" monkeys and apes; persistence of antibodies in colony animals; immunization and vaccines; and general reviews including detailed histopathologic descriptions of tissues derived from man and other animals (Keeble *et al.,* 1958; Nagler and Klotz, 1958; Pierce *et al.,* 1958; W. T. Soper, 1959; Hull and Nash, 1960; Keeble, 1960, 1968; Van Hoosier and Melnick, 1961; Hull *et al.,* 1962; Love and Jungherr, 1962; Appleby *et al.,* 1963; Hartley, 1964, 1966; Kalter *et al.,* 1964b; Gralla *et al.,* 1966; Sumner-Smith, 1966; Zeitlyonok *et al.,* 1966; Cabasso *et al.,* 1967; Cheville and Kluge, 1967; Hull and Peck, 1967; Plummer, 1967; W. C. Cole *et al.,* 1968; Perkins, 1968; Hunt and Melendez, 1969; McCarthy, 1969; Melendez, 1969). It is now well recognized that a close antigenic relationship exists between human herpesvirus (*H. hominis*) and B virus (*H. simiae*) (Burnet *et al.*, 1939b). Examination of the literature indicates conflicting reports as regards this relationship. Shah and Southwick (1965) found antibodies to *H. simiae* in free-living adult rhesus monkeys but not juveniles. Ten percent of freshly captured rhesus monkeys were reported by Hull and Nash (1960) to have antibodies to this virus. Could these antibodies be a reflection of previous contact with *H. hominis* derived from humans?

Because of the importance of this antigenic relationship and its significance as regards immunity to B virus, it may be well to review briefly what is known about the serologic overlapping between simplex and B virus. Sabin (1934a) had originally concluded that the two viruses were distinct entities, although a relationship based upon serum neutralization tests was indicated. Several investigators (Pierce *et al.*, 1958; Nagler and Klotz, 1958) reported individuals with antibody to herpes simplex but not to B virus, even though they had been in contact with monkeys for long periods. A number of investigators (Burnet *et al.*, 1939b; Melnick and Banker, 1954; Van Hoosier and Melnick, 1961; Hull *et al.*, 1962; Cabasso *et al.*, 1967) reported that human sera did have anti-B virus immune globulins. Interpretation of this finding is still not clear. Generally an individual with antibody to B virus also had antibodies to herpes simplex. It is conceded that these antibodies reflect anti-herpes simplex bodies rather than exposure to B virus. Because of this relationship, Love and Jungherr (1962) and Hartley (1966) recommend use of gamma globulin in cases of accidental exposure, although no evidence for its efficacy has been demonstrated.

Other simian herpesviruses have been isolated from different species of Old World and New World monkeys, Table IV. A spontaneous salivary gland virus disease occurring in chimpanzees with inclusions in such tissues as the salivary gland, adrenal cortex, and myocardium was reported by Vogel and Pinkerton (1955). Cytomegalovirus of monkeys was first isolated by Malherbe and Harwin (1957) and by Black *et al.* (1963). This virus, designated SA6, produces typical cytomegalovirus inclusions in the parotid gland of infected animals. Another herpesvirus, but one probably more closely related antigenically to *H. simiae* than SA6, was isolated by Malherbe and Harwin (1963) and designated SA8. This virus does not appear to produce any clinical evidence of infection in its natural host. In 1963, Holmes *et al.* reported on the isolation of a herpesvirus from marmosets. Confirmation of this finding quickly followed with two independent reports describing herpesvirus infections of marmosets (*Saguinus oedipus* and *S. nigricollis*) (Holmes *et al.*, 1964; Melnick *et al.*, 1964). At first called MHV (marmoset herpes virus) or herpes T (*Herpesvirus tamarinus*) then

H. platyrrhinae by Holmes *et al.* (1966), this virus produces a fatal, systemic disseminated disease in cotton-top marmosets, white tamarins, and owl monkeys (Hunt and Melendez, 1966). It appears that the natural host for this virus is the squirrel monkey (*Saimiri sciureus*) in which it exists as a latent infection (Melendez *et al.,* 1966; Melendez and Hunt, 1966). That other species New World monkeys may serve as a natural host for this virus is suggested by the studies of Holmes *et al.* (1966). In a serologic survey of antibodies to this virus, it was found that in addition to the squirrel monkey, the spider (*Ateles* sp.) and cinnamon ringtail (*Cebus albifrons*) monkeys also had herpes T antibody. This virus could also produce disease in the squirrel monkey (Deinhardt and Deinhardt, 1966; Daniel *et al.,* 1967a,b; King *et al.,* 1967). A fatal disease in owl monkeys (*Aotus trivirgatus*) with isolation of herpes T was reported by Emmons *et al.* (1968). Sheldon and Ross (1966) isolated this same virus from an owl monkey in a colony that apparently had experienced an epizootic of the disease. These investigators tested squirrel (*Saimiri*), marmoset (*Callithrix*), spider (*Ateles*), weeper (*Cebus*), and woolly (*Lagothrix*) monkey sera for antibody to herpes T with seropositives found only among the squirrel monkeys. This finding would suggest that there are no survivors among the seronegative species or perhaps they had never been exposed to herpes T. More recently, Burkholder and Soave (1970) were also able to recover herpes T virus from the owl monkey. Because of reported deaths among owl monkeys attributable to *H. platyrrhinae,* it was interesting to note that Burkholder and Soave (1970) reported a mortality rate of 0.8% of a group of 246 animals for this species and that inapparent infection occurred. Hunt and Melendez (1969), however, had indicated that the owl monkey, because of the occurrence of fatal disease, was probably not a reservoir host. The marmoset was shown by B. L. Murphy *et al.* (1971) to maintain a carrier state to herpes T.

In 1967, Clarkson *et al.* reported on a virus disease in vervet (*C. aethiops*) monkeys caused by a new member of the herpesvirus group. This virus produced a fatal exanthematous disease and was isolated from blood and a number of tissues. No relationship could be shown between this virus (Liverpool vervet virus, LVV) and *H. hominis, H. simiae, H. platyrrhinae,* or SA8. The reservoir for this virus was thought to be the squirrel monkey. Another similar outbreak was reported shortly thereafter in patas (*E. patas*) monkeys by McCarthy *et al.* (1968). This disease was clinically and pathologically similar to that seen in *C. aethiops,* but attempts to isolate a virus from necropsy materials in a wide range of laboratory culture media were unsuccessful. Patas monkey inoculation followed by seeding of patas kidney cells proved to be more rewarding. Additional information concerning this virus is necessary for full evalua-

TABLE IV

Herpesviruses Recovered from Simian Tissues

Virus	Originally isolated from
Type A	
H. simiae (B virus)	Rhesus
SA8	African green
H. tamarinus (*platyrrhinae*)	Squirrel monkey
SMV	Spider
H. papio	Baboon
Type B	
SA6	African green
Herpes saimiri (?)	Squirrel monkey
Liverpool vervet agent (?)	African green, patas

tion. Lennette (1968) reported on the recovery of a virus from a spider monkey (SMV) that Hull (1968) temporarily designated as a new virus. As a result of a serologic survey, it is thought that SMV may naturally infect the spider monkey in the wild. SA8 virus had been reported as isolated from the African vervet by Malherbe and Harwin (1957); recently this virus has also been recovered from the baboon (Malherbe and Strickland-Cholmley, 1969a,b; Kalter, 1970). Malherbe and Strickland-Cholmley (1970) also reported on the recovery of an SA15-like herpesvirus from the baboon.

Herpesvirus hominis infection of man is well known and will not be described at this time. It should be emphasized, however, that clinical disease, while common to man (prevalence of antibody in adults is approximately 95%) and considered as a benign disease, may be a serious and sometimes fatal infection. This virus in nonhuman primates produces a disease similar to that caused by herpes T as well as other herpesvirus strains in foreign simian species. Species differences in susceptibility to HSV was reported by McKinley and Douglass (1930). It was noted that the rhesus monkey was resistant to this virus, whereas the capuchin developed an encephalitis after intracerebral inoculation with herpes simplex. These findings confirmed those previously described by Zinsser (1929). Spontaneous encephalitis as well as a natural epizootic resulting from *H. hominis* infection in gibbons had been reported (P. C. Smith *et al.,* 1958, 1969). A natural herpes simplex infection of the gibbon was reported by Emmons and Lennette (1970). L. W. Chu and Warren (1960) reported that HSV produced an encephalitis in cynomolgus monkeys. Fatal disease has been experimentally produced in the owl monkey by Katzin *et al.* (1967) and recovered from natural fatal disease in this same species by Melendez *et al.* (1968b, 1969b). Deinhardt *et al.* (1967b) challenged marmosets with herpes simplex and found that an occasional animal will die but this virus was less pathogenic for the marmoset than *H. platyrrhinae.* The ability to produce an encephalitis in marmosets with herpes simplex, similar to that seen in owl monkeys, was indicated by Hunt and Melendez (1969). As part of studies that will be described in greater detail below, Gerber and Rosenblum (1968) surveyed rhesus monkeys for antibody to herpes simplex virus. Only a small percentage (3%) of these animals, either immediately (1–4 days) following capture or after 4 months in captivity, had antibody. Recently Hampar *et al.* (1969) were able to demonstrate an antigenic relationship between herpes simplex virus and SA8 by studies on human serum. Squirrel monkeys were used by Kaufman and Maloney (1963) to ascertain the effect of various inhibitors on experimental herpetic keratitis.

Melendez *et al.* (1968a) reported the isolation of an apparently new herpesvirus from squirrel monkey (*Saimiri sciureus*) primary kidney cultures. This virus, *H. saimiri* (Melendez *et al.,* 1969c), was considered to produce latent infections in its host animal. Normal squirrel monkeys contain antibody to the agent and are considered, therefore, as the reservoir host (Melendez *et al.,* 1969d). Most interesting was the report that *H. saimiri* when inoculated into marmosets or owl monkeys caused a malignant reticulum cell-type lymphoma with death of the animals in approximately 3–6 weeks (Melendez *et al.,* 1969a). This virus is antigenically distinct from other herpesviruses (Melendez *et al.,* 1969c). Most extensive tissue involvement is seen in the liver, spleen, kidney, lymph nodes, thymus, lung, and adrenal glands. Inclusion bodies *in situ* are not seen, although observed in *H. saimiri*-infected cell cultures. Natural disease has not been reported. The pathology in marmosets and owl monkeys as a result of experimental infections has been reported (Hunt *et al.,* 1970). Further evidence for the oncogenic potential of this virus was described by Melendez *et al.* (1970b) as relates to infection of the ringtail monkey (*Cebus albifrons*). Daniel *et al.* (1970) were able to demonstrate the production of malignant lymphoma in rabbits by *H. saimiri.* The disease, however, could not be produced in stumptail (*M. arctoides*)* nor rhesus (*M. mulatta*) monkeys (Melendez *et al.,* 1970c). African green monkeys (*C. aethiops*) were somewhat more sensitive in that two of the six monkeys demonstrated histopathologic features of this disease. This suggests a possible species susceptibility which is in need of further study. Additional evidence suggesting an etiologic relationship between *H. saimiri* and malignant lymphoma in primates was recently reported by Melendez *et al.* (1970a). It was also demonstrated that *H. saimiri* may produce an acute

* *M. speciosa.*

lymphocytic leukemia in owl monkeys (Melendez *et al.,* 1971). A collaborative study with Dr. L. V. Melendez testing the susceptibility of baboons (*Papio cynocephalus*) to this agent is currently in progress. No evidence of a lymphoproliferative disease has been seen after approximately 1 year.

Of the group B herpesviruses, SA6 has been mentioned. However, other cytomegaloviruses have been also recovered from monkeys and apes. Vogel and Pinkerton (1955) reported the presence of cytomegalovirus inclusions in chimpanzees. It had been previously described in the capuchin (*Cebus fatuellus*) monkey (Cowdry and Scott, 1935), African green monkey (Rowe, 1960; Black *et al.,* 1963; K. O. Smith *et al.,* 1969), and rhesus monkey (Covell, 1932). Recently cytomegalovirus was described in salivary glands of tarsiers (*Tarsius* sp.) dying in captivity (A. A. Smith and McNulty, 1969). Specific antisera to human and monkey cytomegaloviruses were prepared in baboons by Graham *et al.* (1969). A persistent, asymptomatic viruria due to a cytomegalovirus was described by Asher *et al.* (1969) in rhesus monkeys.

A few years ago, Burkitt (1958) recognized a tumor in children of Africa which was histologically defined as a malignant lymphoma. Other epidemiologic studies by Burkitt suggested that some biologic factor, possibly infectious and carried by an arthropod vector, was responsible. Epstein and Barr (1964) were able to cultivate cells from these tumors *in vitro* and also demonstrate the presence of herpes-like particles within these cultivated cells (Epstein *et al.,* 1964a). Space limitations prevent description of the extensive studies attempting to demonstrate the virus etiology of Burkitt's tumor. It was obviously important to try *in vivo* studies. A small experiment was reported by Epstein *et al.* (1964b) in which African green monkeys were inoculated with biopsy material from a patient with this type of lymphoma. In a more extensive report Epstein *et al.* (1966) described development of lesions in the long bones of two of three African green monkeys surviving more than 2 years after inoculation as suckling animals. Henle and Henle (1966) and D. Armstrong *et al.* (1966) using immunofluorescence had indicated a widespread distribution of antibody to these herpes-like particles seen in cultures of Burkitt's lymphoma cells. No positive immunofluorescent reactions were seen to Epstein-Barr virus (EBV) in a variety of simian sera with the possible exception of two experimentally infected baboons (Henle and Henle, 1967). Gerber and Birch (1967) tested five species of nonhuman primates for CF antibodies to EBV. With the exception of baboons, there was a high incidence of antibody among these primates (chimpanzee, cynomolgus, rhesus, and African green), with the chimpanzees having the highest titers. It was suggested that this may reflect cross-reactions with other herpesviruses. More recently, two independent studies (Dunkel, personal communication, Kalter *et al.*, 1972a) found that Old World but not New World monkeys had EBV antibody. In both studies, sera derived from apes were generally found to be 100% seropositive with high titers, whereas only about 50% of the monkey sera were with antibody and then frequently of low titer. C.-T. Chu *et al.* (1971) reported that the prevalence of EBV antibody in *M. cyclopis* was lower than that of normal persons in Taiwan. These investigators also noted that maternal antibody to EBV persisted for some time in the blood of newborn animals. These findings were, in part, similar to those reported by Landon and Malan (1971) that adult *M. mulatta* maintained in close contact with humans and other simians were consistently positive to EBV antibody.

In the course of studies with EBV, it was demonstrated that infectious mononucleosis was related to this virus (Henle *et al.*, 1968; Niederman *et al.*, 1968; Gerber *et al.*, 1968). Further studies regarding herpes-like viruses as related to Burkitt's lymphoma, as well as to infectious mono-nucelosis, prompted Gerber and Rosenblum (1968) to examine rhesus sera for antibody with the finding that 50% of sera from these animals had such globulins. A number of investigators have attempted to demonstrate the susceptibility of nonhuman primates to the causative agent of infectious mononucleosis without success (Wising, 1939, 1942; Bang, 1942; Joncas *et al.*, 1966), but the demonstration of a possible relationship between EBV and this disease suggested renewed investigations. As a result, Gerber *et al.* (1969) attempted to transmit infectious mononucleosis to rhesus monkeys and to marmosets also without success. Landon *et al.* (1968) attempted long-term cultures of chimpanzee leukocytes followed by inoculations with cells and virus derived from cultures of cell lines grown similarly to those derived from Burkitt's tumor. Continuous cultures were established from these infected and normal chimpanzee leukocytes. Electron microscopy demonstrated the presence of herpes-like particles in both normal and infected cultures. The significance of this finding is, however, not clear. Stevens *et al.* (1970) suggest that EBV and those derived from the chimpanzee (Landon *et al.*, 1968) are distinct but share common antigens.

Varicella or herpes zoster (chickenpox) is also included among the herpesviruses. The early experimental work of Rivers (1926, 1927) has been mentioned. Natural disease is also known to occur in anthropoid apes—gorilla, chimpanzee, and orangutan (Eckstein, 1933; Heuschele, 1960).

Perhaps one of the more important observations as relates to herpesviruses and their possible association to cancer, especially cervical cancer, was the observation by Nahmias *et al.* (1971) that the capuchin monkey developed cervical lesions following inoculation with type 2 *H. hominis*.

In the course of their studies it was found that the rhesus and squirrel monkeys were resistant to this virus. These findings have been now substantiated in our laboratory. We have also found that the baboon is resistant, but the marmoset was highly susceptible with 100% fatalities (Kalter *et al.,* 1972).

Discussion of herpesviruses cannot be concluded without some additional mention of the serologic relationships. From the above, it may be seen that a certain amount of antigenic overlapping is recognized. Full extent of these reactions is not known because of lack of sufficient and detailed study. Inasmuch as it is important to detect whether or not an animal has antibody to B virus (the animal must be considered potentially dangerous for life if it does have such antibody), the only mechanism for obtaining this information is a serologic test, preferably serum neutralization. The early studies of Burnet *et al.* (1939b) suggest a one-way cross, i.e., *H. hominis* was neutralized by *H. simiae* antiserum but not vice versa. That this may not be as clearly demarcated as desired is seen by the data reported by Hull (1968) and as found in our laboratory (Kalter *et al.,* 1971). From these data it is quite evident that serologic differentiations between the two viruses may be difficult to interpret. Other studies, still in progress in this laboratory, suggest strong antigenic crossing between *H. hominis, H. simiae,* and SA8. There do not, however, appear to be any cross-reactions between these three viruses and *H. tamarinus.*

D. Myxoviruses

There are apparently only two reported naturally occurring myxoviruses found in monkeys and apes, i.e., SV5 and SV41. Of these, SV5 may not be a true simian virus and SV41 occurs very infrequently. In addition, however, nonhuman primates have been found to be infected with many of the human myxoviruses and have been used for various studies on this group and related viruses. Table V lists various viruses considered to be myxoviruses and paramyxoviruses. Viruses included in this broad category have been referred to as "myxo-like" and "pseudomyxoviruses" or "other RNA-helical-enveloped-viruses."

1. *Simian Myxovirus*

Frequently isolated from uninoculated rhesus monkey kidney cells, SV5 (Hull *et al.,* 1956; Emery and York, 1960; Hsiung and Atoynatan, 1966; Tribe, 1966; Yoshida *et al.,* 1965) has received considerable attention. Most important is the question of origin of this virus, i.e., is it a simian virus or is it derived from another animal source, human, canine, and so on? SV5 virus is most frequently recovered from Old World, but not from

TABLE V

Simian Myxoviruses (Paramyxoviruses, Pseudomyxoviruses, Myxo-like Viruses)

Virus	Originally isolated from
Paramyxoviruses	
SV5	Rhesus (?)
SV41	Cynomolgus
SA10	African green
Pseudomyxoviruses	
Foamy virus type 1	Rhesus, African green
2	African green
3	African green
4	Squirrel monkey
5	Galago (prosimian)
6	Chimpanzee
7	Chimpanzee

New World, monkeys. A number of investigators have reported the isolation or the presence of antibodies to this virus in man (Hsiung et al., 1962; Aulisio et al., 1964; Paine, 1964).

In common with other myxoviruses cellular infection with SV5 may be readily detected by hemadsorption with guinea pig cells (Emery and York, 1960; Yoshida et al., 1965). Serologic cross-reactions with other myxoviruses, especially mumps, makes serologic differentiation difficult. A variant has been designated as SV5a but it is not considered to be a different virus or even a true variant at this time. Animal inoculation with SV5, including monkeys, by Chang and Hsiung (1965), indicated development of inapparent infection. Transmission and disease production in young baboons was reported by Larin et al. (1967). Isolation of SV5 from patas and cynomolgus monkeys by Tribe (1966) has been mentioned. This investigator also suggested a vaccination program and demonstrated its effect on prevention of contamination as well as control of spread in a colony. Although a favorable response was reported, little effective use has been made of this mechanism in preventing infection with SV5. Heath and coworkers (1966) induced SV5 infection in vervet monkeys and suggested that this system may be used as a model for studies of respiratory disease. Immunization of monkeys with SV5 by Hsiung et al. (1965) resulted in a homotypic as well as a heterotypic response. This latter reaction was dependent upon previous viral experiences as well as host susceptibility.

Hsiung et al. (1969) reported that SV5 and other myxoviruses were not isolated from approximately 500 lots of monkey and baboon cultures during a 15-month period. In a subsequent study, Atoynatan and Hsiung

(1969) reported that SV5 infection was acquired during quarantine of the animals. This complements the report of Kalter *et al.* (1967a) and Shah and Morrison (1969), who found the prevalence of SV5 antibody among primates, human and nonhuman (in captivity), to be high. The nonhuman primates were found to become seropositive only after captivity; antibody to SV5 was not found in newly captured baboons (Kalter, 1965, 1969a; Kalter *et al.*, 1967a,b, 1968b). These findings are also interesting when collated with those of Schultz and Habel (1959). These investigators isolated a virus from man (SA virus), now known to be identical with SV5, and also noted that antibody to this virus was present in normal monkeys.

SV41 is less frequently encountered in animal tissues than SV5 (Hull, 1968). This is not only based upon virus-isolation studies but on serologic evidence as well (Kalter *et al.*, 1967a; Kalter and Heberling, 1970). SV41 was originally isolated from cynomolgus kidney cell cultures by R. H. Miller *et al.* (1964). Originally isolated rather frequently, this virus has not been reported recently. Hull (1968) suggests that SV41 is virulent for a number of different animals including monkeys. Cross-reactions with SV5 and other myxoviruses has been noted.

2. Parainfluenza

Both SV5 and SV41 are antigenically related to the parainfluenza viruses. However, viruses of the parainfluenza group have been also recovered from simians. Churchill (1963) isolated a parainfluenza type 3 from patas monkeys with clinically fatal pneumonia. It would appear that parainfluenza types 1 and 2 are rarely encountered in nonhuman primates and this is borne out by the lack of serologic evidence for their presence (Bhatt *et al.*, 1966b; Chang and Hsiung, 1965; Churchill, 1963; Shah and Southwick, 1965; Kalter *et al.*, 1967a; Shah and Morrison, 1969). A small number of nonhuman primates (chimpanzees, baboons) were found with antibody to parainfluenza, type 2 (Kalter *et al.*, 1967a). Akopova and Alekseeva (1968, 1969) described the antigenic structure as well as the various cellular affinities of simian parainfluenza viruses. The hemadsorption procedure was noted by Chanock *et al.* (1961) as helpful for the detection of naturally occurring simian parainfluenza virus infection.

3. Measles

The susceptibility of monkeys to measles viruses have been well documented. Early investigations, primarily those of J. F. Anderson and Goldberger (1911a,b,c), Goldberger and Anderson (1911), Blake and Trask (1921a), Nevin and Bittman (1923), Shaffer *et al.* (1941), and more recently those of Smirnova and Riazantseva (1960), as well as Imaizumi (1966), should be noted. It is of interest that Grunbaum (1904) attempted experimental measles infection in the chimpanzee at the turn of the century. Enders (1940) noted that measles research in monkeys was made difficult by the natural resistance of these animals to infection. The answer to this problem was presented by Peebles *et al.* (1957) in demonstrating the presence of antibody to measles in the majority of monkeys (rhesus) under study. Freshly imported cynomolgus were devoid of antibody. At the same time, Ruckle (1957, 1958a,b) observed that a "monkey intranuclear inclusion agent" (MINIA) was frequently found in monkeys and this agent was serologically indistinguishable from measles virus. These finding led to a study by Meyer *et al.* (1962a) on the prevalence of antibody in monkeys to measles virus as found in animals in the laboratory, at a compound in India, and in "the jungle." The marked increase in prevalence of antibody as associated with human contact was noteworthy. Animals in the jungle did not have antibody to measles virus. These observations were supported by the data of Shishido (1966) and Yamanouchi *et al.* (1969) as determined in Japan, as well as that collected by others elsewhere (Ruckle-Enders, 1962; Shah and Southwick, 1965; Bhatt *et al.*, 1966a,b; Kalter *et al.*, 1967a; Shah and Morrison, 1969). In an effort to develop information that would be helpful in procuring measles-susceptible rhesus monkeys, Vickers (1962) demonstrated that infection will occur when monkeys are maintained in close proximity to infected animals. It was shown that it was virtually impossible to prevent animals in captivity from converting to seropositive. A brief description of naturally occurring disease in cynomolgus monkeys had been provided by Habermann and Williams (1957) and by Sauer and Fegley (1960). Lokhova (1965) described the hemagglutinating properties of a measles-like virus recovered from monkey tissues. An epizootic of measles in a rhesus monkey colony was reported by Potkay *et al.* (1966). Clinical disease was similar to that seen in man, although Koplik spots were absent. Recognition of the failure to see clinical disease in animals was noted. Experimental measles in rhesus and cynomolgus have been reported by Nii *et al.* (1964), Peebles *et al.* (1957), and Shishido (1966). An outbreak occurring in three species of New World marmosets (*Callithrix jacchus, Saguinus oedipus,* and *Saguinus fuscicollis*) was reported by Levy and Mirkovic (1971). Significantly this outbreak was characterized by a high susceptibility and severity to measles virus with numerous deaths. Histopathologic evidence of a giant-cell pneumonitis was also recorded.

Giant cells have been reported in lymphoid tissues of monkeys as a result of measles invasion by a number of investigators (H. Gordon and Knighton, 1941; Ruckle, 1958a,b; Taniguchi *et al.*, 1954; Nii and Kamahova, 1963; Kamahova and Nii, 1965; Manning *et al.*, 1968). Sergiev *et al.* (1966) indicate that vaccine strains produce a milder reaction and different grades of pathologic response. Wild strains induced typical clinical disease. The observations of Yamanouchi *et al.* (1970) in this regard are noteworthy. In general they confirmed the findings of Sergiev *et al.* (1966), but emphasized that they could not definitely correlate pathologic findings with the virulence of strains, even though several of the strains had been attenuated. Sergiev and his collaborators have also employed baboons (*P. hamadryas*) for evaluation and study of measles vaccines (1956, 1959, 1960).

At this time only brief mention will be made of the potential capability of measles to induce disease after induction of a latent infection.* The ability of viruses to persist within a host cell for long periods of time, while recognized for many years, is beginning to assume a greater significance. Perhaps some procedure as that suggested by Charnomordik *et al.* (1965), whereby proteins and 17-hydroxyketosteroids as excreted in the urine may be beneficial in the recognition of infection due to measles. Diseases of unknown etiology are now recognized as possibly associated with certain of the more common agents often presumed only to be responsible for acute disease. In the case of measles, a role in the etiology of subacute sclerosing panencephalitis has been suggested (Adels *et al.,* 1968; Sever and Zeman, 1968). Attempts to induce this disease in chimpanzees, baboons, and rhesus, cynomolgus, and African green monkeys (among other nonhuman primates) have been negative. Serologic data, however, continue to support this suggestion. Pelc *et al.* (1958) induced CNS disease in two cynomolgus monkeys by the intracerebral inoculation of brain suspension from an SSPE patient. Transmission to another monkey was indicated. In view of Ruckle-Enders (1965) finding persisting measles virus in monkey spleen for at least 2 months after clinical disease, there is undoubtedly merit in continuing this type of investigation. The natural occurrence of a similar clinical disease in nonhuman primates has been reported (Kim *et al.,* 1970; Voss *et al.,* 1969).

4. *Influenza*

Influenza virus has not been reported to occur with any regularity among nonhuman primates. Experimental infection has been demonstrated in a number of species (rhesus and cynomolgus), generally without development of apparent clinical disease, although seroconversions and hematologic changes have been described (Woolpert *et al.,* 1941; Burnet, 1941; Saslaw *et al.,* 1946). Mouquet (1926) described the occurrence of "grippale" in three chimpanzees with one death. Deaths were also reported in *Cebus apella* monkeys by Ratcliffe (1942). Panthier *et al.* (1949) de-

* Latent infections and slow viruses will be discussed below.

scribed clinical influenza in cercopithecoid monkeys. Combined infection of influenza virus with streptococci also had minimal clinical effect (Merino *et al.,* 1941). Exposure of rhesus and cynomolgus monkeys to influenza virus in a Henderson apparatus did produce obvious illness in a number of animals (Saslaw and Carlisle, 1965). More recently these investigators (Saslaw and Carlisle, 1969) indicated that rhesus monkeys exposed to the Asian strain of influenza also developed signs of illness.

In our laboratory we have been able to infect baboons (*P. cynocephalus*) with a fresh isolate (A2/Hong Kong/68) of influenza virus (Kalter *et al.,* 1969a). Although clinical disease was minimal, seroconversion occurred and virus-shedding persisted for at least 9 days. Most important was the spread of virus to neighboring control animals. In other experiments with baboons, virus-shedding was found to persist for approximately 25 days. Treatment of inoculated animals with poly-IC delayed or suppressed antibody formation and prevented its development in uninoculated contact animals (Kalter *et al.,* 1970; Heberling and Kalter, 1971). Johnson *et al.* (unpublished data) described experimental infection and colony spread of A2/Hong Kong/68 influenza in gibbons (*Hylobates lar lar*). The effect of intranasal inoculation of various strains of influenza viruses in monkeys was described by Marois *et al.* (1971).

Serologic results with influenza viruses requires further evaluation. Bhatt *et al.* (1966a,b) found rhesus, langur, and bonnet sera to have HI antibody to A2/Jap/305/57. These titers were found to be "mostly 1:10." In our studies (Kalter *et al.,* 1967a) four antigens were employed in the HI test (A_0/PR8/38; A_1/FM1/47; A_2/Jap/305/57; and B/Lee) against sera obtained from man, gorilla, chimpanzee, orangutan, gibbon, baboon, vervet, rhesus, and patas. Large numbers of animals were found with this "antibody," suggesting past infection or the presence of an inhibitor or nonspecific reactor in spite of pretreatment by the usual methods known to remove such materials from human sera. Additional studies (unpublished) suggest that there is need for further evaluation of the HI procedure for serologic surveys of monkey and ape sera. Deinhardt *et al.* (1967a) did not find influenza-soluble antigen types A, B, and C antibody in white-lipped or cotton-top marmoset sera. Questionable serologic results with influenza antigens were reported by Atoynatan and Hsiung (1969). More recently Ohwada *et al.* (1970) also indicated a large number of serologic positives (PR8, FM1, Adachi, Swine, Great Lake) among vervet and cynomolgus sera.

5. *Respiratory Syncytial Virus*

A virus recovered from a chimpanzee with an upper respiratory tract infection and labeled "chimpanzee coryza agent (CCA)" by Morris *et al.*

(1956) was shortly thereafter recovered from infants with a respiratory illness (Chanock *et al.,* 1957). The properties of the two agents were identical. Coates and Chanock (1962) noted that the marmoset and baboon were unreceptive to this virus. In spite of the failure to induce clinical disease, virus was recovered for 10 days postinfection from the upper respiratory tract. Chimpanzees, however, did develop evidence of clinical illness. We have noted the presence of antibody to this agent in a variety of nonhuman primates: chimpanzee, orangutan, vervet, Formosan, and rock macaque. It will be noted, however, that most of the species did not have this antibody. Furthermore, only the chimpanzee was found with a prevalence of any significance.

6. *Rinderpest Virus*

Curasson (1942) failed to infect three species of monkeys (*Cercopithecus*) and one species of baboon with rinderpest virus. However, seroconversion was noted in cynomolgus monkeys inoculated with this virus, although measles virus inoculations did not result in any significant antibody changes (DeLay *et al.,* 1965).

7. *Newcastle Disease Virus*

Fowl plague (fowl pest) was noted by Findlay and Mackenzie (1937a,b) to produce a mild disease in rhesus monkeys.

8. *Foamy Viruses*

Some seven viruses (types 1–7) are presently included among the myxoviruses because of a number of commonly shared properties. Their structure resembles the parainfluenza–mumps viruses more closely than the influenza viruses. Foamy viruses do not grow in chick embryos nor do they hemagglutinate erythrocytes. Cultivation of these viruses is difficult even in rabbit kidney cells, which is considered by many to be the cell system of choice. Differences have been observed in the biologic characteristics of the seven serotypes and it is highly probably that all or part of this group will be separated from the myxoviruses.

First described by Enders and Peebles (1954) and Rustigian *et al.* (1955), these agents are now recognized as some of more common contaminants or indigenous agents present in monkey kidney cell preparations (Henle and Deinhardt, 1955; Hotta and Evans, 1956; Brown, 1957; Lepine and Paccaud, 1957; Falke, 1958; Endo *et al.,* 1959; Carski, 1960; Johnston, 1961; Hsiung and Gaylord, 1961; Plummer, 1962; Stiles, 1968; Stiles *et al.,* 1964, 1966; Rogers *et al.,* 1967; Hooks *et al.,* 1969). Table VI lists these viruses and their original source. For details concerning the fine structure of these viruses see the studies of Jordan *et al.* (1965).

TABLE VI

FOAMY VIRUSES AND THEIR ORIGINAL SOURCE[a]

Type	Species	Reference
1	*M. cyclopis*	Johnston (1961)
2	*M. cyclopis*	Johnston (1961)
3	*C. aethiops*	Stiles *et al.* (1964)
4	*S. sciureus*	Johnston (unpublished data)
5	*Galago crassicaudatus*	Johnston (unpublished data)
6	*Pan* sp.	Rogers *et al.* (1967)
7	*Pan* sp.	Rogers *et al.* (1967)

[a] Several different simian species have since been reported as hosts for foamy viruses.

Malherbe and Harwin (1957) described a foamy virus (SA1) isolated from the African green monkey that is identical with FV1.

Little is known regarding the pathogenesis of those agents or their distribution. A serologic survey attempting to determine monkeys and apes with antibody to these viruses is in progress in this laboratory.

E. PAPOVAVIRUSES

Two viruses, SV40 and SA12, are considered to be simian members of this virus group. The latter virus, isolated by Malherbe and Harwin (1963) was recovered only once from kidney cultures of an African green monkey and its status is not clear. The biologic characteristics of SA12 have been described both by the original investigators and by Hull (1968). The susceptibility of renal cells derived from different species of nonhuman primates was described by Ushijima (1966).

SV40 was first recognized by Sweet and Hilleman (1960) following transfer of viruses grown in rhesus and cynomolgus tissue to African green kidney cells where it produced a characteristic CPE. This distinctive CPE resulted in the term "vacuolating virus." Melnick (1962) suggested the name papovaviruses for a group of viruses with common properties and included SV40. In this regard, it may be important to note that Melnick and Rapp (1965) suggested a possible relationship between human warts (another papovavirus) and SV40. This virus is very frequently encountered in kidney cells derived from *Macaca* sp., especially rhesus and cynomolgus monkeys. Its presence in these cells has resulted in its inclusion in much of the early live, as well as killed, poliomyelitis vaccine.* Growth of SV40 occurs in many cell systems, but CPE is not seen in kidney cells from rhesus and cynomolgus monkeys unless the cells are held for long

* Inactivation procedures employed for polioviruses were ineffectual against SV40.

periods of time (Easton, 1964). In addition to the high incidence of SV40 in kidney cells derived from macaques, this virus is of great concern to investigators because of its oncogenicity for newborn hamsters (Eddy *et al.*, 1961).

Full extent of the natural occurrence of SV40 is not known, primarily because the majority of reports concerning the isolation of this virus from one or another simian species is clouded by the lack of history on the animal providing the isolate. The original reports of Sweet and Hilleman (1960) indicated that this virus was prevalent in approximately 70% of the rhesus monkeys. Hull (1968) suggests that the incidence is closer to 100%, making rhesus monkey kidney cells for production of vaccines useless. Sera from rhesus, cynomolgus, and African green monkeys were assayed by Meyer *et al.* (1962a) with the finding that 69% of the rhesus, 3% of the cynomolgus, and none of the African greens were seropositive. It was also reported by these investigators that animals with antibody were more prone to have positive kidney cell cultures than seronegative animals. Similar findings were obtained by Stiles (1968), who reported SV40 antibody in rhesus but not grivet monkey sera. *Cercopithecus* sp. produced antibody to SV40 according to Rapp *et al.* (1967). Zeitlyonok *et al.* (1966) found cynomolgus monkeys in Indonesia to be free of SV40 antibody. African green monkeys were also reported to be seronegative by Chumakov *et al.* (1963) and Chumakova *et al.* (1963), but converted to positive when maintained in contact with rhesus monkeys (Chumakova *et al.*, 1963). Adult rhesus monkeys were found to be 100% positive, but only 18% of the young animals had SV40 antibody (Shah and Southwick, 1965). In a continuation of this study, Shah and Morrison (1969) reported 34% of the free-living rhesus in India to have SV40 antibody, whereas 81% of laboratory-housed animals and only 1% of the animals at Cayo Santiago were with antibody. Hsiung *et al.* (1969) reported that seronegative rhesus and African green monkeys converted to seropositive upon arrival in their laboratory, although the origin of the infection was not apparent. Thus, the report by Hsiung and Gaylord (1961) on recovery of SV40 from patas monkeys must be viewed carefully as these animals had previous contact with macaques. A number of investigators have now shown that African green monkeys became infected after contact with rhesus (Sweet and Hilleman, 1960; Meyer *et al.*, 1962b; Ashkenazi and Melnick, 1962). Other reports concerning SV40 and infection of different species of primates may be found in studies from the Soviet Union (Kolyaskina, 1963; Lapin *et al.*, 1965a,b; Talash *et al.*, 1969; Yang *et al.*, 1967).

Production of tumors by inoculation of newborn hamsters and mastomys with SV40 (Eddy *et al.*, 1961, 1962; Girardi *et al.*, 1962; Kirschstein and

Gerber, 1962; Rabson, 1962) has resulted in numerous investigations, not within the scope of this report, on the oncogenic potential of this virus. Tumor development in primates by SV40 has *not* been demonstrated. However, the high prevalence of SV40 in monkey kidney cells has made contamination of many vaccine preparations commonplace. Furthermore, SV40 is extremely resistant to inactivation with formaldehyde (Gerber *et al.,* 1961), so live virus is present in presumed inactivated vaccines. As a result of these vaccines, primarily for poliomyelitis, many individuals have received live SV40. Fraumeni *et al.* (1963) have evaluated the carcinogenicity of SV40 for man.

Shah (1966) found neutralizing SV40 antibody in 14 of 161 human sera from residents of North India. The donors of these positive sera gave no history of immunization with monkey-prepared vaccines, but did live in areas of high monkey prevalence. A larger number of seropositive individuals was found among handlers of rhesus monkeys collected for export. Persons receiving formalin-inactivated poliomyelitis vaccines prepared in rhesus kidney cell monolayers produced significant levels of neutralizing antibody to SV40 (Gerber, 1967). Rhesus monkeys infected by a number of different routes developed viremia and neutralizing antibody to SV40 (Shah and Hess, 1968). Antibody to the tumor (T) antigen was also found in experimentally and naturally infected rhesus monkeys. Younger animals (1–2 years of age) were more prone to be seropositive to the T antigen than older animals. Meyer *et al.* (1962b) had previously demonstrated experimental infection of African green monkeys with SV40. Differences in ability to recover virus from these two species of monkeys under experimental conditions may be noted. In other experiments using the African green monkey, Ashkenazi and Melnick (1962) were able to recover SV40 from experimentally infected animals for as long as 6–8 months after inoculation. Inoculation of rhesus monkeys *in utero* at about 90 days gestation did not result in development of tumors on any of the three fetuses (Shah *et al.,* 1969). The effects of intravenous inoculation of SV40 on African green monkeys was also studied by Fabiyi *et al.* (1967). Rapid multiplication of virus occurred in various organs: kidney, lymph nodes, spleen, liver, and lung within 2 days after inoculation. Rapp *et al.* (1967) described development of CF antibody which reacted with the virus tumor and virus capsid antigens following inoculation of African green monkeys with intact simian cells infected with SV40. This demonstrated the production of tumor antibody to tumor antigen in primates without detectable tumors. Of more than passing interest was the recent report by Orsi *et al.* (1969) that low levels of SV40 were present in uninfected African green monkey kidney cells following rapid passage of these cells over a period of a year, even from seronegative animals. Tumor antigen was also detected in these cells. These investigators raise the point that monkeys may have become infected in their natural environment or as a result of shipping.

F. PICORNAVIRUSES

Nonhuman primates have been used in picornavirus (Table VII) studies, primarily as related to the polioviruses ever since Landsteiner and Popper (1909) demonstrated the susceptibility of monkeys and baboons to poliomyelitis virus. Various aspects of the early studies with different species of simians have been mentioned (Flexner and Lewis, 1910a,b; Kraus and Kantor, 1917; Thomsen, 1912; Aycock and Kagan, 1927; Kling *et al.*, 1929, 1934; Stewart and Rhoads, 1929; Amoss, 1928; Levaditi *et al.*, 1926; Aycock and Kramer, 1930; Jungeblut and Engle, 1932; Paul and Trask, 1932; Saddington, 1932; Vignec *et al.*, 1939; Lepine and Sedallian, 1939; Burnet *et al.*, 1939a; Mackay and Schroeder, 1935; Grossman and Kramer, 1936; C. Armstrong, 1939; Kolmer and Rule, 1934; Kolmer, 1934; Goldman, 1935; Burnet, 1940; Aycock, 1940; Faber *et al.*, 1944; Trask and Paul, 1941, 1942; Howe and Bodian, 1940, 1942, 1944; Aisenberg and Grubb, 1943). Numerous additional experimental involvement of simians in enterovirus research should be noted. P. F. Clark *et al.* (1930, 1932) had failed to detect poliovirus in the intestinal contents of 16 sick monkeys presumably following intracerebral inoculations. Muller (1935) reported spontaneous poliomyelitis in the chimpanzee. Grossman and Kramer (1936) were unable to infect marmosets with poliovirus. Isolation of poliovirus from intestinal contents of one of seven rhesus monkeys was recorded by Kramer *et al.* (1939). Intracutaneous inoculation of monkeys with poliomyelitis virus and its excretion in the stool was reported by Trask and Paul (1942). Sabin and Ward

TABLE VII

SIMIAN PICORNAVIRUSES AND THEIR SOURCE

Viruses	Originally isolated from
Entero-	
SV2, 6, 16, 18, 19, 26, 35, 42, 43, 44, 45, 46, 47, 49	Rhesus, cynomolgus
SA5	African green
A13	Baboon
Unclassified	
SV28	Rhesus
SA4	African green

(1942) were able to isolate virus from intestinal contents of five cynomolgus monkeys infected by the oral route. In the chimpanzee, Howe and
Bodian (1941), Bodian and Howe (1945), Bodian (1952, 1956), and
Howe (1962) were repeatedly able to recover poliovirus from feces after
feeding. Additional studies using the chimpanzee for poliovirus and vaccine evaluation have been also reported (Horstmann and Melnick, 1950;
Howe and Bodian, 1945; Howe et al., 1950, 1957; Howe, 1952; Melnick
and Horstmann, 1947). The chimpanzees were unique in that a "healthy"
carrier state developed.

Paul (1942–1943) summarized certain aspects of experimental poliomyelitis in different simian hosts and noted the variations in response. This
suggested to Melnick (1956) to compare the following nonhuman primate
species: M. mulatta, M. cynomolgus, M. mordax, C. mona, C. sabaeus,
C. cephus, C. capucina, and P. satyrus for susceptibility to poliovirus.
All but M. mordax* were found to have virus in their stools. Additional
comparative studies in monkeys were previously described by Melnick
and von Magnus (1948). Reproducible oral infection of rhesus monkeys
with type 1 poliovirus (Maloney) was reported by Gebhardt and Bachtold
(1953). Horstmann et al. (1947) inoculated infant rhesus monkeys orally
and found poliovirus distributed in the intestinal tract, peripheral nerves,
and CNS. A spontaneous occurrence of poliomyelitis in laboratory rhesus
monkeys was noted by Howe and Craigie (1943). Viremia was detected
by Melnick (1945) in paralyzed monkeys after experimental infection.
The spread of poliovirus to uninoculated rhesus monkeys and development of disease following oral administration in animals receiving deoxy-
pridoxine was described by Bodian (1948). Foster et al. (1951) were
able to demonstrate enhanced poliovirus infection in rhesus monkeys
treated with ACTH. The incubation period was shortened and more animals developed paralysis in the treated group than in comparable controls.
Barrera-Oro and Melnick (1961) used cynomolgus monkeys to ascertain
whether or not guanidine might have an effect on naturally occurring
enterovirus infections.

Melnick and Paul (1943) had reported successful transmission of poliomyelitis to the South American ringtail monkey (Cebus capucinus) as well
as to the African mustache monkey (Cercopithecus cebus). It may be recalled that Mackay and Schroeder (1935) had reported on the inability
to transmit poliomyelitis to the South American spider monkey (Ateles
ater). De Rodaniche (1952) also attempted use of South American monkeys (C. capucinus, A. palliata, and A. fusiceps) with successful virus
transmission to the capuchin and howler monkeys. The susceptibility of the

---

* The number of tests done were too limited to be considered significant.

spider monkey to poliovirus infection was described by Jungeblut and De Rodaniche (1954) and Jungeblut and Bautista (1956).

Other avenues of experimental usage of monkeys included attempts to vaccinate chimpanzees with a rodent-adapted strain of poliomyelitis (Koprowski *et al.,* 1954). C. A. Evans *et al.* (1954a,b, 1961) and C. A. Evans and Hoshiwara (1955) noted that various monkey tissues supported the *in vitro* growth of poliovirus: the failure of types 1 and 2 polioviruses to multiply *in vivo* after testicular inoculation, and increased virus proliferation in healing wound tissue of cynomolgus monkeys. The comparative susceptibility of baboon (*M. leucophaeus, P. cynocephalus, P. doguera, P. papio*) kidney cells to enteric viruses was examined by Hsiung and Melnick (1957). Monkey-to-monkey transmission of poliovirus via contact was reported by Craig and Francis (1958).

Development of vaccines (Salk *et al.,* 1954; Sabin, 1959a) resulted in continued expanded employment of nonhuman primates, not only as experimental models, but for evaluation of neurovirulence and vaccine potency. McLean and Taylor (1958) reviewed the use of monkeys for safety testing of poliomyelitis vaccines and this report is suggested for further information. Various studies involving analysis of poliovirus strains continued to involve monkeys especially as concerned the *d* character associated with virulence and reversion after multiplication in vaccinated children (Hsiung and Melnick, 1958b; Melnick, 1959; Melnick and Benyesh-Melnick, 1959; Bonin and Unterharnscheidt, 1964; Unterharnscheidt and Bonin, 1965; Simon, 1965; Simon *et al.,* 1965; Chiappino *et al.,* 1966; Unterharnscheidt, 1966; Smit and Wilterdink, 1966). Similar studies have been carried out in the Soviet Union (Krutyanskaya *et al.,* 1967; Nadaichik and Rozina, 1967; Rogova and Ralf, 1967; Savinov *et al.,* 1967; Adzhighitov and Gordeladze, 1968). The presence of an inhibitor in the central nervous system of the rhesus monkey was reported by Low and Baron (1960). Protective effects of stress were observed by J. T. Marsh *et al.* (1963), who showed that 7 of 11 stressed monkeys survived poliovirus infection, whereas only 1 of 12 controls survived. An outbreak of paralytic poliomyelitis was recorded in higher apes, i.e., gorillas and orangutans (Guilloud and Kline, 1966; Allmond *et al.,* 1967; Guilloud *et al.,* 1969).

D. G. Miller and Horstmann (1966) used proflavin-tagged type 1 poliovirus to study the pathogenesis of this virus in cynomolgus monkeys. By this method these investigators were able to differentiate between parent and progeny virus and the pattern of early replication. Their findings were in agreement with a number of earlier studies in primates (Verlinde *et al.,* 1955; Bodian, 1955, 1956; Sabin, 1956; Wenner *et al.,* 1959–1960), with poliovirus multiplication occurring first in the walls of the oropharynx and of the lower bowel. Using immunofluorescence a number of investigators attempted to determine the pathogenesis of poliovirus infections in different *Macaca* sp. (rhesus, cynomolgus, Japanese macaque). For example, *M. fuscata* resisted oral administration of poliovirus (Kanamitsu *et al.,* 1964, 1967) but others (Majima, 1957; Kaji *et al.,* 1962) have found this species to be susceptible. Various studies (Bodian, 1955; Sabin, 1956; Nathanson and Bodian, 1961) in monkeys have suggested that CNS involvement is the result of viremia and/or neural routes. Examination of peripheral nerves early in the course of infection has indicated the presence of virus (Horstmann *et al.,* 1947; Verlinde *et al.,* 1955; Sabin, 1956; Wenner and Kamitsuka, 1956, 1957). Kovacs *et al.* (1963a,b) demonstrated the presence of poliovirus by immunofluorescence in monkey polymorphonuclear leukocytes and CNS of intracerebrally inoculated monkeys. Similar observations in cynomolgus monkeys following oral infection with poliovirus were reported by Mannweiler and Palacios (1961), Kasahara (1965), and Kanamitsu *et al.* (1964, 1967). Blinzinger (1969) was able to find poliovirus crystals within the endoplasmic reticulum of endothelial and mononuclear cells of monkey spinal cord.

Nonhuman primates were also used for determining the potential pathogenesis of other enteroviruses besides the polioviruses. Melnick and Kaplan (1953) infected chimpanzees with 10 different types of coxsackieviruses, groups A and B. Similarly, the echoviruses (types 2, 3, 4, 6) were given orally to chimpanzees with subsequent development of infection and antibody (Itoh and Melnick, 1957). Clinical disease was not observed following any of these virus infections. Coxsackie A7 was reported by several groups of investigators as producing CNS lesions without overt paralysis in monkeys and chimpanzees (Dalldorf, 1957; Johnsson and Lundmark, 1957; Habel and Loomis, 1957; Chumakov *et al.,* 1956; Horstmann and Manuelidis, 1958). This coxsackievirus A7 had been previously isolated by Chumakov *et al.* (1956) from patients with paralytic disease and it had produced paralysis in monkeys. A viremia with echovirus type 9 was demonstrated in chimpanzees by Yoshicka and Horstmann (1960). Cynomolgus monkeys were found to develop endocarditis as well as initial valvulitis following inoculation with coxsackievirus B4 (Lou *et al.,* 1960, De Pasquale *et al.,* 1966). Sun and co-workers (1967) were able to produce a viral pancarditis with this same virus in cynomolgus monkeys similar to rheumatic pancarditis observed in man. Echovirus 13 was recovered from monkeys with encephalitis according to Prokhorova *et al.* (1967). Association of coxsackievirus B5 with appendicitis as well as enteritis in monkeys was described by Tobe (1967) and Tobe and co-workers (1968).

Theiler's encephalomyelitis virus (of mice) was found to be nonpathogenic for rhesus monkeys (Theiler and Gard, 1940a,b). Helwig and Schmidt (1945) and E. C. H. Schmidt (1948) described an agent (encephalomyocarditis virus) capable of producing interstitial myocarditis in the chimpanzee. A similar agent (meningoencephalomyelitis virus) was described in monkeys by G. W. A. Dick et al. (1958). These investigators also had reported that this agent may induce an encephalitis in monkeys (G. W. A. Dick et al., 1948). Gainer (1967) described the sudden death of two chimpanzees as a result of this virus. Recovery of the virus from heart tissue was accomplished. Recently, Douglas et al. (1970) reported the recovery of poliovirus type 3 from "normal" chimpanzee stools. An extensive study of the intestinal viral flora has been performed on chimpanzees maintained at the Holloman Air Force Base (Day et al., 1966; Soike et al., 1967, 1969a,b, 1971). A large number of enteroviruses, many of them closely related to human strains, have been recovered from these animals.

Rhinoviruses are included among the picornaviruses (Table VII), although the isolation of a simian serotype from nonhuman primates has not been recorded to our knowledge. E. C. Dick (1968) described experimental infection of chimpanzees with human rhinovirus types 14 and 43. It was suggested that the coryza-like syndrome reported in chimpanzees by Dochez et al. (1930) may have been rhinovirus-induced. During those experiments it was noticed that certain of these animals, although asymptomatic, excreted an agent with rhinovirus characteristics, but not the same as those used experimentally (E. C. Dick and Dick, 1968). Identification studies of this isolate suggests it to be closely related to another human rhinovirus, i.e., type 31. Martin and Heath (1969) suggest that the vervet monkey may be used as an in vivo model of human rhinovirus infection. These investigators employed an equine rhinovirus which produced clinical signs in the monkeys associated with virus excretion and antibody development. Human strains of rhinovirus were shown to infect gibbons followed by virus recovery as well as seroconversion to positive (Pinto and Haff, 1969). A patas monkey, similarly inoculated, remained seronegative throughout the experiment.

Extensive serologic information is available on picornavirus, human and simian, infections of primates. Probably because certain of the enteroviruses do not readily invade tissue other than that of the intestinal tract, sera may be obtained from an animal shedding one of these viruses that has little, if any, detectable antibody. Accordingly, enterovirus isolations may frequently be made without detection of serum antibody (Hoffert et al., 1958; Heberling and Cheever, 1966; Hull, 1968). Readily recovered from the intestinal tract of nonhuman primates, the enteroviruses rarely cause latent infections but this is in need of more extensive study. For a review of simian enteroviruses and their properties, see Hull (1968).

In a series of studies using rhesus, cynomolgus, and chimpanzee sera, both CF and SN antibody to various coxsackieviruses were reported in normal animals newly arrived in the laboratory (Kraft and Melnick, 1950, 1953; Kraft, 1952; Melnick and Kaplan, 1953). In our initial serologic studies with baboon sera (Kalter et al., 1964a) examined for evidence of infection with many human enteroviruses (poliovirus types 1–3, cox-sackieviruses B1–6, A9, and echoviruses 21–28), antibody to echovirus type 18 was the only seropositive detected. Shah and Southwick (1965) found 1 of 47 free-living rhesus monkeys had antibody to type 2 poliovirus. Contrasting with these findings were those of Bhatt et al. (1966a) who reported the presence of low-titered HI antibody to echovirus types 3, 7, 11, 12, and 19 in sera from the bonnet, langur, and rhesus monkeys. No antibody was present in 47 rhesus sera examined for a human rhinovirus (CV30). N. J. Schmidt et al. (1965, 1968) did not find preexisting SN or CF antibody to any of the human enteroviruses in rhesus sera. In these investigations the antibody response of rhesus monkeys, infected under conditions simulating those occurring in natural infection of man, were studied. Antibody responses following oral administration with coxsackie-viruses B1–6 and A9 after initial infection, as well as after successive infections with viral heterotypes, were determined. Rapid and type-specific neutralizing responses were obtained. Heterotypic infection elicited recall responses but not to viral types other than those with which the animal had experienced infection. CF antibody responses tended to broaden with subsequent infections. In somewhat similar studies, Kamitsuka et al. (1965) prepared antisera to 26 group A coxsackieviruses by the inocula-tion of rhesus monkeys. These sera were found frequently to cross with other coxsackieviruses. Heterologous crossings such as these have been reported previously in monkeys and chimpanzees (Kraft and Melnick, 1950, 1953). In the course of preparing these antisera, Kamitsuka et al. (1965) also noted development of antibody in these coxsackievirus-inocu-lated rhesus monkeys, to several of the simian enteroviruses: SV2, SV6, SV16, SV18, SV19, SV26, SV35. These antibodies were considered as possibly due to (1) natural infection by these viruses, (2) contamination of coxsackievirus antigens with simian viruses, or (3) sharing of common antigens between the simian viruses and coxsackieviruses. Subsequently these investigators (McMillen et al., 1968) determined that the simian viruses did not share antigens in common with the human group A cox-sackieviruses. Antibody to the simian viruses was therefore considered as probably related to infection either prior to or after arrival in the labora-tory. Much of the early studies on simian enteroviruses were done by

Hoffert *et al.* (1958) and Heberling and Cheever (1966). For information on the characteristics of these simian enteroviruses, consult the works of Heberling and Cheever (1964, 1965a,b, 1966), Kalter (1960, 1964, 1966), and Hull (1968).

As indicated previously, Hoffert *et al.* (1958) and Heberling and Cheever (1966) did not detect antibody to the simian enteroviruses, even though the rhesus monkeys under study were shedding virus. Deinhardt *et al.* (1967b) surveyed newly arrived marmosets as well as animals after a year in captivity without any evidence for antibody to the three polioviruses nor group B coxsackieviruses. In this laboratory, extensive surveys of several different groups of sera representing numerous monkey and ape species indicated large numbers of chimpanzee sera to have antibody to the three polioviruses, coxsackieviruses B1, B2, and A9, echoviruses 1, 7, 12, and SV19, SV49, and A13 (Kalter *et al.,* 1967a, 1971). Orangutans were also found with antibody to echoviruses 7 and 12 as well as to SV19 and SV49. Antibody only to echoviruses 3 and 12 were detected in gibbon sera. Baboon sera, especially those obtained in Africa, and representing newly captured animals, varied considerably in their antibody content depending upon their source and test virus. Examination of the data indicates that epidemics of echoviruses 3 and 7 occurred in Africa to which approximately all of the baboons responded serologically. Many of the vervets were found with antibody to echovirus 3 (15 of 33) but only one was found with antibody to echovirus 7. Numerous rhesus sera had antibody to echovirus 12. Over 50% of the patas sera were found with antibody to coxsackievirus B2.

Other studies with gorilla sera (Kalter *et al.,* 1969c) indicated that most of these animals had antibody as the result of a previous outbreak or vaccination (Guilloud and Kline, 1966; Allmond *et al.,* 1967). The majority of these gorilla sera were seronegative to all six group B, coxsackieviruses, and echoviruses 3, 6, 9, 11, and 13. Antibody was present to coxsackievirus A9 and A20, which represented seroconversions over the 1-year test period and to echoviruses 7 and 12. Additional studies on chimpanzee sera (Kalter and Guilloud, 1970) to these previously reported (Kalter *et al.,* 1967a, 1971) resulted in some noteworthy findings. Antibody to all three poliovirus types (probably as a result of vaccination) (Guilloud and Kline, 1966; Allmond *et al.,* 1967) was found in one group of animals (Lab #1), but epidemics to one or another type, especially poliovirus type 2, occurred at the other laboratories housing these animals. A few chimpanzees were found with antibody to coxsackieviruses A9 and B6. Coxsackievirus A9 antibody was most prevalent in chimpanzees of all five laboratories surveyed. Antibody to A20 was only infrequently present except for the Southwest Foundation for Research and Education (SFRE) colony that showed evidence of converting from 0 of 17 chimpanzees to 4 of 15 animals within a 6–9-month period following arrival. Chimpanzees in Lab #1 also converted from 0 of 16 seropositives in 1963 to 38 of 64 seropositives in 1968. Echovirus antibody was also present in these groups of chimpanzees: types 1, 3, 6, 7, 9, 11, 12, and 13. Evidence for occurrence of scattered "outbreaks" to one or another of these serotypes was also detected. Ohwada *et al.* (1970) reported similar findings for the African green and cynomolgus monkeys. These investigators reported the presence of antibody in varying numbers of these species to the three polioviruses, coxsackieviruses A9, B4, and B5. Antibody to B3 and echoviruses types 4, 6, and 9 were also detected in the African green, but not in the cynomolgus monkeys. The number of animal sera tested, however, were frequently small.

Testing nonhuman primate sera for antibody to the simian enteroviruses brought varying results. Colony differences were frequently noted; for example, one colony of chimpanzees were completely without SV49 antibody, whereas another group had a prevalence of approximately 17%. However, antibody to all six simian enteroviruses was found in all species tested. An antibody to SV4 and SV45 was infrequently present. These results with SV4 are in conflict with those reported by Hull *et al.* (1956). Possible species resistance may exist in that no antibody was ever found for human echovirus type 4.

As indicated above, no simian rhinoviruses have as yet been described, although experimental infection and natural occurrence with human and equine strains have been reported in different simian species (E. C. Dick, 1968; E. C. Dick and Dick, 1968; Martin and Heath, 1969; Pinto and Haff, 1969). Serologic studies of these animals, however, indicated preexisting antibody to one or several of these agents.

G. Poxviruses

The poxviruses have been well characterized and documented in the literature (Downie and Dumbell, 1956; Fenner and Burnet, 1957). These viruses, according to Andrewes and Pereira (1967), may be grouped primarily by their immunologic, biologic, and histopathologic reactivities, into several subgroups (Table VIII). It will be noted that a number of the listed agents are naturally "monkey"-oriented—monkeypox, yaba, yaba-like, and, perhaps, molluscum contagiosum. Hahon (1961) has reviewed the pertinent literature on smallpox and related poxviruses in the simian host and lists the susceptibilities of *M. cynomolgus, M. irus, S. sciureus, M. sinicus, C. callithrix, M. rhesus, M. nemestrinus, C. pathos, Macacus, Rhoesus, Cercopithecus, large African monkey, Java, S. satyrus, C. sebacus,*

TABLE VIII

THE ANIMAL POXVIRUSES[a]

Subgroup 1: Viruses closely related antigenically, biologically, and morphologically to variola	
Alastrim	Monkeypox
Cowpox	Vaccinia
Ectromelia	Variola
Rabbitpox	
Subgroup 2: Viruses related to orf	
Paravaccinia (Milker's nodes)	
Bovine papular dermatitis	
Subgroup 3: Viruses affecting ungulates; some may belong to subgroups 1 and 2	
Sheeppox	Swinepox
Goatpox	Horsepox
Lumpy skin disease	Camelpox
Subgroup 4: Viruses of birds	
Fowlpox	
Canary and other bird poxes	
Subgroup 5: Viruses producing tissue proliferation	
Rabbit myxoma	Hare fibroma
Rabbit fibroma	Squirrel fibroma
Subgroup 6: Viruses that have not been classified	
Molluscum contagiosum	
Yaba monkey virus	
Yaba-like monkey virus	

[a] Andrewes and Pereira (1967).

C. mona, C. albigularis, C. apella, rhesus, cynocephale, and bonnet to variola, alastrim, and vaccinia poxviruses. The usual signs of infection were that of a local lesion but fever, encephalitis, generalized exanthem, orchitis, and death were observed. This review (Hahon, 1961) should be consulted for details as well as for references to the original studies. Hahon (1961) also emphasizes the possible role that monkeys may play in transmitting smallpox to man, although he indicated that there has been "no recorded instance of human infection acquired in this manner." It was also suggested that the susceptibility of primates to the poxviruses may be determined by a serologic survey of existing specific antibodies to the poxviruses.

Hahon (1961) was somewhat prophetic in that shortly after the appearance of his review, a number of reports appeared in the literature involving nonhuman primates and one or another of the poxviruses. In 1958, Bearcroft and Jamieson reported an outbreak of subcutaneous tumors in rhesus monkeys at the laboratories of the West African Council for Medical Research, Lagos, Nigeria. This disease could be transmitted by subcutaneous inoculation to rhesus and West African guenon monkeys (*Cercopithecus aethiops tantalus*). However, transmission attempts to mangabeys (*Cercocebus torquatus torquatos*) and patas monkeys (*Erythrocebus patas patas*) were unsuccessful. The morphologic relationship of this virus (Yaba) to the poxvirus group was readily demonstrated by Andrewes and his collaborators (1959; Andrewes and Pereira, 1967; Niven *et al.*, 1961). While these studies with Yaba virus were underway, von Magnus *et al.* (1959) described the occurrence of a pox disease, now recognized as monkeypox, in their colony of cynomolgus monkeys. As will be described below, a number of such outbreaks have been recorded in various monkey and ape colonies. More recently, cases of monkeypox virus infections have been also reported in a number of humans in West Africa. Thus, this group of viruses assumes a role of importance in primate medicine heretofore not given proper consideration. In this section we will concern ourselves with those poxviruses primarily derived from monkeys, i.e., monkeypox, Yaba, and Yaba-like. Any information regarding other poxviruses and nonhuman primates will be presented where available.

1. *Monkeypox*

The first description of monkeypox as a specific disease entity was presented by von Magnus *et al.* (1959), although Bleyer (1922) described a fatal pox disease in mycetes and *Cebus* monkeys in Brazil. Since then a number of other outbreaks have been recorded (Prier *et al.*, 1960; Sauer *et al.*, 1960; McConnell *et al.*, 1962; Peters, 1966a,b) and their salient features reviewed by Arita and Henderson (1968). While these epidemics are all within the last 10–15 years, it may be noted that naturally occurring pox infections of nonhuman primates date back to 1767 (Arita and Henderson, 1968). Gispen *et al.* (1967) described the virus isolate obtained from the apes and monkeys of the Rotterdam Zoo outbreak (Peters, 1966a,b). Poxvirus was recovered from 7 orangutans, 1 squirrel monkey, 1 white-handed gibbon, and 1 marmoset (*Callithrix jacchus*) that were fatally infected. This outbreak may have resulted from an infected anteater (*Myrmecophaga tridactyla*) from which this virus was also recovered. Biologic properties and cell-culture host range of this virus were described by Rouhandeh *et al.* (1967). The pathogenesis of monkeypox virus (von Magnus strain) for cynomolgus and rhesus monkeys was described by Wenner *et al.* (1968). These investigators noted that the disease was less severe in rhesus as compared with cynomolgus monkeys. All groups of animals, however, seroconverted to positive and resisted challenge with vaccinia virus. However, these animals were found to be susceptible to Yaba virus. The efficacy of vaccinia vaccination in protecting rhesus monkeys and chimpanzees from experimental infection with monkeypox virus was demonstrated by McConnell *et al.* (1964, 1968). Similar studies in the baboon have been reported by Heberling and Kalter (1970, 1971).

Newer methodology and the occurrence of pox diseases in monkeys prompted Wenner and his collaborators to study further the pathogenesis of this disease in nonhuman primates. In an extensive series of experiments (Wenner *et al.,* 1967, 1968, 1969a,b,c; Cho *et al.,* 1970), Wenner *et al.* have attempted to describe monkeypox as regards clinical, virologic, and immunologic responses; dose response and virus dispersion; histopathologic lesions and sites of immunofluorescence; and modification of the disease pattern by antilymphocytic sera and the effects of methisazone on infection. Thus it was demonstrated that both rhesus and cynomolgus monkeys develop a pox eruption. The clinical disease is similar to that seen in man and the histologic reaction was that described for variola and vaccinia. HI antibodies occurred rapidly but started to decline at the time the SN antibody was making its appearance. The HI antibody does not last as long as do CF or SN antibodies. Virus growth is within many different tissues, especially skin and mucous membranes, and such organs as the spleen, tonsils, lymph nodes, testes, and ovary. Treatment with ALS results in death due to either severe pox or bacterial sepsis with the entire course of the disease intensified. It was noted also that methisazone which suppressed plaque development of monkeypox and vaccinia viruses in tissue culture did not modify the disease in cynomolgus monkeys. Rao *et al.* (1968) also noted that immunosuppression (cortisone) enhanced experimental variola in rhesus monkeys.

2. *Yaba and Benign Epidermal Monkeypox (Yaba-like)*

Other investigations include a report by Raghavan and Khan (1968) on a case report of a pox disease in a chimpanzee that appears to be monkeypox. Milhaud *et al.* (1969) also reported clinical and virologic studies on Yaba and Yaba-like outbreaks. As indicated above, occurrence of a tumor disease transmissible to monkeys was first described by Bearcroft and Jamieson (1958) and shown to be of virus etiology by Andrewes and co-workers (1959; Niven *et al.,* 1961). A number of reports emanating from Roswell Park Memorial Institute demonstrated the susceptibility of the rhesus and cynomolgus monkey to Yaba virus. The African green monkey was only slightly susceptible, whereas the mona, vervet, sooty, and squirrel monkeys were not susceptible to this agent (Ambrus *et al.,* 1963). Metzgar *et al.* (1962) had developed a method for measuring the CF response of the host and found that the titer was high during tumor growth and that the animal was resistant to reinfection. Following decline of antibody, susceptibility to reinfection returned. A detailed pathogenesis study of Yaba in monkeys was presented by Sproul *et al.* (1962, 1963). Human disease as a result of accidental inoculation with this virus was described by Grace *et al.* (1962; Grace and Mirand, 1963, 1965). In both monkeys and humans, benign histiocytomas are produced. Back *et al.* (1965) reported detailed studies on Yaba virus infections of rhesus monkeys as well as negative antibody findings in baboon sera, as sampled in Kenya, Africa, to this virus. Kato *et al.* (1965), in addition to studying Yaba virus infection by autoradiography, reported that *M. fuscata* could also be successfully inoculated with Yaba virus. It was then demonstrated by Ambrus and Strandstrom (1966) that African monkeys, previously reported to be resistant (Ambrus *et al.,* 1963), were susceptible to Yaba virus if they were born in the United States. Thus the baboon, vervet, and patas monkey could be successfully inoculated with Yaba virus. More recently, Ambrus *et al.* (1969) described the occurrence of spontaneous Yaba tumors on the hairless areas of the face, palms, and interdigital areas in monkeys at their laboratory. Wolfe *et al.* (1968a,b) were able to demonstrate that Yaba virus could be transmitted to rhesus and cynomolgus monkeys by aerosols with development of pulmonary and nasal tumors. These investigators (Wolfe *et al.,* 1968a,b) were also able to define the immunologic response of monkeys to such infections.

In 1967, several reports appeared in the literature regarding a new pox disease of monkeys that apparently was not monkeypox. Hall and McNulty (1967) described the occurrence in *M. mulatta* and in man of a benign, epidermal disease. *Macaca nemestrina* and *M. fuscata* were also found to be highly susceptible as were, to a lesser extent, *Cynopithecus niger.* These latter species, as well as *Cebus albifrons, Saimiri sciureus, Alouatta caraya, M. speciosa,* and *C. aethiops* in another colony did not develop disease. Electron microscopy of a pox disease occurring in rhesus monkeys and not monkeypox was described by H. W. Casey *et al.* (1967). This report indicates that a similar outbreak (Hall and McNulty, 1967) has occurred at another institution in monkeys obtained from the same distributor. Crandell *et al.* (1969) provided greater details concerning the recovery and characteristics of this virus isolate. A third laboratory was also involved in a similar outbreak with animals derived from the same source (Espana, 1971). This virus was shown to differ from monkeypox and Yaba, although they all share a number of characteristics in common. Nicholas and McNulty (1968) and Nicholas (1970a,b), who originally referred to this agent as "1211," report that this virus resembles vaccinia–monkeypox in thermal stability, growth kinetics, and plaque morphology. Host range is similar to that of Yaba virus. There are sufficient differences to warrant separation from the other poxviruses. Yaba and Yaba-like viruses were also shown to share antigens with each other but not (or only very weakly) with other poxviruses. This agent was also found to produce moderately severe systemic symptoms in man (McNulty *et al.*, 1968).

3. *Miscellaneous Poxvirus Studies*

The pathogenesis of variola in *M. irus* has been described by Hahon and Wilson (1960). Danes and Blazek (1966) reported on the immunization of rhesus monkeys with live variola virus and the failure to immunize with a single dose of inactivated virus. Danes and his collaborators (1966) showed further that single-dose irradiated noninfectious virus induced low levels of antibody in various animals, including the monkey. Exposure of monkeys to aerosols of variola virus resulted in primary lesions developing in the bronchioles and alveoli (Lancaster *et al.*, 1966).

In an attempt to answer a number of questions raised by Arita and Henderson (1968), especially as concerns the possible existence of small-pox in wild monkey populations, Noble and Rich (1969) transmitted variola major virus to cynomolgus monkeys by contact and by the aerosol routes. These investigators were able to consistently transmit the agent by both methods. After six passages, infection could not be continued by con-tact transmission and the number of observed lesions decreased with each contact. These studies do indicate then that infection may be transmitted and it was theoretically possible for monkeys to serve as a reservoir for variola virus. In another study, Noble (1970) has examined New World and Old World monkeys in an attempt to determine whether or not they may serve as a simian reservoir of smallpox. *Cebus apella, Ateles paniscus,* and *Lagothrix lagothricha* were found not to be susceptible to variola minor. An Old World monkey, *Cercopithecus aethiops,* was also not very susceptible to variola major or minor. Serologic studies (HI tests) by this investigator on various sera from monkeys of Africa, Philippines, and South America failed to reveal "significant" levels of antibody to vaccinia virus. It may be noted, however, that 26 of 509 sera examined were found with HI antibody at 1:10 or 1:20 levels. As a result it is concluded that "there is no clear experimental, serological, or epidemiological evidence to support the hypothesis that smallpox can exist in the wild simian popu-lations" (Noble, 1970). It is interesting, however, to note that Mack and Noble (1970) also report on the natural transmission of smallpox from man to monkeys and refer to it as "an ecological curiosity."

Molluscum contagiosum has been reported as a naturally occurring dis-ease of chimpanzees (Douglas *et al.*, 1967). Significantly, while mild, this disease has only been reported as seen in man and chimpanzees. Attempts to produce this clinical entity in *M. mulatta* have not been successful.

Serologic examination of nonhuman primate sera for the presence of antibody to the poxviruses has been complicated by the reliability of the available procedures (complement-fixation, serum-neutralization, and hemagglutination-inhibition). Thus, attempts to demonstrate a simian reservoir of human smallpox has been made difficult because of problems arising from lack of a methodologic capability which would provide unqualified results. As indicated above, Arita and Henderson (1968) have traced the history of monkeypox infections in an attempt to determine whether or not a reservoir to human monkeypox exists in the monkey population. In an attempt to clarify the relationship of simians to the human disease, as well as interpret serologic data obtained by a number of different laboratories, an informal discussion sponsored by the "WHO Smallpox Eradication Unit" was held in Moscow, USSR, March 1969.* Examination of several thousand sera from approximately 20 species of nonhuman primates indicated a small number of HI positives, but not substantiated by SN, to variola and/or monkeypox antigens. In another study, a large number of additional sera obtained from cynomolgus monkeys collected from freshly trapped animals in Malaysia were tested by several of these participating laboratories, again with essentially negative results.

Conclusions drawn from these studies would suggest that the CF test is unsuitable for serologic surveys because of numerous anticomplementary reactions observed as well as the short duration of CF antibody. The HI test was thought to be satisfactory as experimentally infected animals respond with easily detected HI antibody, but there may be difficulties with nonspecific inhibitors in monkey sera. Furthermore, HI antibody is also short-lasting and, like CF antibody, may also not persist for more than 1 year. Neutralizing antibody evidently is more persistent, but only limited information is available on its use for collecting epidemiologic data. Other procedures such as gel diffusion, immunofluoresence, have also not been evaluated under similar circumstances. The more recent studies of Noble (1970) in surveying New World and Old World monkey sera have been mentioned above. Experimentally infected baboons developed clinical disease associated with antibody development to monkeypox virus as do uninoculated control animals maintained in the same room (Heberling and Kalter, 1970, 1971). In these studies good antibody (HI and SN) responses to the infecting agent were noted.

H. REOVIRUSES

Sabin (1959b) described the characteristics of this virus group and also noted the occurrence of common cold symptoms in a chimpanzee as a re-

---

* Participants: I. Arita, F. Fenner, R. Gispen (Chairman), E. B. Gurwich, S. S. Kalter, S. S. Marennikova, G. Meiklejohn, J. Noble, Jr., E. M. Sheluchina, V. D. Soloviev, and I. Tagaya.

sult of infection by a reovirus. Three simian viruses have now been described: SV12 (Hull et al., 1956) and SA3 (Malherbe and Harwin, 1957) recovered from rhesus and African green monkey, respectively, are closely related to reovirus type 1; and SV59 (Hull et al., 1958), also originally isolated from the rhesus, is antigenically similar to reovirus type 2. Isolations of these three serotypes from nonhuman primates is relatively frequent but there does not seem to be a counterpart of reovirus type 3. Deinhardt et al. (1967b) have reported the occurrence of this type in marmosets, but thus far this has not been confirmed, although antibody to this type is very prevalent in primate sera (see below). For discussion of the biologic characteristics of the simian reoviruses, see Hull (1968). Rosen (1960) described the serologic grouping of reoviruses by the HI procedure.

Studies with reoviruses and nonhuman primates has been minimal. Bhatt et al. (1966b) described antibody to all three reovirus types in bonnet, langur, and rhesus monkeys. Antibody to more than one type was found in many of these animal sera. In our previous serologic studies (Kalter, 1965, 1968a, 1969a, 1969b, 1971b; Kalter and Heberling, 1970, 1971a,b; Kalter and Guilloud, 1970; Kalter et al., 1967a, 1969c) we found antibody to these three reoviruses to be commonplace. Antibody to the reoviruses was also frequently encountered in gorilla sera, a finding in contrast with other antibody data for this animal. Chimpanzees were often seen with antibody to all three serotypes, and especially to type 3, even at the time of capture in Africa. A more detailed study of reovirus infections of primates (Kalter and Heberling, 1971a) emphasized the frequent occurrence of reovirus antibodies with type 3 being most prevalent. Multiple infection was also frequently seen with antibody to all 3 reovirus types often observed in many of the different species examined.

George and Feldman (1969) were able to obtain sera from wild and captive bonnet monkeys which were examined for reovirus antibody. Both groups had approximately the same prevalence of antibody to all three serotypes, with type 2 antibody predominant. More of the captive rhesus monkeys were seen with antibody to all three reoviruses than the bonnet monkeys, again with antibody to type 2 most prevalent (approximately 90%). Experimental inoculation of bonnet monkeys with type 3 reovirus resulted in homotypic serologic responses. Only limited experimental infections of nonhuman primates with the reoviruses have been described (Stanley et al., 1954; Rosen, 1962; Lou and Wenner, 1963; Masillamony and John, 1970). Lou and Wenner (1963) noted a febrile response in monkeys infected with type 1. More extensive studies by Masillamony and John (1970) indicate that type 3 reovirus produces infection in bonnet monkeys as evidenced by recovery of virus from throat and rectal samples. A number of animals also excreted virus in their urine. Febrile reactions were seen in all and diarrhea was observed in 13 of 17 animals. The antibody response was specific for the type 3 virus.

I. Miscellaneous

There are a number of viruses still not included in the above-listed families that have been associated with nonhuman primates. These agents, primarily of human origin, have been found or studied in various species monkeys and apes and their inclusion here is, therefore, warranted.

1. *Hepatitis*

Perhaps one of the more important of these viruses is that related to hepatitis. An interesting aspect of monkeys and hepatitis was reported by F. L. Soper and Smith (1938). These investigators, in reporting on their studies with yellow fever vaccine in humans, noted the occurrence of jaundice in 20–30% of individuals receiving vaccine prepared with serum pools from two rhesus monkeys. Vaccines prepared with other sera did not produce similar responses. Reactions of this kind probably represent the first known occurrence of primate-related hepatitis. It was not, however, until 1939 that hepatitis was demonstrated to be of viral etiology (Findlay *et al.*, 1939). Shortly thereafter a number of investigators attempted to use nonhuman primates for experimental infection purposes. In most instances the results were negative or equivocal (Colbert, 1948–1949; A. S. Evans *et al.*, 1953; A. S. Evans, 1954), although Pellisier and Lumaret (1948) as well as Findlay and collaborators (1944) reported successful inoculation of cercopithecus monkeys and chimpanzees. As will be pointed out below, there has been a renewed interest in use of nonhuman primates for studies with this virus (or viruses) as a result of two recent developments: (1) recognition of primate-associated hepatitis in man, and (2) the demonstration by Blumberg *et al.* (1965) that Australia antigen is associated with this disease entity.

Hillis, in the early 1960's (1961, 1963), reported the occurrence of viral hepatitis in chimpanzee handlers, but actual demonstration of the presence of a virus was not documented. Similar occurrences were reported by Riopelle (1963, 1964) and McInerney *et al.* (1968). Bearcroft (1963, 1964, 1969a,b), in an attempt to ascertain whether or not West African monkeys were susceptible to infectious hepatitis, experimentally inoculated *Erythrocebus patas, Cercocebus torquatus torquatus, Cercopithecus aethiops tantalus, C. mona mona, C. erythrotis sclateri, C. erythrotis camerunensis, C. nictitans nictitans, C. nictitans erythrogaster, Mandrillus leucophagus, Papio anubis, Galago demidovii demidovii,* and *Perodicticus potto potto* (a total of 240 animals) with specimens of liver, spleen, colon, and feces from fatal human hepatitis cases. As a result of these studies, it was noted that two separate liver lesions developed in experimentally infected animals. Both parenchymal and stromal elements of the liver were affected with lesions appearing about 3 weeks following inoculation. Disease was maintained in patas monkeys by serial passage. However, no virus-like particles were seen in electron microscopic examination of the liver. Smetana, in a series of reports, also described experimental as well as spontaneous viral hepatitis in primates as associated with patas monkeys and chimpanzees (1965, 1969; Smetana and Felsenfeld, 1969a,b). Here it was suggested that spontaneous hepatitis frequently occurs in chimpanzees and to a lesser extent in the patas monkey. The disease is very mild and usually inapparent, but human disease that results as a consequence of contact with these animals is more severe. Transmission of the agent from man to both species primates occurs if sufficient "potent" virus is given and the animals had no previous exposure to the disease. The histopathology developed in chimpanzees is similar to that seen in man. Similar studies were described by a number of other investigators, including attempts to determine the efficacy of immune globulins on preventing disease in man (Appleby et al., 1963; Douglas and Berge, 1964; Atchley and Kimbrough, 1966; Davenport et al., 1966; Hartwell et al., 1968; Mosley et al., 1967; Ruddy et al., 1967; Krushak, 1970). More recently Bearcroft pursued the status of the patas monkey as relates to susceptibility to infectious hepatitis (1968, 1969a,b,c). Here it was reported that this monkey develops a disease of lymphoid tissue with depression of the bone marrow resulting in abnormal cells. Such animals were capable of reinfection but the disease was ameliorated by the primary infection. Serum glutamic oxalacetic transaminase (SGOT) and serum glutamic pyruvic transaminase (SGPT) values of the normal and hepatitis chimpanzee were determined by Hartwell et al. (1968) with the finding that the infected group had much higher values than normal animals.

Deinhardt et al. (1967a) reported on the successful transmission of human viral hepatitis to marmoset monkeys. These results were questioned by several groups of investigators (Kohler and Apodaca, 1968; Kohler et al., 1968, 1969; Apodaca et al., 1968; Scriba and Oehlert, 1969) with the suggestion that the results in the marmosets could be explained by an adenovirus infection. Parks and Melnick (1969) also were unable to confirm the findings of Deinhardt et al. (1967a) as these investigators found their control marmosets developed the same elevated serum transaminase and histopathologic changes as did the test animals. Other studies by this group of investigators (Parks et al., 1968, 1969) have continued to support the contention that the lesions produced in the marmosets were not that of human hepatitis. These investigators suggested that an indigenous marmoset agent was responsible for the results and thus use of the marmoset would be negated for hepatitis studies. However, in another series of studies by Deinhardt and co-workers (1967a, 1970a,b; Holmes *et al.,* 1965, 1967, 1969; Deinhardt, 1970) it was reported that the disease produced in marmosets was indeed that of human hepatitis. This argument has continued with a joint publication (Deinhardt *et al.,* 1970a; Melnick and Parks, 1970) in which both groups express their view. A cooperative study is currently underway using coded specimens and both cotton-top and white-lipped marmosets in an attempt to resolve the differences reported. Lorenz *et al.* (1970) have also become involved in this problem with the finding that the white-mustached marmoset (*S. mystax*) developed signs of human hepatitis after inoculation with Deinhardt's hepatitis agent or sera from human cases. "Hepatitis-associated antigen" (see below) or partially purified antigen did not produce any changes.

In 1965, Blumberg and co-workers described the presence of an antigen (Australia antigen) in the serum of an Australian aborigine, which is now considered to be closely related to infectious hepatitis, if not the actual etiologic agent itself. A number of investigators have now demonstrated the presence of both antigen and antibody in sera from a variety of simian species (Hirschman *et al.,* 1969; Lichter, 1969; Prince, 1971; Shulman and Barker, 1969; Blumberg, 1972; Blumberg *et al.,* 1968; Deinhardt, 1970; Kalter *et al.,* unpublished data). It is evident from the above that some relationship exists between nonhuman primates and Australia antigen and/or antibody but the full significance is not clear. A number of reviews (Barinskii, 1967; Piazza, 1969; Melnick, 1968; Holmes and Capps, 1966; Hillis, 1967; Koff and Isselbacher, 1968; Deinhardt, 1970) attempt to bring this subject into focus. However, until some clearly defined experimental data become available, the exact relationship of simians to the problem of hepatitis remains obscure.

2. *Rubella*

This virus is responsible for one of the most commonly occurring diseases of man, but its importance was not appreciated until Gregg (1941) pointed out the teratogenic sequelae occurring in infants born of women contracting this disease during their first trimester of pregnancy. A number of investigators had attempted to produce rubella in nonhuman primates. Habel (1942) indicated the successful transmission of the virus to rhesus monkeys by means of patients' blood or throat washings. Clinical disease could not be produced in five rhesus monkeys inoculated intranasally with pools of tissue culture-propagated rubella virus shown to produce clinical disease in human volunteers, although three of the animals developed antibody rises (Sever *et al.,* 1962). Similar inapparent infection in rhesus monkeys was reported by Heggie and Robbins (1963). Parkman *et al.* (1965a,b) demonstrated that the rhesus monkey was highly sensitive to parenteral inoculation of rubella virus with development of subclinical rubella, viremia, and virus-shedding from the throat. Cercopithecus monkeys were shown by Sigurdardottir *et al.* (1963) to develop clinical evidence of infection following inoculation of rubella-infected tissue-culture fluids. This observation was confirmed by Cabasso and Stebbins (1965) in the same species, and Phillips *et al.* (1965) in rhesus monkeys. In these latter studies, Phillips and his collaborators (1965) were also able to demonstrate recovery of rubella virus from the placenta of an infected rhesus monkey whose pregnancy was interrupted 10 days after intravenous inoculation. In another series of pregnant rhesus monkey inoculations, Sever *et al.* (1966) did not produce clinical disease using throat gargle material given during 25–28th day of gestation. Virus-shedding resulted as well as seroconversions. No evidence of congenital infection, malformations, mental retardation, petechiae, or hepatosplenomegaly was noted. One of three live births had antibody at 6 months of age, suggesting undetected transplacental infection. Parkman *et al.* (1965a,b) inoculated pregnant rhesus monkeys during their fourth week of gestation and recovered virus from three fetuses 10–31 days later. These studies strongly suggest uterine infection of the fetus following maternal infection. The patas monkey (*Erythrocebus patas*) was reported to be susceptible to wild and attenuated strains of rubella virus by Draper and Laurence (1969). Overt clinical signs did not occur in this species.

Delahunt (1966) and Delahunt and Rieser (1967) reported on the occurrence of cataracts and other embryopathies in rhesus monkeys after experimental infection of pregnant animals. An increase in abortions were also observed. In our laboratory, we were able to produce clinical disease in pregnant baboons (*Papio cynocephalus*) associated with rash, lymphadenopathy, virus-shedding, and seroconversion following inoculation of throat garglings containing virus. Another group of inoculated pregnant animals showed a higher rate of abortions than generally noted in our normal baboon colony. Cursory examination of fetal tissue did not disclose any evidence of malformations but detailed histopathologic examinations are yet to be done. Elizan *et al.* (1969) reviews the teratogenic effect of rubella virus on experimental animals.

The above results strongly emphasize the need for a good experimental model for the study of malformation capabilities of rubella virus. It would appear that there are suggestions that the baboon and perhaps other nonhuman primates may serve extremely well in this type of study. The numbers of animals employed in most instances have been too few for anyone to make a negative judgment regarding the suitability of this system. In this regard, Cabasso *et al.* (1967) studied rubella passage virus (green monkey kidney cells) in rhesus or African green monkeys in an attempt to ascertain the effect of attenuation on this virus. In these studies, it was found that the monkey and human results as concerns spread in contacts were dissimilar. In view of the past discrepancies reported with this virus and different species of monkeys, these findings must be viewed critically.

Serologic data on various monkeys and apes have indicated marked differences among these species regarding the presence of antibody. At this time there is little uniformity of opinion among the various investigators as to which of these animals had antibody or not. In serologic surveys involving varying numbers of animals and various methods to detect antibody, no preexisting rubella antibody was found in rhesus, African green, or patas monkeys, or baboons (Sever *et al.,* 1962; Sigurdardottir *et al.,* 1963; Heggie and Robbins, 1963; Parkman *et al.,* 1965a,b; Phillips *et al.,* 1965; M. J. Casey *et al.,* 1966; Cabasso *et al.,* 1967; Haas *et al.,* 1969; Amstey, 1969; Draper and Laurence, 1969; Kalter and Heberling, unpublished data). Captive simians found to have rubella antibody were: gorilla, chimpanzee, orangutan, gibbon, baboon, and African green, rhesus, patas, cynomolgus, marmoset, squirrel, and woolly monkeys (Kalter, 1969a,b, 1971b, 1972; Kalter and Guilloud, 1970; Kalter and Herberling, 1970; Kalter *et al.,* 1967a, 1969c). Capuchin, stumptail, Japanese macaque, Formosan rock macaque, white-faced spider, and talapoin monkeys were found to be without rubella antibody. Few of the newly captured baboons or captive baboons were found to have rubella antibody (Kalter and Heberling, 1971b). Other groups of simians were found with antibody prevalences of 100%. Chimpanzees at this institution seroconverted to rubella virus, an event not seen in the baboons. Only small numbers of African green and cynomolgus monkeys were found with rubella antibody according to Ohwada *et al.* (1969, 1970). Horstmann (1969) reported that baboons and chimpanzees all developed antibody following infection with rubella virus derived from humans. These animals also manifested evidence of clinical infection but no rash.

3. *Marburg Virus*

Considerable attention has been given an outbreak of human disease in Frankfurt–Marburg, Germany and Belgrade, Yugoslavia, resulting from contact with blood and tissues of African green monkeys derived from a common source or from patients with the disease. Important in the epidemiology of this disease was the failure to detect any evidence of infection in animal handlers. A number of the original reports and reviews are available and are suggested for information on investigations involving epidemiology, biologic characteristics of the agent, and clinical and laboratory findings (Siegert *et al.,* 1967, 1968a,b; C. E. G. Smith *et al.,* 1967; Hennessen, 1968, 1970; Hennessen *et al.,* 1968; Martini *et al.,* 1968; Luby and Sanders, 1969; G. W. A. Dick and Waterson, 1969; M. Kaplan, 1969; Martini, 1967, 1969; Gedigk *et al.,* 1968; Simpson *et al.,* 1968a,b; Held *et al.,* 1968; Haas, 1968; Haas *et al.,* 1968, 1969; Bonin, 1969; Simpson, 1968, 1969a–d; Zlotnik, 1969; Zlotnik and Simpson, 1968, 1969; Zlotnik *et al.,* 1968a,b; Strickland-Cholmley and Malherbe, 1970; Bowen *et al.,* 1969; Bechteisheimer *et al.,* 1968; Jacob and Solcher, 1968; Hofmann and Kuntz, 1968; Hofmann *et al.,* 1969; Chernyshov, 1969; May and Knothe, 1968; Stille *et al.,* 1968; Carter and Bright, 1968; Maass *et al.,* 1969a,b; May and Herzberg, 1969; May *et al.,* 1968; Saenz, 1969; Slenczka, 1969; Slenczka and Wolff, 1971; Slenczka *et al.,* 1968; Kissling *et al.,* 1968). These reports all provide a comprehensive coverage of the outbreak as it occurred and the attempts to define the agent and understand the mechanism behind the outbreak. It is readily understandable why the sudden occurrence of this disease should produce some consternation among investigators involved in handling monkeys and apes. Not since the discovery of B virus (*Herpes-virus simiae*) had there been a similar situation in which a large number of human deaths occurred. In fact this Marburg outbreak was more devastating since it produced its effect suddenly and at one time.

An attempt was made using serologic surveys to ascertain what monkeys and apes may be responsible for spreading or harboring this agent. Unfortunately the results have not clarified the issue and more studies are necessary before a final evaluation may be made. Kissling and his coinvestigators (1968), Kafuko *et al.* (1970), Stojkovic *et al.* (1971) and studies performed in this laboratory (Kalter, 1971a; Kalter and Heberling, 1970; Kalter *et al.,* 1969b) have demonstrated that a number of simian species possess both CF and SN antibody to this agent. Stojkovic *et al.* (1971), whose laboratory was involved in this outbreak, reported that 90% of the surviving African green monkeys had CF antibody. Our results, using antigen supplied by Dr. R. E. Kissling, suggested a possible focus of infection in African simians. Talapoins, African greens, chimpanzees, and baboons were found with CF antibody (Kalter, 1971a; Kalter and Heberling, 1970, 1971b; Kalter *et al.,* 1969b). Human convalescent sera and infected hamster sera (kindly provided by Dr. D. I. H. Simpson) were also found to be positive. Sera from humans not associated with the disease as well as African simians born in the U. S., South American monkeys, and Asian macaques (there were some exceptions in this later group) were devoid of antibody. We have recently expanded this survey to include sera obtained from elephants, gazelles, wildebeests, zebras, and kongoni bled in Africa with negative results (Kalter *et al.,* 1972c).

Other investigators have not been able to substantiate these findings.

According to Simpson (unpublished data) many of the positive sera provided by us were anticomplementary. Vervet and baboon (chacma) sera examined by Malherbe and Strickland-Cholmley (1968, 1971) were reported to be seronegative. These investigators used an antigen prepared from a surviving vervet monkey previously experimentally infected. Studies are currently underway in our laboratory in collaboration with Dr. W. Slenczka of Marburg, Germany and Dr. R. E. Kissling of C.D.C. to explore the serologic differences reported by the various laboratories. Slenczka (1971) has prepared a tissue-culture antigen which is effective by CF as well as immunofluorescence. One possible explanation for the difference in results may be loss of soluble antigen in the course of preparation. The possibility that some of these antigens are nonspecific must also be considered.

4. Simian Hemorrhagic Fever (SHF)

Another "new" disease with high mortality rates among the involved simian populations has recently been described, primarily as a disease of macaques (*M. mulatta, M. fascicularis, M. nemestrina, M. assamensis,* and *M. speciosa*) in laboratories in the United States, Soviet Union and England (Shevtsova, 1967; Lapin *et al.,* 1967, 1969; Tauraso *et al.,* 1968a,b; Palmer *et al.,* 1968; Krylova, 1968; Krylova and Shevtsova, 1969; Shevtsova *et al.,* 1968; O. Wood *et al.,* 1970; A. M. Allen *et al.,* 1968). Apparently this genus is the only one susceptible to SHF as other primates, including man, in contact with these infected animals, have failed to develop apparent disease. Tauraso *et al.* (1970) has recently reviewed the epidemiology and characteristics of this agent. A serologic survey (Tauraso *et al.,* 1972) for antibodies to this agent in a variety of primate species—man, gorilla, chimpanzee, orangutan, gibbon, baboon, African green, rhesus, cynomolgus, patas, talapoin, stumptail, and marmosets—were consistently negative for CF antibody to SHF. One animal, a rhesus monkey surviving the original outbreak, whose serum served as a control, was positive.

Perhaps one of the more important aspects of this disease, in contrast to the Marburg virus outbreak, is the continuous occurrence. There have been at least six separate epidemics with almost 100% fatalities reported among the animals with clinical disease. The virus is small (less than 50 nm), RNA-type, chloroform-sensitive, labile at pH 3.0, and relatively heat-stable.

5. Tumor Viruses

Monkeys and apes should make an ideal experimental model for the study of a viral causation of cancer. Their phylogenetic relationship allows for extrapolation of data currently not provided by information derived from other animal systems. Perhaps this closeness to man has been the reason for the lack of demonstrative relationships and thus, if a clearly defined system is eventually developed, some definitive interpretation of results could be made. Progress, however, has been encouraging in this direction during the past decade. An attempt will be made to review the pertinent findings.

Munroe and Windle (1963) and, independently, Zilber and his co-workers (1963, 1966), demonstrated the ability of several species of monkeys (*M. mulatta, M. nemestrina,* and *P. hamadryas*) to develop tumors following inoculation with Rous sarcoma virus (RSV). Details of tumor induction by RSV as well as a complete histopathologic report and electron microscopic studies were provided by Munroe *et al.* (1964) and well reviewed in 1969 (Munroe, 1969). The importance of immunosuppression in the successful development of these tumors was indicated (Munroe, 1966). It was also demonstrated that pregnant animals or adults given progesterone were more susceptible to tumor development by RSV than normal or untreated animals (Munroe and Windle, 1967). Deinhardt (1966, 1967) shortly thereafter demonstrated the susceptibility of newborn marmosets (*Saguinus* sp.) to RSV. Newborn cynomolgus monkeys were also reported as susceptible to this virus by Yamanouchi (1966) and Yamanouchi and his co-workers (1967). Cotes (1966) and Berman *et al.* (1967), in a study on fetal rhesus monkeys, found that RSV had produced tumors that were observed at the site of inoculation following delivery. Adenovirus type 12, similarly given, did not induce tumors nor was there any evidence of tumor development 2 years after inoculation *in utero.* As indicated above, we were also unable to produce tumors by inoculating baboons *in utero* with adenovirus type 12 (Kalter *et al.,* 1966c). Tumor development in monkeys as a result of RSV infection elicited CF antibody to both hamster and chicken RSV tumor antigens and to several avian leukosis viruses grown in chicken embryo fibroblast tissue culture (M. J. Casey *et al.,* 1966). Naturally occurring antibody to RSV was not found in sera obtained from baboons, chimpanzees, or African green monkeys according to Morgan (1967). Rabotti *et al.* (1967) were able to demonstrate that rhesus monkeys inoculated with RSV-transformed cells (dura leptomeninges and cerebrum of rhesus monkeys) developed malignant sarcomas. Infectious virus was not recovered but RSV antigen was detected in one monkey tumor. Brain tumors were produced in newborn rhesus monkeys by the intracerebral inoculation of RSV (Janisch *et al.,* 1968). In the marmoset, Deinhardt (1969) indicates that RSV may be recovered from all marmoset tumors but only if the tumor cells are "co-cultivated" with chick embryo cells free of leukosis viruses. These investigators also studied the pathology (R. D. Smith and Deinhardt, 1968a) and noted that the induced RSV tumors produced "unique" cytoplasmic membranes in infected marmosets (R. D. Smith and Deinhardt, 1968b). Levy et al. (1969) were also able to demonstrate the ability of RSV to produce tumors (rhabdomyosarcomas, fibrosarcomas, osteosarcomas) in newborn marmosets. Metastatic tumors were seen in animals with rhabdomyosarcomas and fibrosarcomas but not with osteosarcomas. More recently, Noyes (1970) demonstrated the susceptibility of newborn and adult marmosets (S. oedipus and S. nigricollis) to two different strains of RSV (Carr-Zilber and Schmidt-Ruppin). In our laboratory, in which studies have been carried out using the baboon as the experimental model for RSV tumor production, it was found that infection may result as a consequence of the virus being transmitted via an aerosol (Kalter et al., 1968b). Other studies have demonstrated the importance of age, immunosuppression, and progesterone in influencing the susceptibility of baboons (Kalter and Heberling, unpublished data; Kalter et al., 1968a,c).

In addition to RSV studies in nonhuman primates, these animals have been utilized to detect tumor production by other viruses. Mention has been made of these studies performed with such viruses as adeno-, herpes-, and papovaviruses. Sibal et al. (1967) examined by tanned cell hemagglutination and microimmunodiffusion methods the immune response of rhesus monkeys to murine leukemia virus (Rauscher). Both 19S and 7S antibody were noted, but the latter was higher and more prolonged. Lapin and his co-workers have been reporting on the possibility of a human leukemia virus being transmitted to monkeys (Adzhighitov et al., 1968; Lapin, 1969; Lapin et al., 1968; Yakoleva and Lapin, 1966; Yakoleva, 1969). Thus far, there has been no confirmation of this observation. Needless to say, it would be extremely valuable if this finding could be confirmed. Melnick and his group (1967) have inoculated over 500 baboons with various materials derived from human tumors without any evidence of cancer development. A number of these animals have been inoculated over 5 years ago. A new virus recovered from a mammary tumor of rhesus monkeys has been described by Chopra and Mason (1970). This virus has been studied under the electron microscope and its infectivity for cell cultures recently reported (Chopra et al., 1971). Association of viral particles in breast cancer of animals and possibly man makes this observation worth pursuing. A virus of cats, S7-feline sarcoma virus, has been found by Deinhardt et al. (1970c) to produce multiple sarcomas in marmosets. Both tumor homogenates and cell-free extracts were employed with tumors developing 3–4 weeks postinoculation with the cell-free material. Recently Kinard (1970) has reviewed the status of the nonhuman primate in cancer virus studies providing a general background and rationale for such projects. A listing of many of the nonhuman primate species involved in cancer virus investigations include: *M. mulatta, M. fascicularis, M. radiata, M. nemestrina, P. cynocephalus, C. aethiops, C. albogularis, S. oedipus, E. patas, C. jacchus, Galago crassicaudatus, P. troglodytes, S. fuscicollis,* and *S. mystax.*

A number of investigators have reported on the "natural" occurrence of tumors in various species primates: rhesus monkeys (J. R. Allen *et al.,* 1970); various captive wild animals, including monkeys (Appleby, 1969); Burkitt's lymphoma in the white-handed gibbon (*Hylobates lar*) (Di Giacoma, 1967); breast cancer in a tree shrew (*Tupaia glis*) (Elliot *et al.,* 1966); spontaneous and induced malignant neoplasms in *P. cynocephalus, P. doguera,* and *P. papio* (Kent, 1960); an intrauterine choriocarcinoma in a rhesus monkey (Lindsey *et al.,* 1969); spontaneous blood neoplasms in nonhuman primates (Lingeman *et al.,* 1969); a foreign body granuloma in the rhesus (Lord and Willson, 1968); howler monkeys (Maruffo, 1967); leukemia in *Cynocephalus sphinx* (Massaglia, 1923); uterine tumor in the marmoset (*S. oedipus*) (Moreland and Woodard, 1968); lymphocytic leukemia in the white-cheeked gibbon (*H. concolor*) (De Paoli and Garner, 1968); adenocarcinoma in the intestine of a rhesus monkey (Plentl *et al.,* 1968); reticulum cell sarcoma in the sykes monkey (*Cercopithecus albogularis*) (R. A. Price and Powers, 1969) and basal-cell tumor in the rhesus (Schiller *et al.,* 1969). Occurrence of various neoplastic diseases in simians was described by M. Valerio *et al.* (1968) as well as by Chapman (1968) and Huseby (1969). Allison (1970) raised the question concerning the applicability of utilizing virus tumor data derived from animals. Transplantation of tumors of various sorts to primates was reported by several investigators (Friedman *et al.,* 1968; Godlewski, 1967; Godlewski *et al.,* 1968; Lewis *et al.,* 1968). Herpesviruses and their relationship to development of tumors in monkeys has been described above.

6. Rabies

Little has been done in the way of studying the association of non-human primates to this disease. It would appear that a wider distribution of rabies-infected monkeys would be noted, but this is not the case. Boulger (1966) reported the occurrence of natural rabies in a laboratory (rhesus) monkey. Experimental infection of several species of monkeys with rabies virus was reported in 1926 by Levaditi and his collaborators. In the preparation of rabies-immune globulin and testing in monkeys after virus challenge, G. R. Anderson and Sgouris (1966) did not report antibody in their rhesus monkeys. Karasszon (1969) has also described the clinical symptoms seen in rhesus monkeys following intracerebral inoculation of rabies virus (Högyes fixed strain). The observed incubation period was short (3–4 days) and death occurred in approximately 8–11 days following inoculation. The animals developed all recognized stages of the disease. Recently C. Kaplan (1969) has reviewed the current situation as concerns rabies in nonhuman primates.

7. *Lymphocytic Choriomeningitis* (*LCM*)

This virus was originally isolated from a monkey by C. Armstrong and Lillie (1934) following inoculation with material from a patient with "St. Louis encephalitis." Lillie (1936) described the histopathology of this virus in monkeys. It is now recognized that LCM has a wide host range, including primates. Herrlein *et al.* (1954) suggests that LCM may be a spontaneous disease of laboratory monkeys. In their study on the microbiology of marmosets, Deinhardt *et al.* (1967b) reported that none of the tested animals had antibody to this virus. However, Alice (1951) described a disease in a marmoset that Halloran (1955) classifies as lymphocytic choriomeningitis. Analysis of our serologic data (Kalter *et al.,* 1967a; Kalter and Heberling, 1971b) demonstrated the presence of LCM antibody in sera of a number of species examined: humans, chimpanzees, orangutans, vervets, rhesus, cynomolgus, marmosets, and baboons. No CF LCM antibody was seen in sera from the following monkeys and apes: gorilla, gibbon, Japanese macaque, patas, talapoin, and squirrel.

Antibody was frequently found in sera of baboons sampled immediately following capture in Africa. Seroconversion to positive was also demonstrated in sera of baboons maintained in captivity. The significance of these findings is not clear at this time. Apparent disease due to LCM has not been reported. Inasmuch as simians are known to eat rodents and these animals frequently have access to the feed of simians, it is assumed that infection results from the capture and eating of an infected rodent or from contamination of food supplies by mouse feces and/or urine.

8. *Slow and Other Viruses of the Central Nervous System*

This is an ill-defined group of viruses that potentially may include nearly all viruses known to man. However, the viruses generally considered here are those originally discussed in a conference sponsored by the National Institute of Neurological Diseases and Blindness (Monograph #2), 1965, entitled "Slow, Latent and Temperate Virus Infections" (Gajdusek *et al.,* 1966a). Much credit should be given D. C. Gajdusek and C. J. Gibbs, Jr. for their role in developing our current understanding of the viruses included under this heading. Our present knowledge stems from the realization, followed by the actual demonstration, that "kuru," a disease of the central nervous system limited to the Fore people of New Guinea was transmissible to chimpanzees (Gajdusek, 1967a,b,c; Gajdusek *et al.*, 1966b). This was followed by a series of reports describing the pathology in chimpanzees (Beck *et al.*, 1966); transmission from chimpanzee to chimpanzee but not to "gibbon (black, golden), macaque (rhesus, cynomolgus, Barbary ape, stumptail), African green, patas, spider (black, brown), squirrel, capuchin, woolly, white-lipped marmoset, tree shrew, and slow loris" (Gajdusek *et al.*, 1967); and various comparative aspects of the disease kuru with others that appear to be in this category—scrapie, mink encephalopathy, visna, Aleutian mink disease, maedi, infectious adenomatosis of sheep lungs, and various subacute and chronic neurologic degenerative diseases of man (Gajdusek, 1967a; Gibbs, 1967; Gajdusek and Gibbs, 1968; Beck *et al.*, 1969a,b; Gajdusek *et al.*, 1969). Successful transmission of kuru to the spider monkey following passage in the chimpanzee was described by Gajdusek (1968) and Gajdusek *et al.* (1968) and Lampert *et al.* (1969) described the electron microscopy of experimental kuru in both chimpanzees and spider monkeys.

Fatal spongiform encephalopathy (Creutzfeldt-Jakob disease) was reported as experimentally produced in chimpanzees and transmitted in second passage to another chimpanzee by Gibbs *et al.* (1968; Gibbs and Gajdusek, 1969); the neuropathology was described by Beck *et al.* (1969b). A comprehensive report on attempts to transmit subacute and chronic neurologic diseases to various species of primates and other animals was provided by Gibbs *et al.* (1969). For example, scrapie-infected mouse tissue was inoculated into chimpanzee, rhesus, cynomolgus, and African green monkeys with negative results. Amyotrophic lateral sclerosis, amyotrophic lateral sclerosis–parkinsonism dementia syndrome, Parkinson's disease, multiple sclerosis, Schilder's disease, progressive multifocal leukoencephalopathy, subacute herpes encephalitis, Werding-Hoffmann's disease, spongiform encephalopathy, and myasthenia gravis surgically biopsied or early autopsy material was inoculated into a total of 32 primates with negative findings up to 41 months of observation. Included among the monkeys and apes were: chimpanzee, African green, cynomolgus, rhesus, baboon, and bonnet. Rhesus monkeys were successfully inoculated with transmissible mink encephalopathy virus by R. F. Marsh *et al.* (1969) and this material passaged to a stumptail macaque and squirrel monkeys (Eckroade *et al.*, 1970).

9. *SA11*

This is a simian virus originally isolated by Malherbe and Harwin (1957) and apparently is unrelated to any known virus other than that described by these same investigators (Malherbe and Strickland-Cholmley, 1967)—"O" agent isolated from abattoir wastes. Malherbe and Strickland-

Cholmley (1967) found SN antibody to SA11 in five of six vervet monkeys.

IV. Diseases—Natural and Experimental

A number of texts and monographs are now available on nonhuman primates and many describe various diseases of these animals (Ruch, 1959; R. N. Fiennes, 1967; R. N. T-W Fiennes, 1966; Perkins and O'Donoghue; 1969; Bainer and Beveridge, 1970; D. A. Valerio *et al.,* 1969; Bourne, 1969, 1970a,b, 1971; Hofer, 1969a,b; Carpenter, 1969; Lapin and Yakoleva, 1963; H. H. Reynolds, 1969; Beveridge, 1969a,b; Kratochvil, 1968; Whitney *et al.,* 1967; Starck *et al.,* 1967; Sauer, 1960; Goldsmith and Moor-Jankowski, 1969; Kalter and Heberling, 1971b; World Health Organization, 1971). These offer a diverse background of information as relates to monkeys and apes. Unfortunately many of these reports are inadequate and fail to clearly define the problems associated with nonhuman primates in biomedical research. Mattingly (1966) describes B virus and hepatitis infections; Hartley (1968), in addition to B virus, discusses rabies and "vervet monkey disease," whereas Appleby *et al.* (1963) only discusses B virus as a primate disease infectious for man. In a discussion of common disease problems of laboratory animals, Trum and Routledge (1967) mentions "measles and poxes" but not in the context of being a colony problem. These authors suggested further that *H. simiae* is a serious zoonotic problem, but not for monkeys; *H. tamarinus* is the virus that may be serious for colony monkeys. Habermann *et al.* (1960) included pseudorabies, variola (smallpox), rubeola, varicella, herpes zoster, herpes simplex, Sabin B, giant cell pneumonia, yellow fever, louping ill, salivary gland disease, lymphocytic choriomeningitis, dengue, poliomyelitis, Western and Eastern encephalitis, and St. Louis encephalitis as diseases to which monkeys were susceptible. Necropsies of 708 rhesus and cynomolgus monkeys by these same investigators (Habermann and Williams, 1957) suggested that only two cynomolgus monkeys had a viral disease. Fairbrother and Hurst (1932), in describing spontaneous diseases seen in 600 monkeys (primarily rhesus), do not mention viral diseases. Similarly, Kennard (1941), after 246 necropsies including 218 rhesus, 21 sooty mangebeys, 2 baboons, and 1 each of the brown mangebey, patas, mona, Java, and a hybrid macaque, does not mention any viral diseases in these animals. B virus and cytomegalovirus infections of monkeys were noted as important by Ditchfield (1968). Only brief mention is made of viral diseases as relates to the breeding of macaques by D. A. Valerio *et al.* (1968). Over 500 necropsies were performed by Nelson *et al.* (1966) on imported tamarins (*T. nigricollis*) but little evidence is provided for viral studies on these animals. In another report on these marmosets Gengozian (1969) lists only two virus agents to be of any consequence among marmosets: *H. tamarinus* and yellow fever virus. According to Vickers (1969), B virus, yellow fever, and Rift Valley fever are important in the African green monkey and probably the patas monkey. The only virus disease of marmosets and macaques mentioned was herpesvirus infection. Perhaps the recent review by Eyestone (1968b) should be examined for a more realistic appraisal of infections of nonhuman primates by viruses. This author cites yellow fever, B virus, monkeypox, rabies, infectious hepatitis, rubeola, Kyasanur Forest, Marburg virus, and simian hemorrhagic fever as recent zoonoses associated with nonhuman primates. Wedum and Kruse (1969) also point out the hazards to man of many viruses excreted by monkeys via urine and/or feces. Some 20 viruses are listed as either capable of being excreted by monkeys and apes as well as causing infection in contact monkeys caged with or near the infected animals. Actually, one of the few pathologic studies of nonhuman primates (baboons) done at the time of capture was that of Kim *et al.* (1968).

An interesting commentary has been recently put forth by Cornelius (1970) on the use of animal models. He contends that there are many diseases occurring spontaneously in animals that are similar to human disease that have been only superficially studied. In a very extensive listing, it is noted that only three simian models of virus diseases with human counterparts are given: hepatitis, kuru, and molluscum contagiosum. Jones (1969) had also suggested using animal and avian models for studies of diseases in man, but with only a few monkey and ape examples provided. Previously Koprowski (1958) described virus diseases of man and monkeys that are similar clinically; citing yellow fever and infectious hepatitis as examples.

Most of these reports attempt to place emphasis on clinical comparisons rather than on immunologic and histopathologic responses. We have repeatedly suggested that clinical symptoms fall far short of providing a true indication of host–parasite relationships and demonstrated that serologic information frequently serves as a good indicator of past infection with one or another agent. These sentiments are echoed by Ohwada *et al.* (1969) in their studies on the presence of antibody in African green and cynomolgus monkeys to viruses of human and simian origin.

V. Conclusions and Perspectives

An attempt has been made to describe the use of nonhuman primates in virus research. It became an impossible task and as a result many reports have been omitted. No intent was made to slight these efforts; it was simply a matter of not having a reference to the work or the material had been well reviewed elsewhere. We have tried to draw the attention of investigators to what is available in the literature on viruses as pertains to monkeys and apes and point out that there has been extensive experimentation in certain areas and neglect of others.

A number of misconceptions has crept into the literature, primarily because investigators have failed to consider these animals as living entities with their own biologic flora and fauna and because most data were drawn from animals maintained in captivity. A listing of the recognized simian viruses are provided herein. Descriptions of their isolation, growth, and biologic description may be found in a number of reviews (Hsiung, 1968, 1969; Hull, 1968, 1969a,b; Hull and Minner, 1957; Hull *et al.*, 1956, 1958; Kalter, 1969a,b, 1971b; Kalter and Heberling, 1968, 1971b, 1972). In addition to these 70+ recognized simian virus serotypes, many laboratories have agents recovered from excreta and tissues of normal and sick animals that are in need of identification. As in-depth studies are performed on the different species of simians more isolates are being made. The studies of Melendez and his collaborators (1970c) on New World monkeys emphatically demonstrates the role that these new viruses may play, not as relates to the health of the animal, but in our understanding of human disease. Very few of the recognized simian viruses have proven to be highly pathogenic for either the monkey population or to other animals with which they have had contact. The exceptions, as noted above, have been devastating—herpesviruses of man and monkeys, Marburg virus in man and monkeys, hepatitis in man, Kyasanur Forest virus in man and monkeys, simian hemorrhagic disease in macaques, poxviruses in monkeys and apes, and so on. Infection of an animal other than the natural host is frequently highly invasive and often fatal. Occurrences such as these are generally the result of the quality of animal used as well as poor colony husbandry and management. Cross-infections occur as a result of intermingling of species either by placing the animals directly in contact with each other or as a result of a more subtle mechanism which involves the carrying of the agent(s) by the personnel, their clothes, or instruments. The World Health Organization has recently reviewed this problem and made recommendations for minimizing the situation in one of their reports (1971).

Recommendations for providing a better quality of animal are valueless unless a mechanism for monitoring the animals (serologic surveys, virus isolations) is instituted. In addition, isolation and quarantine of animals are effective only if the laboratory personnel understand the problem and cooperate by limiting their contact and instituting an appropriate protective barrier (clothing, boots, masks) and self-restraint in moving from one group of animals to another. Vaccination is rather limited and effectively includes only yellow fever, poliomyelitis, and smallpox (for monkeypox). One of the major areas in need of study and development involves preparation of vaccines to one or another of the important simian viruses.

Virus data as determined by monitoring of virus-shedding must be interpreted carefully. The history and previous experiences of the animals involved are frequently vague and confused by numerous unknown or unreported contacts with other animals. Shedding of virus is not uniform and very probably governed by many extraneous and unknown factors. In studies performed at this laboratory (Kalter and Heberling, 1968), we have found that the 5-month period following shipment of animals from Africa to be the time of highest virus isolations. This observation has been made on numerous occasions and is attributed to the "stress" of traveling. Hull (1969a) has also reported on the incidence of virus isolations as determined by the frequency of recovering simian viruses over different periods of time. The 1955–1958 isolations were: SV4 (504),* SV12 (173), SV28 (65), SV11 (44), SV15 (42), SV17 (41), SV5 (32), SV23 (25), B virus (20), others (5). During 1958–1962 the following virus isolations were made: SV28 (15), SV23 (13), SV5 (7), SV17 (6), SV40 (5), SA1 (4), SV38 (3), SV32 (2), SV31 (2), others (1). In 1962, emphasis was placed on African green, rather than rhesus monkeys with the following isolations made: SV5 and SA1 (most frequent), SV41 (12 isolations in 1963 only), infrequent recovery of SV5, SV16, SV17, SV18, SV23, SV26, SV40, and SA5.

In our laboratory virus isolations from the baboon indicate that most of the isolates are recognized simian adenovirus serotypes: SV15, SV23, SA7, and V340. Infrequently encountered are: SV1, SV17, SV20, SV25, SV33, SV34, and AA153. Other serotypes are: picornaviruses SV6, SV19, A13; herpesviruses SA8, *H. hominis;* and reovirus type 2. New agents have been infrequently encountered. Soike *et al.* (1967, 1969a,b) have reported similar findings for the chimpanzee. Few of the isolates recovered from these animals were new agents but rather belonged to previously described virus groups. In contrast to our findings with the baboon, many of the chimpanzee isolates are human agents, presumably the result of contact with man. Rogers *et al.* (1967) found that isolates obtained from chimpanzee tissues maintained in culture for extended periods of time were new but could be included among the recognized established virus families—adenovirus, reovirus, foamy virus. The isolates reported by Hsiung and her collaborators (Hsiung and Atoynatan, 1966; Hsiung and Melnick,

* The numbers in parentheses indicate frequency of virus isolations.

1958a) as recovered from various captive monkeys (*M. mulatta, C. aethiops, Papio* sp.) are those common to these species.

A consequence of developing this report has been the realization that there is still a lack of understanding among most investigators using monkeys and apes in their research regarding the magnitude of the problem. Most disconcerting has been the observation that little recognition is given to the potential danger, even though B virus infection of man was described approximately 40 years ago and the danger still persists. This virus, as well as several newly recognized viruses, has continued to produce fatalities and most laboratories have yet to provide adequate protection for their personnel or suitable quarters for their animals in an attempt to minimize the problem. Similar suggestions have recently been expressed by Hunt (1970). Laboratory Animal Facilities and Care of the Institute of Laboratory Animal Resources (1968) does provide information suggesting the need of special facilities for biologic safety in infectious disease units. However, this and other manuals fail to indicate that latent infections, the intermingling of species, or introduction of "new" animals of similar species into a stabilized colony, may have the same effect as working with experimentally infected animals. In this regard is the problem that may develop as breeding colonies of various monkeys and apes are instituted (Kalter *et al.,* 1972b). These animals will be by far cleaner than their native-born counterpart. However, their susceptibility to various agents will also be different than those animals that have survived by virtue of natural selection. Introduction of an infectious agent into a colony of captive-bred animals may result in a high mortality because of the prevalence of numerous susceptibles in the group.

Of developing concern is the use of chimpanzees and baboons (probably other species as well) for transplant or cross-circulation studies with man (Balner, 1969). Very little has been done to examine these donor animals for evidence of viral infection. Perhaps consideration should be given to developing colonies of pathogen-free animals (van Bekkum *et al.,* 1969) or even gnotobiotics (Kalter *et al.,* 1972c). A recent symposium on cross-species transplantation (Reemtsma, 1970) mentions the presence of infectious hepatitis virus, molluscum contagiosum (Douglas, 1970), Marburg virus, "coxsackie BL-34," and simian herpesvirus (G. P. Murphy *et al.,* 1970) as possible agents present in the donor animals. It is urgent that if these types of experiments are to continue and, more importantly, to succeed, a better insight into comparative virology (as well as that of other microorganisms) must be obtained. The similarity of man to these animals necessitates a greater need for study. As pointed out above, representative viruses of every major group are found in most primates. In certain instances the differences between the strains are quite marked; they may, however, be extremely close and at times indistinguishable. Similar clinical diseases have been reported for both man and other primates as a result of infection with one or another agent. Obviously, while distinct and specific viruses exist among all species of animals, many of these agents may be the same. Only by virtue of their perpetuation cycles (Matumoto, 1969) do they assume certain antigenic components of the different host tissues involved thus producing differences in antigenic reactions or perhaps severity of illness.

We have made little mention of the impact that the use of these large numbers of monkeys and apes will make upon ecology and population dynamics. Many of these animals are on the endangered lists and other species are rapidly decreasing in numbers. Consideration must be given to the continued need to use nonhuman primates for biomedical research and all the implications, and if their use is to be continued, then how best to do it. A number of individuals have given voice to this problem (Booth, 1969; Beveridge, 1969a–c; Cass, 1970; Dowling, 1968; Eyestone, 1965, 1968a; Kalter and Heberling, 1971b; Kalter, 1971a,b; Greer, 1969; McPherson, 1970; Southwick *et al.,* 1970). Some suggestions to the breeding of nonhuman primates in captivity in order to supply the scientific needs of the community have been described, but implementation has been far from satisfactory (Epple, 1970; Kalter, 1968b; Derwelis *et al.,* 1969; Moor-Jankowski and Goldsmith, 1969; Davis *et al.,* 1969; Gengozian, 1969; Melby and Baker, 1969; Neurauter, 1969; Booth, 1969; Boiron, 1969; Spiegel, 1969; Beveridge, 1969a–c).

ACKNOWLEDGMENTS

This study was funded in part by USPHS grants RRO5519, RR00278, RR00361, and RR00451 and contracts NIH 69-93 and NIH 71-2348 and WHO grant Z2/181/27 for the World Health Organization Regional Reference Center for Simian Viruses.

REFERENCES

Adels, B. R., Gajdusek, D. C., Gibbs, C. J., Albrecht, P., and Rogers, N. G. (1968). *Neurology* 18, 30.
Adzhighitov, F. I., and Gordeladze, T. D. (1968). *Quest. Physiol. Exp. Pathol.* pp. 368–388.
Adzhighitov, F. I., Lapin, B. A., Yakovleva, L. A., Vasilieva, V. A., and Krivolapchuk, O. Ya. (1968). *Vop. Virusol.* 4, 454.
Aisenberg, M. S., and Grubb, T. C. (1943). *J. Bacteriol.* 46, 311.
Akopova, I. I., and Alekseeva, A. K. (1968). *Vop. Virusol.* 5, 585.
Akopova, I. I., and Alekseeva, A. K. (1969). *Vop. Virusol.* 2, 167.

136   S. S. KALTER

Alice, F. J. (1951). *Rev. Brasil. Biol.* **11**, 85.
Allen, A. M., Palmer, A. E., Tauraso, N. M., and Shelokov, A. (1968). *Amer. J. Trop. Med.* **17**, 413.
Allen, J. R., Houser, W. D., and Carstens, L. A. (1970). *Arch. Pathol.* **90**, 167.
Allen, W. P., Belman, S. G., and Borman, E. R. (1967). *Amer. J. Trop. Med. Hyg.* **16**, 106.
Allison, A. C. (1970). *Proc. Roy. Soc. Med.* **63**, 346.
Allmond, B. W., Froeschle, J. E., and Guilloud, N. B. (1967). *Amer. J. Epidemiol.* **85**, 229.
Altstein, A. D., Tsetlin, E. M., Dodonova, N. N., Levenbuk, I. S., and Chigirinsky, A. E. (1968). *Neoplasma* **15**, 113.
Ambrus, J. L., and Strandstrom, H. V. (1966). *Nature (London)* **211**, 876.
Ambrus, J. L., Feltz, E. T., Grace, J. T., Jr., and Owens, G. (1963). *J. Nat. Cancer Inst.* **10**, 447.
Ambrus, J. L., Strandstrom, H. V., and Kawinski, W. (1969). *Experientia* **25**, 64.
Amoss, H. L. (1928). In *"Filterable Viruses"* (T. M. Rivers, ed.), pp. 159–201. Williams & Wilkins, Baltimore, Maryland.
Amstey, M. S. (1969). *Amer. J. Obstet. Gynecol.* **104**, 573.
Anderson, C. R., and Downs, W. G. (1955). *Amer. J. Trop. Med. Hyg.* **4**, 662.
Anderson, C. R., Spence, L. P., Downs, W. G., and Aitken, T. H. G. (1960). *Amer. J. Trop. Med. Hyg.* **9**, 78.
Anderson, G. R., and Sgouris, J. T. (1966). *Int. Symp. Rabies, Falloires, 1965* Vol. 1, pp. 319–332.
Anderson, J. F., and Goldberger, J. (1911a). *J. Amer. Med. Ass.* **57**, 113.
Anderson, J. F., and Goldberger, J. (1911b). *J. Amer. Med. Ass.* **57**, 1612.
Anderson, J. F., and Goldberger, J. (1911c). *Pub. Health Rep.* **26**, 847.
Andrewes, C. H., and Pereira, H. G. (1967). *"Viruses of Vertebrates,"* 2nd ed. Williams & Wilkins, Baltimore, Maryland.
Andrewes, C. H., Allison, A. C., Armstrong, J. A., Bearcroft, G., Niven, J. S. F., and Pereira, H. G. (1959). *Acta Unio Int. Contra Cancrum* **15**, 760.
Apodaca, J., Lange, W., and Kohler, H. (1968). *Zentralbl. Bakteriol., Parasitenk., Infektionskr. Hyg., Abt. 1, Orig.* **207**, 100.
Appleby, E. C. (1969). *Acta Zool. (Stockholm)* **48**, 77.
Appleby, E. C., Graham-Jones, O., and Keeble, S. A. (1963). *Vet. Rec.* **75**, 81.
Archetti, I., and Bocciarelli, D. S. (1963). *Virology* **20**, 399.
Archetti, I., Bereczky, E., and Bocciarelli, D. S. (1966). *Virology* **29**, 671.
Arita, I., and Henderson, D. A. (1968). *Bull. WHO* **39**, 277.
Armstrong, C. (1939). *Pub. Health Rep.* **54**, 1719.
Armstrong, C., and Lillie, R. D. (1934). *Pub. Health Rep.* **49**, 1019.
Armstrong, C., Wooley, J. G., and Onstott, R. H. (1936). *Pub. Health Rep.* **51**, 298.
Armstrong, D., Henle, G., and Henle, W. (1966). *J. Bacteriol.* **91**, 1257.
Asher, D. M., Gibbs, C. J., and Lang, D. J. (1969). *Bacteriol. Proc.* 191.
Ashkenazi, A., and Melnick, J. L. (1962). *Proc. Soc. Exp. Biol. Med.* **111**, 367.
Atchison, R. W., Castro, B. C., and Hammon, W. M. (1965). *Science* **149**, 754.
Atchley, F. O., and Kimbrough, R. D. (1966). *Lab. Invest.* **15**, 1520.
Atoynatan, T., and Hsiung, G. D. (1969). *Amer. J. Epidemiol.* **89**, 472.
Aulisio, C. G., Wong, D. C., and Morris, J. A. (1964). *Proc. Soc. Exp. Biol. Med.* **117**, 6.
Aycock, W. L. (1940). In *"Virus and Rickettsial Diseases"* (J. E. Gordon *et al.*, eds.), pp. 555–580. Harvard Univ. Press, Cambridge, Massachusetts.

Aycock, W. L., and Kagan, J. R. (1927). *J. Immunol.* 14, 85.

Aycock, W. L., and Kramer, S. D. (1930). *J. Prev. Med.* 4, 189.

Back, N., Ambrus, J. L., and Kalter, S. S. (1965). *In* "The Baboon in Medical Research" (H. Vagtborg, ed.), Vol. I, pp. 421–429. Univ. of Texas Press, Austin.

Balfour, A. (1914). *Lancet* 1, 1176.

Balner, H. (1969). *Ann. N. Y. Acad. Sci.* 162, 437.

Balner, H., and Beveridge, W. I. B., eds. (1970). "Infections and Immunosuppression in Subhuman Primates." Munksgaard, Copenhagan.

Bang, J. (1942). *Acta Med. Scand.* 111, 291.

Bardos, V., Danes, L., Simkova, A., Libich, K., Blazek, K., and Cupkova, E. (1969). *In* "Arboviruses of the California Complex and Bunyamwera Group" (V. Bardos, ed.), pp. 275–284. Akademie Vied, Bratislava, Czeckoslovakia.

Barinskii, I. F. (1967). *Vop. Virusol.* 3, 259.

Barrera-Oro, J. G., and Melnick, J. L. (1961). *Tex. Rep. Biol. Med.* 19, 529.

Bearcroft, W. G. C. (1957). *J. Pathol. Bacteriol.* 24, 295.

Bearcroft, W. G. C. (1963). *Nature (London)* 197, 806.

Bearcroft, W. G. C. (1964). *J. Pathol. Bacteriol.* 88, 511.

Bearcroft, W. G. C. (1968). *J. Med. Microbiol.* 1, 1.

Bearcroft, W. G. C. (1969a). *Brit. J. Exp. Pathol.* 50, 56.

Bearcroft, W. G. C. (1969b). *Brit. J. Exp. Pathol.* 50, 327.

Bearcroft, W. G. C., and Jamieson, M. F. (1958). *Nature (London)* 182, 195.

Bechteisheimer, H., Jacob, H., and Solcher, H. (1968). *Deut. Med. Wochenschr.* 93, 602.

Beck, E., Daniel, P. M., Alpers, M., Gajdusek, D. C., and Gibbs, C. J., Jr. (1966). *Lancet* 2, 1056.

Beck, E., Daniel, P. M., Alpers, M., Gajdusek, D. C., and Gibbs, C. J. (1969a). *Int. Arch. Allergy Appl. Immunol.* 36, 553.

Beck, E., Daniel, P. M., Matthews, W. B., Stevens, D. L., Alpers, M. P., Asher, D. M., Gajdusek, D. C., and Gibbs, C. J. (1969b). *Brain* 92, 699.

Bedson, S. P., and Western, G. T. (1930). *Brit. J. Exp. Pathol.* 11, 502.

Behbehani, A. M., Hiller, M. S., Lenahan, M. F., and Wenner, H. A. (1967). *Amer. J. Trop. Med. Hyg.* 16, 63.

Benda, R., Danes, L., and Fuchsova, M. (1960a). *Acta Virol. (Prague), Engl. Ed.* 4, 160.

Benda, R., Danes, L., and Fuchsova, M. (1960b). *Czech. Epidemiol.* 9, 1.

Benda, R., Fuchsova, M., and Danes, L. (1962). *Acta Virol. (Prague), Engl. Ed.* 6, 46.

Berman, L. D., Cotes, P. M., and Simons, P. J. (1967). *J. Nat. Cancer Inst.* 39, 119.

Beveridge, W. I. B., ed. (1969a). "Primates in Medicine," Vol. 2, Part I. Karger, Basel.

Beveridge, W. I. B., ed. (1969b). "Primates in Medicine," Vol. 3, Part II. Karger, Basel.

Beveridge, W. I. B. (1969c). *Ann. N. Y. Acad. Sci.* 162, 392.

Bhatt, P. N., Goverdhan, M. K., Shaffer, M. F., Brandt, D. C., and Fox, J. P. (1966a). *Amer. J. Trop. Med. Hyg.* 15, 551.

Bhatt, P. N., Brandt, C. D., Weiss, R. A., Fox, J. P., and Shaffer, M. F. (1966b). *Amer. J. Trop. Med. Hyg.* 15, 561.

Bhatt, P. N., Work, T. H., Varma, M. G. R., Trapido, H., Narasimha Myrthy, D. P., and Rodrigues, F. M. (1966c). *Indian J. Med. Sci.* 20, 316.

Binn, L. N., Harrison, V. R., and Randall, R. (1967). *Amer. J. Trop. Med. Hyg.* **16**, 782.

Black, P. H., Hartley, J. W., and Rowe, W. P. (1963). *Proc. Soc. Exp. Biol. Med.* **112**, 601.

Blake, F. G., and Trask, J. D. (1921a). *J. Exp. Med.* **33**, 385.

Blake, F. G., and Trask, J. D. (1921b). *J. Exp. Med.* **33**, 413.

Blanc, G., and Caminopetros, J. (1921). *C. R. Soc. Biol.* **84**, 859.

Bland, J. O. W. (1944a). *J. Pathol. Bacteriol.* **56**, 161.

Bland, J. O. W. (1944b). *J. Pathol. Bacteriol.* **56**, 446.

Bland, J. O. W. (1944c). *Lancet* **247**(6321), 549.

Blaxall, F. R. (1923). *Bull. Acad. Nat. Med., Paris* [3] **89**, 146.

Bleyer, J. G. (1922). *Muenchen. Med. Wochenschr.* **69**, 1009.

Blinzinger, K. (1969). *Nature (London)* **221**, 1336.

Bloch, O., Jr. (1937). *Amer. J. Pathol.* **13**, 939.

Blumberg, B. S. (1972). *In* "Vth International Symposium on Comparative Leukemia Research, Podova/Venice, September, 1971." Karger, Basel (in press).

Blumberg, B. S., Alter, H. J., and Visnich, S. A. (1965). *J. Amer. Med. Ass.* **191**, 541.

Blumberg, B. S., Sutnick, A. I., and London, W. T. (1968). *Bull. N. Y. Acad. Med.* [2] **44**, 1566.

Bodian, D. (1948). *Amer. J. Hyg.* **48**, 87.

Bodian, D. (1952). *Amer. J. Pub. Health* **42**, 1388.

Bodian, D. (1955). *Science* **122**, 105.

Bodian, D. (1956). *Amer. J. Hyg.* **64**, 181.

Bodian, D., and Howe, H. A. (1945). *J. Exp. Med.* **81**, 255.

Boiron, M. (1969). *Ann. N. Y. Acad. Sci.* **162**, 390.

Bonin, O. (1969). *Acta Zool.* **48**, 319.

Bonin, O., and Unterharnscheidt, F. (1964). *Arch. Psychiat. Nervenkr.* **206**, 260.

Boorman, J. P. T., and Draper, C. C. (1968). *Trans. Roy. Soc. Trop. Med. Hyg.* **62**, 269.

Booth, C. (1969). *Ann. N. Y. Acad. Sci.* **162**, 387.

Boshell, J., and Rajagopalan, P. K. (1968). *Indian J. Med. Res.* **56**, 573.

Boulger, L. R. (1966). *Lancet* **1**, 941.

Bourne, G. H., ed. (1969). "The Chimpanzee," Vol. 1. Univ. Park Press, Baltimore, Maryland.

Bourne, G. H., ed. (1970a). "The Chimpanzee," Vol. 2. S. Karger, Basel.

Bourne, G. H., ed. (1970b). "The Chimpanzee," Vol. 3. University Press, Baltimore, Maryland.

Bourne, G. H., ed. (1971). "The Chimpanzee," Vol. 4. University Press, Baltimore, Maryland.

Bowen, E. T. W., Simpson, D. I. H., Bright, W. F., Zlotnik, I., and Howard, D. M. R. (1969). *Brit. J. Exp. Pathol.* **50**, 400.

Bras, G. (1922). *Doc. Med. Geogr. Trop.* **4**, 303.

Breen, G. E., Lamb, S. G., and Otaki, A. T. (1958). *Brit. Med. J.* **2**, 22.

Brown, L. (1957). *Amer. J. Hyg.* **65**, 189.

Bugher, J. C. (1951). *In* "Yellow Fever" (G. K. Strode, ed.), pp. 529–584. McGraw-Hill, New York.

Bullock, G. (1965). *J. Hyg.* **63**, 383.

Burkholder, C. R., and Soave, O. A. (1970). *Lab. Anim. Care* **20**, 186.

Burkitt, D. (1958). *Brit. J. Surg.* **46**, 218.

Burnet, F. M. (1940). *Med. J. Aust.* 1, 325.

Burnet, F. M. (1941). *Aust. J. Exp. Biol. Med. Sci.* 19, 281.

Burnet, F. M., Jackson, A. V., and Robertson, E. G. (1939a). *Aust. J. Exp. Biol. Med. Sci.* 17, 375.

Burnet, F. M., Lush, D., and Jackson, A. V. (1939b). *Aust. J. Exp. Biol. Med. Sci.* 17, 41.

Burnett, J. P., and Harrington, J. A. (1968). *Proc. Nat. Acad. Sci. U. S.* 60, 1023.

Cabasso, V. J., and Stebbins, M. R. (1965). *J. Lab. Clin. Med.* 65, 612.

Cabasso, V. J., Stebbins, M. R., Karelitz, S., Cerini, C. P., Ruegsegger, J. M., and Stillerman, M. (1967). *J. Lab. Clin. Med.* 70, 429.

Carpenter, C. R., ed. (1969). *Proc. Int. Congr. Primatol., 2nd,* 1968 Vol. 1.

Carski, T. R. (1960). *J. Immunol.* 84, 426.

Carter, G. B., and Bright, W. F. (1968). *Lancet* 2, 913.

Casey, H. W., Woodruff, J. M., and Butcher, W. I. (1967). *Amer. J. Pathol.* 51, 431.

Casey, M. J., Turner, H. C., Huebner, R. J., Sarma, P. S., and Miller, R. L. (1966). *Nature (London)* 211, 1417.

Cass, J. S. (1970). *Fed. Proc., Fed. Amer. Soc. Exp. Biol.* 29, 1590.

Causey, O. R., Causey, C. E., Maroja, O. M., and Macedo, D. G. (1961). *Amer. J. Trop. Med. Hyg.* 10, 277.

Chang, P. W., and Hsiung, G. D. (1965). *J. Immunol.* 95, 591.

Chanock, R. M., Roizman, B., and Myers, R. (1957). *Amer. J. Hyg.* 66, 281.

Chanock, R. M., Johnson, K. M., Cook, M. K., Wong, D. C., and Vargosko, A. (1961). *Amer. Rev. Resp. Dis.* 83, 125.

Chapman, W. L., Jr. (1968). *Clin. Exp. Immunol.* 3, 872.

Charnomordik, A. E., Voskresenskaia, G. S., and Kuksova, M. I. (1965). *Pediatriya (Moscow)* 44, 39.

Chernyshov, V. I. (1969). *In* "Primates in Medicine" (W. I. B. Beveridge, ed.), Vol. 3, Part II, pp. 124–128. Karger, Basel.

Cheville, N. F., and Kluge, J. P. (1967). *Veterinarian* 29, 89.

Chiappino, G., Strozzi, F., and Hahn, E. E. A. (1966). *Boll. Ist. Sieroter, Milan.* 45, 1.

Cho, C. T., Bolano, C. R., Kamitsuka, P. S., and Wenner, H. A. (1970). *Amer. J. Epidemiol.* 92, 137.

Chopra, H. C., and Mason, M. M. (1970). *Cancer Res.* 30, 2081.

Chopra, H. C., Zelljadt, I., Jensen, E. M., Mason, M. M., and Woodside, N. J. (1971). *J. Nat. Cancer Inst.* 46, 127.

Chu, C.-T., Yang, C.-S., and Kawamura, P., Jr. (1971). *Appl. Microbiol.* 21, 539.

Chu, L. W., and Warren, G. H. (1960). *Proc. Soc. Exp. Biol. Med.* 105, 396.

Chumakov, M. P., Voroshilova, M. K., Zhevandrova, V. I., Mironova, L. L., Itzelis, F. I., and Robinson, I. S. (1956). *Probl. Virol. (USSR)* 1, 16.

Chumakov, M. P., Dzagurov, S. G., Chumakova, M. Ya., Chernyshov, V. I., and Bardin, M. N. (1963). *In* "Poliomyelit i drugie enterovirusnye infektsii," p. 194. Moskva Meditsina, Moscow.

Chumakova, M. Ya., Chumakov, M. P., Elbert, L. B., Augustinovich, G. I., Ralph, N. M., Voroshilova, M. K., Taranova, G. P., and Tapupere, V. O. (1963). *Vop. Virusol.* 8, 457.

Churchill, A. E. (1963). *Brit. J. Exp. Pathol.* 44, 529.

Clark, H. C. (1952). *Amer. J. Trop. Med. Hyg.* 1, 78.

Clark, P. F., Schindler, J., and Roberts, D. J. (1930). *J. Bacteriol.* 20, 213.

Clark, P. F., Roberts, D. J., and Preston, W. S. (1932). *J. Prev. Med.* 6, 47.

Clarkson, M. J., Thorpe, E., and McCarthy, K. (1967). *Arch. Gesamte Virusforsch.* **22**, 219.

Coates, H. V., and Chanock, R. M. (1962). *Amer. J. Hyg.* **76**, 302.

Colbert, J. W., Jr. (1948–1949). *Yale J. Biol. Med.* **21**, 335.

Cole, R., and Kuttner, A. G. (1925). *J. Exp. Med.* **42**, 799.

Cole, W. C., Bostrom, R. E., and Whitney, R. A., Jr. (1968). *J. Amer. Vet. Med. Ass.* **153**, 894.

Collias, N., and Southwick, C. (1952). *Proc. Amer. Phil. Soc.* **96**, 143.

Cornelius, C. (1970). *N. Engl. J. Med.* **281**, 934.

Cotes, P. M. (1966). *Symp. Zool. Soc. London* **17**, 309.

Covell, W. P. (1932). *Amer. J. Pathol.* **8**, 151.

Cowdry, E. V., and Scott, G. H. (1935). *Amer. J. Pathol.* **11**, 647.

Craig, D. E., and Francis, T., Jr. (1958). *Proc. Soc. Exp. Biol. Med.* **99**, 325.

Crandell, R. A., Casey, H., and Brumlow, W. (1969). *J. Infec. Dis.* **119**, 80.

Curasson, G. (1942). *In* "Traité de pathologie exotique Vétérinaire et Comparée" (G. Curasson, ed.), 2nd ed., Vol. 1, pp. 12–169. Vigot Frères, Paris.

Dalldorf, G. (1957). *J. Exp. Med.* **106**, 69.

Danes, L., and Blazek, K. (1966). *J. Hyg. Epidemiol., Microbiol., Immunol.* **10**, 109.

Danes, L., Hruskova, J., and Blazek, K. (1966). *J. Hyg. Epidemiol., Microbiol., Immunol.* **10**, 210.

Daniel, M. D. Karpas, A., Melendez, L. V., King, N. W., and Hunt, R. D. (1967a). *Arch. Gesamte Virusforsch.* **22**, 324.

Daniel, M. D., King, N. W., Hunt, R. D., and Melendez, L. V. (1967b). *Fed. Proc., Fed. Amer. Soc. Exp. Biol.* **26**, 421.

Daniel, M. D., Melendez, L. V., Hunt, R. D., King, N. W., and Williamson, M. E. (1970). *Bacteriol. Proc.*, 195.

Davenport, F., Hennessy, A. V., Christopher, N., and Smith, C. K. (1966). *Amer. J. Epidemiol.* **83**, 146.

Davis, J. H., McPherson, B. R., and Moor-Jankowski, J. (1969). *Ann. N. Y. Acad. Sci.* **162**, 329.

Day, P. W., Soike, K., Levenson, R. H., and Van Riper, D. C. (1966). *Lab. Anim. Care* **16**, 497.

Deinhardt, F. (1966). *Nature (London)* **210**, 443.

Deinhardt, F. (1967). *Perspect. Virol.* **5**, 183–197.

Deinhardt, F. (1969). *Nat. Cancer Inst., Monogr.* **29**, 327.

Deinhardt, F. (1970). *In* "Infections and Immunosuppression in Subhuman Primates" (H. Balner and W. I. B. Beveridge, ed.), pp. 55–63. Munksgaard, Copenhagen.

Deinhardt, F., and Deinhardt, J. (1966). *In* "Some Recent Developments in Comparative Medicine" (R. N. T-W Fiennes, ed.), pp. 127–152. Academic Press, New York.

Deinhardt, F., Holmes, A. W., Capps, R. B., and Popper, H. (1967a). *J. Exp. Med.* **125**, 673.

Deinhardt, F., Holmes, A. W., Devine, J., and Deinhardt, J. (1967b). *Lab. Anim. Care* **17**, 48.

Deinhardt, F., Holmes, A. W., Wolfe, L. G., Melnick, J. L., and Parks, W. P. (1970a). *J. Infec. Dis.* **121**, 351.

Deinhardt, F., Holmes, A. W., Wolfe, L., and Junge, U. (1970b). *Vox Sang.* **19**, 261.

Deinhardt, F., Wolfe, L. G., Theilen, G. T., and Snyder, S. P. (1970c). *Science* 167, 881.

Delahunt, C. S. (1966). *Lancet* 1, 825.

Delahunt, C. S., and Rieser, N. (1967). *Amer. J. Obstet. Gynecol.* 99, 580.

DeLay, P. D., Stone, S. S., Karzon, D. T., Katz, S., and Enders, J. (1965). *Amer. J. Vet. Res.* 26, 1359.

De Paoli, A., and Garner, F. M. (1968). *Cancer Res.* 28, 2559.

De Pasquale, N. P., Burch, G. E., Sun, S. C., Hale, A. R., and Mogabgab, W. J. (1966). *Amer. Heart J.* 71, 678.

De Rodaniche, E. C. (1952). *Amer. J. Trop. Med. Hyg.* 1, 205.

Derwelis, S. K., Douglas, J. D., Fineg, J., and Butler, T. M. (1969). *Ann. N. Y. Acad. Sci.* 162, 311.

Dick, E. C. (1968). *Proc. Soc. Exp. Biol. Med.* 127, 1079.

Dick, E. C., and Dick, C. R. (1968). *Amer. J. Epidemiol.* 88, 267.

Dick, G. W. A., and Waterson, A. P. (1969). *Trans. Roy. Soc. Trop. Med. Hyg.* 63, 324.

Dick, G. W. A., Best, A. M., Haddow, A. J., and Smithburn, K. C. (1948). *Lancet* 2, 286.

Dick, G. W. A., Smithburn, K. C., and Haddow, A. J. (1958). *Brit. J. Exp. Pathol.* 29, 547.

Di Giacomo, R. F. (1967). *Cancer Res.* 27, 1178.

Ditchfield, W. J. B. (1968). *Can. Med. Ass. J.* 98, 903.

Dochez, A. R., Shibley, G. S., and Mills, K. C. (1930). *J. Exp. Med.* 52, 701.

Douglas, J. D. (1970). *Transplant. Proc.* 2, 539.

Douglas, J. D., and Berge, T. O. (1964). *Amer. Vet. Med. Ass. Sci., Proc. Annu. Meet., 1964,* p. 243.

Douglas, J. D., Tanner, K. N., Prine, J. R., Van Riper, D. C., and Derwelis, S. K. (1967). *J. Amer. Vet. Med. Ass.* 151, 901.

Douglas, J. D., Soike, K., and Raynor, J. (1970). *Lab. Anim. Care* 20, 265.

Dowling, H. F. (1968). *In* "Conference on Nonhuman Primate Toxicology at Airlee House, 1966" (C. O. Miller, ed.), pp. 5–7. U. S. Govt. Printing Office, Washington, D. C.

Downie, A. W., and Dumbell, K. R. (1956). *Annu. Rev. Microbiol.* 10, 237.

Draper, C. C., and Laurence, G. D. (1969). *J. Med. Microbiol.* 2, 249.

Dunkel, V. C. Personal communication.

Easton, J. M. (1964). *J. Immunol.* 93, 716.

Eckroade, R. J., ZuRhein, G. M., Marsh, R. F., and Hanson, R. P. (1970). *Science* 169, 1088.

Eckstein, A. (1933). *Z. Gesamte Neurol. Psychiat.* 149, 176.

Eddy, B. E., Borman, G. S., Berkeley, W., and Young, R. D. (1961). *Proc. Soc. Exp. Biol. Med.* 107, 191.

Eddy, B. E., Borman, G. S., Grubbs, G. E., and Young, R. S. (1962). *Virology* 17, 65.

Elford, W. J., and Galloway, I. A. (1933). *J. Pathol. Bacteriol.* 37, 381.

Elizan, T. S., Fabiyi, A., and Sever, J. L. (1969). *J. Mt. Sinai Hosp., New York* 38, 108.

Elliot, O. S., Elliot, M. W., and Lisco, H. (1966). *Nature (London)* 211, 1105.

Elton, N. W. (1952). *Mil. Surg.* 111, 157.

Elton, N. W. (1953). *Mil. Surg.* 112, 424.

Emery, J. B., and York, C. J. (1960). *Virology* 11, 313.
Emmons, R. W., and Lennette, W. H. (1970). *Arch. Gesamte Virusforsch.* 31, 215.
Emmons, R. W., Gribble, D. H., and Lennette, E. H. (1968). *J. Infec. Dis.* 118, 153.
Enders, J. F. (1940). *In* "Virus and Rickettsial Diseases" (J. E. Gordon *et al.,* eds.), pp. 237–267. Harvard Univ. Press, Cambridge, Massachusetts.
Enders, J. F. (1952). *J. Immunol.* 69, 639.
Enders, J. F., and Peebles, T. C. (1954). *Proc. Soc. Exp. Biol. Med.* 86, 277.
Enders, J. F., Weller, T. H., and Robbins, F. C. (1949). *Science* 109, 85.
Endo, M., Aoyama, Y., Kaminara, T., Hayshida, T., and Kinjo, T. (1959). *Jap. J. Exp. Med.* 29, 355.
Endo, M., Kamimura, T., Aoyama, Y., Hayashida, T., Kinjo, T., Ono, Y., Kotera, S., Suzuki K., Tajimaet, Y., and Ando, K. (1960a). *Jap. J. Exp. Med.* 30, 227.
Endo, M., Kamimura, T., Kusano, N., Kawai, K., Aoyama, Y., Tajima, Y., Suzuki, K., and Kotera, S. (1960b). *Jap. J. Exp. Med.* 30, 385.
Epple, G. (1970). *Folia Primatol.* 12, 56.
Epstein, M. A., and Barr, Y. M. (1964). *Lancet* 1, 252.
Epstein, M. A., Achong, B. G., and Barr, Y. M. (1964a). *Lancet* 1, 702.
Epstein, M. A., Woodall, J. P., and Thomson, A. D. (1964b). *Lancet* 1, 288.
Epstein, M. A., Thomson, A. D., and Woodall, J. P. (1966). *In* "Some Recent Developments in Comparative Medicine" (R. N. T-W Fiennes, ed.), pp. 323–335. Academic Press, New York.
Espana, C. (1971). *In* "Medical Primatology 1970" (E. I. Goldsmith and J. Moor-Jankowski, eds.) pp. 694–708. Karger, Basel.
Eugster, A. K., Kalter, S. S., Kim, C. S., and Pinkerton, M. E. (1969). *Arch. Gesamte Virusforsch.* 26, 260.
Evans, A. S. (1954). *Nat. Acad. Sci.—Nat. Res. Counc., Publ.* 322, 58.
Evans, A. S., Evans, B. K., and Sturtz, V. (1953). *Proc. Soc. Exp. Biol. Med.* 82, 437.
Evans, C. A., and Hoshiwara, I. (1955). *Proc. Soc. Exp. Biol. Med.* 89, 86.
Evans, C. A., Byatt, P. H., Chambers, V. C., and Smith, W. M. (1954a). *J. Immunol.* 72, 348.
Evans, C. A., Chambers, V. C., Smith, W. M., and Byatt, P. H. (1954b). *J. Infec. Dis.* 94, 273.
Evans, C. A., Graham, C. B., Jr., Hoshiwara, I., and Oh, J. O. (1961). *Proc. Soc. Exp. Biol. Med.* 107, 97.
Eyestone, W. H. (1965). *J. Amer. Vet. Med. Ass.* 147, 1482.
Eyestone, W. H. (1968a). *In* "Conference on Nonhuman Primate Toxicology at Airlee House, 1966" (C. O. Miller, ed.), pp. 155–156. U. S. Govt. Printing Office, Washington, D. C.
Eyestone, W. H. (1968b). *J. Amer. Vet. Med. Ass.* 153, 1767.
Faber, H. K., Silverberg, R. J., and Dong, L. (1944). *J. Exp. Med.* 80, 39.
Fabiyi, A., Calcagno, P. L., Sever, J. L., Antonovych, T., and Wolman, F. (1967). *Nature (London)* 215, 88.
Fairbrother, R. W., and Hurst, E. W. (1932). *J. Pathol. Bacteriol.* 35, 867.
Falke, D. (1958). *Zentrabl. Bakteriol., Parasitenk., Infektionskr Hyg., Abt. 1. Orig.* 170, 377.
Fenner, F., and Burnet, F. M. (1957). *Virology* 4, 305.
Fiennes, R. (1967). "Zoonoses of Primates." Cornell Univ. Press, Ithaca, New York.

Fiennes, R. N. T-W, ed. (1966). "Some Recent Developments in Comparative Medicine." Academic Press, New York.

Findlay, G. M. (1932). *Trans. Roy. Soc. Trop. Med. Hyg.* **25,** 229.

Findlay, G. M., and Clarke, L. P. (1935). *J. Pathol. Bacteriol.* **40,** 55.

Findlay, G. M., and MacCallum, F. O. (1939). *J. Pathol. Bacteriol.* **49,** 53.

Findlay, G. M., and Mackenzie, R. D. (1937a). *Brit. J. Exp. Pathol.* **18,** 146.

Findlay, G. M., and Mackenzie, R. D. (1937b). *Brit. J. Exp. Pathol.* **18,** 258.

Findlay, G. M., Stefanopoulo, G. J., Davis, T. H., and Mahaffy, A. F. (1936). *Trans. Roy. Soc. Trop. Med. Hyg.* **29,** 419.

Findlay, G. M., MacCallum, F. O., and Murgatroyd, F. (1939). *Trans. Roy. Soc. Trop. Med. Hyg.* **32,** 575.

Findlay, G. M., Martin, N. H., and Mitchell, J. B. (1944). *Lancet* **2,** 365.

Flexner, S. (1937). *J. Exp. Med.* **65,** 497.

Flexner, S., and Lewis, P. A. (1910a). *J. Exp. Med.* **12,** 227.

Flexner, S., and Lewis, P. A. (1910b). *J. Amer. Med. Ass.* **55,** 662.

Fong, C. K., Bensch, K. G., and Hsiung, G. D. (1968a). *Virology* **35,** 297.

Fong, C. K., Hsiung, G. D., and Bensch, K. G. (1968b). *Virology* **35,** 311.

Foster, C., Siegel, M. M., Henle, W., and Stokes, J., Jr. (1951). *J. Lab. Clin. Med.* **38,** 359.

Fox, J. P., and Penna, H. A. (1943). *Amer. J. Hyg.* **38,** 152.

Fox, J. P., Lennette, E. H., Manso, C., and Sousa-Agniar, J. R. (1942). *Amer. J. Hyg.* **36,** 117.

Francis, T., Jr., and Magill, T. P. (1938). *J. Exp. Med.* **68,** 147.

Fraumeni, J. F., Jr., Ederer, F., and Miller R. W. (1963). *J. Amer. Med. Ass.* **185,** 713.

Friedman, M., Moldovanu, G., Moore, A. E., and Miller, D. G. (1968). *Progr. Exp. Tumor Res.* **10,** 1.

Fuentes-Marins, R. A., Rodriguez, A. R., Kalter, S. S., and Hellman, A. (1963). *J. Bacteriol.* **85,** 1045.

Gainer, J. H. (1967). *J. Amer. Vet. Med. Ass.* **151,** 421.

Gaitan, L. (1938). *Actas Decima Conf. Sanit. Panamer., 1938* p. 84.

Gajdusek, D. C. (1967a). *Acad. Med. Sci. USSR, 1967.*

Gajdusek, D. C. (1967b). *Curr. Top. Microbiol. Immunol.* **40,** 59.

Gajdusek, D. C. (1967c). *N. Engl. J. Med.* **276,** 392.

Gajdusek, D. C. (1968). *Science* **162,** 693.

Gajdusek, D. C., and Gibbs, C. J., Jr. (1968). *Res. Publ., Ass. Res. Nerv. Ment. Dis.* **44,** 254–280.

Gajdusek, D. C., Gibbs, C. J., Jr., and Alpers, M. (1966a). *U. S. Pub. Health Serv., Publ.* **1378.**

Gajdusek, D. C., Gibbs, C. J., Jr., and Alpers, M. (1966b). *Nature (London)* **209,** 794.

Gajdusek, D. C., Gibbs, C. J., Jr., and Alpers, M. (1967). *Science* **155,** 212.

Gajdusek, D. C., Gibbs, C. J., Jr., Asher, D. M., and David, E. (1968). *Science* **162,** 693.

Gajdusek, D. C., Rogers, N. G., Basnight, M., Gibbs, C. J., Jr., and Alpers, M. (1969). *Ann. N. Y. Acad. Sci.* **162,** 529.

Galindo, P., and Rodaniche, E. (1964). *Amer. J. Trop. Med. Hyg.* **13,** 844.

Galindo, P., and Srihongse, S. (1967). *Bull. WHO* **36,** 151.

Galindo, P., Srihongse, S., De Rodaniche, E., and Grayson, M. (1966). *Amer. J. Trop. Med. Hyg.* **15,** 385.

144   S. S. KALTER

Gebhardt, L., and Bachtold, J. G. (1953). *Proc. Soc. Exp. Biol. Med.* **83**, 807.
Gedigk, P., Bechtelsheimer, H., and Korb, G. (1968). *Deut. Med. Wochenschr.* **93**, 590.
Gengozian, N. (1969). *Ann. N. Y. Acad. Sci.* **162**, 336.
George, S., and Feldman, R. A. (1969). *Indian J. Med. Res.* **57**, 2001.
Gerber, P. (1967). *Proc. Soc. Exp. Biol. Med.* **125**, 1284.
Gerber, P., and Birch, S. M. (1967). *Proc. Nat. Acad. Sci., 1967* **58**, 478.
Gerber, P., and Rosenblum, E. N. (1968). *Proc. Soc. Exp. Biol. Med.* **128**, 541.
Gerber, P., Hotle, G. A., and Grubbs, R. E. (1961). *Proc. Soc. Exp. Biol. Med.* **108**, 205.
Gerber, P., Hamre, D., Moy, R. A., and Rosenblum, E. N. (1968). *Science* **161**, 173.
Gerber, P., Branch, J. W., and Rosenblum, E. N. (1969). *Proc. Soc. Exp. Biol. Med.* **130**, 14.
Gerloff, R. K., and Larson, C. L. (1959). *Amer. J. Pathol.* **35**, 1043.
Gibbs, C. J., Jr. (1967). *Curr. Top. Microbiol.* **40**, 44.
Gibbs, C. J., Jr., and Gajdusek, D. C. (1969). *Science* **165**, 1023.
Gibbs, C. J., Jr., Gajdusek, D. C., Asher, D. M., Alpers, M. P., Beck, E., Daniel, P. M., and Matthews, W. B. (1968). *Science* **161**, 388.
Gibbs, C. J., Jr., Gajdusek, D. C., and Alpers, M. P. (1969). *Int. Arch. Allergy* **36**, 519.
Gilden, R. V., Kern, J., Heberling, R. L., and Huebner, R. J. (1968). *Appl. Microbiol.* **16**, 1015.
Girardi, A. J., Sweet, B. H., Slotnick, V. B., and Hilleman, M. R. (1962). *Proc. Soc. Exp. Biol. Med.* **109**, 649.
Gispen, R., Verlinde, J. D., and Zwart, P. (1967). *Arch. Gesamte Virusforsch.* **21**, 205.
Gleiser, C. A., Gochenour, W. S., Jr., and Berge, T. O. (1961). *J. Immunol.* **87**, 504.
Godlewski, H. G. (1967). *Folia Histochem. Cytochem.* **5**, 325.
Godlewski, H. G., Masironi, R., and Bourne, G. H. (1968). *Neoplasma* **15**, 157.
Goldberger, J., and Anderson, J. F. (1911). *J. Amer. Med. Ass.* **57**, 476.
Goldman, K. (1935). *Berlin Tieraerztl. Wochenschr.* **51**, 497.
Goldsmith, E. I., and Moor-Jankowski, J. (1969). *Ann. N. Y. Acad. Sci.* **162**, 1–704.
Goldsmith, E. I., and Moor-Jankowski, J., eds. (1971). "Medical Primatology 1970." Karger, Basel.
Goodwin, W. J. (1970). *Lab. Anim. Care* **20**, 329.
Gordon, H., and Knighton, H. I. (1941). *Amer. J. Pathol.* **17**, 165.
Gordon, M. H. (1914). *Rep. Loc. Gov. Bd., New Ser. No. 96.*
Gordon, M. H. (1925). *Med. Res. Counc. (Gt. Brit.), Spec. Rep. Ser.* SRS-98.
Gordon, M. H. (1930). *Lancet* **1**, 1174.
Grace, J. T., Jr., and Mirand, E. A. (1963). *Ann. N. Y. Acad. Sci.* **108**, 1123.
Grace, J. T., Jr., and Mirand, E. A. (1965). *Exp. Med. Surg.* **23**, 213.
Grace, J. T., Jr., Mirand, E. A., Millian, S. J., and Metzgar, R. S. (1962). *Fed. Proc., Fed. Amer. Soc. Exp. Biol.* **21**, 32.
Graham, B., Minamishima, Y., and Benyesh-Melnick, M. (1969). *Bacteriol. Proc.* **69**, 198.
Gralla, E. J., Ciecura, S. J., and Delahunt, C. S. (1966). *Lab. Anim. Care* **16**, 510.
Gravrilov, V. I., Dodonova, N. N., Kuborina, L. N., Voronin, E. S., and Schekochikhina, E. A. (1963). *Probl. Virol.* **4**, 475.

Greer, W. E. (1969). *PAHO Sci. Publ.* **182**, 107.

Gregg, N. M. (1941). *Trans. Ophthalmol. Soc. Aust.* **3**, 35.

Gresikova, M., Rajcani, J., and Huizik, J. (1966). *Acta Virol. (Prague)*, *Engl. Ed.* **10**, 420.

Groot, H. (1962). *Bull. WHO* **27**, 709.

Grossman, L. H., and Kramer, S. D. (1936). *Proc. Soc. Exp. Biol. Med.* **35**, 345.

Grunbaum, A. (1904). *Brit. Med. J.* **1**, 817.

Guilloud, N. B., and Kline, I. C. (1966). *J. Amer. Phys. Ther. Ass.* **46**, 516.

Guilloud, N. B., Allmond, B. W., Froeschle, J. E., and Fitz-Gerald, F. L. (1969). *J. Amer. Vet. Med. Ass.* **155**, 1190.

Haas, R. (1968). *J. Embryol. Exp. Morphol.* **20**, 210.

Haas, R., Maass, G., Muller, J., and Oehlert, W. (1968). *Z. Med. Mikrobiol. Immunol.* **154**, 210.

Haas, R., Maass, G., and Ochlert, W. (1969). *In* "Primates in Medicine" (W. I. B. Beveridge, ed.), Vol. 3, pp. 138–139. Karger, Basel.

Habel, K. (1942). *Pub. Health Rep.* **57**, 1126.

Habel, K., and Loomis, L. N. (1957). *Proc. Soc. Exp. Biol. Med.* **95**, 597.

Haber, P. (1935). *C. R. Soc. Biol.* **119**, 136.

Habermann, R. T., and Williams, F. T. (1957). *Amer. J. Vet. Res.* **18**, 419.

Habermann, R. T., Williams, F. P., and Fite, G. L. (1960). *J. Amer. Vet. Med. Ass.* **137**, 161.

Haddow, A. J. (1952). *Ann. Trop. Med. Parasitol.* **46**, 135.

Haddow, A. L. (1968). *Proc. Roy. Soc. Edinburgh* **70**, 191.

Haddow, A. J., and Ellice, J. M. (1964). *Trans. Roy. Soc. Trop. Med. Hyg.* **58**, 521.

Haddow, A. J., Smithburn, K. C., Mahaffy, A. F., and Burgher, J. C. (1947). *Trans. Roy. Soc. Trop. Med. Hyg.* **40**, 677.

Hahon, N. (1961). *Bacteriol. Rev.* **25**, 459.

Hahon, N., and Wilson, B. J. (1960). *Amer. J. Hyg.* **71**, 69.

Halberstaedter, L., and von Prowazek, S. (1907). *Arb. Gesundh.* **26**, 44.

Hall, A. S., and McNulty, W. P., Jr. (1967). *J. Amer. Vet. Med. Ass.* **151**, 833.

Halloran, P. O'C. (1955). *Amer. J. Vet. Res.* **16**, 465.

Hammon, W. McD., Reeves, W. C., and Sother, G. (1952). *J. Immunol.* **69**, 493.

Hamper, B., Stevens, D. A., Martos, L. M., Ablashi, D. V., Burroughs, M. A. K., and Wells, G. A. (1969). *J. Immunol.* **102**, 397.

Harrison, V. R., Binn, L. N., and Randall, R. (1967a). *Amer. J. Trop. Med Hyg.* **16**, 786.

Harrison, V. R., Marshall, J. D., and Guilloud, N. B. (1967b). *J. Immunol.* **98**, 979.

Hartley, E. G. (1964). *Vet Rec.* **76**, 555.

Hartley, E. G. (1966). *Brit. Vet. J.* **122**, 46.

Hartley, E. G. (1968). *Vet. Rec.* **83R**, 6.

Hartwell, W. V., Kimbrough, R. D., and Love, G. J. (1968). *Amer. J. Vet. Res.* **29**, 1449.

Heath, R. B., El Falaky, I., Stark, J. E., Herbst-Laier, R. H., and Larin, N. M. (1966). *Brit. J. Exp. Pathol.* **47**, 93.

Heberling, R. L. (1972). *In* "Pathology of Simian Primates" (R. N. T-W Fiennes, ed.), 572–591. Karger, Basel.

Heberling, R. L., and Cheever, F. S. (1960). *Ann. N. Y. Acad. Sci.* **85**, 942.

Heberling, R. L., and Cheever, F. S. (1964). *Amer. J. Epidemiol.* **81**, 106.

Heberling, R. L., and Cheever, F. S. (1965a). *Proc. Soc. Exp. Biol. Med.* **118**, 151.

Heberling, R. L., and Cheever, F. S. (1965b). *Proc. Soc. Exp. Biol. Med.* **120**, 825.

Heberling, R. L., and Cheever, F. S. (1966). *Amer. J. Epidemiol.* **83**, 470.

Heberling, R. L., and Kalter, S. S. (1970). *Proc. Soc. Exp. Biol. Med.* **135**, 717.

Heberling, R. L., and Kalter, S. S. (1971). *J. Infec. Dis.* **124**, 33.

Heggie, A. D., and Robbins, F. C. (1963). *Proc. Soc. Exp. Biol. Med.* **114**, 750.

Held, J. R., Richardson, J. H., and Mosley, J. W. (1968). *J. Amer. Vet. Med. Ass.* **153**, 881.

Helwig, F. C., and Schmidt, E. C. H. (1945). *Science* **102**, 31.

Henderson, B. E., Cheshire, P. P., Kirya, G. B., and Lule, M. (1970). *Amer. J. Trop. Med. Hyg.* **19**, 110.

Henle, G., and Deinhardt, F. (1955). *Proc. Soc. Exp. Biol. Med.* **89**, 556.

Henle, G., and Henle, W. (1966). *J. Bacteriol.* **91**, 1248.

Henle, G., and Henle, W. (1967). *Cancer Res.* **27**, 2442.

Henle, G., Henle, W., and Diehl, V. (1968). *Proc. Nat. Acad. Sci. U. S.* **59**, 94.

Hennessen, W. (1968). *Nat. Cancer Inst., Monogr.* **29**, 161.

Hennessen, W. (1970). *Lab. Anim. Handb.* **4**, 137.

Hennessen, W., Bonin, O., and Mauler, R. (1968). *Deut. Med. Wochenschr.* **93**, 582.

Herrlein, H. G., Coursen, G. B., Randall, R., and Slanetz, C. A. (1954). *Nat. Acad. Sci.—Nat. Res. Counc., Publ.* **317**, 77.

Heuschele, W. P. (1960). *Amer. J. Vet. Med. Ass.* **136**, 256.

Hillis, W. D. (1961). *Amer. J. Hyg.* **73**, 316.

Hillis, W. D. (1963). *Transfusion* **3**, 445.

Hillis, W. D. (1967). *Johns Hopkins Med. J.* **120**, 176.

Hillis, W. D., and Goodman, R. (1969). *J. Immunol.* **103**, 1089.

Hillis, W. D., Holmes, A. W., and Davison, V. (1968). *Proc. Soc. Exp. Biol. Med.* **129**, 366.

Hillis, W. D., Garner, A. C., and Hillis, A. I. (1969). *Amer. J. Epidemiol.* **90**, 344.

Hindle, E. (1929a). *Trans. Roy. Soc. Trop. Med. Hyg.* **22**, 405.

Hindle, E. (1929b). *Proc. Roy. Soc. Med.* **22**, 823.

Hirschman, R. J., Shulman, R. N., Barker, L. F., and Smith, K. O. (1969). *J. Amer. Med. Ass.* **208**, 1667.

Hofer, H. O. (1969a). *Proc. Int. Congr. Primatol., 2nd, 1968* Vol. 2.

Hofer, H. O. (1969b). *Proc. Int. Congr. Primatol., 2nd, 1968* Vol. 3.

Hoffert, W., Bates, M. E., and Cheever, F. S. (1958). *Amer. J. Hyg.* **68**, 15.

Hofmann, H., and Kunz, C. (1968). *Zentralbl. Bakteriol., Parasitenk., Infektionskr. Hyg., Abt. 1. Orig.* **208**, 344.

Hofmann, H., Kunz, C., and Kriwanek, M. (1969). *Zentralbl. Bakteriol., Parasitenk., Infektionskr. Hyg., Abt. 1. Orig.* **209**, 288.

Holmes, A. W., and Capps, R. B. (1966). *Medicine (Baltimore)* **45**, 553.

Holmes, A. W., Dedmon, R. E., and Deinhardt, F. (1963). *Fed. Proc., Fed. Amer. Soc. Exp. Biol.* **22**, 334.

Holmes, A. W., Caldwell, R. G., Dedmon, R. E., and Deinhardt, F. (1964). *J. Immunol.* **92**, 602.

Holmes, A. W., Capps, R. B., and Deinhardt, F. (1965). *J. Lab. Clin. Med.* **66**, 879.

Holmes, A. W., Devine, J., Nowakowski, E., and Deinhardt, F. (1966). *J. Immunol.* **96**, 668.

Holmes, A. W., Capps, R. B., and Deinhardt, F. (1967). *J. Clin. Invest.* **46**, 1072.

Holmes, A. W., Wolfe, L., Rosenblate, H., and Deinhardt, F. (1969). *Science* **165**, 816.

Hooks, J., Rogers, N., Gibbs, C. J., Jr., and Gajdusek, D. C. (1969). *Bacteriol. Proc.,* 179.

Horstmann, D. M. (1969). *Ann. N. Y. Acad. Sci.* **162,** 594.

Horstmann, D. M., and Manuelidis, E. E. (1958). *J. Immunol.* **81,** 32.

Horstmann, D. M., and Melnick, J. L. (1950). *J. Exp. Med.* **91,** 573.

Horstmann, D. M., Melnick, J. L., Ward, R., and Fleitas, J. S. (1947). *J. Exp. Med.* **86,** 309.

Hotta, S., and Evans, C. A. (1956). *J. Infec. Dis.* **98,** 88.

Howe, H. A. (1952). *Amer. J. Hyg.* **56,** 265.

Howe, H. A. (1962). *Proc. Soc. Exp. Biol. Med.* **110,** 110.

Howe, H. A., and Bodian, D. (1940). *Proc. Soc. Exp. Biol. Med.* **43,** 718.

Howe, H. A., and Bodian, D. (1941). *Bull. Johns Hopkins Hosp.* **69,** 149.

Howe, H. A., and Bodian, D. (1942). *J. Pediat.* **21,** 713.

Howe, H. A., and Bodian, D. (1944). *J. Exp. Med.* **80,** 383.

Howe, H., and Bodian, D. (1945). *J. Exp. Med.* **81,** 247.

Howe, H. A., and Craigie, J. (1943). *J. Bacteriol.* **45,** 87.

Howe, H. A., Bodian, D., and Morgan, I. M. (1950). *Amer. J. Hyg.* **51,** 85.

Howe, H. A., O'Leary, W., Bender, W., Kiel, M., and Fontanella, A. (1957). *Amer. J. Pub. Health* **47,** 871.

Hsiung, G. D. (1968). *Bacteriol. Rev.* **32,** 185.

Hsiung, G. D. (1969). *Ann. N. Y. Acad. Sci.* **162,** 483.

Hsiung, G. D., and Atoynatan, T. (1966). *Amer. J. Epidemiol.* **83,** 38.

Hsiung, G. D., and Gaylord, W. H. (1961). *J. Exp. Med.* **114,** 975.

Hsiung, G. D., and Melnick, J. L. (1957). *J. Immunol.* **78,** 137.

Hsiung, G. D., and Melnick, J. L. (1958a). *Ann. N. Y. Acad. Sci.* **70,** 342.

Hsiung, G. D., and Melnick, J. L. (1958b). *J. Immunol.* **80,** 282.

Hsiung, G. D., Isacson, P., and McCollum, R. W. (1962). *J. Immunol.* **88,** 284.

Hsiung, G. D., Chang, R. W., Cuadrado, R. R., and Isacson, P. (1965). *J. Immunol.* **94,** 67.

Hsiung, G. D., Atoynatan, T., and Lee, C. W. (1969). *Amer. J. Epidemiol.* **89,** 464.

Huang, C. H. (1943). *J. Exp. Med.* **78,** 111.

Hughes, J. H. (1967). *In* "The Baboon in Medical Research" (H. Vagtborg, ed.), Vol. II, pp. 99–113. Univ. of Texas Press, Austin.

Hull, R. N. (1968). *In* "Virology Monographs," No. 2, pp. 1–66. Springer Publ. New York.

Hull, R. N. (1969a). *Nat. Cancer Inst., Monogr.* **29,** 173.

Hull, R. N. (1969b). *Ann. N. Y. Acad. Sci.* **162,** 472.

Hull, R. N., and Minner, J. R. (1957). *Ann. N. Y. Acad. Sci.* **67,** 413.

Hull, R. N., and Nash, J. C. (1960). *Amer. J. Hyg.* **71,** 15.

Hull, R. N., and Peck, F. B., Jr. (1967). *PAHO Sci. Publ.* **147,** 266.

Hull, R. N., Minner, J. R., and Smith, J. W. (1956). *Amer. J. Hyg.* **63,** 204.

Hull, R. N., Minner, J. R., and Mascoli, C. C. (1958). *Amer. J. Hyg.* **68,** 31.

Hull, R. N., Peck, F. B., Jr., Ward, T. G., and Nash, J. C. (1962). *Amer. J. Hyg.* **76,** 239.

Hull, R. N., Johnson, I. S., Culbertson, C. G., Reimer, C. B., and Wright, H. F. (1965). *Science* **150,** 1044.

Hunt, R. D. (1970). *Lab. Anim. Care* **20,** 1007.

Hunt, R. D., and Melendez, L. V. (1966). *Pathol. Vet.* **3,** 1.

Hunt, R. D., and Melendez, L. V. (1969). *Lab. Anim. Care* **19,** 221.

Hunt, R. D., Melendez, L. V., King, N., Gillmore, C., Daniel, M., Williamson, M.,

Jones, T. C., Donawick, W., Johnstone, C., Martens, J., Dodd, D., Martin, J., and Marshak, R. (1970). *J. Nat. Cancer Inst.* 44, 447.

Hurst, E. W. (1931). *J. Comp. Pathol. Ther.* 44, 231.

Hurst, E. W. (1936). *J. Exp. Med.* 63, 449.

Huseby, R. A. (1969). *Fed. Proc., Fed. Amer. Soc. Exp. Biol.* 28, 211.

Ilyenko, V. I., and Pokroskaya, O. A. (1959). *Acta Virol. (Prague), Engl. Ed.* 4, 75.

Imaizumi, K. (1966). *Jap. J. Med. Sci. Biol.* 19, 215.

Itoh, H., and Melnick, J. L. (1957). *J. Exp. Med.* 106, 677.

Iyer, C. G. S., Work, T. H., Narasimha, M. D. P., Napido, H., and Rajagopalan, P. K. (1960). *Indian J. Med. Res.* 48, 276.

Jacob, H., and Solcher, H. (1968). *Acta Neuropathol.* 11, 29.

Janisch, W., Horn, K. H., Scholtze, P., and Schreiber, D. (1968). *Exp. Pathol.* 2, 226.

Johnson, C. D., and Goodpasture, E. W. (1935). *Amer. J. Hyg.* 21, 46.

Johnson, C. D., and Goodpasture, E. W. (1936). *Amer. J. Hyg.* 23, 329.

Johnson, C. D., *et al.* Personal communication.

Johnsson, T., and Lundmark, C. (1957). *Lancet,* 1148.

Johnston, P. B. (1961). *J. Infec. Dis.* 109, 1.

Johnston, P. B. Unpublished data.

Joncas, J., Lussier, G., and Pavilanis, V. (1966). *Can. Med. Ass. J.* 95, 151.

Jones, T. C. (1969). *Fed. Proc., Fed. Amer. Soc. Exp. Biol.* 28, 162.

Jordan, L. E., Plummer, G., and Mayor, H. D. (1965). *Virology* 25, 156.

Jungeblut, C. W., and Bautista, G., Jr. (1956). *J. Infec. Dis.* 99, 103.

Jungeblut, C. W., and De Rodaniche, E. C. (1954). *Proc. Soc. Exp. Biol. Med.* 86, 604.

Jungeblut, C. W., and Engle, E. T. (1932). *Proc. Soc. Exp. Biol. Med.* 29, 879.

Kafuko, G. W., Henderson, B. E., Williams, M. C., and Kissling, R. E. (1970). *In* "Infection and Immunosuppression in Subhuman Primates" (H. Balner and W. I. B. Beveridge, eds.), pp. 45–48. Munksgaard, Copenhagen.

Kaji, M., Kamiya, S., Takewaki, E., and Nagabuchi, J. (1962). *Kyushu J. Med. Sci.* 13, 199.

Kalter, S. S. (1960). *Bull. WHO* 22, 319.

Kalter, S. S. (1964). *In* "Institute of Occupational Diseases Acquired from Animals, Jan. 7–9, 1964," pp. 126–159. Sch. Pub. Health, University of Michigan, Ann Arbor.

Kalter, S. S. (1965). *In* "The Baboon in Medical Research" (H. Vagtborg, ed.), Vol. I. pp. 407–420. Univ. of Texas Press, Austin.

Kalter, S. S. (1966). *In* "Basic Medical Virology" (J. E. Prier, ed.), pp. 207–245. Williams & Wilkins, Baltimore, Maryland.

Kalter, S. S. (1968a). *Bibl. Haematol.* 28, 338.

Kalter, S. S. (1968b). *In* "Primates in Medicine" (W. I. B. Beveridge, ed.), Vol. 2, pp. 45–61. Karger, Basel.

Kalter, S. S. (1969a). *Ann. N. Y. Acad. Sci.* 162, 499.

Kalter, S. S. (1969b). *PAHO Sci. Publ.* 182, 86.

Kalter, S. S. Unpublished data.

Kalter, S. S. (1970). *In* "Infection and Immunosuppression in Subhuman Primates" (H. Balner and W. I. B. Beveridge, eds.), pp. 119–120. Munksgaard, Copenhagen.

Kalter, S. S. (1971a). *In* "Marburg Virus Disease" (G. A. Martini and R. Siegert, eds.), pp. 177–187. Springer-Verlag, Berlin and New York.

Kalter, S. S. (1971b). *In* "Defining the Laboratory Animal," pp. 481–527. Nat. Acad. Sci., Washington, D. C.

Kalter, S. S. (1972). *In* "Pathology of Simian Primates" (R. N. T-W Fiennes, ed.) pp. 469–496. Karger, Basel.

Kalter, S. S., and Guilloud, N. B. (1970). *In* "The Chimpanzee" (G. H. Bourne, ed.), Vol. 2, pp. 361–389. Karger, Basel.

Kalter, S. S., and Heberling, R. L. (1968). *Nat. Cancer Inst., Monogr.* 29, 149.

Kalter, S. S., and Heberling, R. L. (1969). *World Health Organ., Chron.* 23, 112.

Kalter, S. S., and Heberling, R. L. (1970). *In* "Infection and Immunosuppression in Subhuman Primates" (H. Balner and W. I. B. Beveridge, eds.), pp. 121–137. Munksgaard, Copenhagen.

Kalter, S. S., and Heberling, R. L. (1971a). *Amer. J. Epidemiol.* 93, 403.

Kalter, S. S., and Heberling, R. L. (1971b). *Bacteriol. Rev.* 35, 310–364.

Kalter, S. S., and Heberling, R. L. (1971c). *In* "Medical Primatology 1970" (E. I. Goldsmith and J. Moor-Jankowski, eds.) pp. 272–280. Karger, Basel.

Kalter, S. S., and Heberling, R. L. Unpublished data.

Kalter, S. S., and Jeffries-Klitch, H. (1969). *Amer. J. Trop. Med. Hyg.* 18, 466.

Kalter, S. S., Fuentes-Marins, R., Crandell, R. A., Rodriguez, A. R., and Hellman, A. (1964a). *J. Bacteriol.* 87, 744.

Kalter, S. S., Rodriguez, A. R., and Ratner, J. J. (1964b). *Bacteriol. Proc.* 127.

Kalter, S. S., Kuntz, R. E., Al-Doory, Y., and Katzberg, A. (1966a). *Lab. Anim. Care* 16, 161.

Kalter, S. S., Ratner, I. A., Britton, H. A., Vice, T. E., Eugster, A. K., and Rodriguez, A. R. (1966b). *Nature (London)* 213, 610.

Kalter, S. S., Ratner, I. A., Britton, H. A., Vice, T. E., Eugster, A. K., and Rodriguez, A. R. (1966c). *In* "Some Recent Developments in Comparative Medicine" (R. N. T-W Fiennes, ed.), pp. 303–307. Academic Press, New York.

Kalter, S. S., Ratner, J., Kalter, G. V., Rodriguez, A. R., and Kim, C. S. (1967a). *Amer. J. Epidemiol.* 86, 552.

Kalter, S. S., Ratner, J. J., Rodriguez, A. R., and Kalter, G. V. (1967b). *In* "The Baboon in Medical Research" (H. Vagtborg, ed.), Vol. II, pp. 757–773. Univ. of Texas Press, Austin.

Kalter, S. S., Eugster, A. K., Vice, T. E., Kim, C. S., and Ratner, J. J. (1968a). *Fed. Proc., Fed. Amer. Soc. Exp. Biol.* 27, 715 (2800).

Kalter, S. S., Eugster, A. K., Vice, T. E., Kim, C. S., and Ratner, I. A. (1968b). *Nature (London)* 218, 884.

Kalter, S. S., Eugster, A. K., Vice, T. E., Kim, C. S., and Ratner, I. A. (1968c). *Bacteriol. Proc.* p. 175.

Kalter, S. S., Kuntz, R. E., Myers, B. J., Eugster, A. K., Rodriguez, A. R., Benke, M., and Kalter, G. V. (1968d). *Primates* 9, 123.

Kalter, S. S., Heberling, R. L., Vice, T. E., Lief, F. S., and Rodriguez, A. R. (1969a). *Proc. Soc. Exp. Biol. Med.* 132, 357.

Kalter, S. S., Ratner, J. J., and Heberling, R. L. (1969b). *Proc. Soc. Exp. Biol. Med.* 130, 10.

Kalter, S. S., Ratner, J. J., Rodriguez, A. R., Heberling, R. L., and Guilloud, N. B. (1969c). *Lab. Anim. Care* 19, 63.

Kalter, S. S., Heberling, R. L., Rodriguez, A. R., and Lief, F. S. (1970). *Bacteriol. Proc.* p. 166.

Kalter, S. S., Heberling, R. L., Helmke, R. J., and Rodriguez, A. R. (1971). *Bacteriol. Proc.* p. 168.

Kalter, S. S., Heberling, R. L., and Claussen, B. (1971b). *Lab. Anim. Sci.* 21, 829.

Kalter, S. S., Heberling, R. L., and Ratner, J. J. (1972a). *Bibl. Haematologica* 39, Karger, Basel.

Kalter, S. S., Felsburg, P. J., Heberling, R. L., Nahmias, A. J., and Brack, M. (1972b). *Proc. Soc. Exper. Biol. Med.* 139, 964.

Kalter, S. S., Heberling, R. L., Rodriguez, A. R., and Helmke, R. J. (1972c). *Proc. Int. Congr. Primatol., 3rd,* 2, 173–180.

Kamahora, J., and Nii, S. (1965). *Arch. Gesamte Virusforsch.* 16, 161.

Kamitsuka, P. S., Lou, T. Y., Fabiyi, A., and Wenner, H. A. (1965). *Amer. J. Epidemiol.* 81, 283.

Kanamitsu, M., Kasamaki, A., Ogana, M., Shimpo, K., and Kasahara, S. (1964). *Rep. Nat. Congr. Jap. Virol. Soc., 12th,* pp. 25–26.

Kanamitsu, M., Kasamaki, A., Ogawa, M., Kasahara, S., and Imanura, M. (1967). *Jap. J. Med. Sci. Biol.* 20, 175.

Kaplan, C. (1969). *Lab. Anim. Handb.* 4, 117.

Kaplan, M. (1969). *In* "Primates in Medicine" (W. I. B. Beveridge, ed.), Vol. 3, pp. 140–145. Karger, Basel.

Karasszon, D. (1969). *Acta Vet. Acad. Sci. (Hungary)* 19, 299.

Kasahara, S. (1965). *Sapporo Med. J.* 27, 219.

Kato, S., Tsuru, K., and Miyamoto, H. (1965). *Biken's J.* 8, 45.

Katzin, D. S., Connor, J. D., Wilson, L. A., and Sexton, R. S. (1967). *Proc. Soc. Exp. Biol. Med.* 125, 391.

Kaufman, H. E., and Maloney, E. D. (1963). *Arch. Ophthalmol.* 69, 626.

Keeble, S. A. (1960). *Ann. N. Y. Acad. Sci.* 85, 960.

Keeble, S. A. (1968). *In* "Some Diseases of Animals Communicable to Man in Britain" (O. Graham-Jones, ed.), pp. 183–188. Pergamon, Oxford.

Keeble, S. A., Christofinis, G. T., and Wood, W. (1958). *J. Pathol. Bacteriol.* 76, 189.

Kennard, M. A. (1941). *Yale J. Biol. Med.* 13, 701.

Kent, S. P. (1960). *Ann. N. Y. Acad. Sci.* 85, 819.

Kim, C. S., Sueltenfuss, E. A., and Kalter, S. S. (1967). *J. Infec. Dis.* 117, 292.

Kim, C. S., Eugster, A. K., and Kalter, S. S. (1968). *Primates* 9, 93.

Kim, C. S., Kriewaldt, F. H., Hagino, N., and Kalter, S. S. (1970). *J. Amer. Vet. Med. Ass.* 157, 730.

Kinard, R. (1970). *Science* 169, 828.

King, N. W., Hunt, R. D., Daniel, M. D., and Melendez, L. V. (1967). *Lab. Anim. Care* 17, 413.

Kirschstein, R. L., and Gerber, P. (1962). *Nature (London)* 195, 299.

Kissling, R. E., Robinson, R. Q., Murphy, F. A., and Whitfield, S. G. (1968). *Science* 160, 888.

Kling, C., Levaditi, C., and Lepine, L. (1929). *Bull. Acad. Nat. Med., Paris* [3] 102, 158.

Kling, C., Levaditi, C., and Hornus, G. (1934). *Bull. Acad. Nat. Med., Paris* [3] 111, 709.

Koff, R. S., and Isselbacher, K. J. (1968). *N. Engl. J. Med.* 278, 1371.

Kohler, H., and Apodaca, J. (1968). *Zentralbl. Bakteriol., Parasitenk., Infektionskr. Hyg., Abt. 1. Orig.* 206, 1.

Kohler, H., Lange, W., Apodaca, J., and Eggert, E. (1968). *Fbl. Baks.* 208, 207.

Kohler, H., Apodaca, J., and Lange, W. (1969). *Zentralbl. Bakteriol., Parasitenk., Infektionskr. Hyg. Abt. 1. Orig.* 209, 423.

Kokernot, R. H., McIntosh, B. M., Worth, C. B., DeMorais, T., and Weinbren, M. P. (1962). *Amer. J. Trop. Med. Hyg.* 11, 678.

Kolmer, J. A. (1934). *Amer. J. Med. Sci.* 188, 510.

Kolmer, J. A., and Rule, A. M. (1934). *J. Immunol.* 26, 505.

Kolyaskina, G. I. (1963). *Probl. Virol.* 4, 450.

Koprowski, H. (1958). *Ann. N. Y. Acad. Sci.* 70, 369.

Koprowski, H., Jervis, G. A., and Norton, T. W. (1954). *Arch. Gesamte Virusforsch.* 5, 413.

Kovacs, E., Baratawidjaja, R. K., and Labzoffsky, N. A. (1963a). *Nature (London)* 2, 497.

Kovacs, E., Baratawidjaja, R. K., Wolmsley-Hewson, A., and Labzoffsky, N. A. (1963b). *Arch. Gesamte Virusforsch.* 14, 143.

Kraft, L. M. (1952). *Proc. Soc. Exp. Biol. Med.* 80, 498.

Kraft, L. M., and Melnick, J. L. (1950). *J. Exp. Med.* 92, 483.

Kraft, L. M., and Melnick, J. L. (1953). *J. Exp. Med.* 97, 401.

Kramer, S. D., Hoskirth, B., and Grossman, L. H. (1939). *J. Exp. Med.* 69, 49.

Kratochvil, C. H., ed. (1968). "Primates in Medicine," Vol. 1. Karger, Basel.

Kraus, R., and Kantor, C. (1917). *Rev. Inst. Bacteriol. Carlos G. Malbrian* 1, 43.

Krushak, D. H. (1970). *Lab. Anim. Care* 20, 52.

Krutyanskaya, G. L., Elbert, L. B., Ralph, N. M., and Tyufanov, A. V. (1967). *Vop. Virusol.* 12, 680.

Krylova, P. I. (1968). *Quest. Physiol. Exp. Pathol.* pp. 311–316.

Krylova, P. I., and Shevtsova, Z. V. (1969). *Arkh. Patol.* 31, 65.

Laboratory Animal Facilities and Care of The Institute of Laboratory Animal Resources. (1968). *U. S., Pub. Health Serv., Publ.* 1024.

Laemmert, H. W., and de Castro Ferreira, L. (1945). *Amer. J. Trop. Med. Hyg.* 25, 231.

Laemmert, H. W., and Kumm, H. W. (1950). *Amer. J. Trop. Med. Hyg.* 30, 723.

Lampert, P. W., Earle, K. M., Gibbs, C. J., Jr., and Gajdusek, D. C. (1969). *J. Neuropathol. Exp. Neurol.* 28, 353.

Lancaster, M. C., Boulter, E. A., Westwood, J. C. N., and Randles, J. (1966). *Brit. J. Exp. Pathol.* 47, 466.

Landon, J. C., and Bennett, D. G. (1969). *Nature (London)* 222, 683.

Landon, J. C., and Malan, L. B. (1971). *J. Nat. Cancer Inst.* 46, 881.

Landon, J. C., Ellis, L. B., Zeve, V. H., and Fabrizio, D. P. A. (1968). *J. Nat. Cancer Inst.* 40, 181.

Landsteiner, K., and Popper, E. (1909). *Z. Immunitaetsforsch.* 2, 377.

Lange, W., Apodaca, J., and Kohler, H. (1969). *Zentralbl. Bakteriol., Parasitenk., Infektionskr. Hyg. Abt. 1. Orig.* 210, 337.

Lapin, B. A. (1969). "Primates on Medicine" (W. I. B. Beveridge, ed.), Vol. 3, pp. 23–27.

Lapin, B. A., and Yakoleva, L. A. (1963). "Comparative Pathology in Monkeys." Thomas, Springfield, Illinois.

Lapin, B. A., Djikidze, E. K., Yakovleva, L. A., Chumakova, M. Y., and Adjigitov, F. I. (1965a). *Vop. Virusol.* 2, 226.

Lapin, B. A., Dzhikidze, E. K., and Yakovleva, L. A. (1965b). *Vop. Virusol.* 10, 226.

Lapin, B. A., Pekerman, S. M., Yakovleva, L. A., Dzhikidze, E. K., Shevtsova, Z. V., Kuksova, M. I., Danko, L. V., Krilova, R. I., Akbiroit, E. Ya., and Agraba, V. Z. (1967). *Vop. Virusol.* 12, 168.

Lapin, B. A., Yakovleva, L. A., Deichman, G. I., Kuksova, M. I., Krivoshein, I. S., Antonichev, A. V., Ivanov, M. T., and Akaeva, E. A. (1968). *In* "Meditsinskaia

Primatologiia" (B. A. Lapin, ed.), pp. 167–180. Akad. Med. Nauk USSR, Tbilisi.

Lapin, B. A., Shevtosova, Z. V., and Krylova, R. I. (1969). *Proc. Int. Cong. Primatol., 2nd, 1968* Vol. 3, pp. 120–128.

Larin, N. M., Herbst-Laier, R. H., Copping, M. P., and Wenham, R. B. M. (1967). *Nature (London)* 213, 827.

Lee, G. C. Y., Grayston, J. T., and Wang, S. (1967). *Proc. Soc. Exp. Biol. Med.* 125, 803.

Lee, H. W., and Scherer, W. F. (1961). *J. Immunol.* 86, 151.

Lennette, E. W. (1968). "Workshop on Viral Diseases which Impede Colonization of Nonhuman Primates." Nat. Center for Primate Biol., University of California, Davis.

Lepine, P., and Paccaud, M. (1957). *Ann. Inst. Pasteur, Paris* 92, 289.

Lepine, P., and Sedallian, P. (1939). *C. R. Acad. Sci.* 208, 129.

Levaditi, C., Nicolau, S., and Schoen, R. (1926). *Ann. Inst. Pasteur, Paris* 40, 973.

Levy, B. M., and Mirkovic, R. R. (1971). *Lab. Anim. Sci.* 21, 33.

Levy, B. M., Taylor, A. C., Hampton, S., and Thoma, G. W. (1969). *Cancer Res.* 29, 2237.

Lewis, J. L., Brown, W. E., Hertz, R., Davis, R. C., and Johnson, R. H. (1968). *Cancer Res.* 28, 2032.

Libikova, H., and Albrecht, P. (1959). *Vet. Cos.* 8, 461.

Libikova, H., Tesarova, J., and Rajcani, J. (1970). *Acta Virol. (Prague), Engl. Ed.* 14, 64.

Lichter, E. A. (1969). *Nature (London)* 244, 810.

Lillie, R. D. (1936). *Pub. Health Rep.* 51, 303.

Lindsey, J. R., Wharton, L. R., Woodruff, J. D., and Baker, H. J. (1969). *Pathol. Vet.* 6, 378.

Lingeman, C. H., Reed, R. E., and Garner, F. M. (1969). *Nat. Cancer Inst., Monogr.* 32, 157.

Livesay, H. R. (1949). *J. Infec. Dis.* 84, 306.

Lokhova, S. V. (1965). *Vop. Virusol.* 10, 243.

Lord, G. H., and Willson, J. E. (1968). *J. Amer. Vet. Med. Ass.* 153, 910.

Lorenz, D., Barker, L., and Stevens, D. (1970). *Proc. Soc. Exp. Biol. Med.* 135, 348.

Lou, T. Y., and Wenner, H. A. (1963). *Amer. J. Hyg.* 77, 293.

Lou, T. Y., Wenner, H. A., and Kamitsuka, P. S. (1960). *Arch. Gesamte Virusforsch.* 10, 451.

Love, F. M., and Jungherr, E. (1962). *J. Amer. Med. Ass.* 179, 804.

Low, R. J., and Baron, S. (1960). *Science* 132, 622.

Luby, J. P., and Sanders, C. V. (1969). *Ann. Intern. Med.* 71, 657.

Lucas, W. P., and Prizer, E. C. (1912). *J. Med. Res.* 26, 181.

Maass, G., Haas, R., and Oehlert, W. (1969a). *Lab. Anim. Handb.* 4, 155.

Maass, G., Muller, J., Seemayer, N., and Haas, R. (1969b). *Amer. J. Epidemiol.* 89, 681.

McCarthy, K. E. (1969). *Lab. Anim. Handb.* 4, 121.

McCarthy, K. E., Thorpe, E., Laursen, A. C., Heymann, C. S., and Beale, A. J. (1968). *Lancet* 2, 856.

McConnell, S. J., Herman, Y. F., Mattson, D. E., and Erickson, L. (1962). *Nature (London)* 195, 1128.

McConnell, S. J., Herman, Y. F., Mattson, D. E., Huxsoll, D. L., Lang, C. M., and Yager, R. H. (1964). *Amer. J. Vet. Res.* **25**, 192.

McConnell, S. J., Hickman, R. L., Wooding, W. L., Jr., and Huxsoll, D. L. (1968). *Amer. J. Vet. Res.* **29**, 1675.

McInerney, T., Condon, F., Barr, T. M., and Coohon, D. B. (1968). *Morbid. Mortal. Week. Rep.* **17**, 271.

McIntosh, B. M., Paterson, H. E., McGillivray, G., and DeSousa, J. (1964). *Ann. Trop. Med. Parasitol.* **58**, 45.

Mack, T. M., and Noble, J., Jr. (1970). *Lancet* **1**, 752.

Mackay, C. M., and Schroeder, C. R. (1935). *Proc. Soc. Exp. Biol. Med.* **33**, 373.

McKinley, E. B., and Douglass, M. (1930). *J. Infec. Dis.* **47**, 511.

McLean, I. W., Jr., and Taylor, A. R. (1958). *Progr. Med. Virol.* **1**, 122.

McMillen, J. Macasaet, F., Lenahan, M., Kamitsuka, P., and Wenner, H. A. (1968). *Amer. J. Epidemiol.* **88**, 126.

McNamara, F. N. (1953). *Brit. J. Exp. Pathol.* **34**, 392.

McNulty, W. P., Lobitz, W. C., Hu, F., Maruffo, C. A., and Hall, A. S. (1968). *Arch. Dermatol.* **97**, 286.

McPherson, C. (1970). *J. Amer. Vet. Med. Ass.* **157**, 1959.

Majima, E. (1957). *Osaka Daigaku Igaku Zasshi* **9**, 1011.

Malherbe, H., and Harwin, R. (1957). *Brit. J. Exp. Pathol.* **38**, 539.

Malherbe, H., and Harwin, R. (1963). *S. Afr. Med. J.* **37**, 407.

Malherbe, H., and Strickland-Cholmley, M. (1967). *Arch. Gesamte Virusforsch.* **22**, 235.

Malherbe, H., and Strickland-Cholmley, M. (1968). *Lancet* **1**, 1434.

Malherbe, H., and Strickland-Cholmley, M. (1969a). *Lancet* **2**, 1300.

Malherbe, H., and Strickland-Cholmley, M. (1969b). *Lancet* **2**, 1427.

Malherbe, H., and Strickland-Cholmley, M. (1970). *Lancet* **7**, 785.

Malherbe, H., and Strickland-Cholmley, M. (1971). *In* "Marburg Virus Disease" (G. A. Martini and R. Siegert, eds.), pp. 188–194. Springer-Verlag, Berlin and New York.

Malherbe, H., Harwin, R., and Ulrich, M. (1963). *S. Afr. Med. J.* **37**, 407.

Manning, P. J., Banks, K. L., and Lehner, N. D. M. (1968). *J. Amer. Vet. Med. Ass.* **153**, 899.

Mannweiler, K., and Palacios, O. (1961). *Z. Naturforsch.* **16S**, 705.

Marois, P., Boudreault, A., DiFranco, E., and Pavilanis, V. (1971). *Can. J. Comp. Med.* **35**, 71.

Marsh, J. T., Lavender, J. F., Chang, S. S., and Rasmussen, A. F. (1963). *Science* **140**, 1414.

Marsh, R. F., Burger, D., Echroade, R. J., ZuRhein, G. M., and Hanson, R. P. (1969). *J. Infec. Dis.* **120**, 713.

Martin, G. V., and Heath, R. B. (1969). *Brit. J. Exp. Pathol.* **50**, 516.

Martini, G. A. (1967). *Lancet* **2**, 1363.

Martini, G. A. (1969). *Trans. Roy. Soc. Trop. Med. Hyg.* **63**, 295.

Martini, G. A., Knauff, H. G., Schmidt, H. A., Mayer, G., and Baltzer, G. (1968). *Deut. Med. Wochenschr.* **93**, 559.

Maruffo, C. A. (1967). *Nature (London)* **213**, 521.

Masillamony, R., and John, T. J. (1970). *Amer. J. Epidemiol.* **91**, 446.

Massaglia, A. C. (1923). *Lancet* **1**, 1056.

Mattingly, S. F. (1966). *J. Amer. Vet. Med. Ass.* **149**, 1677.

Matumoto, M. (1969). *Bacteriol. Rev.* **33**, 404.

May, G., and Herzberg, K. (1969). *Zentralbl. Bakteriol., Parasitenk., Infektionskr. Hyg., Abt. 1. Orig.* **211**, 133.

May, G., and Knothe, H. (1968). *Deut. Med. Wochenschr.* **93**, 620.

May, G., Kuothe, H., Hulser, D., and Herzberg, K. (1968). *Zentralbl. Bakteriol., Parasitenk., Infektionskr. Hyg., Abt. 1. Orig.* **207**, 145.

Mayer, V. (1966). *Acta Virol. (Prague), Engl. Ed.* **10**, 561.

Mayer, V., and Rajcani, J. (1967). *Acta Virol. (Prague), Engl. Ed.* **11**, 321.

Mayer, V., and Rajcani, J. (1968). *Acta Virol. (Prague), Engl. Ed.* **12**, 403.

Mayor, H. D., and Ito, M. (1967). *Proc. Soc. Exp. Biol. Med.* **126**, 723.

Mayor, H. D., Jamison, R. M., Jordan, L. E., and Melnick, J. L. (1965). *J. Bacteriol.* **90**, 235.

Melby, E. C., and Baker, H. J. (1969). *Ann. N. Y. Acad. Sci.* **162**, 373.

Melendez, L. V. (1969). *PAHO Sci. Publ.* No. 182, pp. 103–106. Pan Amer. Health Org.

Melendez, L. V., and Hunt, R. D. (1966). *Proc. Pan-Amer. Congr. Vet. Med. Zootech., 5th, 1966.*

Melendez, L. V., Hunt, R. D., Garcia, F. G., and Trum, B. F. (1966). *In* "Some Recent Developments in Comparative Medicine" (R. N. T-W Fiennes, ed.), pp. 393–397. Academic Press, New York.

Melendez, L. V., Daniel, M. D., Hunt, R. D., and Garcia, F. G. (1968a). *Lab. Anim. Care* **18**, 374.

Melendez, L. V., Hunt, R. D., Daniel, M. D., and Garcia, F. G. (1968b). *Fed. Proc., Fed. Amer. Soc. Exp. Biol.* **27**, 664.

Melendez, L. V., Daniel, M. D., Hunt, R. D., and Garcia, F. G. (1969a). *Lab. Primate Newslett.* **8**, 1.

Melendez, L. V., Espana, C., Hunt, R. D., Daniel, M. D., and Garcia, F. G. (1969b). *Lab. Anim. Care* **19**, 38.

Melendez, L. V., Garcia, F. G., Fraser, C. E. O., Hunt, R. D., and King, N. W. (1969c). *Lab. Anim. Care* **19**, 372.

Melendez, L. V., Hunt, R. D., Garcia, F. G., and Fraser, C. E. O. (1969d). *Lab. Anim. Care* **19**, 378.

Melendez, L. V., Daniel, M. D., Hunt, R. D., Fraser, C. E. O., Garcia, F. G., King, N. W., and Williamson, M. E. (1970a). *J. Nat. Cancer Inst.* **44**, 1175.

Melendez, L. V., Hunt, R. D., Daniel, M. D., Fraser, C. E. O., Garcia, F. G., and Williamson, M. E. (1970b). *Int. J. Cancer* **6**, 431.

Melendez, L. V., Hunt, R. D., Daniel, M. D., and Trum, B. F. (1970c). *In* "Infections and Immunosuppression in Subhuman Primates" (H. Balner and W. I. B. Beveridge, eds.), pp. 111–117. Munksgaard, Copenhagen.

Melendez, L. V., Hunt, R. D., Daniel, M. D., Blake, B. J., and Garcia, F. G. (1971). *Science* **171**, 1161.

Melnick, J. L. (1945). *Proc. Soc. Exp. Biol. Med.* **58**, 14.

Melnick, J. L. (1956). *J. Immunol.* **53**, 277.

Melnick, J. L. (1959). *Live Poliovirus Vaccines* **44**, 65.

Melnick, J. L. (1962). *Science* **138**, 1128.

Melnick, J. L. (1968). *Perspec. Virol.* **6**, 99–103.

Melnick, J. L., and Banker, D. D. (1954). *J. Exp. Med.* **100**, 181.

Melnick, J. L., and Benyesh-Melnick, M. (1959). *J. Amer. Med. Ass.* **171**, 1165.

Melnick, J. L., and Horstmann, D. M. (1947). *J. Exp. Med.* **85**, 287.

Melnick, J. L., and Kaplan, A. S. (1953). *J. Exp. Med.* **97**, 367.

Melnick, J. L., and Parks, W. P. (1970). *J. Infec. Dis.* **121**, 351.

Melnick, J. L., and Paul, J. R. (1943). *J. Exp. Med.* **78**, 273.

Melnick, J. L., and Rapp, F. (1965). *J. Nat. Cancer Inst.* **34**, 529.

Melnick, J. L., and von Magnus, H. (1948). *Amer. J. Hyg.* **48**, 107.

Melnick, J. L., Midulla, M., Wimberly, I., Barrerra-Oro, J. G., and Levy, B. M. (1964). *J. Immunol.* **92**, 596.

Melnick, J. L., Benyesh-Melnick, M., Fernback, D. J., Phillips, C. F., and Mirkovic, R. R. (1967). *In* "The Baboon in Medical Research" (H. Vagtborg, ed.), Vol. II, pp. 669–682. Univ. of Texas Press, Austin.

Merino, C., Doan, C. A., Woolpert, O. C., Schwab, J. L., and Saslaw, S. (1941). *Proc. Soc. Exp. Biol. Med.* **48**, 563.

Metzgar, R. S., Grace, J. T., Jr., and Sproul, E. E. (1962). *Ann. N. Y. Acad. Sci.* **101**, 192.

Meyer, H. M., Jr., Brooks, B. E., Douglas, R. D., and Rogers, N. G. (1962a). *Amer. J. Dis. Child.* **103**, 307.

Meyer, H. M., Jr., Hopps, H. E., Rogers, N. G., Bivoks, B. E., Bernheim, B. C., and Jones, W. P. (1962b). *J. Immunol.* **88**, 796.

Milhaud, C., Klein, M., and Virat, J. (1969). *Exp. Anim.* **2**, 121.

Miller, D. G., and Horstmann, D. M. (1966). *Virology* **30**, 319.

Miller, R. H., Pursell, A. R., and Mitchell, F. E. (1964). *Amer. J. Hyg.* **80**, 365.

Moor-Jankowski, J., and Goldsmith, E. I. (1969). *Ann. N. Y. Acad. Sci.* **162**, 324.

Moreland, A. F., and Woodard, J. C. (1968). *Pathol. Vet.* **5**, 193.

Morgan, H. R. (1967). *J. Nat. Cancer Inst.* **39**, 1229.

Morris, J. A., O'Connor, J. R., and Smadel, J. E. (1955). *Amer. J. Hyg.* **62**, 327.

Morris, J. A., Blount, R. E., Jr., and Savage, R. E. (1956). *Proc. Soc. Exp. Biol. Med.* **92**, 544.

Mosley, J. W., Reinhardt, H. P., and Hassler, F. R. (1967). *J. Amer. Med. Ass.* **199**, 695.

Mouquet, A. D. (1926). *Soc. Cent. Med. Vet.* p. 46.

Muckenfuss, R. S., Armstrong, C., and McCordock, H. A. (1933). *Pub. Health Rep.* **48**, 1341.

Muller, W. (1935). *Monatsschr. Kinderheilk.* **63**, 134.

Munroe, J. S. (1966). *In* "Some Recent Developments in Comparative Medicine" (R. N. T-W Fiennes, ed.), pp. 229–250. Academic Press, New York.

Munroe, J. S. (1969). *Ann. N. Y. Acad. Sci.* **162**, 556.

Munroe, J. S., and Windle, W. F. (1963). *Science* **140**, 1415.

Munroe, J. S., and Windle, W. F. (1967). *Nature (London)* **216**, 811.

Munroe, J. S., Shipkey, F., Erlandson, R. A., and Windle, W. F. (1964). *Nat. Cancer Inst., Monogr.* **17**, 365.

Murphy, B. L., Krushak, D. H., and Fields, R. M. (1970). *Amer. J. Vet. Res.* **31**, 947.

Murphy, B. L., Maynard, J. E., Krushak, D. H., and Fields, R. M. (1971). *Appl. Microbiol.* **21**, 50.

Murphy, G. P., Brede, H. D., Cohen, E., and Grace, J. T., Jr. (1970). *Transplant. Proc.* **2**, 546.

Nadaichik, L. V., and Rozina, E. E. (1967). *Viral Infec. Antiviral Prep.* pp. 48–55.

Nagler, F. P., and Klotz, M. (1958). *Can. Med. Ass. J.* **79**, 743.

Nahmias, A. J., London, W. T., and Catalono, L. W., Jr. (1971). *Science* **171**, 3968.

Nathanson, N., and Bodian, D. (1961). *Bull. Johns Hopkins Hosp.* **108**, 308.

Nathanson, N., and Cole, G. A. (1970). *Clin. Exp. Immunol.* **6**, 161.

Nathanson, N., and Harrington, B. (1966). *Amer. J. Epidemiol.* **84**, 541.

Nathanson, N., and Harrington, B. (1967). *Amer. J. Epidemiol.* **85**, 494.

Nathanson, N., Goldblatt, D., Davis, M., Thind, I. S., and Price, W. H. (1965). *Amer. J. Epidemiol.* **82**, 359.

Nathanson, N., Davis, M., Thind, I. S., Price, W. S., and Rivet-Moulin, H. (1966). *Amer. J. Epidemiol.* **84**, 524.

Nathanson, N., Gittelsohn, A. M., Thind, I. S., and Price, W. H. (1967). *Amer. J. Epidemiol.* **85**, 503.

Nathanson, N., Thind, I. S., O'Leary, W., and Price, W. H. (1968). *Amer. J. Epidemiol.* **88**, 103.

Nathanson, N., Cole, G. A., and Hodous, J. (1969a). *Amer. J. Epidemiol.* **89**, 480.

Nathanson, N., Stolley, P. D., and Boolukos, P. J. (1969b). *J. Comp. Pathol.* **79**, 109.

Nelson, B., Cosgrove, G. E., and Gengozian, N. (1966). *Lab. Anim. Care* **16**, 255.

Neurauter, L. J. (1969). *Ann. N. Y. Acad. Sci.* **162**, 378.

Nevin, M., and Bittman, F. R. (1923). *J. Infec. Dis.* **32**, 33.

Nicholas, A. H. (1970a). *J. Nat. Cancer Inst.* **45**, 897.

Nicholas, A. H. (1970b). *J. Nat. Cancer Inst.* **45**, 907.

Nicholas, A. H., and McNulty, W. P. (1968). *Nature (London)* **217**, 745.

Nicolau, S., Cruveilhier, L., Truche, C., Kopciowska, L., and Viala, C. (1938). *C. R. Soc. Biol.* **129**, 176.

Nicolle, C., and Conseil, E. (1911). *C. R. Acad. Sci.* **153**, 1522.

Nicolle, C., and Conseil, E. (1920). *C. R. Soc. Biol.* **83**, 56.

Niederman, J. C., McCollum, R. W., Henle, G., and Henle, W. (1968). *J. Amer. Med. Ass.* **203**, 205.

Nii, S., and Kamahora, J. (1963). *Biken's J.* **6**, 229.

Nii, S., Kamahora, J., Mori, Y., Takahashi, M., Nishimura, S., and Okuno, Y. (1964). *Biken's J.* **6**, 271.

Niven, J. S. F., Armstrong, J. A., Andrewes, C. H., Pereira, H. G., and Valentine, R. C. (1961). *J. Pathol. Bacteriol.* **81**, 1.

Noble, J. (1970). *Bull. WHO* **42**, 509.

Noble, J., and Rich, J. A. (1969). *Bull. WHO* **40**, 279.

Noguchi, H. (1927). *J. Amer. Med. Ass.* **89**, 739.

Noguchi, H. (1928). *J. Exp. Med.* **48**, Supple. No. 2.

Noyes, W. F. (1970). *J. Nat. Cancer Inst.* **45**, 579.

Ohwada, Y., Yamane, Y., Nagashima, T., Takada, M., and Asahara, T. (1969). *Kitasato Arch. Exp. Med.* **42**, 41.

Ohwada, Y., Yamane, Y., Nagashima, T., Takada, M., Asahara, T., and Kashara, S. (1970). *Jap. J. Vet. Sci.* **32**, 69.

Olitsky, P. K. (1930). *Trans. Amer. Acad. Ophthamol.* **35**, 225.

Olitsky, P. K., and Tyler, J. R. (1933). *J. Exp. Med.* **57**, 229.

Olitsky, P. K., Knutti, R. E., and Tyler, J. R. (1931a). *J. Exp. Med.* **53**, 753.

Olitsky, P. K., Knutti, R. E., and Tyler, J. R. (1931b). *J. Exp. Med.* **54**, 31.

Olitsky, P. K., Knutti, R. E., and Tyler, J. R. (1931c). *J. Exp. Med.* **54**, 557.

Olitsky, P. K., Syverton, J. T., and Tyler J. R. (1933). *J. Exp. Med.* **57**, 871.

O'Reilly, K. J., Smith, C. E. G., McMahon, D. A., Wilson, A. L., and Robertson, I. M. (1965). *J. Hyg.* **63**, 213.

Orsi, E. V., Franko, M., Rodriguez, L., and Holden, H. T. (1969). *Experientia* **25**, 181.

Osterrieth, P. M., and Deleplanque-Biegeois, P. (1961). *Ann. Soc. Belge Med. Trop.* **41**, 63.

Osterrieth, P. M., Deleplanque-Biegeois, P., and Renoirte, R. (1960). *Ann. Soc. Belge Med. Trop.* **40**, 205.

Paine, T. F., Jr. (1964). *Bacteriol. Rev.* **28**, 472.

Palmer, A. E., Allen, A. M., Tauraso, N. M., and Shelokov, A. (1968). *Amer. J. Trop. Med. Hyg.* **17**, 404.

Panthier, R., Coteigne, G., and Hammoun, C. (1949). *Bull. Inst. Nat. Hyg.* **4**, 109.

Parkman, P. D., Phillips, P. E., Kirschstein, R. L., and Meyer, H. M., Jr. (1965a). *J. Immunol.* **95**, 743.

Parkman, P. D., Phillips, P. E., and Meyer, H. M. (1965b). *Amer. J. Dis. Child.* **110**, 390.

Parks, W. P., and Melnick, J. L. (1969). *J. Infec. Dis.* **120**, 539.

Parks, W. P., Voss, W. R., and Melnick, J. L. (1968). *Bacteriol. Proc.* **68**, 150.

Parks, W. P., Melnick, J. L., Voss, W. R., Singer, D. B., Rosenberg, H. S., Alcot, J., and Casazza, A. M. (1969). *J. Infec. Dis.* **120**, 548.

Pasteur, L., Chamberland, M. M., and Roux, M. E. (1884a). *C. R. Acad. Sci.* **98**, 457.

Pasteur, L., Chamberland, M. M., and Roux, M. E. (1884b). *C. R. Acad. Sci.* **98**, 1229.

Paul, J. R. (1942–1943). *Harvey Lect.* **38**, 104.

Paul, J. R., and Trask, J. D. (1932). *J. Exp. Med.* **56**, 319.

Paul, J. R., Melnick, J. L., and Sabin, A. B. (1948). *Proc. Soc. Exp. Biol. Med.* **68**, 193.

Pedreira, F. A., Tauraso, N. M., and Palmer, A. E. (1968). *Proc. Soc. Exp. Biol. Med.* **129**, 472.

Peebles, T. C., McCarthy, K., Enders, J. F., and Holloway, A. (1957). *J. Immunol.* **78**, 63.

Pelc, S., Perier, J.-O., and Quersin-Thiry, L. (1958). *Rev. Neurol.* **98**, 3.

Pellissier, A., and Lumaret, R. (1948). *Ann. Inst. Pasteur, Paris* **74**, 507.

Pereira, H. G., Huebner, R. J., Ginsberg, H. S., and Van der Veen, J. (1963). *Virology* **20**, 613.

Perkins, F. T. (1968). *In* "Some Diseases of Animals Communicable to Man in Britain" (O. Graham-Jones, ed.), pp. 189–194. Pergamon, Oxford.

Perkins, F. T., and O'Donohue, P. N., eds. (1969). *Lab. Anim. Handb.* **4**.

Peters, J. C. (1966a). *Int. Zool. Yearbk.* **6**, 274.

Peters, J. C. (1966b). *Kleintier-Prax.* **11**, 65.

Pettit, A., and Aguessy, C. D. (1932). *Bull. Soc. Pathol. Exot.* **25**, 190.

Phillips, P. E., Parkman, P. D., Kirschstein, R. L., and Meyer, H. M., Jr. (1965). *Fed. Proc., Fed. Amer. Soc. Exp. Biol.* **24**, 570.

Piazza, M. (1969). *In* "Experimental Viral Hepatitis," p. 195. Thomas, Springfield, Illinois.

Pierce, E. C., Pierce, J. D., and Hull, R. N. (1958). *Amer. J. Hyg.* **68**, 242.

Pinto, C. A., and Haff, R. F. (1969). *Nature (London)* **224**, 1310.

Plentl, A. A., Dede, J. A., and Grey, R. M. (1968). *Folia Primatol.* **8**, 307.

Plummer, G. (1962). *J. Gen. Microbiol.* **29**, 703.

Plummer, G. (1967). *Progr. Med. Virol.* **9**, 302.

Pond, W. L., Russ, S. B., and Warren, I. (1953). *J. Infec. Dis.* **93**, 294.

Potkay, S., Ganaway, J. R., Rogers, N. G., and Kinard, R. (1966). *Amer. J. Vet. Res.* **27**, 331.

Price, R. A., and Powers, R. D. (1969). *Pathol. Vet.* **6**, 369.

Price, W. H., Lee, R. W., Gunkel, W. F., and O'Leary, W. (1961). *Amer. J. Trop. Med. Hyg.* 10, 403.

Price, W. H., Parks, J. J., Ganaway, J., Lee, R., and O'Leary, W. (1963a). *Amer. J. Trop. Med. Hyg.* 12, 624.

Price, W. H., Parks, J. J., Ganaway, J., O'Leary, W., and Lee, R. (1963b). *Amer. J. Trop. Med. Hyg.* 12, 787.

Prier, J. E., and Sauer, R. M. (1960). *Ann. N. Y. Acad. Sci.* 85, 951.

Prier, J. E., Sauer, R. M., Malsberger, R. G., and Sillaman, J. M. (1960). *Amer. J. Vet. Dis.* 21, 381.

Prince, A. M. (1971). *In* "Medical Primatology 1970" (E. I. Goldsmith and J. Moor-Jankowski, eds.) pp. 731–739. Karger, Basel.

Prokhorova, I. A., Zhevandrova, V. I., and Sobol', A. V. (1967). *In* "Materials of XIIIth Sessions of Institute on Present Problems of Virology and Specific Prophylaxis of Viral Infection," pp. 40–41. Institute of Poliomyelitis and Viral Encephalitis, Moscow.

Rabotti, G. F., Landon, J. C., Pry, T. W., Beadle, L., Doll, J., Fabrizio, D. P., and Dalton, A. J. (1967). *J. Nat. Cancer Inst.* 38, 821.

Rabson, A. S. (1962). *J. Nat. Cancer Inst.* 29, 765.

Raghavan, R. S., and Khan, G. A. (1968). *Indian Vet. J.* 45, 75.

Rake, G., and Shaffer, M. F. (1940). *J. Immunol.* 38, 177.

Rao, A. R., Sukumar, M. S., Kamalakshi, S., Paramasivam, T. V., Parasuraman, A. R., and Shantha, M. (1968). *Indian J. Med. Res.* 56, 1855.

Rapoza, N. P. (1967). *Amer. J. Epidemiol.* 86, 736.

Rapoza, N. P., and Atchinson, R. W. (1967). *Nature (London)* 215, 1186.

Rapp, F., Tevethia, S., Rawls, W. E., and Melnick, J. L. (1967). *Proc. Soc. Exp. Biol. Med.* 125, 794.

Ratcliffe, H. L. (1942). *Rep. Penrose Res. Lab.* pp. 11–25.

Reemtsma, K. (1970). *Transplant. Proc.* 2, 431.

Reynolds, H. H., ed. (1969). "Primates in Medicine," Vol. 4. Karger, Basel.

Reynolds, S. L., Craig, C. P., Whitford, H. W., Airhart, J., and Staab, E. V. (1968). *Radiat. Res.* 35, 451.

Riggs, J. L., Takemori, N., and Lennette, E. H. (1968). *J. Immunol.* 100, 348.

Riopelle, A. J. (1963). *In* "Conference on Research with Primates" (D. E. Pickering, ed.), pp. 19–26.

Riopelle, A. J. (1964). "Hepatitis Associated with Chimpanzees," Final Tech. Rep. U. S. Army Med. Res. Develop. Command, Dept. of the Army, Washington, D. C.

Rivers, T. M. (1926). *J. Exp. Med.* 43, 275.

Rivers, T. M. (1927). *J. Exp. Med.* 45, 961.

Rivers, T. M., and Berry, G. P. (1931a). *J. Exp. Med.* 54, 105.

Rivers, T. M., and Berry, G. P. (1931b). *J. Exp. Med.* 54, 119.

Rivers, T. M., and Berry, G. P. (1931c). *J. Exp. Med.* 54, 129.

Rivers, T. M., and Berry, G. P. (1931d). *Trans. Ass. Amer. Physicians* 46, 197.

Rivers, T. M., Berry, G. P., and Sprunt, D. H. (1931). *J. Exp. Med.* 54, 91.

Robbins, F. C., and Enders, J. F. (1950). *Amer. J. Med. Sci.* 220, 316.

Rodaniche, E., and Galindo, P. (1957). *Amer. J. Trop. Med. Hyg.* 6, 232.

Rogers, N. G., Basnight, M., Gibbs, C. J., Jr., and Gajdusek, D. C. (1967). *Nature (London)* 216, 446.

Rogova, V. M., and Ralf, N. M. (1967). *In* "Materials of XIIIth Sessions of Institute

on Present Problems of Virology and Specific Prophylaxis of Viral Infection,"
pp. 52–54. Institute of Poliomyelitis and Viral Encephalitis, Moscow.
Rosen, L. (1958a). *Amer. J. Trop. Med. Hyg.* 7, 406.
Rosen, L. (1958b). *Amer. J. Trop. Med. Hyg.* 12, 924.
Rosen, L. (1960). *Amer. J. Hyg.* 71, 242.
Rosen, L. (1962). *Ann. N. Y. Acad. Sci.* 101, 461.
Rouhandeh, H., Engler, R., Fouad, M. T. A., and Sells, L. L. (1967). *Arch. Gesamte Virusforsch.* 20, 363.
Rowe, W. P. (1960). In "Viral Infections of Infancy and Childhood" (H. M. Rose, ed.), pp. 205–214. Harper (Hoeber), New York.
Ruch, T. C. (1959). "Diseases of Laboratory Primates." Saunders, Philadelphia, Pennsylvania.
Ruckle, G. (1957). *Ann. N. Y. Acad. Sci.* 67, 355.
Ruckle, G. (1958a). *Arch. Gesamte Virusforsch.* 8, 139.
Ruckle, G. (1958b). *Arch. Gesamte Virusforsch.* 8, 167.
Ruckle-Enders, G. (1962). *Amer. J. Dis. Child.* 103, 297.
Ruckle-Enders, G. (1965). *Arch. Gesamte Virusforsch.* 16, 182.
Ruddy, S. J., Mosley, J. W., and Held, J. R. (1967). *Amer. J. Epidemiol.* 86, 634.
Russell, P. K., Yuill, Y. M., Nisalak, A., Udomsakdi, S., Gould, D. J., and Winter, P. E. (1968). *Amer. J. Trop. Med. Hyg.* 17, 600.
Rustigian, R., Johnston, P. B., and Reihart, H. (1955). *Proc. Soc. Exp. Biol. Med.* 88, 8.
Sabin, A. B. (1934a). *Brit. J. Exp. Pathol.* 15, 248.
Sabin, A. B. (1934b). *Brit. J. Exp. Pathol.* 15, 321.
Sabin, A. B. (1949). *J. Clin. Invest.* 28, 808.
Sabin, A. B. (1956). *Science* 123, 1151.
Sabin, A. B. (1959a). *Brit. Med. J.* 1, 663.
Sabin, A. B. (1959b). *Science* 130, 1387.
Sabin, A. B., and Hurst, E. W. (1935). *Brit. J. Exp. Pathol.* 16, 133.
Sabin, A. B., and Ward, R. (1942). *J. Bacteriol.* 43, 86.
Sabin, A. B., and Wright, A. M. (1934). *J. Exp. Med.* 59, 115.
Saddington, R. S. (1932). *Proc. Soc. Exp. Biol. Med.* 29, 838.
Saenz, A. C. (1969). In "Primates in Medicine" (W. I. B. Beveridge, ed.), Vol. 3, pp. 129–134. Karger, Basel.
Salk, J. E., Krech, U., Youngner, J. S., Bennett, B. L., Lewis, L. J., and Bazeley, P. L. (1954). *Amer. J. Pub. Health* 44, 563.
Sanchez, F. G. (1939). In "Conference in Tropical Medicine." Cienc. Medis. Universidad de San Carlos, Guatamala.
Saslaw, S., and Carlisle, H. N. (1965). *Proc. Soc. Exp. Biol. Med.* 119, 838.
Saslaw, S., and Carlisle, H. N. (1969). *Ann. N. Y. Acad. Sci.* 162, 568.
Saslaw, S., Wilson, H. E., Doan, L. A., Woolpert, O. C., and Schwab, J. L. (1946). *J. Exp. Med.* 84, 113.
Sather, G. E., and Hammon, W. McD. (1970). *Proc. Soc. Exp. Biol. Med.* 135, 573.
Sauer, R. M. (1960). *Ann. N. Y. Acad. Sci.* 85, 735–992.
Sauer, R. M., and Fegley, H. C. (1960). *Ann. N. Y. Acad. Sci.* 85, 866.
Sauer, R. M., Prier, J. E., Buchanan, R. S., Creamer, A. A., and Fegley, H. C. (1960). *Amer. J. Vet. Res.* 21, 377.
Savinov, A. P., Zubri, G. A., Goldfarb, M. M., Lavrova, I. K., and Voroshilova, M. K. (1967). In "Materials on Current Problems of Commission of Academy

of Medical Sciences, USSR," pp. 101–103. Institute of Poliomyelitis and Viral Encephalitis. Moscow.

Scherer, W. F., and Muira, T. (1965). *Proc. Soc. Exp. Biol. Med.* **118**, 1167.

Schiller, A. L., Hunt, R., and Di Giacomo, R. F. (1969). *J. Pathol.* **99**, 327.

Schmidt, E. C. H. (1948). *Amer. J. Pathol.* **24**, 97.

Schmidt, N. J., Dennis, J., Lennette, E. H., Ho, H. H., and Shinomoto, T. T. (1965). *J. Immunol.* **95**, 54.

Schmidt, N. J., Lennette, E. H., and Dennis, J. (1968). *J. Immunol.* **100**, 99.

Schultz, E. W., and Habel, K. (1959). *J. Immunol.* **82**, 274.

Schwartz, A., and Allen, W. P. (1970). *Infec. Immunity* **2**, 762.

Scriba, M., and Oehlert, W. (1969). *Z. Med. Mikrobiol. Immunol.* **155**, 16.

Semenov, B. F., and Lapin, B. A. (1967). *Probl. Virol.* **6**, 755.

Sergiev, P. G., Ryazantseva, N. E., and Smirnova, E. V. (1956). *Zh. Mikrobiol., Epidemiol. Immunobiol.* **11**, 88.

Sergiev, P. G., Ryazantseva, N. E., and Smirnova, E. V. (1959). *Probl. Virol.* **5**, 558.

Sergiev, P. G., Ryazantseva, N. E., and Shroit, I. G. (1960). *Acta Virol. (Prague), Engl. Ed.* **4**, 265.

Sergiev, P. G., Shroit, I. G., Chelysheva, K. M., Smirnova, E. V., Kuskova, M. I., Shevtsova, Z. V., Levinshtam, M. A., Chernomorkik, A. E., Kozlyuk, A. S., Stromova, G. I., Manjko, T. G., Yefinov, E. E., Shamryeva, S. A., and Voskresenskaya, G. S. (1966). *Acta Virol. (Prague), Engl. Ed.* **10**, 430.

Sever, J. L., and Zeman, W. (1968). *Neurology* **18**, 95.

Sever, J. L., Schiff, G. M., and Traub, R. G. (1962). *J. Amer. Med. Ass.* **182**, 663.

Sever, J. L., Meier, G. W., Windle, W. F., Schiff, G. M., Monif, G. R., and Fabiyi, A. (1966). *J. Infec. Dis.* **116**, 21.

Shaffer, M. F., Rake, G., Stokes, J., and O'Neil, G. C. (1941). *J. Immunol.* **41**, 241.

Shah, K. V. (1965). *Acta Virol. (Prague), Engl. Ed.* **9**, 71.

Shah, K. V. (1966). *Proc. Soc. Exp. Biol. Med.* **121**, 303.

Shah, K. V., and Hess, D. M. (1968). *Proc. Soc. Exp. Biol. Med.* **128**, 480.

Shah, K. V., and Morrison, J. A. (1969). *Amer. J. Epidemiol.* **89**, 308.

Shah, K. V., and Southwick, C. H. (1965). *Indian J. Med. Res.* **53**, 488.

Shah, K. V., Willard, S., Myers, R. E., Hess, D. M., and Di Giacomo, R. (1969). *Proc. Soc. Exp. Biol. Med.* **130**, 196.

Sheldon, W. G., and Ross, M. A. (1966). *U. S. Army Med. Res. Lab., Rep.* **670**, 1.

Shevtsova, Z. V. (1967). *Vop. Virusol.* **12**, 47.

Shevtsova, Z. V., Kuksova, M. I., Agrba, V. Z., and Krylova, R. I. (1968). *Questions Physiol. Exp. Pathol.* pp. 307–310.

Shishido, A. (1966). *Jap. J. Med. Sci. Biol.* **19**, 221.

Shope, R. E., Causey, C. E., and Causey, O. R. (1961). *Amer. J. Trop. Med. Hyg.* **10**, 264.

Shortt, H. E., Pandit, C. G., Anderson, W. M. E., and Rao, R. S. (1940). *Indian J. Med. Res.* **27**, 847.

Shulman, R. N., and Barker, L. F. (1969). *Science* **165**, 304.

Sibal, L. R., Fink, M. A., McCune, C. L., and Coroles, E. A. (1967). *J. Immunol.* **98**, 368.

Siegert, R., Shu, H. L., Slenczka, W., Peters, D., and Muller, G. (1967). *Deut. Med. Wochenschr.* **92**, 2341.

Siegert, R., Shu, H. L., and Slenczka, W. (1968a). *Deut. Med. Wochenschr.* **93**, 604.

Siegert, R., Shu, H. L., and Slenczka, W. (1968b). *Deut. Med. Wochenschr.* **93**, 616.
Sigurdardottir, B., Givan, K. F., Rozee, K. R., and Rhodes, A. J. (1963). *Can. Med. Ass. J.* **88**, 128.
Simkova, A., and Bardos, V. (1969). *In* "Arboviruses of the California Complex and Bunyamwera group." (V. Bardos, ed.), pp. 269–274. Publ. House Slav. Acad. Sci., Bratislavia.
Simkova, A., and Danes, L. (1968). *Acta Virol. (Prague), Engl. Ed.* **12**, 474.
Simkova, A., and Danielova, V. (1969). *Folia Parasitol. (Praha)* **16**, 255.
Simmons, J. S. (1940). *In* "Virus and Rickettsial Diseases" (J. E. Gordon *et al.,* eds.), pp. 349–364. Harvard Univ. Press, Cambridge, Massachusetts.
Simmons, J. S., St. John, J. H., and Reynolds, F. H. K. (1931). *Manila Bur. Sci., Monogr.* **29**, 19–77, 112–146, and 189–247.
Simon, J. (1965). *Arch. Gesamte Virusforsch.* **15**, 220.
Simon, J., Vonka, V., and Janda, Z. (1965). *Arch. Gesamte Virusforsch.* **15**, 681.
Simon, J., Slonim, D., and Zavadova, H. (1967a). *Acta Neuropathol.* **8**, 24.
Simon, J., Slonim, D., and Zavadova, H. (1967b). *Acta Neuropathol.* **8**, 35.
Simpson, D. I. H. (1968). *Brit. J. Exp. Pathol.* **49**, 458.
Simpson, D. I. H. (1969a). *In* "Premates in Medicine" (W. I. B. Beveridge, ed.), Vol. 3, pp. 135–137. Karger, Basel.
Simpson, D. I. H. (1969b). *Brit. J. Exp. Pathol.* **50**, 389.
Simpson, D. I. H. (1969c). *Trans. Roy. Soc. Trop. Med. Hyg.* **63**, 303.
Simpson, D. I. H. (1969d). *Lab. Anim. Handb.* **4**, 149.
Simpson, D. I. H. Personal communication.
Simpson, D. I. H., Bowen, E. T. W., and Bright, W. F. (1968a). *Lab. Anim. Handb.* **2**, 75.
Simpson, D. I. H., Zlotnik, I., and Rutter, D. A. (1968b). *Brit. J. Exp. Pathol.* **49**, 458.
Slenczka, W. (1969). *Lab. Anim. Handb.* **4**, 143.
Slenczka, W., and Wolff, G. (1971). *In* "Marburg Virus Disease" (G. A. Martini and R. Siegert, eds.), pp. 105–111. Springer-Verlag, Berlin and New York.
Slenczka, W., Shu, H-L, Piepenburg, G., and Siegert, R. (1968). *Deut. Med. Wochenshr.* **93**, 612.
Slifkin, M., Merkow, L. P., Pardo, M., and Rapoza, N. P. (1969). *J. Nat. Cancer Inst.* **43**, 423.
Slonim, D., Simon, J., and Zavadova, H. (1966). *Acta Virol. (Prague), Engl. Ed.* **10**, 413.
Smetana, H. F. (1965). *Lab. Invest.* **14**, 1366.
Smetana, H. F. (1969). *Amer. J. Pathol.* **55**, 65A.
Smetana, H. F., and Felsenfeld, A. D. (1969a). *Virchows Arch., A* **348**, 309.
Smetana, H. F., and Felsenfeld, A. D. (1969b). *Gastroenterology* **56**, 1222.
Smirnova, E. V., and Riazantseva, N. E. (1960). *In* "Theoretical and Practical Problems of Medicine and Biology in Experiments on Monkeys" (I. A. Utkin, ed.), pp. 243–248. Pergamon, Oxford.
Smit, G. L., and Wilterdink, J. B. (1966). *Arch. Gesamte Virusforsch.* **18**, 261.
Smith, A. A., and McNulty, W. P. (1969). *Lab. Anim. Care* **19**, 479.
Smith, C. E. G., Simpson, D. I. H., Bowen, E. T. W., and Zlotnick, I. (1967). *Lancet* **2**, 1119.
Smith, K. O., Thiel, J. F., Newman, J. T., Harvey, E., Trousdale, M. D., Gehle, W. D., and Clark, G. (1969). *J. Nat. Cancer Inst.* **42**, 489.

Smith, P. C., Yuill, T. M., and Buchanan, R. D. (1958). *Annu. Progr. Rep. SEATO Med. Res. Lab. SEATO Clin. Res. Cent., Bangkok, Thailand* pp. 258–261.

Smith, P. C., Yuill, T. M., Buchanan, R. D., Stanton, J. S., and Chaicumpa, V. (1969). *J. Infec. Dis.* **120**, 292.

Smith, R. D., and Deinhardt, F. (1968a). *Amer. J. Pathol.* **52**, 58A.

Smith, R. D., and Deinhardt, F. (1968b). *J. Cell Biol.* **37**, 819.

Smithburn, K. C. (1952). *J. Immunol.* **68**, 441.

Smithburn, K. C. (1954). *J. Immunol.* **72**, 376.

Smithburn, K. C., and Haddow, A. J. (1944). *J. Immunol.* **49**, 141.

Smithburn, K. C., and Haddow, A. J. (1949). *Amer. J. Trop. Med. Hyg.* **29**, 389.

Smithburn, K. C., Hughes, T. P., Burke, A. W., and Paul, J. H. (1940). *Amer. J. Trop. Med. Hyg.* **20**, 471.

Smithburn, K. C., Haddow, A. J., and Mahoffy, A. F. (1946). *Amer. J. Trop. Med. Hyg.* **26**, 189.

Smorodintsev, A. A. (1958). *Progr. Med. Virol.* **1**, 210.

Soike, K. F., Coulston, F., Day, P., Deibel, R., and Plager, H. (1967). *Exp. Mol. Pathol.* **7**, 259.

Soike, K. F., Coulston, F., and Douglas, J. D. (1969a). *Exp. Pathol.* **11**, 323.

Soike, K. F., Douglas, J. D., Plager, H., and Coulston, F. (1969b). *Exp. Mol. Pathol.* **11**, 333.

Soike, K. F., Krushak, D. H., Douglas, J. D., and Coulston, F. (1971). *Exp. Mol. Pathol.* **14**, 373.

Soper, F. L., and Smith, H. H. (1938). *Amer. J. Trop. Med. Hyg.* **18**, 111.

Soper, F. L., Rickard, E. R., and Crawford, P. J. (1934). *Amer. J. Hyg.* **19**, 549.

Soper, W. T. (1959). Technical Study 17. U. S. Army Biol. Warfare Lab., Ft. Detrick, Maryland.

Southwick, C. H., Siddiqi, M. R., and Siddiqi, M. F. (1970). *Science* **170**, 1051.

Spiegel, A. (1969). *Ann. N. Y. Acad. Sci.* **162**, 391.

Sproul, E. E., Metzgar, R. S., and Grace, J. T., Jr. (1962). *Proc. Amer. Ass. Cancer Res.* **3**, 363.

Sproul, E. E., Metzgar, R. S., and Grace, J. T., Jr. (1963). *Cancer Res.* **23**, 671.

Stanley, N. F., Dorman, D. C., and Ponsford, J. (1954). *Aust. J. Exp. Biol. Med. Sci.* **32**, 543.

Starck, D., Schneider, R., and Kuhn, H. J., eds. (1967). "Progress in Primatology. 1st Congress of the International Primatological Society." Fischer, Stuttgart.

Stevens, D. A., Pry, T., Blackham, E., and Manaker, R. (1970). *Proc. Soc. Exp. Biol. Med.* **133**, 678.

Stewart, F. W., and Rhoades, C. P. (1929). *J. Exp. Med.* **49**, 959.

Stiles, G. E. (1968). *Proc. Soc. Exp. Biol. Med.* **127**, 225.

Stiles, G. E., Bittle, J. L., and Cabasso, V. J. (1964). *Nature (London)* **201**, 1350.

Stiles, G. E., Vasington, P. J., Bittle, J. L., and Cabasso, V. J. (1966). *Bacteriol. Proc.* **66**, 137.

Stille, W., Bohle, E., Helm, E., Van Rey, W., and Siede, W. (1968). *Deut. Med. Wochenschr.* **93**, 572.

Stojkovic, Lj., Bordjoski, M., Gligic, A., and Stefanovic, Z. (1971). *In* "Marburg Virus Disease" (G. A. Martini and R. Siegert, eds.), pp. 24–33. Springer-Verlag, Berlin and New York.

Stokes, A., Bauer, J. H., and Hudson, N. P. (1928). *Amer. J. Trop. Med. Hyg.* **8**, 103.

Strickland-Cholmley, M., and Malherbe, H. (1970). *Lancet* **1**, 476.

Sumner-Smith, G. (1966). *Primate Newslett.* **5**, 1.

Sun, S. C., Sohal, R. S., Burch, G. E., Chu, K. C., and Colcolough, H. L. (1967). *Brit. J. Exp. Pathol.* **48**, 655.

Swan, C., and Mawson, J. (1943). *Med. J. Aust.* **1**, 411.

Sweet, B. H., and Hilleman, M. R. (1960). *Proc. Soc. Exp. Biol. Med.* **105**, 420.

Sweet, B. H., Hatgi, J., and Polise, F. (1969). *Bacteriol. Proc.* **69**, 160.

Talash, M., Chumakov, M. P., and Zavodova, T. I. (1969). *Vop. Virusol.* **3**, 301.

Taniguchi, T., Kamahora, S., Sato, S., and Hagiwara, K. (1954). *Med. J. Osaka Univ.* **5**, 367.

Tauraso, N. M., Shelokov, A., Allen, A. M., and Palmer, A. E. (1968a). *Nature (London)* **218**, 876.

Tauraso, N. M., Shelokov, A., Palmer, A. E., and Allen, A. M. (1968b). *Amer. J. Trop. Med. Hyg.* **17**, 422.

Tauraso, N. M., Myers, M. G., McCarthy, K., and Tribe, G. W. (1970). *In* "Infection and Immunosuppression in Subhuman Primates" (H. Balner and W. I. B. Beveridge, eds.), pp. 101–109. Munksgaard, Copenhagen.

Tauraso, N. M., Kalter, S. S., Ratner, J. J., and Heberling, R. L. (1971). *In* "Medical Primatology 1970" (E. I. Goldsmith and J. Moor-Jankowski, eds.), pp. 660–670. Karger, Basel (in press).

Taylor, R. M. (1967). *U. S., Pub. Health Serv., Publ.* **1760**.

Taylor, R. M., Haseeb, M. A., and Work, T. H. (1955). *Bull. WHO* **12**, 711.

Teissier, P., Gastinel, P., and Reilly, J. (1925). *C. R. Soc. Biol.* **92**, 1015.

Teissier, P., Reilly, J., Rivalier, E., and Stefanesco, V. (1931). *C. R. Soc. Biol.* **108**, 1039.

Theiler, M., and Casals, J. (1958). *Amer. J. Trop. Med. Hyg.* **7**, 584.

Theiler, M., and Gard, S. (1940a). *J. Exp. Med.* **72**, 49.

Theiler, M., and Gard, S. (1940b). *J. Exp. Med.* **72**, 79.

Thind, I. S., and Price, W. H. (1966a). *Amer. J. Epidemiol.* **84**, 193.

Thind, I. S., and Price, W. H. (1966b). *Amer. J. Epidemiol.* **84**, 214.

Thomas, H. W. (1907). *Brit. Med. J.* **1**, 138.

Thomsen, O. (1912). *Z. Immunitaetsforsch. Exp. Ther.* **14**, 198.

Tobe, T. (1967). *Jap. J. Clin. Med.* **25**, 1263.

Tobe, T., Horikoshi, Y., Hamada, C., and Hamashima, Y. (1968). *Gastroenterol. Jap.* **3**, 423.

Torres, C. M. (1928). *C. R. Soc. Bull.* **99**, 1344.

Trapido, H., and Galindo, P. (1956). *Exp. Parasitol.* **5**, 285.

Trask, J. D., and Paul, J. R. (1941). *J. Exp. Med.* **73**, 453.

Trask, J. D., and Paul, J. R. (1942). *Ann. Intern. Med.* **17**, 975.

Tribe, G. W. (1966). *Brit. J. Exp. Pathol.* **47**, 472.

Trum, B. F., and Routledge, J. K. (1967). *J. Amer. Vet. Med. Ass.* **151**, 1886.

Tyrrell, D. A. J., Buckland, F. E., Lancaster, M. C., and Valentine, R. C. (1960). *Brit. J. Exp. Pathol.* **41**, 610.

Unterharnscheidt, F. (1966). *Arb. Ehrlich Inst.* **62**, 1.

Unterharnscheidt, F., and Bonin, O. (1965). *Arch. Psychiat. Nervenkr.* **206**, 454.

Ushijima, R. N. (1966). *Proc. Soc. Exp. Biol. Med.* **122**, 673.

Vagtborg, H., ed. (1965). "The Baboon in Medical Research," Vol. I. Univ. of Texas Press, Austin.

Vagtborg, H., ed. (1967). "The Baboon in Medical Research," Vol. II. Univ. of Texas Press, Austin.

Vagtborg, H., ed. (1968). "Use of Nonhuman Primates in Drug Evaluation." Univ. of Texas Press, Austin.

Valerio, D. A., Courtney, K. D., Miller, R. L., and Pallotta, A. J. (1968). *Lab. Anim. Care* **18**, 589.

Valerio, D. A., Miller, R. L., Innes, J. R. M., Courtney, K. D., Pallotta, A. J., and Guttmacher, R. M. (1969). *"Macaca mulatta* (Management of a Laboratory Breeding Colony)." Academic Press, New York.

Valerio, M., Landon, J. C., and Innes, J. R. M. (1968). *J. Nat. Cancer Inst.* **40**, 751.

van Bekkum, D. W., van der Waay, D., and van Putten, L. M. (1969). *Ann. N. Y. Acad. Sci.* **162**, 363.

van Hoosier, G. L., Jr., and Melnick, J. L. (1961). *Tex. Rep. Biol. Med.* **19**, 376.

van Rooyen, C. E., and Rhodes, A. J. (1948). *In* "Virus Diseases of Man," 2nd ed., pp. 1160–1163. Thomas Nelson & Sons, New York.

Vargas-Mendez, O., and Elton, N. W. (1953). *Amer. J. Trop. Med. Hyg.* **2**, 850.

Verlinde, J. D. (1968). *Trop. Geogr. Med.* **20**, 385.

Verlinde, J. D., Kret, A., and Wyler, R. (1955). *Arch. Gesamte Virusforsch.* **6**, 175.

Vickers, J. H. (1962). *Amer. J. Dis. Child.* **103**, 342.

Vickers, J. H. (1969). *Ann. N. Y. Acad. Sci.* **162**, 659.

Vienchange, J. (1935). *C. R. Soc. Biol.* **118**, 512.

Vignec, A. J., Paul, J. R., and Trask, J. D. (1939). *Proc. Soc. Exp. Biol. Med.* **41**, 246.

Vogel, F. S., and Pinkerton, H. (1955). *Arch. Pathol.* **60**, 281.

von Magnus, P., Anderson, E. K., Petersen, K. B., and Birch-Anderson, A. (1959). *Acta Pathol. Microbiol. Scand.* **46**, 156.

Voss, W. R., Benyesh-Melnick, M., Singer, D. B., and Nora, A. H. (1969). *Lab. Primate Newslett.* **8**, 10.

Webb, H. E. (1969). *Lab. Anim. Handb.* **4**, 131.

Webb, H. E., and Burston, J. (1966). *Trans. Roy. Soc. Trop. Med. Hyg.* **60**, 325.

Webb, H. E., and Chatterjea, J. B. (1962). *Brit. J. Haematol.* **8**, 401.

Webb, H. E., and Laxmana, R. R. (1961). *Trans. Roy. Soc. Trop. Med. Hyg.* **55**, 284.

Wedum, A. G., and Kruse, R. H. (1969). Misc. Publ. 30. Dept. of the Army, Fort Detrick, Frederick, Maryland.

Wenner, H. A., and Kamitsuka, P. (1956). *Virology* **2**, 83.

Wenner, H. A., and Kamitsuka, P. (1957). *Virology* **3**, 429.

Wenner, H. A., Kamitsuka, P., Lenahan, P. M., and Archetti, I. (1960). *Arch. Gesamte Virusforsch.* **9**, 537.

Wenner, H. A., Kamitsuka, P., Macasaet, F., and Kidd, P. (1967). *Antimicrobiol. Ag. Chemother.* **7**, 40.

Wenner, H. A., Macasaet, F. D., and Kamitsuka, P. S. (1968). *Amer. J. Epidemiol.* **87**, 551.

Wenner, H. A., Cho, C. T., Bolano, C. R., and Kamitsuka, P. S. (1969a). *Arch. Gesamte Virusforsch.* **27**, 166.

Wenner, H. A., Bolano, C. R., and Cho, C. T. (1969b). *Arch. Gesamte Virusforsch.* **27**, 179.

Wenner, H. A., Bolano, C., Cho, C. T., and Kamitsuka, P. S. (1969c). *J. Infec. Dis.* **120**, 318.

Whitehead, R. H., Chaicumpa, V., Olson, L. C., and Russell, P. K. (1970). *Amer. J. Trop. Med. Hyg.* **19**, 94.

Whitelock, O. V. St. (1958). *Ann. N. Y. Acad. Sci.* **70**, 277.

Whitelock, O. V. St. (1960). *Ann. N. Y. Acad. Sci.* **85**, 735.

Whitney, R. A., Johnson, D. J., and Cole, W. C. (1967). "The Subhuman Primate: A Guide to the Veterinarian" Med. Res. Lab., Res. Lab., Edgewood Arsenal, Maryland.

Williams, M. C., Woodall, J. P., and Simpson, D. I. H. (1965). *Trans. Roy. Soc. Trop. Med. Hyg.* **59**, 444.

Wilson, R. P. (1928). *3rd. Annu. Rep. Giza Mem. Ophthal. Lab., Cairo* p. 78.

Wising, P. J. (1939). *Acta Med. Scand.* **98**, 328.

Wising, P. J. (1942). *Acta Med. Scand., Suppl.* **133**, 1.

Wolfe, L. G., Griesemer, R. A., and Farrell, R. L. (1968a). *J. Nat. Cancer Inst.* **41**, 1175.

Wolfe, L. G., Adler, A., and Griesemer, R. A. (1968b). *J. Nat. Cancer Inst.* **41**, 1197.

Wood, O., Tauraso, N., and Liebhaber, H. (1970). *J. Gen. Virol.* **7**, 129.

Wood, W., and Shimada, F. T. (1954). *Can. J. Pub. Health* **45**, 509.

Woodall, J. P. (1969). *Ann. Trop. Med. Parasitol.* **62**, 522.

Woodall, J. P., Dykes, J. R. W., and Williams, M. C. (1969). *Ann. Trop. Med. Parasitol.* **62**, 528.

Woolpert, O. C., Schwab, J. L., Saslaw, S., Merino, C., and Doan, C. A. (1941). *Proc. Soc. Exp. Biol. Med.* **48**, 558.

Work, T. H., and Trapido, H. (1957). *Indian J. Med. Sci.* **11**, 340.

World Health Organization. (1971). *World Health Organ., Tech. Rep. Ser.* **470**.

Yakovleva, L. A. (1969). *Bibl. Haematol.* **36**, 1.

Yakovleva, L. A., and Lapin, B. A. (1966). *Vestn. Akad. Med. Nauk SSSR* **21**, 65.

Yamanouchi, K. (1966). *Jap. J. Med. Sci. Biol.* **19**, 226.

Yamanouchi, K., Fukuda, A., Kobune, F., Uchida, N., and Tsuruhara, T. (1967). *Jap. J. Med. Sci. Biol.* **20**, 443.

Yamanouchi, K., Fukuda, A., Kobune, F., Hikita, M., and Shishido, A. (1969). *Jap. J. Med. Sci. Biol.* **22**, 117.

Yamanouchi, K., Egashira, Y., Uchica, N., Kodama, H., Kobune, F., Hayami, M., Fukuda, A., and Shishido, A. (1970). *Jap. J. Med. Sci. Biol.* **23**, 131.

Yang, C. S., Kuo, C., and Chen, C. (1967). *J. Formosan, Med. Ass.* **66**, 143.

Yoshicka, I., and Horstmann, D. M. (1960). *N. Engl. J. Med.* **262**, 224.

Yoshida, E. H., Yamamoto, H., and Shimojo, H. (1965). *Jap. J. Med. Sci. Biol.* **18**, 151.

Zeitlyonok, N. A., Chumakova, M. Y., Ralph, N. M., and Goen, L. S. (1966). *Acta Virol. (Prague), Engl. Ed.* **10**, 537.

Zilber, L. A., Lapin, B. A., and Adgighytov, F. I. (1963). *Nature (London)* **205**, 1123.

Zilber, L. A., Lapin, B. A., Adgighytov, F. I., Shevljaghyn, V. J., and Jakovleva, L. A. (1966). *Int. J. Cancer* **1**, 395.

Zinsser, S. (1929). *J. Exp. Med.* **49**, 661.

Zlotnik, I. (1969). *Trans. Roy. Soc. Trop. Med. Hyg.* **63**, 310.

Zlotnik, I., and Simpson, D. I. H. (1968). *Lancet* **1**, 205.

Zlotnik, I., and Simpson, D. I. H. (1969). *Brit. J. Exp. Pathol.* **50**, 393.

Zlotnik, I., Simpson, D. I. H., Bright, W. F., Bowen, E. T. W., and Balter-Hatton, D. (1968a). *Brit. J. Exp. Pathol.* **49**, 311.

Zlotnik, I., Simpson, D. I. H., and Howard, D. M. R. (1968b). *Lancet* **2**, 26.

Zlotnik, I., Smith, C. E. G., Grant, D. P., and Peacock, S. (1970). *Brit. J. Exp. Pathol.* **51**, 434.

CHAPTER 6

Models for Investigation in Parasitology*

ROBERT E. KUNTZ

I. Introduction

At an early time several species of nonhuman primates arrived in the port cities of Europe and the Mediterranean, accompanying the crews of ships which had made exploratory or trading expeditions to different parts of Africa. Initially, these animals were items of curiosity. In the minds of some people they were representatives of the supernatural. They were accorded such a reception since they previously were unknown to the

* Host names are those given by authors from which information has been taken. There has been no attempt to synonymize names.

This work has been supported in part by grant nos. RR-00278, RR-00451, RR-05519, and AI-08207 from the National Institutes of Health.

167

Europeans, and obviously possessed certain basic biologic features which resembled man.

With the development of schools of learning, and especially the introduction of medicine, these vertebrates whetted the imagination of philosophical minds and soon found their way into the laboratories of practicing anatomists. Public scorn for the use of human cadavers made it difficult for early scholars to obtain first-hand information in the principles of human anatomy. Galen (120–200 AD), a man of considerable scientific repute, and his students saw in the nonhuman primates a reasonable substitute for their studies. Thus it was in the first century that the Barbary ape, or macaque, became one of the earliest animals to be recognized as a biomedical model to satisfy the requisites of the anatomists.

The macroscopic parasites (lice, ticks, large cestodes and nematodes, and various encysted forms) of man were noted in the writings of Aristotle, Celsus, Pliny, Aristophanes, and other scientists, and there were incidental references to common parasites in some of the early classics. Galen recognized several species of helminths in man, and knew in which organs they were likely to occur. The Hippocratic collection of treatises described a variety of cestode cysts, and it may be assumed that Galen called to the attention of his students the presence of worms in the Barbary ape, *Macaca sylvanus*. Many centuries passed, however, before the scientific professions considered or accepted any of the nonhuman primates as possible models in which to study conditions and biologic phenomena relating to the health of man. The different disciplines of microbiology developed and promoted the use of nonhuman primates as reasonable models to serve man on a large scale long before these vertebrates gained wide recognition in parasitologic laboratories. It must be realized, however, that parasitologists, working in areas where some of the more common nonhuman primates were indigenous, did employ these animals as incidental hosts in life-cycle and experimental studies on a variety of parasites. The use of nonhuman primates in parasitology, however, was limited.

The biologic requisites for a biomedical model are basically similar, regardless of the purpose for which an animal is intended. The use of a nonhuman primate for a given discipline, or even for a specific purpose, frequently comes at the end of a series of trials in which a number of other animals representing different phylogenetic groups have been considered. Some of these have failed and others have not satisfactorily fulfilled investigational requirements. Admittedly, with the increased availability of many species of nonhuman primates in the past few years, some of the intermediate steps in the search for a model may have been bypassed. For years it seemed reasonable to accept different rodents, cats, dogs, and even birds, as animals which would yield pertinent information to the experimentalist studying the disease processes of man. Development of greater respect and sophistication for biomedical research led to the realization that ordinary mice, rats, and hamsters would not suffice. As a consequence, special strains were developed to allow more critical and acceptable evaluation of specific diseases, various anomalies, and pathologic conditions. With expansion in the overall field of biomedical research, investigators were confronted with a demand for a more precise understanding of a greater diversity of diseases and how these infections in nonhuman hosts related to diseases in man. Also, there were growing demands for better prevention and treatment of many of the poorly understood illnesses of man.

The time came for the parasitologist, as well as for other investigators representing different fields of specialization, when it was realized that a number of the lower vertebrates did not entirely satisfy the requisites of experimental biology and medicine. It was only natural that the next step was a serious consideration for the use of nonhuman primates, animals which, due to their anatomic and physiologic similarities with man, should be more amenable to the growing list of infectious agents. Even though it was assumed that the rhesus monkey and certain other available nonhuman primates would serve as a panacea to several specific fields of biomedical research, limitations were recognized, and as a consequence many lower vertebrates still maintain a strong position in the laboratory. As a matter of fact, Homburger and Bajusz (1970) have recently called attention to the fact that the Syrian hamster still serves a very important function for studies of serious disease entities, including inherited susceptibilities to teratogenic agents (thalidomide), poisoning by cyclamates, and carcinogenic activity of polycylic hydrocarbons. The hamster persists as a prominent host in experimental parasitology.

Selection of nonhuman primate models to serve the multidisciplines of biomedicine is an expensive task involving much effort, time, and funding. And, obviously, careful consideration should be given to the conservation aspects of the vertebrates selected as candidates (Hill, 1969). Models for studies in parasitology depend upon the species of protozoa or helminths in concern and on the goals to be attained. Even though the selection of models is dependent upon numerous factors, ranging from the biologic to the mundane aspects of scientific investigations, the works of many investigators adequately describe the situation. In a recent symposium representing different biologic and medical disciplines, Benirschke (1969), Frenkel (1969a), Jones (1969), and others have given a careful consideration to the definition, development, evaluation, and acceptance of animal models for a multiplicity of uses. Although the symposium included fields other than microbiology and parasitology, attention was given to animal models for a study of different disease processes, the criteria being applicable to many biomedical problems, including parasitology (Frenkel, 1969b).

Human parasitology, owing to the rather benign nature of many parasite infections, has not encouraged, supported, or demanded the extensive use of nonhuman primates required for investigations in virology, bacteriology, or for the preparation of vaccines. Limited numbers of nonhuman primates have been employed with the more serious parasite diseases (malaria, schistosomiasis, sparganosis, etc.) of man, but many voids exist in our knowledge of the detailed life-cycle, and especially the potential pathology associated with some of the more common parasites (*Ascaris, Trichuris, Enterobius,* etc.) which, under ordinary circumstances, seemingly render only minor or moderate medical inconveniences to man. Parasitism may be responsible for great morbidity and reduction of working potentials of sizeable human populations, but parasite infections in underdeveloped or even well-advanced countries seldom elicit more than secondary concern since deaths directly attributable to parasites may not be statistically significant in the annual mortality records.

It is difficult to assess the current use of nonhuman primates for biomedical research in parasitology since records are widely scattered in the literature, and parasites, with few exceptions, are not responsible for epidemic-type situations and do not arouse great public concern which demands immediate solutions. The value and potential use of these vertebrates and the role they may play in a better understanding of the parasite diseases of man are based upon: (1) records of natural infection; (2) records of parasites (often human) acquired by these animals when introduced into captivity; (3) records of zoonotic situations; and (4) data obtained by investigators with an academic regard for more basic knowledge on the biology of parasites. The field of nonhuman primate parasitology and the application of these vertebrates for experimental purposes to follow disease processes have not been sufficiently explored to determine their full usefulness. Nonhuman primates demonstrate a great diversity in their susceptibility to infection by the parasites of man, and it is not entirely safe to infer that information gained from infections in monkeys, baboons, or apes may be extrapolated to man.

A brief review of the nature of parasitism in the nonhuman primates and a presentation of some of the uses to which these animals have been put in the laboratory will indicate, to a degree, the potential which this group of vertebrates offers. Successful employment of the nonhuman primates for biomedical research depends, in part, upon an appreciation for the parasitism to which these animals may be subjected. These hosts, as a rule, harbor protozoa and/or helminths which, although seemingly benign in nature, probably influence the physiologic state, nutritional processes, and other basic biologic phenomena of monkeys or apes employed in varied experimental endeavors. Little is known of the effects of the so-called commensals and parasites on the basic physiology or of the synergistic relationship of parasites in the presence of bacteria, fungi, viruses, and newly introduced parasite fauna.

II. Protozoa

A. INTESTINAL AND VAGINAL PROTOZOA

Blood protozoa in most of the nonhuman primates are somewhat limited, but these mammals host a long list of intestinal protozoa and commensals. Parasite checklists, compiled from the literature, have indicated the potential range of species which may be expected for the African baboon (Myers and Kuntz, 1965), the Taiwan macaque (Kuntz and Myers, 1969b) and the chimpanzee (Myers and Kuntz, 1972) examined shortly after capture or after varying periods in captivity.

Trends in the study of pathogenic protozoa have fluctuated greatly since Flexner's work (Ruch, 1959) on bacillary dysentery in the Philippines in 1900 and realization that some of the amoebas in macaques were similar to those associated with amebiasis in man. Several decades passed after Flexner's description of *Entamoeba histolytica*-like organisms in *Macaca philippinensis* before other investigators made attempts to explore the biologic and pathogenic characteristics of these amoebas (Dobell, 1931; Hegner, 1929, 1934; Hegner and Chu, 1930; Kessel, 1924, 1928; M. J. Miller and Bray, 1966; and others).

Repeated observation on a number of protozoa in different nonhuman primates has led to the consensus that these organisms display little host specificity, and that the same organisms probably occur naturally in wild primates as well as in man. Actually, there is a paucity of information on the intestinal protozoa and commensals, except for a few of the more commonly recognized hosts which are used in routine biomedical research. This has recently led Dunn (1966), in an attempt to correlate primate infections, to state that the intestinal protozoa are well known only for the chimpanzee. Fairly extensive studies, however, have been made for the African baboons, *Papio cynocephalus* (Myers and Kuntz, 1968; Myers *et al.,* 1971).

The majority of intestinal protozoa observed in primates are nonpathogens. Experience and parasitologic records, however, indicate that amebiasis, on occasion, may be a serious infection in some primates (Eichhorn and Gallagher, 1916; M. J. Miller and Bray, 1966; Ruch, 1959). Fremming *et al.* (1955) reported fatal amebiasis due to liver infection and ulcerative colitis, and M. J. Miller and Bray (1966) have described how amoebic infections may persist for long periods without clinical manifestations, then active amebiasis may result in acute pathologic conditions and, ultimately, death. Collected information suggests that there are great differences in the pathogenicity in different geographic or host strains of *E. histolytica*. Furthermore, it seems that pathogenicity may vary from one individual or one situation to another, and may depend in part on circumstances surrounding a given case. As an example, it is recognized that the infection potentials of *E. histolytica* in New World and Old World monkeys vary considerably (M. J. Miller and Bray, 1966; Ratcliffe, 1931) pointed to the danger of mixing New World and Old World primates in captivity. Thus, there is evidence that South American *Ateles* and *Lagothrix* develop serious amoebic infections subsequent to contact with Asiatic monkeys known to be carriers of *E. histolytica*. Rewell (1969) has presented the pathologic situation and various host–parasite relationships of the intestinal protozoa found in different lower primates.

Through the years parasitologists have exposed a considerable number of rodents, kittens, dogs, and other mammals, including monkeys and marmosets, to *E. histolytica*. These experimental infections have demonstrated wide ranges of severity and have led to the assumption by some workers that conditions characteristic to the large intestine of the dog resemble most closely those found in man (Geiman, 1964). It is a well-accepted opinion that *E. histolytica* and several of the common commensals (*Entamoeba coli, Endolimax nana, Iodamoeba bütschlii*) of nonhuman primates are morphologically identical with those found in man. Accumulated evidence indicates that basic parasitologic requisites being appropriate, a number of the lower primates may be satisfactorily employed for clinical and pathologic aspects of amebiasis, and may in some respects yield more in-depth information than is available from lower vertebrates alone. Possibly the use of appropriate nonhuman primates would provide information on the recurring question of why there may be long periods of latency in amebiasis, with only minimal pathologic assault to the host. Then, as already demonstrated in experimental hosts as in man, due perhaps to altered physiologic conditions or to synergistic-type reactions in the presence of other organisms, there may be a "flare-up" leading to critical clinical manifestations.

Trichomonas, a flagellate common to many nonhuman primates and to man, has been reported many times, especially accompanying dysentery and episodes of diarrhea. Even though these protozoa are considered as nonpathogens, they have unwittingly been incriminated since there is a tendency for multiplication and establishment of enormous populations in intestinal tracts temporarily altered by bacterial infections or due to other circumstances. It seems logical to assume that *Trichomonas*, along with experimentally induced alterations in the physiology or microbiologic state of the intestine, could be employed as a model system for investigation of a similar situation which, in man, apparently arises during travel into foreign environments, or may be associated with other circumstances leading to an imbalance of intestinal physiology.

Trichomonas vaginalis, a flagellate which resides in the female vagina, on occasion elicits symptomatology. The infection has been surveyed in many populations, but its biology is only poorly understood. Because of its rather benign nature, this protozoa has been given only secondary attention and its introduction into nonhuman primates has, for the most part, been neglected. Occurrence of this species in monkeys and related primates is not known. We have failed to find it in several hundred cercopithecids from East Africa. Recently, however, Grewal (1966) has been able to establish a light infection in a monkey (rhesus?) in India.

The ciliate, *Balantidium coli,* frequents both man and lower primates, but it is only on rare occasions that it leads to clinical manifestations in either. *Balantidium* occurs in localized human populations (Kuntz and Lawless, 1966) where unusual epidemiologic factors lead to infection. In baboons (*P. cynocephalus*) *Balantidium* infections are lost within a few weeks after transfer of hosts from the field into captivity. Even though the stress of captivity discourages infection, it is likely that different nonhuman primates, especially cercopithecids, could be employed as research models if conditions suitable to parasite propagation were established.

B. Blood and Somatic Protozoa and Related Genera

Different monkeys have been used extensively as experimental hosts of malarial parasites, and the use of Western Hemisphere monkeys has been recently discussed by Young (1970). Through the years there has been vacillation on the study and employment of nonhuman primates for some of the other blood protozoa. Bullock *et al.* (1967), Deane (1964), Dunn *et al.* (1963), Marinkelle (1966), and others, have called attention to *Trypanosoma cruzi* and related trypanosomes in the monkeys of South America. Recent reports by Kuntz *et al.* (1970a), Siebold and Wolf (1970), and Weinman and Wiratmadja (1969) on the occurrence of *T. cruzi*-like protozoa of several species of monkeys and the slow loris lend support to earlier assertions (Malamos, 1935) that *T. cruzi* probably occurs in Asia. Trypanosomes have not been reported commonly in Old World hosts, and we have failed to detect them in moderately large samples of *P. cynocephalus* and *Cercopithecus aethiops* from Kenya. Experimental endeavors with different species of the parasite, however, have proven that trypanosomes can be established in different genera of hosts. Corson (1936) experimentally transmitted *Trypanosoma gambiense* to

Cercopithecus by *Glossina morsitans,* and the same author, in 1938, demonstrated *T. rhodesiense* in the cerebrospinal fluid of a laboratory-infected *Cercopithecus.* Godfrey and Killick-Kendrick (1967) transmitted *T. rhodesiense, T. gambiense,* and *T. brucei* (nonhuman parasite) to the chimpanzee. Even more recently, Baker (1970) has been able to demonstrate nervous system lesions in a chimpanzee infected with *T. brucei.* Limbos and Jadin (1963) employed a monkey to study trypanosomiasis isolated from a European, previously a resident of the Belgian Congo. *Trypanosoma simiae,* which leads to fatal infections in monkeys and porcines, presents an interesting parasitologic situation in which the parasite has been successfully transmitted to rabbits. The trypanosome was introduced into the latter host since Desowitz and Watson (1953) realized that pigs and monkeys were too expensive for basic research.

Cercopithecus aethiops and *Erythrocebus patas* have been poor hosts for *Leishmania donovani,* but the bushbaby, *Galago s. senegaliensis,* has shown promising use for investigations (Sati, 1963) with this parasite which is of great concern in parts of Africa. Previously it has been studied primarily in hamsters and other rodents.

Sarcocystis and *Toxoplasma,* parasites of debatable taxonomic status, occur extensively in wild mammals, but reports of their presence in the primates are not numerous, and the former is found only occasionally in isolated human populations. *Sarcocystis* is not regarded as a serious disease of man, but the fact that it probably is available in several species of nonhuman primates provides a model system for basic research. *Sarcocystis* has been observed with some frequency in African baboons by Strong *et al.* (1965) and by C. S. Kim *et al.* (1968). Mandour (1969) recorded *S. kortei* from the rhesus monkey, *E. patas,* and *Cercopithecus mitis,* and at the same time described *S. nesbitti* as a new species from *Macaca mulatta.* B. Nelson *et al.* (1966) found *Sarcocystis* in white-lipped tamarins (*Tamarinus nigricollis*), the first record for New World monkeys.

Toxoplasma has been recorded for baboons, chimpanzees, and macaques as well as in the white-faced capuchin (*Cebus capucinus*), the marmoset (*Oedipomidas oedipus*), and the squirrel monkey (*Saimiri sciureus*) (Levine, 1961) from South America, and the slow loris (*Nycticebus coucang*) in Malaya (Zaman and Goh, 1968). Toxoplasmosis may be a serious infection for man and nonhuman primates. It has been responsible for epizootics, with death in captive animals (McKissick *et al.,* 1968; Ruch, 1959). Cowen and Wolf (1945) made the statement that monkeys in general appear to be refractive to *Toxoplasma* infection, in spite of the fact that they produced fatal infection with this parasite in a young *M. mulatta.* Later, Rodaniche (1954) demonstrated that marmosets (*Marikina geoffroyi*) and night monkeys (*Aotis zonalis*) were highly susceptible, with fatal infections following inoculation. Benirschke and

Richart (1960) observed spontaneous acute toxoplasmosis in an immature marmoset (*Oedipomidas oedipus*) 5 days after importation from South America, and Benirschke and Low (1970) reported infection in *Lagothrix*.

The host–parasite relationships and basic biology of *Sarcocystis* and *Toxoplasma* are far from clear. It is likely that critical survey type of samplings along with conscientious surveillance would show these parasites to be much more extensive in distribution, and possibly more serious to animal health, than current opinion would indicate. Even though experimental infection has varied from one situation to another, it is reasonably safe to assume that different nonhuman primates would provide ample potential for more critical studies on different aspects of taxoplasmosis, some of which would enlighten knowledge of the parasitism of man.

There is a heterogeneous group of parasites which does not fit conveniently into the conventional taxonomic scheme of parasitologists or protozoologists, yet they may be a potential hazard to the health of man. Ruch (1959) has presented a compilation of host records and general parasitology with reference to *Babesia* and piroplasmosis. Garnham (1966) has further discussed the piroplasms and their relationship to the malarial parasites. Bray and Garnham (1961) tried unsuccessfully to induce infection by allowing *Theileria parva-* (causative agent for East Coast fever, Africa) infected ticks (*Rhipicephalus appendiculatus*) to feed upon splenectomized primates, two chimpanzees and a sooty mangabey (*Cercocebus torquatus atys*). In other work (Garnham and Bray, 1959), however, they demonstrated that *Babesia divergens* var. *bovis* behaved similarly in man and in chimpanzees. Intact chimpanzees were refractive, but splenectomized animals developed fulminating infections characterized by blackwater.

Baboons and other cercopithecids are subject to bites by ticks which carry *Theileria* in indigenous areas of Africa, yet little attention has been given to these parasites. We have found *Theileria*-like organisms* in intact baboons from Kenya and others have made similar observations. With the recent discovery of babesiosis in man in Ireland (Garnham et al., 1969) along with the realization that splenectomized persons may be vulnerable to infection, these parasites take on renewed significance. Although unproven at this time, it seems that latent piroplasmosis may be present in man in indigenous areas of infection. The possible role of nonhuman primates, and especially cercopithecids, as models for badly needed investigations, is obvious.

Besnoitia, a parasite of questionable pathogenicity, has been seen in a number of mammals, and a new species (*Besnoitia panamensis*), described

---

* Recently designated as *Entopolypoides*.

by Schneider (1965) from Panamanian lizards, produced large numbers of trophozoites in the peritoneal fluid of a marmoset (*Saguinus geoffroyi*) several days after inoculation. A new perspective for this disease in man and domestic animals may be gained if the causative organism is introduced into different species of primates for observation.

III. Helminths

The nonhuman primates are subject to parasitism by a wide representation of worms ranging in size from the microscopic invading larvae of nematodes and encysted forms of trematodes (diplostomatid mesocercariae) to the impressive *Ascaris*-like *Abbreviata,* large cestodes (*Bertiella*), and the voluminous cysts of *Echinococcus.* In general there is a fairly well-established host–parasite compatibility, but certain types of helminth disease (oesophagostomiasis, echinococcosis, sparganosis) in some individuals lead to serious pathology and debilitation. In some instances organ or tissue invasion by parasites (*Oesophagostomum*) provides avenues for infection by bacteria and other microorganisms.

Epidemiologically and epizootologically, nonhuman primates, especially peripatetic types with omnivorous eating habits, come into contact with biologic and parasitologic factors which set the stage for parasite propagation and transmission. These hosts frequent habitats in which arthropod vectors are present, if not common, yet for some unexplained reason a number of these mammals do not harbor infections expected. As an example, African baboons and vervets may be rather closely associated with human populations, but only infrequently become infected with the common helminths of man. In antithesis, checklists (Kuntz and Myers, 1969b; Myers and Kuntz, 1965, 1972; Yamashita, 1963) of the parasites of several nonhuman primates supports the contention that this group of vertebrates, under appropriate conditions, will accept a number of the parasites of man. It is difficult, with available records, to determine which hosts are infected with the parasites of man or other mammals under natural conditions. Some of the host–parasite records obviously represent infections in animals under biologic stress or animals subjected to unusual biologic circumstances. Yamashita (1963) has listed 228 species of helminths from nonhuman primates, 34 of which have also been found in man.

A. TREMATODA

Although an impressive collection of trematodes (representing 13 families) can establish infections in nonhuman primates (Kuntz, 1972), paragonimiasis and schistosomiasis are the diseases of greatest concern as far as biomedical research host potentials apply. Other trematodes

(*Clonorchis,* and several heterophyids), which depend upon fish as intermediate or carrier hosts, accidentally parasitize monkeys in Asia and it is
likely that several species of monkeys have been employed in the laboratories of the Orient to study different phases in the biology and treatment of
these parasites. The large intestinal fluke, *Fasciolopsis buski,* has been
reported from *M. mulatta* (Hartman, 1961) and Kuntz and Lo (1967)
demonstrated that the squirrel monkey could be infected. Several types
of diplostomatid and strigeid larvae have been observed in the African
baboon and several monkeys.

Paragonimiasis, a debilitating helminth disease with unusual epidemiologic and epizootologic features, is prevalent in many areas of Asia. The
parasite has a broad spectrum of host infectivity and, as a consequence, has
been studied biologically in several species of carnivores and other lower
vertebrates (M. Yokogawa, 1965; S. Yokogawa *et al.,* 1960). Natural
infections in nonhuman primates are unusual, but Sandosham (1954) and
Hashimoto and Honjo (1966) have reported adults from *Macaca irus* in
Malaya and in a Japan primate facility, respectively. M. Yokogawa (1965)
presented general information on the use of the Taiwan macaque (*M.
cyclopis*), Japanese macaque, and the Malaya crab-eating macaque (*M.
cynomolgus*) in investigations with *Paragonimus westermani,* and indicated
that there were differences in susceptibility to infection. *Mandrillus leucophaeus* was employed as an experimental host for *P. africana* (Vogel and
Crewe, 1965). Even though the parasite jeopardizes the health of thousands
of people, the biomedical professions have relied almost entirely on rodents,
dogs, and cats for evaluation of the parasitologic aspects of infection and
as hosts for studies in experimental chemotherapy. Antigens for skin-
testing purposes have been derived from parasites grown in the lower
vertebrates. Undoubtedly, the expense associated with maintenance of nonhuman primates has been an influencing factor. Much of the work presenting the details of worm migration, pulmonary and neurologic complications, extrapulmonary host–parasite relationships, and even treatment has
been conducted with recognized cases of paragonimiasis in man. Nonhuman primates seldom have been used for basic biomedical approaches
or treatment (D. C. Kim *et al.,* 1964) of *Paragonimus* infection. With
numerous cases of cerebral involvement and associated neuropathology
in patients in China, Japan, and Korea, it is obvious that the medical
profession should have a model which would allow critical host–parasite
studies in all phases of the infection.

Monkeys or other nonhuman primates are likely to yield beneficial
biologic information which previously may have been overlooked and may
not have been obtainable from lower vertebrates. The Taiwan macaque
(*M. cyclopis*) (Fig. 2), the grivet (*Cercopithecus aethiops*), and the
African baboon (*P. cynocephalus*) (Fig. 3) accommodate *Paragonimus*

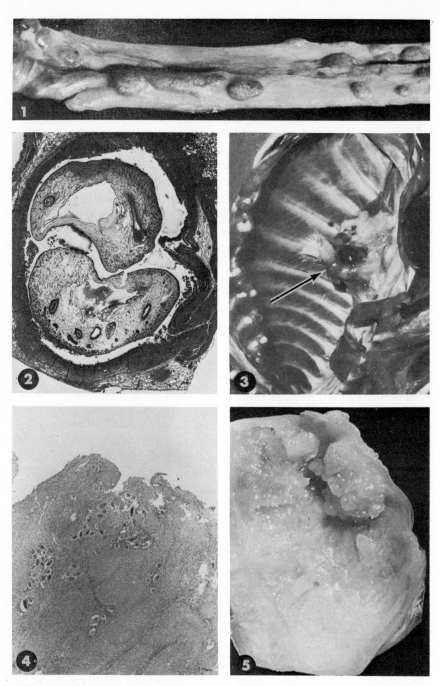

Fig. 1. Schistosomiasis haematobia in capuchin monkey (*Cebus apella*). Large intestine with marked fibrosis and massive deposits of *S. haematobium* (Iran strain) eggs, a manifestation in common with man in endemic areas.

well, and the Taiwan macaque, which has been used extensively in the writer's laboratory, shows most of the clinical manifestations which occur in man. In this host the parasite displays phenomenal migrating propensities and pathologic complications which may be related to superimposed infections and other host–parasite peculiarities. Although unproven, it seems probable that some of the unusual clinical aspects of lung fluke parasitism in man may be explained by observations made in nonhuman primates.

There have been scattered reports on the erratic occurrence of amphistome trematodes, *Gastrodiscoides hominis* and *Watsonius watsoni,* in man and in nonhuman primates (Buckley, 1964; Graham, 1960; Ruch, 1959). Although scarcely recognized as a worm of medical concern, Mukerji (in Buckley, 1964) states that it does lead to ill health in some peoples of India, and untreated children have been known to die. This little-known disease entity probably should be reinvestigated in an appropriate nonhuman primate host. Buckley (1964) has noted that monkeys are included in the list of natural hosts, although the ability of the parasite to infect different species under proper parasitologic conditions in the laboratory is unknown.

Schistosomiasis is currently accepted as the most serious parasite disease influencing the health and general welfare of many peoples in Africa, the Middle East, South America, and parts of Asia. Far-reaching effects on mankind and its burden upon the development of different geographic areas is easily ascertained in the voluminous literature of the past five decades. The schistosomiases are dependent upon infection by one or more of the three common species of *Schistosoma* in man, i.e., *S. haematobium, S. japonicum,* and *S. mansoni.* Schistosomiasis haematobia has received less general attention than schistosomiasis japonica and schistosomiasis mansoni, although all three forms of the disease render substantial medical and economic impact upon extensive indigenous populations.

The first indication that primates other than man would figure in the history of schistosomiasis is found in a report by Cobbold (1872) in

---

Fig. 2. Adult *Paragonimus westermani* (Taiwan strain) in section of lung.

Fig. 3. *Paragonimus westermani* (Taiwan) in African baboon (*Papio cynocephalus*). Extrapulmonary location of parasites (arrow) and extensive adhesions also characterize the infection in man in the Orient.

Fig. 4. Section of urinary bladder of *Cebus apella* with eggs of *S. haematobium* (Iran). Parasites and eggs elicit pathologic involvement and carcinoma of the bladder similar to that common in man in parts of Africa.

Fig. 5. Urinary bladder of squirrel monkey (*Saimiri sciureus*) 32 weeks after exposure to 1000 cercariae of *S. haematobium* (Iran). Parasites and eggs lead to hyperplasia, tissue alterations, and malfunction of bladder.

which he made reference to his finding of *Bilharzia* (presumably *Schistosoma haematobium*) in the portal vein of a sooty monkey (*Cercopithecus fuliginosus*) in South Africa. Since that time, different species of nonhuman primates have played a fairly significant role in investigations on the biologic and pathologic aspects of schistosomiasis. Recent reports on natural infections in these hosts have intensified interests and lead to further recognition of the nonhuman primates as possible important models. Natural infections of *Schistosoma mansoni* have been reported in monkeys (*Cercopithecus sabaeus*) in the West Indies (Cameron, 1928), in baboons (Fenwick, 1969; J. H. Miller, 1960; G. S. Nelson *et al.*, 1962; Strong *et al.*, 1965), in the squirrel monkey (Swellengrebel and Rijpstra, 1965), and in the chimpanzee (S. Y. Hsü and Hsü, 1968a). DePaoli (1965) found a chimpanzee from West Africa infected with *S. haematobium*. There are additional references to natural infections and to the use of nonhuman primates as subjects for experimental schistosomiasis in different areas of the world. A resumé indicating the general recognition of primates in the development of basic biologic knowledge of the schistosomes may be found in an earlier review (Kuntz, 1955). Investigations including these mammals in schistosomiasis research in recent years are far too numerous to recite. A few, however, indicate the prominence attained by the nonhuman primates as models for the schistosomiases.

Although a long list of primates has been subjected to human and animal schistosomes in most of the major areas of the world where the parasites are found, some of the earliest uses of nonhuman primates occurred in different parts of Africa in which schistosomiasis haematobia and schistosomiasis mansoni were endemic. Fairley (1920) used a series of monkeys as hosts for *S. haematobium* and *S. mansoni* in an attempt to contrast lesions in these animals with his observations on the same diseases in the Egyptian population. As the need for a better understanding of schistosomiasis increased, there was a realization that better laboratory models would greatly enhance endeavors in experimental schistosomiasis. As a consequence, many mammals, including nonhuman primates (mostly African species), were tested for their susceptibility potentials (Kuntz and Malakatis, 1955).

Additional indications of nonhuman vertebrates as hosts for the schistosomes were given by Malek (1961) and Martins (1958). Reference to and evaluation of nonhuman primates for research on the schistosomes of man was further discussed in the works of Edwards and McCullough (1954), Jordan and Goatly (1966), Jordan *et al.* (1967), Meisenhelder and Thompson (1963), Sadun *et al.* (1966a,b, 1970a,b), Standen (1949), Vogel (1967), Webbe and Jordan (1966), and others. Pellegrino and Katz (1968) described the prerequisites for ideal vertebrate hosts in drug-screening programs and referred to methods of infection, suscepti-bility of nonhuman primates, and handicaps of these hosts in experimental chemotherapy of schistosomiasis. With greater sophistication in technology and a more critical need for research models for the schistosomes, re-searches in recent years have pursued more specific parameters.

Gear (1967), working with South African schistosomes, has concluded that the South African vervet (*Cercopithecus aethiops*) closely simulates the parasitologic characteristics common in man. Pathologic changes sub-sequent to infection with *S. mansoni* give rise to "pipe-stem" fibrosis of portal tracts and von Lichtenberg and Sadun (1968) also produced this condition in chimpanzees experimentally infected with the Puerto Rican strain of *S. mansoni*. The same parasite was used by Warren and Jane (1967) to determine susceptibilities of the squirrel monkey (*Saimiri sciureus*), a slow loris (*Nycticebus coucang*), and the Malay tree shrew (*Tupaia glis*). The squirrel monkey was susceptible, but the slow loris was relatively resistant, and the "preprimate" (*Tupaia*) was virtually non-susceptible. Both Cheever (1969) and Warren (1964) have made com-parisons and correlations between experimental schistosome infections in various animals and in man. The latter author called attention to the early work of Fairley (1920) in Egypt in which monkeys were employed for acute schistosomiasis. These investigations have led into more in-tensified study on a broad representation of nonhuman primates subjected to different schistosomes, but emphasis has definitely leaned to the char-acterization of schistosomiasis mansoni (Bruce *et al.*, 1966; Cheever and Powers, 1971; Naimark *et al.*, 1960; Sadun *et al.*, 1966a,b, 1970a; and many others).

With a somewhat limited host-infectivity spectrum, *S. haematobium*, parasitologically, is quite different from *S. mansoni* and *S. japonicum*. In man, *S. haematobium* is responsible for extensive and serious damage to the urogenital system with stress on remarkable involvement of the blad-der. Schistosomiasis haematobia is one of the few parasite diseases in which there is a statistical correlation between infection and the presence of carcinoma, i.e., in bladder. Conventional pathologic and clinical mani-festations of schistosomiasis haematobia in lower animals are rare, even though the hamster and other rodents may develop fairly substantial in-fections and be used as hosts for maintenance of the parasite cycle. With an increased concern over the spread of schistosomiasis haematobia, and a lack of suitable hosts for detailed pathobiologic investigations, greater effort has been given to the development of nonhuman primates as candi-date models. Recent researches have been most promising (Kuntz *et al.*, 1971a; Myers *et al.*, 1970a,b; Sadun *et al.*, 1970b; Vogel, 1967) and it now appears that selected species may provide the sought-for model.

Significant bladder lesions due to induced *S. haematobium* infections were reported for West African baboons by Edwards and McCullough (1954), in hamadryad baboons by Kuntz and Malakatis (1955), in mangabeys (*Cercocebus*), and chimpanzees (*Pan*) by Vogel (1967), and in vervets (*Cercopithecus aethiops*) by Obuyu (1970). Comprehensive studies on chimpanzees by Sadun *et al.* (1970b) have shown without doubt that this ape can be recommended since overall infection features are comparable to those observed in man. Since long-term, in-depth investigations will require larger series of smaller, cheaper, and more easily managed hosts, other investigators (Kuntz *et al.*, 1971a,b; Myers *et al.*, 1970a,b) have explored the feasibility for consideration of other species.

Available host–parasite and pathobiologic data, information and observations suggest that several of the smaller, more common primates (Kuntz *et al.*, 1971a,b) accommodate *S. haematobium* well and produce pathology comparable to that in the chimpanzee (Sadun *et al.*, 1970b) and in man (Fig. 1). Impressive lesions have developed in the bladder of the African baboon (*Papio cynocephalus*), gelada baboon (*Theropithecus gelada*), talapoin monkey (*Cercopithecus talapoin*), capuchin monkey (*Cebus apella*) (Fig. 4), squirrel monkey (*Saimiri sciureus*) (Fig. 5), and langur (*Presbytis cristatus*). Individualism is a feature of schistosomiasis haematobia. There has been variable involvement of the ovaries and oviducts, uterus and vagina, testes and epididymis. The infection in some hosts has given rise to hydroureters, unilateral or bilateral hydronephrosis, and prominent cauliflower-like, papillomatous proliferations of the bladder wall. Moderate-to-heavy egg deposits in several males have resulted in external lesions on the penis. In theory, lesions witnessed in some individuals (talapoin, capuchin, and squirrel monkey) indicate that a model for a study of parasite oncogenic agents may be near to reality.

At an early date, Fairly (1927) gave consideration to immunity in schistosomiasis as demonstrated by *Schistosoma spindalis* infection in Indian primates, and Meleney and Moore (1954) studied immunity in monkeys with superinfections. For Nelson and his group (Amin *et al.*, 1968; G. S. Nelson, 1970; G. S. Nelson and Saoud, 1968; etc.) the rhesus and other monkeys served as a means of recognizing geographic strain differences in schistosomes. The macaques of Asia [Japanese (*M. fuscata*), Taiwan (*M. cyclopis*), and Philippines (*M. philippinensis*)] have occupied a pertinent position in the outstanding researches of H. F. Hsü and Hsü (1956, 1958, 1959, 1960, 1962; S. Y. Li Hsü, 1968b), delineating the true nature of geographic strains of *S. japonicum*. Other reports on pathology and varied host–parasite relationships (Swanson and Williams, 1963;

Williams and Swanson, 1963) of *S. japonicum* in macaques supported the strain-differences concept.

In recent years nonhuman primates have contributed extensively to the exploration of other parameters (acquired immunity, resistance, use of irradiated parasites, zooprophylaxis, serology, special pathology, and many others) of experimental schistosomiasis:

1. Acquired immunity, immunity to heterologous infection, immunization, challenge infections, immunity related to suppressive therapy (S. Y. Li Hsü, 1969, 1970; S. Y. Li Hsü and Hsü, 1965, 1968a; H. F. Hsü et al., 1965; S. Y. Li Hsü et al., 1966; Smithers, 1962; Smithers and Terry, 1965, 1967, 1969).

2. Drug susceptibility and use in chemotherapy (S. Y. Li Hsü et al., 1963; Pellegrino and Katz, 1968, 1969).

3. Schistosome removal by portal filtration (Goldsmith and Kean, 1966).

4. Fluorescent antibody reaction, complement-fixation tests, serum protein changes, immunoprecipitins (Hillyer and Ritchie, 1967; S. Y. Li Hsü, 1970; S. Y. Li Hsü and Hsü, 1966; H. F. Hsü et al., 1962; Mc-Mahon, 1967; Sadun et al., 1962).

5. Reagin-like antibodies (Sadun and Gore, 1970).

There is little doubt that the trematodes, in general, and the schistosomes, in particular, will be united with a greater number of the nonhuman primates in research endeavors of the future. Even a cursory review of the use of these mammals in this facet of parasitology indicates that a better knowledge of nonhuman primate–trematode parasitism might have spared different investigators the necessity of self-infection (*Fasciolopsis buski*, Barlow, 1925; *S. haematobium*, Barlow and Meleney, 1949; *Plagiorchis*, McMullen, 1937; and others) and the use of human volunteers.

B. CESTODA

Heavy infections with the large fleshy adults of *Bertiella* in the intestine or the cystic forms of *Echinococcus* or *Multiceps* in different sites in the body are impressive. Even though representatives of eight families of cestodes or tapeworms have been listed for the nonhuman primates (Myers, 1972), adults are not considered common. Members of three (Diphyllobothriidae, Mesocestoididae, Taeniidae) of the eight families are larval stages which, from the standpoint of parasitology, are classed as accidental infection, i.e., the parasite does not depend routinely upon these hosts for completion of the life cycle and propagation. Host–parasite relationships, pathogenic potentials, and other aspects of parasitism, which characterize the cestodes in nonhuman primates, contrast markedly with those described for the trematodes. Pertinent information on infection patterns among different hosts, including nonhuman primates, in diverse habitats may be found in the works of Dunn and co-workers (Dunn, 1963, 1966, 1968, 1970).

The life cycle of cestodes depend in great part on eating and social habits, and upon host contacts with different biologic elements in the ecosystem. Little is known of the biology of the tapeworms of this group of vertebrates, but it is safe to assume, based upon available parasite records, that only a few of the species represented depend entirely upon the nonhuman primate population for propagation and transmission from one host to another. Primates, which are basically arboreal, probably would not harbor numerous cestodes, and it is likely that a good proportion of worms reported were acquired as arboreal hosts made terrestrial excursions. Chances for infection with the so-called accidental, larval stages are greatly increased as nonhuman primates frequent the habitats of man or have varying contact with wild and domestic carnivores. As a consequence, the macaques of Asia and some cercopithecids of Africa possess a greater potential for cestode infections.

Two species of *Bertiella* (*B. studeri* and *B. mucronata*) parasitize primates other than man (Faust *et al.*, 1970). Infection, presumably, is acquired by ingestion of free-living mites which serve as intermediate hosts. Although it is assumed that light-to-moderate infections render only mild symptomatology, the presence of large numbers of worms in some hosts, e.g., African baboons examined in the field, may on occasion lead to occlusion of the intestinal tract, and it is likely there may be toxic effects in some individuals. This tapeworm obviously is of secondary interest to the medical profession, but it is obvious that the baboon and other cercopithecids may serve as excellent hosts for biomedical researches if occasions demand. The life cycle, however, is poorly understood and would demand unusual parasitologic expertise if considered as a model.

The adults of taeniid cestodes such as *Taenia saginata* and *T. solium*, which infect man under special epidemiologic situations, seldom parasitize nonhuman primates as adult worms. Nonetheless, adult *T. solium* developed in the lar gibbon (*Hylobates lar lar*) 52 days subsequent to feeding of cysticerci from "measly pork" (Cadigan *et al.*, 1967). Verster (1965) failed in attempts to infect *Cercopithecus aethiops*, but obtained immature *T. solium* from chacma baboons (*Papio ursinus*) 7 days after fed cysticerci. There are, however, a number of instances in which the larval stages of taeniids have been found in different species of monkeys (Fiennes, 1967; Graham, 1960; Kuntz and Myers, 1967, 1969a; Ruch, 1959). Identity of such cysts may be uncertain or difficult. A good proportion of reports are based on sectioned materials examined by pathologists or discovered by others at postmortem examination. Sectioned worms may not present adequate characterization to allow a determination. Many cysts undoubtedly may be *T. hydatigena* (Kuntz and Myers, 1967) or stages of related species found as adults in nonhuman vertebrates. Taeniid larvae such as the hydatids, coenuri, and some forms of cysticerci give cause for concern in man and lower primates as well as other animals.

Cysticercus cellulosae, the larval stage of the pork tapeworm, *T. solium*, may cause deleterious effects upon hosts since it has the potential to gain entry into all parts of the body, including the central nervous system. This parasite has been reported in several nonhuman primates (Myers, 1972; Vickers and Penner, 1968; Walker, 1936) and cerebral involvement has been achieved in the Taiwan macaque (*M. cyclopis*) by Hsieh (1960) and in the spider monkey (*Ateles geoffroyi*) by Mazzotti *et al.* (1965).

Echinococcus has long been recognized as a parasite characterized by its capability to infect many vertebrate hosts. Hydatid disease or echinococcosis in the past 2 decades has been observed in many African and Asiatic nonhuman primates (Crosby *et al.*, 1968; Fiennes, 1967; Ilievski and Esber, 1969; Myers, 1972; Myers and Kuntz, 1965; Myers *et al.*, 1965, 1970c), and Hutchison (1966) suggested the rhesus as a laboratory host for extended investigations in hydatid disease. Although Myers *et al.* (1970c) detected the parasites in baboons shortly after being transported to laboratory from Kenya, G. S. Nelson and Rausch (1963) failed to find cysts in 271 nonhuman primates, including baboons examined in Kenya. Actually, in many instances it is difficult to determine whether infections were natural or acquired after introduction of hosts into infection-prone habitats. *Echinococcus* is a serious infection which may drastically alter the physical condition of its host, man or lower primates, and lead to death.

Coenurus or *Multiceps* infection may account for serious consequences in nonhuman primates or in man, but the association with these hosts is entirely accidental and the records, while presenting interesting cases, do not include a broad spectrum of primate genera (Fiennes, 1967; Myers, 1972; Ruch, 1959). Coenurus cysts varying in size and appearance may be found in any part of the body. As an example, Clark (1969) found coenurus cysts in the gelada baboon (*Theropithecus gelada*). The biology of *Multiceps* has been one of some confusion, but it is a well-established fact that the adult taeniid is an intestinal parasite of domestic and wild canines. Rodents, primates, and other mammals may become infected by ingestion of eggs through the medium of contamination.

The incidence of infection for "sparganum"- and "tetrathyridium"-

type cestodes is variable and, in all probability, far from accurate. For purposes of discussion, these can be lumped together, even though they represent larval stages of adult tapeworms which occur in different orders. Sparganums are the larval stages of the diphyllobothriid, *Spirometra*, while tetrathyridia are associated with the taeniid, *Mesocestoides*. Adults occur in the intestinal tracts of a broad spectrum of mammals, with preference frequently for felines and canines. The life cycles of these parasites may be complicated, but it is assumed that, in the majority of instances, nonhuman primates become infected by consumption of other vertebrates which may harbor more immature larval stages.

Larval stages in primates may appear as single, fleshy, ribbonlike worms, or as congregations of worms in the body cavity or in the musculature (Figs. 6, 7). Some may be detected as viable cysts, or as cysts with parasites in varying stages of degradation, depending upon the duration of infection and host–parasite reactions. These larvae may render considerable damage in their migrations, and in some instances give rise to abscesses which allow for increased complications due to contamination and bacterial infection. Macroscopically, the worms are similar in morphology, and coloration. Differentiation of these larvae is dependent upon microscopic examination which will allow for observation of the presence of suckers for the tetrathyridium stage of *Mesocestoides* and only bothrial grooves for sparganum.

Sparganum and sparganum-like helminths have been recognized for years in the nonhuman primates (Fiennes, 1967; Morton, 1969; G. S. Nelson *et al.*, 1965; Ruch, 1959) but it was not until 1965 that G. S. Nelson *et al.* called attention to the close resemblance of mesocestoid and diphylobothriid larvae. G. S. Nelson *et al.* (1965) stated that records for spargana in Africa are few. They found larvae which tentatively were recognized as "spargana," but subsequently proved to be tetrathyridia, in the baboon (*Papio doguera*) and *Cercopithecus mitis*. Myers and Kuntz (1967) identified spargana in 42% of 50 baboons taken by Fison on the Cambridge Expedition to Mwanza. These authors (Kuntz *et al.*, 1970b) also described conventional spargana in Kenya baboons and proliferative-type spargana in *Cercopithecus aethiops* of East Africa. The significance of sparganosis in man in the Orient is recognized, and Mueller (1938) established experimental infections of the Asiatic *Diphyllobothrium mansonoides* (=*Spirometra mansonoides*) in the rhesus monkey.

Some of the larvae of cestodes under proper circumstances have unusual capabilities for infection of man, many other mammals, and especially for the lower primates. Therefore, it is relatively safe to predict that the rhesus, baboon, vervet, and other easily obtainable nonhuman primates may be considered as models to support this particular aspect

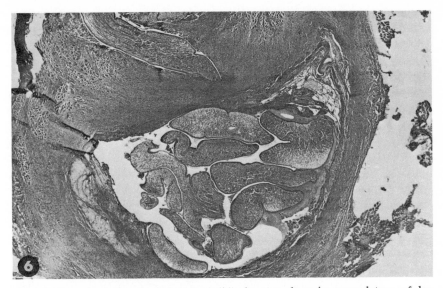

Fig. 6. Spargana (larval diphyllobothriid) in cyst deep in musculature of leg of vervet (*Cercopithecus aethiops*).

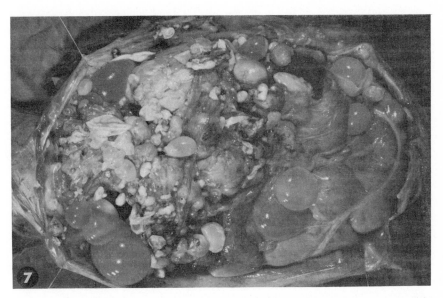

Fig. 7. Abdominal viscera are traumatized by presence of numerous hydatid (*Echinococcus*) cysts which fill body cavity of *Colobus* monkey.

of medical parasitology. Little is known of the relationship of invading parasites as vehicles for introduction of various microorganisms into the body. Possibly some of the cestodes in favorable combination with monkeys or baboons would provide a system for basic investigations with bacteria, fungi, viruses, and other agents.

C. NEMATODA

The Nematoda comprise a diverse congregation of helminths which range greatly in size, exhibit marked differences in life-cycle patterns, show a broad capability for infection of vertebrates, and may be characterized by their unusual potential for tissue invasion and the resulting pathology. Thus, some of the nematodes of the nonhuman primates require intermediate hosts for transmission of infection, some display moderate host specificity, and some are responsible for variable pathologic injury to the host. Any compilation of information on the helminths of nonhuman primates (Fiennes, 1967; Graham, 1960; Kuntz et al., 1968; Ruch, 1959) and parasite checklists (Kuntz and Myers, 1969b; Myers and Kuntz, 1965, 1972) show a long listing of nematodes encompassing several orders and a number of families. Orihel (1970) estimated that no less than 250 species of helminths have been described for the nonhuman primates, the majority being nematodes.

Since there are possible zoonotic implications and a number of the parasites in concern have been reported from man as well as from nonhuman primates, and vice versa (Chitwood, 1970; Orihel, 1970), it is quite obvious that the different nonhuman primates may offer varied, but in many instances untested, potential as models in medical parasitology. Host–parasite model systems, however, depend upon the goal in mind, and a recommendation of nonhuman primates for specific problems with the nematodes must rely upon careful pretesting and experimental evaluations. With somewhat limited exploration in this field of specialization, it is difficult to assess the use of nonhuman primate hosts for specific endeavors.

Since nonhuman primate parasitology is still in its infancy, there is a lack of agreement relative to the use of these animals as models for some of the nematode diseases of man. At a recent workshop (Workshop on Nonhuman Primate Parasites, 1970), Chitwood (1970) and Orihel (1970) provided pertinent information on general biologic relationships of a number of nematode parasites to disease in the nonhuman primates with insight on the potential of these hosts for experimental parasitoses. In the same workshop, Hunt et al. (1970) pointed to mechanisms of parasite damage and host response due to nematodes and other helminths.

In these proceedings Chitwood presented a synoptic table indicating host–parasite relationships, i.e., infection potentials of a broad spectrum of nematodes reported from New World and Old World primates and man. Although the fairly common practice of parasite-sharing was discussed, emphasis was given to basic parasitologic concepts, which, other factors being equal, determine whether a given host may be accepted as a biologic or biomedical model. It was a consensus that the full value of nonhuman primates to medical parasitology could not be predicted or realized until more critical explorations had been made in the use of specific hosts for designated disease entities, and this is especially applicable to the nematodes.

Checklists on the parasites of nonhuman primates as well as other compilations on helminths (Chitwood, 1970; Dunn, 1968; Fiennes, 1967; Ruch, 1959; Yamashita, 1963) give numerous references and indicate the broad representation of nematodes harbored by these mammals. Further scrutiny also reveals that many of the nematodes are species which also parasitize man. On the other side of the coin, as a review of parasitologic records likewise will show, there are many instances where the nematodes of man, e.g., different filariae, hookworms, and even *Ascaris*, are found in several species of lower primates. Some of these represent situations of theoretical zoonotic consequence but most probably are localized biologic anomalies in which the parasites, due to epidemiologic or epizootologic factors, have crossed host barriers. In some instances it is surmised that a given species of parasite, which apparently occurs in man as well as nonhuman primates in the same area, represent strains associated with the hosts in concern. Survey and surveillance-type studies on the parasite fauna of the nonhuman primates undoubtedly would reveal many additional examples of parallel parasitism. As an example, it has been assumed that the whipworm (*Trichuris*) of African and Asiatic monkeys is the *Trichuris trichiura* of man. Study of pertinent morphologic details, however, indicates that there are discrete differences, even though, macroscopically, the whipworms appear identical.

The strongyles are responsible for disease in man in numerous areas where living and working conditions allow infection. Such parasitoses frequently pose problems to localized populations in underdeveloped countries. Ternideniasis unquestionably is a health problem in certain localities of Africa, and oesophagostomiasis, although recognized as a disease common in nonhuman primates, appears in human populations. The common occurrence of *Oesophagostomum* in macaques and other cercopithecids presents ample evidence for the use of these vertebrates for clinical as well as parasitologic investigations. Hookworms, such as *Necator americanus* have been reported from several nonhuman primates, and Orihel (1971) has recently removed adults from the patas monkey (*E. patas*) and the chimpanzee (*Pan*). These apparently are identical with the same parasite in man. Under laboratory conditions, Orihel (1971) has demonstrated that the patas and chimpanzee are much more suitable than other laboratory hosts, even though all are susceptible to *N. americanus* derived from man. In spite of individual host–parasite variabilities, persistence of infections and the development of severe anemias and other clinical manifestations clearly indicate that some of the lower primates may be employed for investigations which will lead to a better understanding of this group of nematodes. T. A. Miller (1968), in his investigations on pathogenesis and immunity in hookworm infections, expressed his lament over the want of a suitable nonhuman primate host to fortify his field of specialization.

Other nematode groups have received variable attention. Studies with *Strongyloides simiae* from Panamanian monkeys have allowed experimentation which could be extrapolated to basic parasitologic aspects of *S. stercoralis* infection in man (Beach, 1936). Grétillat and Vassiliadès (1968) have evaluated the symptoms of trichinosis by introduction of *Trichinella spiralis* into *Erythrocebus patas* and *Papio papio*.

Except for unusual situations in school children or other minors, enterobiasis (*Enterobius vermicularis*) as a rule is considered an infection of little consequence. Some of the oxyurids are recognized for their host specificity, yet there is ample evidence to indicate that even *E. vermicularis* occasionally finds its way into lower primates. Sandosham (1950) has recorded *E. vermicularis* from the chimpanzee, the lar gibbon (*Hylobates lar*), and the lion marmoset (*Leontocebus rosalia*), and Dunn (1966) has referred to the oxyurid situation in his discussion on patterns of parasitism in primates. Nonhuman primates harbor a rather extensive complex of oxyurids (Dunn, 1966; Hugghins, 1969; Pope, 1966; Sandosham, 1950) and these nematodes, on occasion, give rise to problems in captive animals (Schmidt and Prine, 1970). With even a limited availability of host–parasite information, it seems likely that there are oxyurid–nonhuman primate combinations which would allow for basic research on enterobiasis and would be applicable to the infection in man as well as in other lower primates.

Minor parasitologic problems face populations exposed to parasite infection-prone habitats. Lê-Van-Hoa *et al.* (1963) reported a nematode cutaneous infection in a Vietnamese soldier. The invader apparently was a member of the genus *Anatrichosoma*, a trichuroid recognized in several macaques (Allen, 1960; Chitwood, 1970). Since medical facilities are not available to many peoples in underdeveloped regions of the world, it is possible that such infections may be more common than realized. The *Anatrichosoma*–macaque host–parasite system, offers a model for a problem of academic and possibly medical concern.

In recent years the medical profession has been forced to recognize several species of nematodes which previously were assigned almost without question to the domain of animal parasitology. Accidental infections of man by the rodent lungworm (*Angiostrongylus cantonensis*), by *Anisakis*, a parasite of marine mammals, by *Toxocara cati* and *T. canis*, from cats and dogs, respectively, and by *Gnathostoma* renders serious symptomatology and clinical manifestations (Belding, 1965; Faust *et al.*, 1970). Punyagupta *et al.* (1970) have recently reported on a long series of human cases of eosinophilic meningitis in Thailand due to infection by *A. cantonensis*. The pathogenicity of migrating larval forms of these nematodes is recognized and respected, but there are many details in need of elucidation for a complete knowledge of these infections foreign in man. Loison *et al.* (1962) demonstrated the tropism of *A. cantonensis* for the brain of an experimentally infected rhesus and produced a fatal eosinophilic meningitis, and Alicata *et al.* (1963) reported similar complications. Weinstein *et al.* (1963) also established experimental infection of *Angiostrongylus* in rhesus monkeys. Lesions developed in the central nervous system even though there was no overt symptomatology. The exposure of primates to *A. cantonensis*, however, is not entirely predictable, as indicated by the work of Bisseru (1969a) in which larvae of this parasite failed to infect *Macaca irus*.

Our knowledge of the nature of *Toxocara* and *Anisakis* infections is far from satisfactory, and it seems that future endeavors along this line would benefit academic as well as medical parasitology. *Toxocara* infections have been followed in different lower vertebrates as well as in *Macaca irus* (Bisseru, 1969b). Wiseman (1969) introduced *T. canis* into rhesus monkeys to evaluate allergic and eosinophilic responses then, in continued investigations, Wiseman *et al.* (1969) turned attention to skin sensitivity testing. Somewhat related parameters were explored by Aljeboori and Ivey (1970) in white-cell and serum-protein responses in baboons (*P. anubis*) subsequent to per os introduction of *T. canis*. In this study it was concluded that baboons may display a different white-cell response than man. Unfortunately, other primates have not been subjected to these nematodes to determine whether conditions comparable to those in man develop. However, with evaluation of additional nonhuman primates, it seems likely that optimal models may evolve.

The filariases for several decades have presented significant medical problems in many areas of the tropics, especially in islands of the Pacific and parts of Asia. Pertinent studies on various biologic parameters of the species which parasitize man have been drastically curtailed for want of experimental host–parasite systems and models. Examination of non-human primates in regions endemic for human filariasis and the introduction of a number of species of these mammals to interested specialists has revitalized this facet of medical parasitology. The nonhuman primates undoubtedly harbor species of filariae which are closely related to the common species in man, and some probably can be used as acceptable models after experimentation has delineated biologic requisites of host–parasite systems. Although variable biologic features are recognized, one can point to *Brugia tupaiae* of tree shrews (a preprimate) and *Dirofilaria aethiops* and *Macacanema formosana* in monkeys.

Investigations on the filariae have been somewhat exploratory in nature, but work with several species in different hosts has indicated the probable use of nonhuman primates as models for basic biomedical considerations. In 1968 Orihel briefly discussed the filarial situation relative to the filariae of macaques in Asia and disclosed that laboratory-born primates were usually susceptible to the simian strain of *Brugia malayi*, but wild hosts were not. Dissanaike and Niles (1965) failed to establish *Wuchereria bancrofti* in a toque monkey. Hawking and Gammage (1968) used *B. malayi*-infected rhesus to follow periodic migrations of microfilariae, while Wong *et al.* (1969) used this host–parasite system for immunization studies.

Loiasis and onchocerciasis, due to biologic characteristics, have been the subject of minimal experimental scrutiny. The mandrill and several other cercopithecids have been noted as hosts for *Loa loa* and Duke and Wijers (1958) have described the biologic differences in the *Loa* of man and lower primates. Hawking *et al.* (1967) employed drills (*Mandrillus leucophaeus*) infected with *Loa loa* as a means of correlating microfilariae activities in drills and man. Duke and co-workers (Duke, 1962; Duke *et al.*, 1967) have infected chimpanzees with a Cameroon and Guatemalan strain of *Onchocerca volvulus*, and Neumann and Gunders (1963) produced experimental ocular lesions with the same filarid in chimpanzees. Guinea worm- (*Dracunculus medinensis*) infection, or dracunculiasis, is another of the helminth diseases of man which has received scant attention in the laboratory. Muller (1970b), however, has successfully established dracunculiasis in rhesus monkeys subsequent to maintenance of larvae of *D. medinensis* in the frozen state and then passaged through *Cyclops*, its natural intermediate host. This has introduced a new concept into parasitology, one which, with the use of nonhuman primates, will allow more sophisticated experimentation with a disease rarely seen outside the area of endemicity.

D. ACANTHOCEPHALA

The Acanthocephala, on rare occasion, parasitize man in societies where native customs include unusual living and eating habits. The thorny-headed worms, or the Acanthocephala, pose no real problems to the health of man. Members of this group, however, are found in a number of non-human primates, and are especially prevalent in South America. *Prosthenorchis elegans* commonly parasitizes *Ateles, Callicebus, Cebus, Saimiri,* and many of the marmosets, and *Macrocanthorhynchus hirudinaeus,* the large species occasionally found in man, has been recorded from *Cebus, Cebuella, Callithrix, Leontideus,* and *Saimiri* (Dunn, 1968). Even though of minor medical concern, a host–parasite model system for basic parasitology, pathology, and allied aspects of helminth damage may be found in the *Prosthenorchis*–South American monkey complex.

IV. Pentastomida

Pentastomiasis or porocephalosis occurs in widely separated areas of the world (Belding, 1965; Faust *et al.,* 1970) where customs of certain ethnic groups account for consumption of a wide variety of unusual food including raw or semicooked fish, amphibians, and reptiles. Even though the incidence of infection is not great, it is of special importance to note that *Porocephalus* was recently detected in 45% of aborigines sampled in Malaysia (Prathap *et al.,* 1969). This group of parasites has been removed from the tissues of a number of lower primates (Cosgrove *et al.,* 1970) which serve in the capacity of intermediate hosts. *Armillifer* has been reported from several species of cercopithecids (Graham, 1960; Heuschele, 1961; Whitney and Kruckenberg, 1967), and *Porocephalus* from New World marmosets (*Saguinis*) (B. Nelson *et al.,* 1966) and *Saimiri sciureus* (Cosgrove *et al.,* 1970).

Tissue reactions and varied host–parasite relationships of the nymphal stages of the pentastomids in man and in lower vertebrates is poorly documented, and this parameter of medical parasitology is essentially untouched. Even though associated with the health of man in rather localized populations, it would be a distinct advantage to the medical profession to understand more completely the pathology and clinical manifestations due to these infections. Observations on the pentastomids in a series of primates, suggest that a number of species may be used as laboratory hosts. Successful experimentation, however, will depend upon great expertise in biologic handling of the life cycle of the parasite and appropriate introduction into a nonhuman primate host–parasite system.

V. Status of Nonhuman Primate Parasitology

As the literature and available host–parasite records indicate, the non-human primates naturally harbor many protozoa and helminths. They are also susceptible to additional parasite infections under experimental conditions. Information on the true potential of different species of lower primates to support investigations on parasite diseases of man, however, are somewhat limited, and there are only a few instances in which these mammals have been considered for their true value in this field of bio-medical research. Myers (1968) discussed the use of nonhuman primates for investigations with the intestinal parasites of man. However, recognition and acceptance of nonhuman primates as animals of choice for specific types of study has developed slowly, and consideration of a host–parasite system to satisfy a given situation has frequently depended upon preliminary testing coupled with the available accumulation of parasitologic data and information.

Certain aspects of medical parasitology have also matured rather slowly, and as a consequence there are many unexplored parameters in this field of specialization which must rely upon prudent and judicial use of non-human primates. Some investigators in the fields of immunology, serology, and certain aspects of pathobiology are just beginning to appreciate the potentials offered by the nonhuman primates. The use of tissue cultures has scarcely entered the field of nonhuman primate parasitology. Wood and Suitor (1966) successfully explored the possibility of using mosquito-cell cultures for development of *Macacanema formosana,* a filarid easily obtainable from the Taiwan macaque (*M. cyclopis*). Herman (1968) and Zaman and Yin-Murphy (1969) have successfully cultivated Leishman-Donovan bodies and *Toxoplasma,* respectively, in monkey kidney cells. Methods for fluorescent antibody tests, various sensitivity evaluations, and allied aspects of helminth infection have been slow to develop, but currently are evolving with support offered by the use of nonhuman primates (Malley *et al.,* 1968; Muller, 1970a). Patterson (1969) and Weiszer *et al.* (1968) checked different aspects of reagin hypersensitivity in rhesus monkeys introduced to *Ascaris lumbricoides* antigens, and heterophile antibodies and immunoglobulins have been evaluated in *M. mulatta* given intravenous inoculations of trypanosomes (Houba *et al.,* 1969). Strannegård (1967) has used sera of cynomolgus monkeys in studies of *Toxoplasma*-hostile factor. Immunologic responses to reagin-like antibodies in chimpanzees, rhesus, and *Cercopithecus sabeus* infected with *Schistosoma mansoni* and *S. haematobium* have been followed by Sadun and Gore (1970). Synergistic relationships of different categories of parasites (pathogens and nonpathogens) to viruses, bacteria, rickettsiae, and other microorganisms are essentially unknown. Also, a consideration of the possibility that protozoa and helminths may serve as carriers or vehicles for introduction of microorganisms into nonhuman primates is a field ripe for investigation.

ACKNOWLEDGMENTS

The author is indebted to Dr. B. J. Myers, Southwest Foundation for Research and Education, for assistance in compilation of information and for reading the manuscript, and to Dr. S. S. Kalter, Chairman, Division of Microbiology and Infectious Diseases, for general support in studies on the parasites of nonhuman primates.

REFERENCES

Alicata, J. E., Loison, G., and Cavallo, A. (1963). *J. Parasitol.* **49**, 156–157.
Aljeboori, T., and Ivey, M. (1970). *Amer. J. Trop. Med. Hyg.* **19**, 249–254.
Allen, A. M. (1960). *Amer. J. Vet. Res.* **21**, 389–392.
Amin, M. A., Nelson, G. S., and Saoud, M. F. A. (1968). *Bull. WHO* **38**, 19–27.
Baker, J. R. (1970). *Trans. Roy. Soc. Trop. Med. Hyg.* **64**, 14.
Barlow, C. H. (1925). *Amer. J. Hyg., Monogr.* No. 4, pp. 1–98.
Barlow, C. H., and Meleney, H. E. (1949). *Amer. J. Trop. Med.* **29**, 79–87.
Beach, T. D. (1936). *Amer. J. Hyg.* **23**, 243–277.
Belding, D. L. (1965). "Textbook of Parasitology," 3rd ed. Appleton, New York.
Benirschke, K. (1969). *Fed. Proc., Fed. Amer. Soc. Exp. Biol.* **28**, 170–178.
Benirschke, K., and Low, R. J. (1970). *Comp. Pathol. Bull.* **11**, 3–4.
Benirschke, K., and Richart, R. (1960). *Amer. J. Trop. Med. Hyg.* **9**, 269–273.
Bisseru, B. (1969a). *In* "Proceedings of Seminar on Filariasis and Immunology of Parasitic Infections and Laboratory Meeting, Singapore, 1968," pp. 115–124. Malaysian Society of Parasitology and Tropical Medicine, Singapore.
Bisseru, B. (1969b). *J. Helminthol.* **43**, 267–272.
Bray, R. S., and Garnham, P. C. C. (1961). *J. Parasitol.* **47**, 538.
Bruce, J. I., von Lichtenberg, F., Schoenbechler, M. J., and Hickman, R. L. (1966). *J. Parasitol.* **52**, 831–832.
Buckley, J. J. C. (1964). *J. Helminthol.* **38**, 1–6.
Bullock, B. C., Wolf, R. H., and Clarkson, T. B. (1967). *J. Amer. Vet. Med. Ass.* **151**, 920–921.
Cadigan, F. C., Jr., Stanton, J. S., Tanticharoenyos, P., and Chaicumpa, V. (1967). *J. Parasitol.* **53**, 844.
Cameron, T. W. M. (1928). *J. Helminthol.* **6**, 219–222.
Cheever, A. W. (1969). *Trans. Roy. Soc. Trop. Med. Hyg.* **63**, 781–795.
Cheever, A. W., and Powers, K. G. (1971). *Amer. J. Trop. Med. Hyg.* **20**, 69–76.
Chitwood, M. (1970). *Lab. Anim. Care* **20**, 389–394.
Clark, J. D. (1969). *J. Amer. Vet. Med. Ass.* **155**, 1258–1263.
Cobbold, T. S. (1872). *Brit. Med. J.* **2**, 89–92.
Corson, J. F. (1936). *Ann. Trop. Med. Parasitol.* **30**, 389–400.
Corson, J. F. (1938). *Ann. Trop. Med. Parasitol.* **32**, 437–443.

Cosgrove, G. E., Nelson, B. M., and Self, J. T. (1970). *Lab. Anim. Care* **20**, 354–360.
Cowen, D., and Wolf, A. (1945). *J. Infec. Dis.* **77**, 144–157.
Crosby, W. M., Ivey, M. H., Shaffer, W. L., and Holmes, D. D. (1968). *Lab. Anim. Care* **18**, 395–397.
Deane, L. M. (1964). *Rev. Brasil. Malariol. Doencas Trop.* **16**, 27–48.
DePaoli, A. (1965). *Amer. J. Trop. Med. Hyg.* **14**, 561–565.
Desowitz, R. S., and Watson, H. J. C. (1953). *Ann. Trop. Med. Parasitol.* **47**, 324–334.
Dissanaike, A. S., and Niles, W. J. (1965). *Ann. Trop. Med. Parasitol.* **59**, 189–192.
Dobell, C. C. (1931). *Parasitology* **23**, 1–72.
Duke, B. O. L. (1962). *Trans. Roy. Soc. Trop. Med. Hyg.* **56**, 271.
Duke, B. O. L., and Wijers, D. J. B. (1958). *Ann. Trop. Med. Parasitol.* **52**, 158–175.
Duke, B. O. L., Moore, P. J., and de Leon, J. R. (1967). *Ann. Trop. Med. Parasitol.* **61**, 332–337.
Dunn, F. L. (1963). *J. Parasitol.* **49**, 717–722.
Dunn, F. L. (1966). *Folia Primatol.* **4**, 329–345.
Dunn, F. L. (1968). *In* "The Squirrel Monkey" (L. A. Rosenblum and R. W. Cooper, eds.), pp. 31–68. Academic Press, New York.
Dunn, F. L. (1970). *Lab. Anim. Care* **20**, 383–388.
Dunn, F. L., Lambrecht, F. L., and du Plessis, R. (1963). *Amer. J. Trop. Med. Hyg.* **12**, 524–534.
Edwards, E. E., and McCullough, F. S. (1954). *Ann. Trop. Med. Parasitol.* **48**, 164–177.
Eichhorn, A., Gallagher, B. A. (1916). *J. Infec. Dis.* **19**, 395–407.
Fairley, N. H. (1920). *J. Pathol. Bacteriol.* **23**, 289–314.
Fairley, N. H. (1927). *Indian J. Med. Res.* **14**, 685–700.
Faust, E. C., Russell, P. F., and Jung, R. C. (1970). "Craig and Faust's Clinical Parasitology," 8th ed. Lea & Febiger, Philadelphia, Pennsylvania.
Fenwick, A. (1969). *Trans. Roy. Soc. Trop. Med. Hyg.* **63**, 557–567.
Fiennes, R. N. (1967). "Zoonoses of Primates." Cornell Univ. Press, Ithaca, New York.
Fremming, B. D., Vogel, F. S., Benson, R. E., and Young, R. J. (1955). *J. Amer. Vet. Med. Ass.* **126**, 406–407.
Frenkel, J. K. (1969a). *Fed. Proc., Fed. Amer. Soc. Exp. Biol.* **28**, 160–161.
Frenkel, J. K. (1969b). *Fed. Proc., Fed. Amer. Soc. Exp. Biol.* **28**, 179–190.
Garnham, P. C. C. (1966). "Malaria Parasites and Other Haemosphoridia." Blackwell, Oxford.
Garnham, P. C. C., and Bray, R. S. (1959). *J. Protozool.* **6**, 352–353.
Garnham, P. C. C., Donnelly, J., Hoogstraal, H., Kennedy, C. C., and Walton, G. A. (1969). *Brit. Med. J.* **4**, 768–770.
Gear, J. H. S. (1967). *In* "Bilharziasis" (F. K. Mostofi, ed.), Int. Acad. Pathol., Spec. Monogr. pp. 248–258. Springer Publ., New York.
Geiman, Q. M. (1964). *Lab. Anim. Care* **14**, 441–454.
Godfrey, D. C., and Killick-Kendrick, R. (1967). *Trans. Roy. Soc. Trop. Med. Hyg.* **6**, 781–791.
Goldsmith, E. I., and Kean, B. H. (1966). *Gastroenterology* **50**, 805–807.
Graham, G. L. (1960). *Ann. N. Y. Acad. Sci.* **85**, 842–860.
Grétillat, S., and Vassiliadès, G. (1968). *Rev. Elevage Med. Vet. Pays Trop.* **21**, 85–99.
Grewal, M. S. (1966). *Indian Practitioner* **19**, 403–410.

Hartman, H. A. (1961). *Amer. J. Vet. Res.* **22**, 1123–1125.
Hashimoto, I., and Honjo, S. (1966). *Jap. J. Med. Sci. Biol.* **19**, 215–227.
Hawking, F., and Gammage, K. (1968). *Amer. J. Trop. Med. Hyg.* **17**, 724–729.
Hawking, F., Moore, P., Gammage, K., and Worms, M. J. (1967). *Trans. Roy. Soc. Trop. Med. Hyg.* **61**, 674–683.
Hegner, R. (1929). *Science* **70**, 539–540.
Hegner, R. (1934). *Amer. J. Hyg.* **19**, 480–501.
Hegner, R., and Chu, H. J. (1930). *Amer. J. Hyg.* **12**, 62–108.
Herman, R. (1968). *J. Protozool.* **15**, 35–44.
Heuschele, W. P. (1961). *J. Amer. Vet. Med. Ass.* **139**, 911–912.
Hill, W. C. O. (1969). *Ann. N. Y. Acad. Sci.* **162**, 7–14.
Hillyer, G. V., and Ritchie, L. S. (1967). *Exp. Parasitol.* **20**, 326–333.
Homburger, F., and Bajusz, E. (1970). *J. Amer. Med. Ass.* **212**, 604–610.
Houba, V., Brown, K. N., and Allison, A. C. (1969). *Clin. Exp. Immunol.* **4**, 113–123.
Hsieh, H. C. (1960). *Formosan Sci.* **14**, 66–80.
Hsü, H. F., and Hsü, S. Y. Li (1956). *Amer. J. Trop. Med. Hyg.* **5**, 136–144.
Hsü, H. F., and Hsü, S. Y. Li (1958). *Trans. Roy. Soc. Trop. Med. Hyg.* **52**, 363–367.
Hsü, H. F., and Hsü, S. Y. Li (1959). *Proc. Int. Congr. Trop. Med. Malar., 6th, 1900* Vol. 2, pp. 58–66.
Hsü, H. F., and Hsü, S. Y. Li (1960). *J. Parasitol.* **46**, 228.
Hsü, H. F., and Hsü, S. Y. Li (1962). *Exp. Parasitol.* **12**, 459–465.
Hsü, H. F., Hsü, S. Y. Li, and Osborne, J. W. (1962). *Nature (London)* **194**, 98–99.
Hsü, H. F., Hsü, S. Y. Li, and Osborne, J. W. (1965). *Nature (London)* **206**, 1338–1340.
Hsü, S. Y. Li (1969). *Exp. Parasitol.* **25**, 202–209.
Hsü, S. Y. Li (1970). *Trans. Roy. Soc. Trop. Med. Hyg.* **64**, 597–600.
Hsü, S. Y. Li, and Hsü, H. F. (1965). *Z. Tropenmed. Parasitol.* **16**, 83–89.
Hsü, S. Y. Li, and Hsü, H. F. (1966). *Z. Tropenmed. Parasitol.* **17**, 264–278.
Hsü, S. Y. Li, and Hsü, H. F. (1968a). *Trans. Roy. Soc. Trop. Med. Hyg.* **62**, 901–902.
Hsü, S. Y. Li, and Hsü, H. F. (1968b). *Z. Tropenmed. Parasitol.* **19**, 43–59.
Hsü, S. Y. Li, Chu, K. Y., and Hsü, H. F. (1963). *Z. Tropenmed. Parasitol.* **14**, 37–40.
Hsü, S. Y. Li, Hsü, H. F., Chu, K. Y., Tsai, C. T., and Eveland, L. K. (1966). *Z. Tropenmed. Parasitol.* **17**, 407–412.
Hugghins, E. J. (1969). *J. Parasitol.* **55**, 680.
Hunt, R. D., Jones, T. C., and Williamson, M. (1970). *Lab. Anim. Care* **20**, 345–353.
Hutchison, W. F. (1966). *J. Parasitol.* **52**, 416.
Ilievski, V., and Esber, H. (1969). *Lab. Anim. Care* **19**, 199–204.
Jones, T. C. (1969). *Fed. Proc., Fed. Amer. Soc. Exp. Biol.* **28**, 162–169.
Jordan, P., and Goatly, K. D. (1966). *Ann. Trop. Med. Parasitol.* **60**, 63–69.
Jordan, P., von Lichtenberg, F., and Goatly, K. D. (1967). *Bull. WHO* **37**, 393–403.
Kessel, J. F. (1924). *Proc. Soc. Exp. Biol. Med.* **22**, 206–208.
Kessel, J. F. (1928). *Univ. Calif., Berkeley, Publ. Zool.* **31**, 275–306.
Kim, C. S., Eugster, A. K., and Kalter, S. S. (1968). *Primates* **9**, 93–104.
Kim, D. C., Sun, S. C., and Bergner, J. F., Jr. (1964). *Rep. Nat. Inst. Health Korea* **1**, 153–166.
Kuntz, R. E. (1955). *Amer. J. Trop. Med. Hyg.* **4**, 383–413.

Kuntz, R. E. (1972). *In* "Pathology of Primates" (R. N. Fiennes, ed.) **2**, 104–123. Karger, Basel.

Kuntz, R. E., and Lawless, D. K. (1966). *J. Formosan Med. Ass.* **65**, 287–293.

Kuntz, R. E., and Lo, C. T. (1967). *Trans. Amer. Microsc. Soc.* **86**, 163–166.

Kuntz, R. E., and Malakatis, G. M. (1955). *Amer. J. Trop. Med. Hyg.* **4**, 75–89.

Kuntz, R. E., and Myers, B. J. (1967). *Primates* **8**, 83–88.

Kuntz, R. E., and Myers, B. J. (1969a). *Proc. Int. Congr. Primatol., 2nd, 1969* Vol. 3, pp. 184–190.

Kuntz, R. E., and Myers, B. J. (1969b). *Primates* **10**, 71–80.

Kuntz, R. E., Myers, B. J., Bergner, J. F., Jr., and Armstrong, D. E. (1968). *Formosan Sci.* **22**, 120–136.

Kuntz, R. E., Myers, B. J., and McMurray, T. S. (1970a). *Trans. Amer. Microsc. Soc.* **89**, 304–307.

Kuntz, R. E., Myers, B. J., and Katzberg, A. A. (1970b). *J. Parasitol.* **56**, 196–197.

Kuntz, R. E., Myers, B. J., Moore, J. A., and Huang, T. C. (1971a). *Exp. Parasitol.* **29**, 33–41.

Kuntz, R. E., Myers, B. J., Huang, T. C., and Moore, J. A. (1971b). *Proc. Int. Congr. Primatol., 3rd, 1970,* **2**, 162–172.

Lê-Van-Hoa, Duong-Hong-Mo, and Nguyên-Luu-Viên (1963). *Bull. Soc. Pathol. Exot.* **56**, 121–126.

Levine, N. D. (1961). "Protozoan Parasites of Domestic Animals and of Man." Burgess, Minneapolis, Minnesota.

Limbos, P., and Jadin, J. (1963). *Ann. Soc. Belge Med. Trop.* **43**, 739–746.

Loison, G., Cavallo, A., and Vervent, G. (1962). "An Experimental Study on the Monkey of the Role of *'Angiostrongylus cantonensis'* in the Etiology of Eosinophilic Meningitis," Tech. Inform. Circ. No. 53, pp. 1–11. South Pacific Commission, New Caledonia.

McKissick, G. E., Ratcliffe, H. L., and Koestner, A. (1968). *Pathol. Vet.* **5**, 538–560.

McMahon, J. E. (1967). *East Afr. Med. J.* **44**, 250–255.

McMullen, D. B. (1937). *J. Parasitol.* **23**, 113–115.

Malamos, B. (1935). *Arch. Schiffs- Trop.-Hyg.* **39**, 156–171.

Malek, E. A. (1961). *Bull. Tulane Univ. Med. Fac.* **20**, 181–207.

Malley, A., Amkraut, A. A., Strejan, G., and Campbell, D. H. (1968). *J. Immunol.* **101**, 292–300.

Mandour, A. M. (1969). *J. Protozool.* **16**, 353–354.

Marinkelle, C. J. (1966). *Trans. Roy. Soc. Trop. Med. Hyg.* **60**, 109–116.

Martins, A. V. (1958). *Bull. WHO* **18**, 931–944.

Mazzotti, L., Dávalos, A., and Martínez Marañón, R. (1965). *Rev. Inst. Salubr. Enferm. Trop., Mex.* **25**, 151–162.

Meisenhelder, J. E., and Thompson, P. E. (1963). *J. Parasitol.* **49**, 567–570.

Meleney, H. E., and Moore, D. V. (1954). *Exp. Parasitol.* **3**, 128–139.

Miller, J. H. (1960). *Trans. Roy. Soc. Trop. Med. Hyg.* **54**, 44–46.

Miller, M. J., and Bray, R. S. (1966). *J. Parasitol.* **52**, 386–388.

Miller, T. A. (1968). *Trans. Roy. Soc. Trop. Med. Hyg.* **62**, 473–489.

Morton, H. L. (1969). *Lab. Anim. Care* **19**, 253–255.

Mueller, J. F. (1938). *Amer. J. Trop. Med.* **18**, 41–66.

Muller, R. L. (1970a). *Exp. Parasitol.* **27**, 357–361.

Muller, R. L. (1970b). *Nature (London)* **226**, 662.

Myers, B. J. (1968). *Int. Congr. Trop. Med. Malar., 8th, 1968* Abstracts and reviews, pp. 180–181.

Myers, B. J. (1972). *In* "Pathology of Primates" (R. N. Fiennes, ed.) **2**, 123–144. Karger, Basel.

Myers, B. J., and Kuntz, R. E. (1965). *Primates* **6**, 137–194.

Myers, B. J., and Kuntz, R. E. (1967). *East Afr. Med. J.* **44**, 322–324.

Myers, B. J., and Kuntz, R. E. (1968). *J. Protozool.* **15**, 363–365.

Myers, B. J., and Kuntz, R. E. (1972). To be published.

Myers, B. J., Kuntz, R. E., and Vice, T. E. (1965). *J. Parasitol.* **51**, 1019–1020.

Myers, B. J., Kuntz, R. E., Huang, T. C., and Moore, J. A. (1970a). *Lab. Anim. Care* **20**, 1004–1006.

Myers, B. J., Kuntz, R. E., Huang, T. C., and Moore, J. A. (1970b). *Proc. Helminthol. Soc. Wash.* **37**, 189–192.

Myers, B. J., Kuntz, R. E., Vice, T. E., and Kim, C. S. (1970c). *Lab. Anim. Care* **20**, 283–286.

Myers, B. J., Kuntz, R. E., and Malherbe, H. (1971). *Trans. Amer. Microsc. Soc.* **90**, 80–83.

Naimark, D. H., Benenson, A. S., Oliver-Gonzalez, J., McMullen, D. B., and Ritchie, L. S. (1960). *Amer. J. Trop. Med. Hyg.* **9**, 430–435.

Nelson, B., Cosgrove, G. E., and Gengozian, N. (1966). *Lab. Anim. Care* **16**, 255–275.

Nelson, G. S. (1960). *Trans. Roy. Soc. Trop. Med. Hyg.* **54**, 301–324.

Nelson, G. S. (1970). *H. D. Srivastava Commemoration Vol.* pp. 19–25.

Nelson, G. S., and Rausch, R. L. (1963). *Ann. Trop. Med. Parasitol.* **57**, 136–149.

Nelson, G. S., and Saoud, M. F. A. (1968). *J. Helminthol.* **62**, 339–362.

Nelson, G. S., Teesdale, C., and Highton, R. B. (1962). *Bilharziasis, Ciba Found. Symp., 1962* pp. 127–156.

Nelson, G. S., Pester, F. R. N., and Richman, R. (1965). *Trans. Roy. Soc. Trop. Med. Hyg.* **59**, 507–524.

Neumann, E., and Gunders, A. E. (1963). *Amer. J. Ophthalmol.* **56**, 573–588.

Obuyu, C. K. A. (1970). *Ann. Trop. Med. Parasitol.* **64**, 395–398.

Orihel, T. C. (1968). *Int. Congr. Trop. Med. Malar., 8th, 1968* Abstracts and reviews, p. 87.

Orihel, T. C. (1970). *Lab. Anim. Care* **20**, 395–401.

Orihel, T. C. (1971). *J. Parasitol.* **57**, 117–121.

Patterson, R. (1969). *Progr. Allergy* **13**, 332–407.

Pellegrino, J., and Katz, N. (1968). *Advan. Parasitol.* **6**, 233–290.

Pellegrino, J., and Katz, N. (1969). *Ann. N. Y. Acad. Sci.* **160**, 429–460.

Pope, B. L. (1966). *J. Parasitol.* **52**, 166–168.

Prathap, K., Lau, K. S., and Bolton, J. M. (1969). *Amer. J. Trop. Med. Hyg.* **18**, 20–27.

Punyagupta, S., Bunnag, T., Juttijudata, P., and Rosen, L. (1970). *Amer. J. Trop. Med. Hyg.* **19**, 950–958.

Ratcliffe, H. L. (1931). *Amer. J. Hyg.* **14**, 337–352.

Rewell, R. E. (1969). *In* "The Chimpanzee" (G. H. Bourne, ed.), Vol. 1, pp. 425–458. Karger, Basel.

Rodaniche, E. de (1954). *Amer. J. Trop. Med. Hyg.* **3**, 1026–1032.

Ruch, T. C. (1959). "Diseases of Laboratory Primates." Saunders, Philadelphia, Pennsylvania.

Sadun, E. H., and Gore, R. W. (1970). *Exp. Parasitol.* **28**, 435–449.

Sadun, E. H., Anderson, R. I., and Williams, J. S. (1962). *Bull. WHO* **27**, 151–159.

Sadun, E. H., von Lichtenberg, F., and Bruce, J. I. (1966a). *Amer. J. Trop. Med. Hyg.* **15**, 705–718.
Sadun, E. H., von Lichtenberg, F., Hickman, R. L., Bruce, J. I., Smith, J. H., and Schoenbechler, M. J. (1966b). *Amer. J. Trop. Med. Hyg.* **15**, 496–506.
Sadun, E. H., von Lichtenberg, F., Cheever, A. W., and Erickson, D. G. (1970a). *Amer. J. Trop. Med. Hyg.* **19**, 258–277.
Sadun, E. H., von Lichtenberg, F., Cheever, A. W., Erickson, D. G., and Hickman, R. L. (1970b). *Amer. J. Trop. Med. Hyg.* **19**, 427–458.
Sandosham, A. A. (1950). *J. Helminthol.* **24**, 171–204.
Sandosham, A. A. (1954). *Stud. Inst. Med. Res. F. M. S.* No. 26, pp. 213–226.
Sati, M. H. (1963). *Exp. Parasitol.* **14**, 52–53.
Schmidt, R. E., and Prine, J. R. (1970). *Pathol. Vet.* **7**, 56–59.
Schneider, C. R. (1965). *J. Parasitol.* **51**, 340–344.
Siebold, H. R., and Wolf, R. H. (1970). *Lab. Anim. Care* **20**, 514–517.
Smithers, S. R. (1962). *Bilharziasis, Ciba Found. Symp., 1962* pp. 239–265.
Smithers, S. R., and Terry, R. J. (1965). *Parasitology* **55**, 695–700.
Smithers, S. R., and Terry, R. J. (1967). *Trans. Roy. Soc. Trop. Med. Hyg.* **61**, 517–533.
Smithers, S. R., and Terry, R. J. (1969). *Ann. N. Y. Acad. Sci.* **160**, 826–840.
Standen, O. D. (1949). *Ann. Trop. Med. Parasitol.* **43**, 268–283.
Strannegård, O. (1967). *Acta Pathol. Microbiol. Scand.* **69**, 465–476.
Strong, J. P., Miller, J. H., and McGill, H. C., Jr. (1965). *In* "The Baboon in Medical Research" (H. Vagtborg, ed.), Vol. I, pp. 503–512. Univ. of Texas Press, Austin.
Swanson, V. L., and Williams, J. E. (1963). *Amer. J. Trop. Med. Hyg.* **12**, 748–752.
Swellengrebel, N. H., and Rijpstra, A. C. (1965). *Trop. Geogr. Med.* **17**, 80–84.
Verster, A. J. M. (1965). *J. S. Afr. Vet. Med. Ass.* **36**, 580.
Vickers, J. H., and Penner, L. R. (1968). *J. Amer. Vet. Med. Ass.* **153**, 868–871.
Vogel, H. (1967). *Ann. Soc. Belge Med. Trop.* **47**, 107–116.
Vogel, H., and Crewe, W. (1965). *Z. Tropenmed. Parasitol.* **16**, 109–125.
von Lichtenberg, F., and Sadun, E. H. (1968). *Exp. Parasitol.* **22**, 264–278.
Walker, A. E. (1936). *J. Comp. Pathol. Ther.* **49**, 141–145.
Warren, K. S. (1964). *Nature (London)* **201**, 899–901.
Warren, K. S., and Jane, J. A. (1967). *Trans. Roy. Soc. Trop. Med. Hyg.* **61**, 534–537.
Webbe, G., and Jordan, P. (1966). *Trans. Roy. Soc. Trop. Med. Hyg.* **60**, 279–306.
Weinman, D., and Wiratmadja, N. S. (1969). *Trans. Roy. Soc. Trop. Med. Hyg.* **63**, 497–506.
Weinstein, P. P., Rosen, L., Laqueur, G. L., and Sawyer, T. K. (1963). *Amer. J. Trop. Med. Hyg.* **12**, 358–377.
Weiszer, I., Patterson, R., and Pruzansky, J. J. (1968). *J. Allergy* **41**, 14–22.
Whitney, R. A., and Kruckenberg, S. M. (1967). *J. Amer. Vet. Med. Ass.* **151**, 907–908.
Williams, J. E., and Swanson, V. L. (1963). *Amer. J. Trop. Med. Hyg.* **12**, 753–757.
Wiseman, R. A. (1969). *Proc. Roy. Soc. Med.* **62**, 1046–1048.
Wiseman, R. A., Woodruff, A. W., and Pettitt, L. E. (1969). *Trans. Roy. Soc. Trop. Med. Hyg.* **63**, 246–250.
Wong, M. M., Fredericks, H. J., and Ramachandran, C. P. (1969). *Bull. WHO* **40**, 493–501.
Wood, D. E., and Suitor, E. C., Jr. (1966). *Nature (London)* **211**, 868–870.

Workshop on Nonhuman Primate Parasites. (1970). *Lab. Anim. Care* **20**, 319–409.
Yamashita, J. (1963). *Primates* **4**, 1–97.
Yokogawa, M. (1965). *Advan. Parasitol.* **3**, 99–158.
Yokogawa, S., Cort, W. W., and Yokogawa, M. (1960). *Exp. Parasitol.* **10**, 81–137.
Young, M. D. (1970). *Lab. Anim. Care* **20**, 361–367.
Zaman, V., and Goh, T. K. (1968). *Ann. Trop. Med. Parasitol.* **62**, 52–53.
Zaman, V., and Yin-Murphy, M. (1969). *Singapore Med. J.* **10**, 64–65.

CHAPTER 7

Primates as Organ Donors in Transplantation Studies in Man

KEITH REEMTSMA

I. Introduction

Two central problems now limit the widespread application of organ transplantation. The first is the immunologic problem of rejection; the second is the scarcity of suitable donor organs. Although the first problem generally is greater when nonhuman primate organs are used than when human organs are used, there is considerable overlap in the severity of reaction in so-called homografts and heterografts. Furthermore, some recent studies suggest that overcoming the heterograft reaction may be feasible. If means could be found to control the immunologic reaction of the human to transplanted organs from nonhuman primates, the problem of the supply of suitable donors might be greatly ameliorated.

The modern era of organ heterotransplantation began in 1963. Prior to this, it was assumed generally that any transplant across the species barrier would undergo rapid, inevitable, and irrevocable rejection. The present

203

evidence suggests that transplants between individuals of closely related species may, with immunologic suppression, behave in a qualitatively similar fashion to transplants between individuals of the same species. However, the quantitative response in most instances appears to be considerably more vigorous and difficult to control. Although the ferocity of the immunologic response usually appears to be related to the degree of genetic disparity, there are enough exceptions to make a firm generalization unwarranted in this regard.

The purpose of this report is to summarize the clinical, immunologic, and pathologic data available in clinical renal heterotransplantation. Emphasis is placed on the Tulane University series of chimpanzee-to-man transplants (Reemstma et al., 1964a) and the University of Colorado series of baboon-to-man transplants (Starzl, 1964b) because of the relatively large experience of these two groups. Excellent studies on other clinical cases with heterotransplants have been made by Hitchcock et al. (1964), Hume (1964), Traeger, Goldsmith, and Cortesini and Stefanini (personal communication). Much of the immunologic work referred to in this report is derived from the studies of DeWitt (1965, 1972) and of Kirkpatrick and Wilson (1964). Extensive pathologic studies in heterotransplanted kidneys have been reported by Porter (1964).

II. Historical Perspective

At the beginning of this century the demonstration that blood vessels could be connected with continuing patency was followed by a spate of work in organ transplantation. These pioneering studies involved a wide range of work, including transplants within the same animal (autograft), between animals of the same species (homograft), and across species lines (heterograft).* One of the earliest studies on experimental renal transplantation, reported by Ullmann in 1902, included reference to a dog-to-goat transplant.

Reports on renal heterotransplantation into man appeared early in the century. In 1905 Princeteau inserted slices of rabbit kidney into a nephrotomy on a child with renal insufficiency. "The immediate results were excellent," he wrote. "The volume of urine increased; vomiting stopped. . . . On the 16th day the child died of pulmonary congestion. . . ." In the following year, 1906, Jaboulay on two occasions attempted renal

---

* The word "heterograft" in this report applies to transplants between closely related species, such as chimpanzee and man, and also for grafts across lines of species, genus, and family. The term "xenograft" is applied to transplants between individuals of greater genetic disparity, for reasons previously discussed (Reemtsma, 1965).

heterotransplantation into man using vascular anastomoses. The hetero-grafts, one from a pig and another from a goat, were inserted into the antecubital spaces. Neither graft functioned, and failure was attributed to vascular thromboses. Unger, in 1910, described his attempt at trans-plantation of kidneys from nonhuman primate with man. The patient died 32 hr following operation, and at autopsy thromboses were found in the veins.

In 1923, Neuhof reported an experience in clinical renal heterotrans-plantation. He wrote: "In a recent case of poisoning by bichloride of mercury, I attempted a heterotransplantation of the kidney. There had been almost complete anuria for three days and the prognosis appeared hopeless. I believed that a transplanted kidney that would function tem-porarily might aid in tiding the patient over the period of urinary suppres-sion. Unfortunately, a human kidney could not be obtained and an animal graft had to be employed. The patient's blood was tested against the blood of several lambs and the animal that showed relatively little hemolysis for the patient's red blood cells was selected."

The patient died 9 days following transplantation, and, in commenting on the findings at autopsy, Neuhof wrote: "[This case] proves, however, that a heterografted kidney in a human being does not necessarily become gangrenous and the procedure is, therefore, not necessarily a dangerous one, as had been supposed. It also demonstrates that thrombosis or hemorrhage at the anastomosis is not inevitable. I believe this case report should turn attention anew. . . ."

In the following year, 1924, Avramovici reported renal transplants from cat to dog following bilateral nephrectomy in the recipient with survival, in two instances, of 49 and 58 days. However, detailed studies, such as histologic sections, are lacking in this report.

The failure to achieve long-term success in renal transplantation be-tween individuals of the same species and between individuals of different species led to a waning of interest, and during the next two decades studies were directed largely toward observations on physiologic studies and the early postoperative course following auto- and homotransplantation. More recent studies on the immediate response to cross-species grafting in animals have been reported by Brull, Louis-Bar and Marson (1956) and by Calne (1963).

The work of Medawar (1944, 1945) established that the rejection of skin homografts was basically an immunologic reaction. That a similar mechanism was involved in whole organ grafts was established by Demp-ster (1953) and Simonsen et al. (1953).

The first extensive experience with renal transplantation into unmodified recipients was reported in 1955 by Hume et al. This study provided the background for subsequent developments in the field of renal transplantation. Attempts to modify the ability of the host to reject the kidney initially focused on total body irradiation. This method was applied in this country by Murray *et al.* (1960) and in France by Hamburger *et al.* (1962) and Küss *et al.* (1951). Despite occasional successes, this method proved difficult to apply because of severe marrow depression that frequently accompanied total body irradiation.

In 1959 Schwartz and Dameshek (1959, 1960) demonstrated the specific immunosuppressive effect of a purine analog, 6-mercaptopurine, administered to animals receiving foreign antigens. The following year, 1960, Calne and Murray (1961) and Zukoski, Lee, and Hume (1960) demonstrated prolongation of renal homograft function in dogs receiving 6-mercaptopurine. These studies ushered in the modern era of clinical renal homotransplantation. Historic developments in the field of renal transplantation have been reviewed by Hume (1959), Woodruff, (1960), Calne (1963), and Starzl (1964a).

Early in the present decade encouraging results were reported from several centers using immunosuppressive therapy with renal transplantation. Paradoxically, improvement in clinical results with renal homografts served to highlight an increasingly critical problem in renal transplantation: the supply of suitable donor organs.

The decision of the Tulane group to explore clinical heterotransplantation was prompted in part by clinical urgency and in part by the belief that the pessimism surrounding heterotransplantation, derived largely from studies between widely disparate species, might not be applicable among higher primates, including man. Furthermore, it was impossible to test this basic conjecture in the laboratory, as there was no nonhuman system which would be a valid parallel to chimpanzee-to-man grafting.

In practice, all patients were in terminal uremia, maintained on dialysis, who were presented the following alternatives: (1) supporting treatment only; (2) homograft from a human volunteer, with the work "volunteer" defined in the strictest sense; (3) cadaveric homograft, if and when available; or (4) heterograft.

The risks, the uncertainties, and the experimental nature of the work were discussed with the patients and their families. If they chose to proceed with transplantation and had no volunteer donor, a search was made for a cadaveric kidney. If no suitable cadaver kidney became available, a heterograft was used, with the patient's understanding and consent.

The chimpanzee offered several advantages, principally his close relationship to man. Taxonomically the chimpanzee is classified in the category of great apes, with a number of characteristics suggesting similarity to humans. Of particular interest are the major blood groups of chimpanzees,

A and O, although only approximately 12% of all chimpanzees are blood type O (Wiener et al., 1942; Wiener and Moor-Jankowski, 1963), and several studies (Smith and Clarke, 1938; Gagnon and Clarke, 1957) have demonstrated similarity in renal function. Other studies suggesting close relationship between chimpanzee and man involve serum proteins, hemoglobin structure, and comparative parasitology. In addition, the size of the chimpanzee approximates that of man.

Serious limiting factors in the use of chimpanzees are their scarcity and expense. For this reason the University of Colorado group explored the use of the baboon as a donor. Although a more distant relative of the human than the chimpanzee, the baboon is much more common and available animal.

III. Clinical Aspects of Renal Heterotransplantation

On the basis of existing studies, it is difficult to define with precision clinical or biochemical features which would distinguish heterografts from homografts.

With the exception of one instance in which major blood group incompatibility occurred in a chimpanzee-to-man heterograft, prompt function occurred in all instances. Marked diuresis was often seen, in one instance 52 liters of urine appearing in 24 hr (Hume, 1964). The urinary loss of sodium and, particularly, potassium often appeared large in patients receiving heterografts, but whether water or electrolyte excretion differ qualitatively in cross-species transplantation compared with homografts remains conjectural.

Overall survival figures show that most heterografts failed within 2 months, and the maximal period of survival has been 9 months. However, initial function of the grafts, occurrence of episodes of rejection, and reversibility of rejection have not differed qualitatively from homografts. Immunosuppressive measures used in the New Orleans (Reemtsma et al., 1964a) and Denver (Starzl, 1964b) series were similar to protocols used in patients receiving homografts. Generally patients were treated prior to transplant with azathioprine, steroids, and actinomycin C. Postoperative treatment included these agents and, in addition, local irradiation to the graft administered either routinely or when indicated by appearance of rejection.

The occurrence of rejection episodes and the degree of reversibility of rejection followed varied patterns. Occasionally no rejection episode was observed, but generally such episodes occurred at intervals similar to those seen with homotransplants. Treatment of rejection usually included adjustment of immunosuppressive drugs and, often, local radiation to the graft. In the first six chimpanzee-to-man grafts rejection appeared to be a less serious problem than infection. However, in the subsequent six transplants performed by the same group, early and irreversible rejection was observed in all. In the Denver series of baboon-to-man grafts, at least two rejection episodes were seen in five of the six cases.

Detection of rejection episodes followed similar patterns to those seen in homografts. In most instances, sudden deterioration of renal function was seen, but occasionally slow and insidious rejection was observed. In addition to the usual determinations, serial renograms proved useful in assessing rejection.

IV. Immunologic Studies in Renal Heterotransplantation

Immunologic aspects of renal heterotransplantation have been studied extensively by DeWitt (1965, 1972) in the chimpanzee-to-man grafts and by Kirkpatrick and Wilson (1964) in the baboon-to-man grafts. These two species of donors differ in that chimpanzees may be typed as either human blood group A or O on the basis of response with anti-A and anti-B typing sera; baboons, however, do not show much cross-reaction with type-specific antisera and can be typed as A, B, and AB on the basis of blood group substances in the saliva and the appearance of the reciprocal serum antibody, as described by Wiener and Moor-Jankowski (1963). No blood type O baboons have been described, nor have blood type B or AB chimpanzees.

Detailed serologic studies of the responses of isohemagglutinin and heterohemagglutinin titers following transplantation have been made in both the chimpanzee-to-man and baboon-to-man series. Transplants between donors and recipients mismatch with respect to the ABO system were performed on two occasions in the chimpanzee-to-man series and on three occasions in the baboon-to-man series. These cases were associated with postoperative decreases in the serum levels of preformed anti-A and/or anti-B isoantibodies, presumably the result of absorption by the transplanted organs. In the baboon-to-man series, increases in these antibodies were usually seen during rejection episodes.

The heterohemagglutination response similarly has been studied in both the chimpanzee-to-man series and baboon-to-man series. A human anti-chimpanzee hemagglutinin has been found in all patients receiving chimpanzee grafts and also has been found widely distributed in the population (49 of 50 positive). The antibody is a typical heat-stable, γ-M-globulin molecule, sensitive to 2-mercaptoethanol. Following heterotransplantation a monophasic, anamnestic response occurs, apparently stimulated by residual erythrocytes in the donor organ. The heterohemagglutinin is not absorbed from human serum by chimpanzee spleen and kidney tissues, and the antigen responsible is believed to be present only on the erythrocyte. In the chimpanzee-to-man studies this system does not, therefore, reflect tissue immunity.

In the baboon-to-man series, however, there was a rapid fall in heterohemagglutination activity immediately following transplantation, and a subsequent rise was observed in one case coincident with the first rejection episode and a rise in titer was seen in the subsequent rejection crises in five patients. The reappearance of the heterohemagglutinin response in five of the six patients indicates that the transplant stimulated production of antibody, suggesting that, in the baboon, unlike the chimpanzee, the antigen is present on the kidneys as well as on the erythrocytes.

The appearance of a serum cytotoxin has been observed in all patients except one receiving chimpanzee renal heterografts (DeWitt, 1972). The response is variable in time of onset, degree, duration, and reversibility. It appears to be closely related to rejection crises and probably is a measure of immune response to transplantation. The antibody is heat-stable and complement-dependent and appears in both the γ-M and γ-G fractions. It apparently is a true heterospecific antibody as judged by cross-absorption tests with cells from six chimpanzees, but isospecificities also may be present. The reverse heteroantibody (chimpanzee antihuman) shows definite isospecificity.

Cytotoxin responses in 12 patients have shown consistent correlation with the timing of rejection episodes. The degree of response and reversibility likewise have been closely correlated with the clinical outcome.

V. Pathology of Renal Heterotransplantation

The pathologic changes in the transplants following cross-species grafting have been reviewed extensively by Porter (1964). His study provides a detailed analysis of kidneys of various species transplanted into man.

The first chimpanzee-to-man heterograft, examined 9 weeks after implantation, showed interstitial edema and tubular necrosis and repair, but no cellular infiltration and no glomerular or vascular changes. Porter commented: "The lack of any histological stigmata of rejection in these chimpanzee kidneys is most striking. . . ." Sections of the chimpanzee kidneys which functioned in a patient for 9 months similarly showed absence of cellular infiltration but marked intimal thickening of the interlobular arteries. In the recent series of six chimpanzee-to-man transplants prompt rejection occurred, and pathologic changes of severe rejection was evident: gross swelling, hemorrhagic infarcts, cellular infiltration, and vascular lesions.

The pathologic changes in the baboon-to-man renal heterotransplants have been described in detail by Porter. These kidneys showed more uniform changes than did the chimpanzee-to-man transplants. Porter commented, "All were swollen, edematous, and contained many scattered hemorrhages and hemorrhagic infarcts. . . . The baboon kidneys showed a very heavy cellular infiltration."

Although the severity of pathologic changes appears to be related to the degree of genetic disparity between donor and host, there have been exceptions which make a firm generalization unwarranted on the basis of existing evidence.

VI. Implications of Heterotransplantation

The supply of suitable organs for clinical transplantation has imposed serious restrictions, for both logistic and ethical reasons. The use of nonhuman sources removes the ethical concern, but the supply of the higher nonhuman primates is precarious. Furthermore the cross-species transplantation of organs imposes a high biologic penalty, as demonstrated by the results of these studies, performed with strenuous immune suppression.

Nevertheless the demonstration that such grafts may respond similarly to human organs suggests that the field of heterotransplantation deserves close scrutiny and continued effort. In the long-range development of transplantation the use of nonhuman organs offers several intriguing possibilities. The transplantation of nonpaired vital organs, such as heart and liver, would be greatly facilitated. The recent studies of Starzl and associates (1965) and Lower, Dong, and Shumway (1965) suggest that immunologic problems associated with transplantation of these organs may differ from those associated with renal transplantation. The development of tests to predict donor–host compatibility might be applied with particular effectiveness in heterografts, as a pool of donors might be screened on this basis. Additionally, the use of a preselected nonhuman donor would facilitate the preparation of the donor and/or host by specific immunologic methods, should such techniques prove feasible in achieving improved graft acceptance.

Finally, the specific breeding of selected nonhuman donors, including blood type O chimpanzees, should receive serious attention, as the supply of these animals is extremely limited.

REFERENCES

Avramovici, A. (1924). *Lyon Chir.* 21, 734.
Brull, L., Louis-Bar, D., and Marson, F. G. (1956). *Arch. Int. Physiol. Biochim.* 64, 269.
Calne, R. Y. (1963). "Renal Transplantation." Williams & Wilkins, Baltimore, Maryland.
Calne, R. Y. (1960). *Lancet* 1, 417.
Calne, R. Y., and Murray, J. E. (1961). *Surg. Forum* 12, 118.
Cortesini, R., and Stefanini, P. Personal communication.
Dempster, W. J. (1953). *Brit. J. Surg.* 40, 447.
DeWitt, C. W. (1965). *Fed. Proc., Fed. Amer. Soc. Exp. Biol.* 24, 573.
DeWitt, C. W. (1972). In preparation.
Gagnon, J. A., and Clarke, R. W. (1957). *Amer. J. Physiol.* 190, 117.
Goldsmith, E. A. Personal communication.
Hamburger, J. *et al.* (1962). *Amer. J. Med.* 32, 854.
Hitchcock, C. R. *et al.* (1964). *J. Amer. Med. Ass.* 189, 934.
Hume, D. M. (1959). *In* "Transplantation of Tissues" (L. A. Peer, ed.). Williams & Wilkins, Baltimore, Maryland.
Hume, D. M. (1964). Discussion of Reemtsma *et al.* (1964a).
Hume, D. M., Merrill, Miller, and Thorn. (1955). *J. Clin. Invest.* 34, 327.
Jaboulay, M. (1906). *Lyon Med.* 39, 575.
Kirkpatrick, C. H., and Wilson, W. E. C. (1964). *In* "Experience in Renal Transplantation" (T. E. Starzl, ed.), 1st ed., p. 000. Philadelphia, Pennsylvania.
Kuss, R., Teinturier, J., and Milliez, P. (1951). *Mem. Acad. Chir.* 77, 775.
Lower, R. R., Dong, E., and Shumway, N. E. (1965). *Surgery* 58, 110.
Medawar, P. B. (1944). *J. Anat.* 78, 176.
Medawar, P. B. (1945). *J. Anat.* 79, 157.
Murray, J. E. *et al.* (1960). *Surgery* 48, 272.
Neuhof, H. (1923). "The Transplantation of Tissues." Appleton, New York.
Porter, K. A. (1964). *In* "Experience in Renal Transplantation" (T. E. Starzl, ed.), 1st ed. Saunders, Philadelphia, Pennsylvania.
Princeteau, M. (1905). *J. Med. Bordeaux Sud-Ouest* 26, 549.
Reemtsma, K. (1965). *N. Engl. J. Med.* 272, 380.
Reemtsma, K. *et al.* (1964a). *Ann. Surg.* 160, 409.
Reemtsma, K. *et al.* (1964b). *J. Amer. Med. Ass.* 187, 691.
Reemtsma, K. *et al.* (1964c). *Science* 143, 700.
Schwartz, R., and Dameshek, W. (1959). *Nature (London)* 183, 1682.
Schwartz, R., and Dameshek, W. (1960). *J. Clin. Invest.* 39, 952.
Simonsen, M. *et al.* (1953). *Acta Pathol. Microbiol. Scand.* 40, 480.
Smith, H. W., and Clarke, R. W. (1938). *Amer. J. Physiol.* 122, 132.
Starzl, T. E., ed. (1964a). "Experience in Renal Transplantation," 1st ed. Saunders, Philadelphia, Pennsylvania.
Starzl, T. E. (1964b). *Transplantation* 2, 752.
Starzl, T. E. *et al.* (1965). *Surgery* 58, 131.
Traeger, J. Personal communication.
Ullmann, E. (1902). *Wien. Klin. Wochenschr.* 15, 281.
Unger, E. (1910). *Berlin. Klin. Wochenschr.* 47, 573.
Wiener, A. S., and Moor-Jankowski, J. (1963). *Science* 142, 67.

Wiener, A. S., Candela, P. B., and Goss, L. J. (1942). *J. Immunol.* **45**, 229.
Wilson, W. E. C., and Kirkpatrick, C. H. (1964). *In* "Experience in Renal Trans-
plantation" (T. E. Starzl, ed.), 1st ed. Saunders, Philadelphia, Pennsylvania.
Woodruff, M. F. A. (1960). "The Transplantation of Tissues and Organs," 1st ed.
Thomas, Springfield, Illinois.
Zukoski, C. F., Lee, H. M., and Hume, D. M. (1960). *Surg. Forum* **11**, 470.

CHAPTER 8

The Importance of Monkeys for the Study of Malignant Tumors in Man

B. A. LAPIN

Attempts to produce malignant tumors in monkeys were unsuccessful for a long time. The negative results of the experiments were associated with a small number of available "spontaneous" tumors in primates. Not more than a hundred cases involving the formation of neoplasms in monkeys are described in the literature, although the number of different monkey species employed in biomedical experiments and in the production of vaccines and tissue cultures has increased enormously. We can assume that single cases of tumors in monkeys observed by veterinarians and animal keepers are not reported; however, we believe that most tumors diagnosed in monkeys are mentioned in the literature.

We have attempted to set up a table (Table I) of malignant tumors in a number of monkey species based on the literature available. The table is made up in such a way that all benign tumors and cases involving experimentally induced malignant tumors are excluded, and it records primarily data on those animals which developed tumors after a relatively long period following total irradiation.

As far as nonmalignant tumors (or states which could be treated as pretumor) are concerned, we did not think it feasible to enumerate them. Benign tumors, such as gastric polyps (often associated with nematode (*Nochtia nochti*) invasion), are common in Java monkeys, rhesus monkeys, and *Macaca speciosa* (Bonne and Sandgroud, 1939; Lapin and Yakovleva, 1960; Smetana and Orihel, 1969). Cases of pronounced hyperplasia of the gastric mucosa of the pyloric area, sometimes with formation of mucous polyps, are observed in hamadryad baboons (Vadova and Gel'shtein, 1956b, Lapin and Yakovleva, 1960). Simple or papillar ovarian cysts occur in females of rhesus monkeys, hamadryad baboons, and South American marmosets (Lapin and Yakovleva, 1960; Flinn, 1967; Martin et al., 1970; and others). Old rhesus and South American monkeys show polyps of the cervix uteri (Lapin and Yakovleva, 1960; Moreland and Woodard, 1968).

A tumor-like hyperplasia of the thyroid gland (Lapin and Yakovleva, unpublished data) occurs in young *Papio hamadryas*. Fibromas, papillomas, angiomas of various localization (Lapin and Yakovleva, unpublished data, 1964; Ruch, 1959; and many others), and endometriosis (observed frequently in young adult and old *M. mulatta*) should be mentioned among diseases which could be related to pretumor processes (Lapin and Yakovleva, 1960; Palotay, 1971; and others). If we consider monkeys from the point of view of their sensitivity to cancerogenic effects on the basis of the frequency of described spontaneous malignant tumors, an opinion can be formed that monkeys are animals with a low tumor incidence; as a matter of fact only 70 cases among many thousands of experimental animals have been described. The opinion expressed by a great number of primatologists on the low frequency of tumors in monkeys seems to be supported by research. Further investigation of tumors in monkeys shows convincingly that the greatest number of tumors are found in old and very old monkeys, over 10 years of age. A few tumors are found in sexually mature and adult monkeys, from 3 to 10 years of age. (See Table I.)

Only one case of tumor in a monkey under 3 years of age is recorded in our table. Thus, data from the literature strongly suggest that most tumors occur in monkeys at an age corresponding to human old age. However, taking into consideration the fact that young sexually immature rhesus and green monkeys are generally employed in the production of vaccines and tissue culture as well as other experimental work, we should bear in mind that studies of neoplasm frequency in monkeys are concerned with the occurrence of tumors in monkeys in general, without regard for species specificity. However, it is known that sensitivity to cancerogenic factors can differ sharply not only within the same monkey species, but also in different populations within that species.

Thus it follows that the opinion that tumors occur rarely in monkeys is wrong. We have studied the incidence of tumors in certain monkey species at the "sensitive age," and preliminary calculations suggest that the frequency of tumors in monkeys is very much the same as the frequency of tumor diseases in man.

Leukemia, lymphogranulomatosis, and myeloid disease in green monkeys can serve as examples supporting this suggestion. We have recorded and described four cases of such diseases since 1953. It was a very high percentage (there were about 40 green monkeys in the colony at that period), surpassing considerably that observed in man and leukemic cattle.

Similar examples might be presented from the experience of maintaining squirrel monkeys at the Sukhumi colony. Two cases of epidermoid carcinoma of the cheek mucosa were detected during 1 year in a group of 30 animals that had lived 3.5 years in Sukhumi and had been imported in the sexually mature state.

Everything mentioned above and also other data show the great value of monkeys kept in special colonies in which it is possible to control breeding and to observe the animals for a long period of time. This possibility seems to be very important for experimental studies in oncology.

The absence of positive results in attempts to produce tumors in monkeys over a long period of time confirmed the experimenter's view about the low sensitivity of these animals to cancerogenic agents, just the opposite of the high sensitivity of laboratory rodents.

Petrov (1957, Sukhumi) was the first to produce malignant bone tumors in monkeys with chemical (methyl holantren) and radioactive (radium ore, radium bromide) cancerogens. An extremely long latent period following cancerogen incorporation until the time when the tumor appeared was observed. However, from our point of view, this latent period correlated with the life span of monkeys (in contrast to small laboratory animals in which the life span and latent periods are shorter), and probably better reflected the regularities characterizing human oncogenesis. However, experimenters Melnikov and Barabadze (1960) demonstrated that a considerable increase in the dose of the irradiated compound reduced the latent period between the inoculation of cancerogens and the appearance of the first signs of tumor. The consequent studies demonstrated that monkey cancerogenesis, allowing for certain peculiarities, did not differ significantly from that of small laboratory animals. It is well known that a similar carcinogenic sensitivity to coal-tar derivatives, X-rays, and radioactive phosphorus has been observed in man.

Thus it is possible to say that monkeys filled the gap between laboratory rodents and man in the experiments mentioned above.

Monkeys became of great importance for oncologic experiments when it was shown that they can be highly sensitive to the effect of tumor viruses. The publication of Munroe and Windle (1963) on the pathogenicity of Rous sarcoma virus for newborn rhesus monkeys was the first report about the possibility of producing "virus tumors" in monkeys. Shortly after this study we together with Zilber and Adjigitov (see Zilber

TABLE I

MALIGNANT TUMORS IN MONKEYS

No.	Tumor localization	Tumor characteristics	Monkey species	Sex	Age (years)	Reference
1	Upper jaw	Fibrosarcoma	M. mulatta	M	6	Rankow (1947)
2	Upper jaw	Osteosarcoma	M. mulatta	M	3	Kent and Pickering (1958)
3	Lower jaw	Sarcoma	M. mulatta	F	Adult	Bagg (1931)
4	Jaw	Periosteal tumor	Cebus (albifrons)	M	Adult	Benjamin and Lang (1969)
5	Jaw	Epithelial tumor	Ateles geofroyi		Adult	Cran (1969)
6	Humerus	Sarcoma	M. mulatta	F	3	Vadova, Gel'shehin (1956)
7	Cubit	Fibrosarcoma with metastases	P. comatus	M	9	Ratcliffe (1930)
8	Cubit	Osteosarcoma	Lemur catta	F	5	Warwick (1951)
9	Right hind limb	Sarcoma of soft tissues	Cercopithecus aethiops sabaeus	F	—	Tufanov (1968)
10	Under the skin on the abdomen	Sarcoma	Galago crassicaudatus	F	13	
11	Skin of the leg	Fibrosarcoma	Saimiri sciureus	M	Adult	Zuckerman (1930)
12	Mouth	Epidermoid carcinoma	M. mulatta	M	Adult	Sukhumi (unpublished data)
13	Mouth (mucosa)	Epidermoid cancer	Saimiri sciureus	M	Adult	Sukhumi (unpublished data)
14	Mouth (mucosa)	Epidermoid cancer	Saimiri sciureus	F	Adult	G. Hemmens (unpublished data)
15	Tongue	Epidermoid carcinoma	Macaque		—	Steiner et al. (1942)
16	Tongue	Carcinoma	M. irus	M	16	Steiner et al. (1942)
17	Tongue	Carcinoma	M. mulatta	M	16	Klüver and Brunschwig (1947)
18	Tongue	Carcinoma	M. mulatta	M	14	Klüver and Brunschwig (1947)
19	Tongue	Carcinoma	M. mulatta	F	8	Krotkina (1956)
20	Tongue	Epidermoid carcinoma	Saimiri sciureus	M	25	

	Tissue	Tumor	Species	Sex	Age	Reference
21	Nasal tissue	Undifferentiated adenocarcinoma	Cercopithecus aethiops sabaeus	M	11	Fox (1932)
22	Esophagus	Epidermoid carcinoma	M. mulatta	M	15	McCarrison (1919–1920)
23	Stomach	Carcinoma	M. fuscata	F	Adult	Kliiver (1900)
24	Stomach	Carcinoma	M. sinica	M	Adult	Schmey (1914)
25	Stomach	Adenocarcinoma	M. mulatta		—	O'Connor (1947)
26	Small intestine	Carcinoma				
27	Duodenum	Adenocarcinoma	Cercopithecus aethiops pigerithrus	F	8	Lapin (1958)
28	Pancreatic gland	Adenocarcinoma	Pigtail monkey	F	18	Ratcliffe (1933)
29	Pancreatic gland	Adenocarcinoma	P. cynocephalus	F	13	Ratcliffe (1933)
30	Papilla Fateri	Cancer	P. hamadryas	M	23	Sukhumi monkey nursery, unpublished
31	Papilla Fateri	Cancer	P. hamadryas	F	29.5	Sukhumi monkey nursery, unpublished
32	Papilla Fateri	Cancer	P. hamadryas	F	14	Sukhumi monkey nursery, unpublished
33	Gall bladder	Adenocarcinoma	Cercopithecus aethiops sabaeus	F	—	Lombard and Witte (1959)
34	Gall bladder	Adenocarcinoma	P. papio	M	23	Fox (1938)
35	Rectum	Adenocarcinoma	M. sinica		—	Ratcliffe (1940)
36	Mammary gland	Fibrosarcoma	Cercopithecus aethiops sabaeus		—	O'Connor (1947)
37	Mammary gland	Adenocarcinoma	M. mulatta	M	Old	Lombard and Witte (1959)
38	Mammary gland	Cancer	M. mulatta	F	10	Vadova and Gel'shtein (1956)
39	Mammary gland	Cancer	M. mulatta	F	Old	Ruch (1900)
40	Mammary gland (?)	Adenocarcinoma	Mandrillus sphinx	F	5	Fox (1936)
41	Cervix uteri	Carcinoma	M. sinica	F	—	
42	Cervix uteri	Epidermoid carcinoma	M. mulatta	F	Old	Hisaw and Hisaw (1958)
43	Ovary	Adenocarcinoma	M. irus	F	13	Fox (1937)

(Continued)

TABLE I (Continued)

No.	Tumor localization	Tumor characteristics	Monkey species	Sex	Age (years)	Reference
44	Kidneys	Carcinoma	M. mulatta	M	28	Fox (1934)
45	Kidneys	Carcinoma with metastases	M. mulatta	M	16	Fox (1935)
46	Kidneys	Carcinoma	M. mulatta	M	10	Fox (1935)
47	Kidneys	Carcinoma	M. mulatta	M	10	Fox (1935)
48	Kidneys	Carcinoma with metastases	Cebus apella		—	Scott (1928)
49	Kidneys	Malignant adenoma	M. mulatta	M	27	Vadova and Gel'shtein (1956)
50	Kidneys	Adenocarcinoma	Cacajao sp.		—	Plimmer (1915)
51	Kidneys	Carcinoma	M. mulatta	F	16	Antonov (1956)
52	Kidneys	Carcinoma	M. mulatta		—	Ratcliffe (1940)
53	Kidneys	Carcinoma	Aotus trivirgatus	F	—	Lund et al. (1970)
54	Brain	Gliosarcoma	P. doguera	M	6	Vadova and Gel'shtein (1956)
55	Lungs	Carcinoma	Saimiri sciureus	F	—	Lombard and Witte (1959)
56	Lungs	Alveolar carcinoma			—	Maruffo (1967)
57	Epicardium	Carcinoma	Cebus apella	F	2	Iglesias and Lipschütz (1947)
58	Skin	Pigment-free melanoma (?)	M. mulatta	F	6	Blochin et al. (1955)
59	Scalp	Carcinoma	M. maura		9	Ratcliffe (1956)
60	Scalp	Sarcoma	Cercopithecus		—	Plimmer (1914)
61	Skin	Epidermoid carcinoma	M. maura		9	Ratcliffe (1956)
62	Skin	Epithelioma	M. mulatta	F	11	Schiller et al. (1969)
63	Hypophysis	Carcinoma	Baboon		—	Goodhart (1885)
64	Adrenal	Hypernephroma	Cebus apella	M	4	Fox (1923)

				Sex	Age	Reference
65	Prostate gland	Carcinoma	*M. mulatta*	M	Very old	Engle and Stout (1940)
66	Ovary	Sarcoma	*P. cynocephalus*	F	20	Lapin and Yakovleva (1957)
67	Left testicle	Seminoma	*M. mulatta*	M	—	Palotay (1971)
68	Testicle	Seminoma	Howler monkey	M	5	Maruffo (1967)
69	Ovary	Papillar cystoadenoma	*M. mulatta*	F	—	Martin *et al.* (1970)
70	Sexual skin	Cancer	*P. hamadryas*	F	19	Lapin and Yakovleva (1964)
71	Thyroid gland	Cancer	*Saguinus nigricollis*	F	Adult	Williamson and Hunt (1970)
72		Acute lymphatic leukemia	*Hylobates concolor*	M	3.5	De Paoli and Garner (1968)
73		Lymphoma (malignant)	*Hylobates concolor*	M	3.5	Lingeman (1969)
74		Burkitt's lymphosarcoma	*Hylobates lar*	M	4.5	Di Giacomo (1967)
75		Reticulosis	*Cercopithecus aethiops sabaeus*	F	10	Lapin and Yakovleva (1960, 1964)
76		Hemocytoblastosis	*Cercopithecus aethiops sabaeus*	F	8	Lapin and Yakovleva (1960, 1964)
77		Myelomatosis	*Cercopithecus aethiops sabaeus*	M	5	Lapin and Yakovleva (1964)
78		Lymphogranulomatosis	*Cercopithecus aethiops sabaeus*	F	10	Lapin (1956)

et al., 1964, 1965), managed to produce tumors not only in newborn macaques, but also in hamadryad baboons and pigtail macaques. It appeared that the sensitivity to the virus was demonstrated not only in newborns, but also in 1-month-old animals. Deinhart (1967) induced tumors in South American marmosets which, in contrast to tumors in our and Munroe and Windle's experiments, showed widespread metastases. It is difficult to overestimate the significance of the experiments by Munroe and Windle because (1) they demonstrated for the first time the possibility of inducing virus tumors in a member of the order of Primates; and (2) showed the ability of the tumor virus to overcome the species barrier. The sensitivity of primates to an adapted tumor virus belonging to another host was also demonstrated. Later on it was shown by Obukh that there existed certain Rous sarcoma strains that were highly pathogenic, not only for newborns (or infants at the age of a few days or weeks), but also for adult animals. Our experiments, and especially the experiments carried out by Obukh, proved that tumors, including very large ones, can be subjected to partial and complete regression. These observations on immunologic dependence of tumors are of the utmost importance and stimulate our search for similarity in man.

Numerous experiments carried out by researchers on different monkey species involving inoculation of various viruses oncogenic for other species of animals (e.g., SV40, polyoma virus, different types and strains of human and monkey adenoviruses) have not led to positive results up to the present time. Perhaps it would be more precise to speak in terms of the absence of positive results of such studies, but not the absence of the oncogenic effect of the viruses on primates.

Very interesting tests on the inoculation of green monkeys with cellular material from patients with Burkitt's lymphoma were carried out by Epstein *et al.* (1964). These animals showed atypical growth in the long-bone marrow. This enabled the author of the study to suggest a possible relationship between the inoculation of material and the appearance of pathologic growth in the bone marrow. In spite of the fact that the studies and suggestions of this author were subjected to severe criticism (Wright and Bell, 1964) we believe that the relation of the observed changes to "spontaneous" phenomena remain unproved.

Given the virus concept of malignant tumors and the phylogenetic similarity of monkeys and man, we think it very important to test different neoplasms of man in different monkey species. We have studied the tumor activity of material from leukemic patients for several years (1967, 1969, 1970, 1971). We established that the inoculation of two monkey species, *M. speciosa* and *P. hamadryas*, with human leukemic blood produced a disease which could be characterized as "leukemia-like," in some cases acquiring the character of true leukemia. This disease, appearing once, could be passed later to other monkeys many times by both whole blood and cell-free material, including filtrated material. Virus particles, 850–1200 Å, were detected in sick monkeys after primary inoculation as well as in the process of passages in blood plasma after ultracentrifugation at 40,000g and negative contrast. C-type virus particles were regularly detected in ultrathin leukocyte sections of sick monkeys after short-term cultivation. These virus particles propagated by means of budding, had the same size, and did not differ in morphology from viruses detected in leukemic mice, birds, and cats. These experiments were most successful in *P. hamadryas* of the Sukhumi breeding colony in which pregnant females were inoculated *in utero*. After the birth of the infant the mother was inoculated again with material from a leukemic patient.

A considerable outbreak of leukemia was recorded during the above-mentioned experiment in hamadryad baboons of the Sukhumi stock which were not subjected to experimental procedures but kept in close contact with the experimental animals. The manner of infection is not known; however, it was established that the urine of leukemic baboons contained an active virus, which, when ultracentrifuged and then injected into animals of the same species or *M. speciosa,* caused a similar disease. The question whether the virus causing a "leukemia-like" disease or true leukemia in *P. hamadryas* and *M. speciosa* is a human leukemic virus or a monkey leukemic virus activated by human leukemic blood, is still being disputed. In spite of the fact that the blood from healthy people injected into the monkey did not induce the disease and, although human leukemic blood lost a similar ability after heating (2 hr, 60°C), the question about the origin of the virus is still open.

The successful production of the disease in animals of a certain breeding stock shows again how important it is to carry out investigations in colonies, in animals with a known genealogy where it is possible to conduct genetic and epizoological studies, and the conditions of the laboratory vivarium do not restrict the duration of the experiment.

The importance of monkeys for the study of the oncologic problem in its great variety, from our point of view, cannot be disputed. There is a gap between increasingly numerous data obtained from different laboratory animals and human practice. The creative ideas and theoretical hypotheses are related to the animal world and actually do not involve human beings.

Thus, for example, the virus etiology of most of the tumors in animals gives rise to doubt in no one, while the majority of experimenters reject the idea that even certain human tumors are provoked by viruses. A great number of such instances might be given. Thus it is clear that experimental studies in man lag behind experimental studies in animals. In such a situation the obtaining of data from monkeys, phylogenetically similar to man, seems to fill the gap between experimental oncology and human oncology. If it is possible to confirm that the virus discovered by Chopra *et al.* (1971) in a spontaneous tumor of the mammary gland in an *M. mulatta* has an etiologic importance for the development if this tumor, it will be possible to extend a chain from Bittner's mammary gland cancer in mice to that in primates and will be more likely to prove the probability of the virus origin of the mammary gland cancer in women.

The importance of monkeys for experimental studies in oncology is also related to the evolutionary similarity of monkeys and man. The anatomophysiologic similarity of a number of physiologic systems allows us to expect in monkeys a sensitivity to cancerogenic effects of the environment similar to man and close development peculiarities of pathologic processes.

REFERENCES

Antonov, A. M. (1956). *Vop. Onkol.* [N.S.] 2, 198.
Bagg, H. J. (1931). *Amer. J. Cancer* 15, 2143–2148.
Benjamin, S. A., and Lang, C. M. (1969). *J. Amer. Vet. Med. Ass.* 155, 1236–1240.
Blochin, N. N., Vasil'ev, Ju.M., and Pogosjanc, E. E. (1955). *Vop. Onkol.* [N.S.] 1, 91.
Chopra, H. C., Zelljadt, I., Jensen, E. M., Mason, M. M., and Woodside, N. Y. (1971). *J. Nat. Cancer Inst.* 46, 127–137.
Cran, I. A. (1969). *Oral Surg., Oral Med. Oral Pathol.* 27, 494–498.
Deinhardt, F. (1967). Rous sarcoma virus induced neoplasms in New World non-human primates. *Perspect. Virol.* 5, 183–197.
De Paoli, A., and Garner, F. M. (1968). *Cancer Res.* 28, 2559–2561.
Di Giacomo, R. F. (1967). *Cancer Res.* 27, 1178–1179.
Engle, E., and Stout, A. P. (1940). *Amer. J. Cancer* 39, 334–337.
Epstein, M. A., Woodall, I. P., and Thompson, A. O. (1964). *Lancet* i, 288.
Flinn, R. M. (1967). *J. Pathol. Bacteriol.* 94, 451–452.
Fox, H. (1923). "Disease in Captive Wild Mammals and Birds. Incidence, Description, Comparison." Lippincott, Philadelphia, Pennsylvania.
Fox, H. (1932). *Rep. Lab. Comp. Pathol. Philadelphia* pp. 16–22.
Fox, H. (1934). *Rep. Lab. Comp. Pathol. Philadelphia* pp. 17–30.
Fox, H. (1935). *Rep. Lab. Comp. Pathol. Philadelphia* pp. 12–18.
Fox, H. (1936). *Rep. Penrose Res. Lab.* pp. 14–19.
Fox, H. (1937). *Rep. Penrose Res. Lab.* pp. 7–12.
Fox, H. (1938). *Rep. Penrose Res. Lab.* pp. 17–26.
Goodhart, L. F. (1885). *Trans. Pathol. Soc. London* 36, 36.
Hisaw, F. L., and Hisaw, F. L. (1958). *Cancer* 11, 810–816.
Iglesias, R., and Lipschütz, A. (1947). *J. Endocrinol.* 5, 88–98.
Kent, S. P., and Pickering, J. E. (1958). *Cancer* 11, 138–147.
Klüver, H. Unpublished data.

Knauer, K. W., Vice, T. E., Kim, C. S., and Kalter, S. S. (1969). *Primates* **10**, 285–293.

Krotkina, N. A. (1956). *Vop. Onkol.* [N.S.] **2**, 748–749.

Lapin, B. A. (1956). *In* "Theoretical and Practical Problems of Medicine and Biology in Experiments on Monkeys," pp. 123–125. Medgiz, Moscow.

Lapin, B. A., and Yakovleva, L. A. (1957). *Revolutsii i 30-letiu Sukhumskoi Med.-Biol. Stantsii AMN SSSR, Sukhumi*, pp. 55–59.

Lapin, B. A., and Yakovleva, L. A. (1960). "Comparative Pathology of Monkeys." Medgiz, Moscow.

Lapin, B. A., and Yakovleva, L. A. (1964). "Vergleichende Pathologie der Affen." Fischer, Jena.

Lapin, B. A., and Yakovleva, L. A. Unpublished data.

Lingeman, C. H. (1969). *Nat. Cancer Inst., Monogr.* **32**, 157–170.

Lombard, L. S., and Witte, E. J. (1959). *Cancer Res.* **19**, 127–141.

Lund, J. E., Burkholder, C., and Soave, D. (1970). *Pathol. Vet.* **7**, 270–274.

McCarrison, R. (1919–1920). *Indian J. Med. Res.* **7**, 342–345.

Martin, C. B., Misenhimer, H. R., and Ramsey, E. M. (1970). *Lab. Anim. Care* **20**, 682–692.

Maruffo, C. A. (1967). *Nature (London)* **213**, 521.

Melnikov, R. A., and Barabadze, E. M. (1960). *Vop. Onkol.* [N.S.] **6**, 69–72.

Melnikov, R. A., and Barabadze, E. M. (1968). "Malignant Tumours in Monkeys." Izdat. "Nauka," Leningrad.

Moreland, A. F., and Woodard, J. C. (1968). *Pathol. Vet.* **5**, 193–198.

Munroe, J. S., and Windle, W. F. (1963). *Science* **140**, 1415–1416.

O'Connor, H. (1947). *Anim. Zoo Mag.* **14**, 2.

Palotay, J. (1971). *Primate News* **9**, 1.

Petrov, N. N., Krotkina, N. A., Vadova, A. V., and Postnikova, Z. A. (1951). "Dynamics of Malignant Growth Appearance and Development in Experiments in Monkeys." AMN SSSR, Moscow.

Plimmer, H. G. (1914). *Proc. Zool. Soc. London* **1**, 181–190.

Plimmer, H. G. (1915). *Proc. Zool. Soc. London* **1**, 123–130.

Rankow, R. M. (1947). *J. Dent. Res.* **26**, 333–336.

Ratcliffe, H. L. (1930). *J. Cancer Res.* **14**, 453–460.

Ratcliffe, H. L. (1933). *Amer. J. Cancer* **17**, 116–135.

Ratcliffe, H. L. (1940). *Amer. J. Pathol.* **16**, 619–624.

Ratcliffe, H. L. (1954). *Rep. Penrose Res. Lab.* pp. 6–16.

Ruch, T. C. (1959). "Diseases of Laboratory Primates." Saunders, Philadelphia, Pennsylvania.

Ruch, T. C. Unpublished data.

Schiller, A. L. (1969). *J. Pathol.* **99**, 327–329.

Schmey, M. (1914). *Deut. Tierärztl. Wochenschr.* **22**, 377–380.

Scott, H. H. (1928). *Proc. Zool. Soc. London*, 173–198.

Smetana, H. F., and Orihel, T. C. (1969). *J. Parasitol.* **55**, 349–351.

Steiner, P. E., Klüver, H., and Brunschwig, A. (1942). *Cancer Res.* **2**, 704–709.

Sukhumi Primate Center. Unpublished data.

Tufanov, A. V. (1968). *Mater. Nauch. Sess. Inst. Polyomielita i Virus. Entsefalitov,* *15th* pp. 94–95.

Vadova, A. V., and Gel'shtein, V. I. (1956a). *Vop. Onkol.* [N.S.] **2**, 391.

Vadova, A. V., and Gel'shtein, V. I. (1956b). *In* "Theoretical and Practical Problems of Medicine and Biology in Experiments on Monkeys," pp. 107–122. Medgiz, Moscow.

Warwick, R. A. (1951). *J. Pathol. Bacteriol.* **63**, 499–501.

Williamson, M. E., and Hunt, R. D. (1970). *Lab. Anim. Care* **20**, 6.

Wright, D. H., and Bell, T. M. (1964). *Lancet* **7**, 7366.

Yakovleva, L. A. (1969). *In* "Comparative Leukemia Research," pp. 761–772. Karger, Basel.

Yakovleva, L. A., and Lapin, B. A. (1960). *Int. Symp. Pathol. Anim. Zoos, 2nd., 1960* pp. 201–208.

Zilber, L. A., Lapin, B. A., and Adzigitov, F. I. (1964). *Vop. Virusol.* **4**, 498–499.

Zilber, L. A., Lapin, B. A., and Adzigitov, F. I. (1965). *Nature (London)* **205**, 1123–1124.

Zuckerman, S. (1930). *Proc. Zool. Soc. London* **1**, 59–61.

CHAPTER 9

The Use of Primates in Cardiovascular Research

G. A. GRESHAM

I. Introduction

Cardiovascular disease is, nowadays, one of the major problems that confronts man; this is especially true for people living in organized, stressful, overfed societies. The prime cause of death is occlusive vascular disease affecting coronary vessels in the folk of many western countries and the cerebral vessels in those who live in eastern parts of the world. Contributory factors seem to be such things as stress, hypertension, obesity, and hyperlipidemia and many of these are clearly interrelated. There is no doubt that longevity and better diagnosis are not the sole explanations for this upsurge of vascular disease (Morris, 1951) and that others must be sought by various means available to us. Studies of man himself have not been especially fruitful; epidemiologic and social studies are often fraught with the difficulties of recognizing the presence of arterial disease in the human subject (Gresham and Howard, 1965a). Prospective dietary studies in man abound with similar difficulties. It is not therefore surprising that experimental animals have been much-used in this field from the end of the first decade of this century.

When one reviews the whole field of the use of experimental animals in cardiovascular research it is interesting to observe how certain species of animal have become, as it were, earmarked for particular interests. Rabbits were first used (Anitschkow, 1933) to study atherosclerosis because Ignatowski discovered the disease accidentally when he was studying the effect of unnatural diets containing meat, milk, and eggs. Similar results were not so easily obtained in guinea pigs and rats so that little was done with such animals until the 1950's. Surgical expediency dictated the use of dogs in experimentally induced hypertension (Goldblatt et al., 1934); rats were used later but the picture was complicated by the fact that these animals often became hypertensive spontaneously (Wilson and Byrom, 1939).

This hit-and-miss method, often dictated by financial circumstances, has in recent times been replaced by a more systematic approach to the study of problems in cardiovascular disease research. Many have criticized the alleged similarity between induced cardiovascular disease in the conventional laboratory animal and that found in man. Though not entirely justifiable, such comments have prompted a search for experimental models that bear a closer resemblance to the human being. Primates have been the subject of such scrutiny over the past 20 years and this chapter is concerned with a consideration of their merits and failings in this regard.

Acute experiments, such as the surgical induction of hypertension, cardiac transplantation, and studies on mechanical interference with cardiac blood supply present few problems. The results are usually clearly defined and easy to measure. It is the chronic experiments that span years, such as a study of atherogenesis, that create difficulties in supplies of adequate controls and in the mensuration of the extent of disease as compared with controls. The establishment of primate centers has already gone far to solve these problems and if a case can be made for the use of primates in cardiovascular research, and I think it can, then such centers need to be developed further throughout the scientific world and particularly in Europe.

II. Historical Aspects of Primate Research

Primates are new arrivals in the field of vascular research and little work was done with them before the 1950's. Some early experiments were made in 1937 by Goldblatt who used giant macaques to study hypertension. He found the procedure more difficult than in dogs because of the small renal arteries but succeeded in producing hypertension in three out of five animals (Goldblatt, 1937). As early as 1927 Kawamura attempted to produce atherosclerosis in monkeys by feeding diets containing cholesterol but he found these animals to be most resistant to this procedure so that little more was done along these lines with rhesus monkeys until Taylor and his colleagues instituted dietary studies some 20 years later (1959). Meanwhile a number of other avenues were explored. Attempts to accentuate arterial disease by thyroidectomy coupled with cholesterol-feeding were not successful (Sperry *et al.,* 1944). Nor was the work, in 1949 (Rinehart and Greenberg, 1949), that claimed to produce fibrous plaques in pyridoxine-deficient rhesus monkeys subsequently confirmed (Mann *et al.,* 1953b). At one time (Mann *et al.,* 1953a) methionine deficiency was thought to be necessary if cholesterol-feeding were to result in atherosclerosis. Once more this view was not subsequently sustained. It did appear that primates were not likely to be useful in the study of atherosclerosis, particularly because lesions were sparse and were most often found in the aorta but not in the coronary arteries that were the vessels of especial interest to workers in the field. Furthermore it was seemingly possible to produce thrombosis in diseased vessels and this is one of the major events that affects the atherosclerotic arteries in man.

Part of the difficulty lay in the failure to appreciate that many human cardiovascular diseases develop over a long period of time and in the expectation that they might be replicated in young agile primates in a few years. Lack of knowledge about the occurrence of cardiovascular diseases in wild primates and a dearth of information about their normal vascular physiology and structure also impeded an understanding of many of the experiments that had been made. Today much of this basic work has been and is being done in primate centers and in smaller collections throughout the world and prospects for the meaningful use of these animals in a study of cardiovascular disorder are much more optimistic.

III. The Selection of a Suitable Primate

To some extent the type of primate that is used depends upon the nature of the experiments that are to be made. Organ transplantation, such as the grafting of hearts, demands large primates and a supply of similar animals of related blood-group types to act as blood donors. Studies of cardiovascular hemodynamics are easiest made in docile animals that are amenable to handling and are readily restrained. Work on arterial disease usually demands large numbers of animals that are cheap, easily housed, and not too costly to feed. In addition to all of this, however, the first aim should be to use animals that resemble man as closely as possible. Similarities of blood groups, lipid metabolism, myocardial polarization, and blood pressure and flow are but a few of the desirable resemblances that should be sought. In many fields, such as tissue-typing, much awaits discovery in the future.

The first major transition in using primates as experimental subjects is their removal from the wild state. Increasing evidence is accumulating to show how different these animals are socially, biochemically, and physiologically when observed in their natural habitat, and every effort should be made to mimic this as closely as possible in the experimental situation. A solitary baboon in a cage is a very different creature from the same animal roaming wild in a troop.

Washburn *et al.* (1965) have recently reviewed field studies on primate behavior; among the most notable pieces of work is that of Goodall in 1965. The importance of a thorough understanding of such problems cannot be too strongly emphasized when studying vascular disease in these animals. Despite much work, however, the definition of stress remains unclear, but if stress implies frustration then this is certainly one of the features of captivity. This, in turn, may lead to excessive outpouring of catecholamines, release of free fatty acids with consequent widespread effects upon the vascular system that may confuse attempts to interpret experiments such as those made to study mechanisms of atherogenesis. A variety of workers have illustrated the need to mimic the wild state as closely as is possible and this can be done (Cooper, 1964; Eyestone, 1965; Hummer, 1965; Van Riper *et al.*, 1967).

Any studies on arterial disease must be preceded by a detailed knowledge of the prevalence of arterial disorders in the particular primate in the wild state. Extensive studies have now been made on many primates and lesions have been demonstrated in *Cebus, Saimiri,* Macaques, Pongidae, *Papio,* and others (Gresham and Howard, 1965; Goldsmith and Moor-Jankowski, 1969). Generally speaking, lesions were most often found in the aorta, less often in coronary arteries. Lesions were similar in appearance in both New World and Old World monkeys and consisted essentially of intimal thickenings composed of proliferations of fibroblasts and smooth-muscle cells together with mucopolysaccharide and fragmentation of the adjacent internal elastic lamella (Lindsay and Chaikoff, 1966). Lipid was not always found; indeed the naturally occurring plaques seemed to be more in the nature of a proliferative intimal process rather than a lipid infiltration. This is in contrast to many experimentally induced lesions in primates and in other animals where lipid infiltration often precedes the proliferative changes in the intima.

Nevertheless the lesions that have been found, for example, in the coronary arteries of free-ranging howler monkeys, are very similar to the early stages of atherosclerosis in man (Malinow and Maruffo, 1966; Malinow and Storvick, 1968). McGill *et al.* (1960) have described a similar process in the arteries of free-ranging baboons in Kenya. These lesions are not related to the sex of the animal and occur in association with cholesterol levels that are low compared to man. They are ideal for the study of the evolution of atherosclerosis. Similar lesions in children may well be the substrate upon which lipids collect under the influence of diet and rising blood pressure in later life, resulting in occlusive athero-matosis which is the disease par excellence that requires further study and elucidation of the mechanisms involved in its production. There seems little doubt that occlusive atheroma develops in two stages: first comes the intimal proliferative response, then lipids accumulate and narrow the vessel. In some animals, such as dogs and cats, it is difficult to develop this pro-gression; in primates it is easy and this encourages the need to use these animals for studies of atherogenesis.

Other changes may be found in the arteries of wild and captive primates: some are clearly parasitic in origin; others, such as medial and intimal elastic calcifications, are less easy to explain. An excessive intake of vitamin D_3 will cause elastic-tissue calcification in primates such as the baboon (Gresham and Howard, 1969). We do not, as yet, know enough about the levels of and needs for this vitamin in different primate species and we certainly need to learn more of the variable susceptibilities of the arteries of different primates when exposed to the effects of vitamin D_3. A good deal of confusion has occurred in the past because of failure to appreciate the possibility that differential responses might occur in primates given this substance as part of the "natural" diet. Vitamin D_3 also seems to have a "lipid-loading" effect (Norman, 1968). The major classes of lipids have all been shown to be increased in rats fed upon large amounts of this vitamin. This effect alone can add considerable confusion to studies of the effects of dietary lipids on atherogenesis.

For example, vitamin D_3 has been shown to be biologically more active than vitamin D_2 in New World monkeys using the absorption of calcium as one of the measures of activity. Examination of animals that had received large doses of vitamin D_3 revealed nephrocalcinosis in squirrel monkeys but much more extensive calcification involving other organs, such as the heart and aorta, in *Cebus* monkeys (Hunt *et al.,* 1969). Here there is a clear demonstration of the point, namely, that a detailed careful study of a wide variety of parameters needs to be made in any primate that is to be used for vascular research.

So far as dietary experiments are concerned it is important to establish that the metabolism of various food substances is similar to that of man. One aspect of this that has been extensively studied is lipid metabolism. The principal fatty acids that esterify lipids vary a good deal from one animal to another. Linoleic acid is one of the principal fatty acids in man and this has been shown to be true for other primates such as the baboon (Kritchevsky *et al.*, 1962). Furthermore it can be shown experimentally that ^{14}C-labeled linoleic acid is readily incorporated into the phospholipids of slices of aortic wall studied *in vitro*. The establishment of the close similarity between these fundamentally important structural lipids in the vasculature of man and other primates is a strong recommendation for the use of these animals for vascular studies that involve lipids. Further studies using the technique of arterial perfusion that has been popularized by Lofland and Clarkson (1965) and Bowyer *et al.* (1968) are continuing. These experiments will, we hope, increase our understanding of the ways in which the normal lipid metabolism of the vessel wall proceeds. When this has been adequately recorded further studies of different arteries, such as coronary, cerebral, and aorta, and indeed of different parts of the aorta, may explain the differential responses of these vessels in disease states.

It is important, in studies of this sort that are made upon captive primates, not to forget the possibility that captivity itself may have caused some change in the lipid metabolism of blood vessels. The effect of an abnormal environment has already been discussed and cannot be stressed too frequently. Meier *et al.* (1963) studied the effect of restraint on the serum-cholesterol levels of male and female rhesus monkeys and demonstrated variations in levels of cholesterol that could be statistically related to the periods of restraint. St. Clair *et al.* (1967) also showed alterations in the levels of serum cholesterol in squirrel monkeys kept in the laboratory as compared to the levels in the same animals after they had been newly trapped. The cholesterol levels rose after captivation, though they tended to fall again when the animals became acclimatized to the environment. These changes were entirely accounted for by a rise in the β-lipoprotein cholesterol and there was no demonstrable change in that transported by the α-lipoproteins. It is interesting to note that this altered ratio of β- to α-lipoproteins approximates to the condition that obtains in man. The precise way in which the stress of captivity achieves this result is not as yet clear but further studies may go far to explain some of the unsolved problems of atherogenesis.

Alterations in the blood lipid levels of animals in captivity, that may or may not be fed on atherogenic diets, vary within a species. In several species it can be demonstrated that some animals are "responders"; that is to say, their lipid levels rise following the appropriate stimulus. Others do not respond at all well. We have observed this phenomenon in several experiments using baboons, squirrel monkeys, rabbits, and rats (Bowyer and Gresham, 1971). Sometimes a responding species can be recognized by its anatomic characteristics but often, as, for example, among the New Zealand strain of rabbit, there is no external feature that enables a distinction to be made between responders and nonresponders. It is important to be fully aware of this possibility and to weed out nonresponders early in an experimental program. It may be possible to breed responders selectively just as one can breed strains of hypertensive rats. Here again is another aspect of primate research that could best be done in a well-equipped primate center with all the necessary equipment for the measurement of normal and abnormal parameters.

There is clearly a need for a thorough knowledge of the basic properties of any primate that is used for cardiovascular research, and much remains to be done. This point may be the more clearly illustrated by consideration of animals that have been studied in the past.

IV. The Rhesus Monkey

Any hope of solving the problem of atherogenesis can only be achieved by a study of the earliest stages of the disease. There is still disagreement about the initial stages. There are those who support the early occurrence of subendothelial edema (Dahme, 1968), those who regard fragmentation of the internal elastic lamella as the primary event, and finally there is the continued debate about the significance of diffuse musculoelastic intimal thickening that develops in man with increasing age. It is at present difficult to ascertain whether the latter process is a unique feature of the primate *Homo sapiens* that occurs because of progressively rising blood pressure from childhood to adult life. Even more difficult is the relationship of diffuse intimal thickening to atherosclerosis.

Vlodaver *et al.* (1968) studied the coronary arteries of immature rhesus monkeys and were able to demonstrate that focal changes occurred in the vessels but not diffuse proliferation. These focal changes consisted first of fragmentation of the internal elastic membrane followed by fibroblastic proliferation in the vicinity. Macrophages, collagen, and new elastic fibers later appeared. These various changes were not related to the fat content of the diet nor to the blood lipoid levels. Because such histological appearances did not appear in immature wild monkeys they concluded that some factor, such as stress, associated with captivity, may have contributed to their development. They also postulated that diffuse intimal thickening was a feature of normal man, though the evidence for this statement is not clearly given. Any discussion about young and old primates must involve a knowledge of the precise determinants for aging, and there is often no agreement about these. Haigh and Scott (1965) have laid down useful criteria for age determination and these and other authors should be consulted for such information.

Other workers take a different view about the earliest stage of atherosclerosis, and Shimamoto (1969), working with the rhesus monkey and other animals, has emphasized the so-called "edematous reaction" that is also a feature of some German writings. The view is that fluid first collects in the subendothelial space either by increased permeability of endothelial cells or by increased porosity of terminal capillaries of the vasa vasorum. This process can be exaggerated by the intravenous use of catecholamines and other atherogenic agents.

Some workers claim that rhesus monkeys do not develop atherosclerosis unless they are subjected to a dietary regime of some sort or another. Portman and Alexander (1966) did not find any lesions in a study of fetal and adult animals but they did record changes in the lipid composition of the aorta with increasing size. These changes occurred mainly in the inner layers of the vessel and were not paralleled by lipid changes from a variety of other tissues and organs. The alterations observed in the aorta consisted of an increase in free and esterified cholesterol and in sphingomyelin. It is of considerable interest that similar lipid changes occur in the human aorta as it becomes atherosclerotic; they had therefore defined a biochemical rather than a morphologic criterion for the recognition of atherosclerosis. Work along these lines is likely to be of considerable profit in the study of atherogenesis, for it is always difficult to quantitate morphologic processes precisely. Detailed analysis of various sorts of human lesions by Bowyer (1967) and Smith (1965) support the work of Portman and Alexander and attention is being diverted to the various metabolic pathways that might be altered in order to produce this effect. It is most likely that sphingomyelin is synthesized, rather than derived from the membranes of cells, including those that circulate in the blood. As it is a relatively inert substance metabolically, this may explain accumulation with increasing age.

Attempts to reverse lipid accumulation in atherosclerosis may follow a variety of lines of thought. Some advocate experiments designed to lower the plasma load of lipids and lipoproteins. Others consider it important to try to impede lipid synthesis in the arterial wall, and still others have worked on schemes to promote more rapid removal. A variety of arterial and plasma enzymes have been studied with these points in mind. Cholesterol ester hydrolase and sphingomyelin hydrolases seem likely proteins to study. If their activity can be specifically enhanced there is good reason to believe that lipid accumulation may be reversed. The Glomsett enzyme, which is principally concerned in the synthesis of cholesterol esters [lecithin cholesterol acyl transferase (LCAT)], may be another avenue of study. Here the aim would be to reduce synthesis rather than to promote removal of lipid. It is important to realize that many enzymes involved in lipid anabolism and catabolism by the arterial wall, other than the few that have been mentioned, may provide better clues to an understanding of atherogenesis. In this particular field much more needs to be done with primate arteries.

No section about rhesus monkeys would be complete without reference to the extensive studies of Taylor and his colleagues (1962), who were able to reproduce a variety of atherosclerotic lesions by dietary means. The more intensive measures led to the appearance of xanthomatosis as has been reported by Armstrong et al. (1967). They reported the development of xanthomas that were distributed in much the same way as in man following periods of cholesterol-feeding of as little as 10 months. Such xanthomatosis was closely correlated with hypercholesterolemia but not with hypertriglyceridemia. It was notable that, despite the appearance of extensive xanthomas, xanthelasma and corneal arcus senilis were always absent. This is interesting in relationship to the frequently debated suggestion that arcus senilis, in man, is one of the indices of coronary atherosclerosis. Most workers now agree that this is not so. Cholesterol, phospholipids, and neutral fats are the main constituents of the arcus lesion. It increases in incidence with age, but is only occasionally significant when it appears in young men with hypercholesterolemia, i.e., more than 250 mg/100 ml (McAndrew and Ogston, 1965).

Detailed studies of the arterial lesions induced by diet have been made by means of electron microscopy (Scott et al., 1967). They now confirm the widely accepted view that the fundamental process ab initio seems to be a proliferation of smooth-muscle cells in the intima.

The rhesus monkey is now well established as a tool for the study of atherogenesis and dietary regimes of various sorts will produce arterial lesions (Younger et al., 1969). The most satisfactory reproducible model is achieved with thyroid ablation by ^{131}I joined with a diet containing cholesterol. When feeding of fat is coupled with experimentally induced hypertension the percentage of the intimal surface that is involved with atherosclerotic lesions increases. McGill et al. (1961) showed that normotensive rhesus monkeys had about 1–3% aortic involvement, whereas the extent of the disease increased from 3 to 30% in animals made hypertensive. The lesions in the hypertensive monkeys were nearly all fibrous plaques, and it looked as though fatty streaks and fibrous plaques are sequential lesions in the rhesus monkey. The important aggravating factor that precipitates conversion is hypertension and points to the need for detailed investigation of changes in blood pressure in growing human beings in relation to the extent of atherosclerosis found at necropsy. Such a study would need to be on a nationwide cooperative basis as the amount of material available is scanty.

V. The Baboon

These primates are more amenable and less liable to infections that are a danger to man than is the rhesus monkey. The recognition of precise species within the genus *Papio* is not as easy as some would have us believe and there are undoubted differences in the responses to dietary induction of hyperlipidemia from one species to another (Bowyer and Gresham, 1971). Nevertheless the baboon has been much-used in atherosclerosis research and has an advantage that the pattern of lipid metabolism is similar to that of man. The structure of blood vessels also resembles that of man. Katzberg (1966) showed that the amount of elastic tissue in the abdominal aorta was less than that in the thoracic part in the baboon. The distal part of the vessel might be expected to respond differently both physiologically and pathologically. This is important work that stresses the need for a study of differences in structure and function between different blood vessels and indeed between different parts of the same vessel. Only with this knowledge shall we start to understand the basic process of atherogenesis, for there is no doubt that hemodynamic factors are important in pathogenesis. The discrepancy between degrees of atherosclerosis in various parts of the human vascular tree is well recognized. The amount of disease in coronary arteries tends to run parallel with that in the aorta, though this is not always so. Cerebrovascular disease, however, is only severe when hypertension exists. A similar situation was observed by D. A. Lapin and Yakovleva (1963), who examined baboons aged 5–27 years. They found fibrous plaques and streaks in the aortas, fewer in the coronary arteries, and scanty lipid deposits in the intima of cerebral arteries. This sort of pattern is the usual feature in most animals that show atherosclerosis when captured in the wild state and links up with the fact that, in most experimental animals and others that have been studied, hypertension is not usually found.

The existence of naturally occurring atherosclerosis has been observed by a variety of workers (Gresham and Howard, 1965b). As we have said before this makes control experiments difficult to interpret. Lindsay and Chaikoff in 1957 wrote that atherosclerosis had rarely been observed in subhuman primates, which illustrates how recent much of the work is that is concerned with primate arterial disease. They described lesions in the doguera baboon (*Papio anubis*) and emphasized the proliferative nature of the lesions in the aorta, coronary, and iliac arteries, and concluded that lipid infiltration was a secondary event and not the primary cause of the disease. Their animals came from the San Diego Zoo and were all about 20 years of age. All of the lesions were of mild degree consisting either of fatty streaks or fibrous plaques. The earliest stage of the disease was thought to be a fragmentation of the inner elastic tissue of the vessel followed by an accumulation of mucopolysaccharide, suggesting that atherosclerosis is caused by vascular injury followed by a proliferative stage of repair. This view was supported by the occurrence of collagen and reticulum fibers in the more elevated, older lesions.

Writing again in 1966, Lindsay and Chaikoff reviewed the occurrence of arterial lesions in a range of primates, including eight wild baboons from Africa. They emphasized again that the lesion was a proliferative response and that lipid accumulation was a secondary event. As compared to other animals, however, they did agree that lipid accumulation was more often seen in primate atherosclerosis than in the disease in other creatures. Occasionally they found fibrin on the intimal surface and even within the lesions but concluded that, so far as these animals were concerned, fibrin incorporation did not play an important role in the genesis of the atherosclerotic lesion as has been suggested in man (von Rokitansky, 1846; Duguid, 1949).

Gillman and Gilbert (1957) and McGill et al. (1960) also studied a number of baboons and noted essentially the same appearance as described by Lindsay and Chaikoff. Once more the emphasis was that lipid deposition was a secondary event and, though it might be an important factor in arterial occlusion at a later stage, it was not the initiating factor in atherosclerosis. Most people working with different animals, including man, now agree with this point of view.

Attempts to elucidate further the steps in atherogenesis in the baboon fatty streak have been made by Geer et al. (1968). They fed diets containing various levels of fat, protein, and cholesterol to baboons for 2 years. At the end of the experiment the animals were killed, the aorta opened and split longitudinally; one-half was used for lipid analysis, the other for morphological studies. This is a common practice in atherosclerosis work.

Electron microscopy revealed three types of cells in the lesions: smooth-muscle cells, cells that resemble blood monocytes, and foam cells. The smooth-muscle cell with its myofilaments, numerous pinocytotic vesicles, and limiting basement membrane was the cell most frequently seen in the intima of these baboons. Smooth-muscle cells have been described in atherosclerotic lesions from many animals, including bird, pig, rat, rabbit, and man (Roberts and Straus, 1965). The precise origin of these cells and their fate and function remain debatable. Indeed it is pertinent to consider whether they are smooth muscle at all as they have not been shown to possess the cardinal feature of all muscle cells, namely, contractility. The so-called smooth-muscle cells of the baboon lesions were smaller and more pleomorphic than medial smooth-muscle cells, had fewer myofilaments, and, on the whole, contained more cytoplasmic organelles. Mitochondria and endoplasmic reticulum were conspicuous and the general appearance suggested enhanced metabolic activity. The precise name that should be given to these intimal cells is immaterial so long as their function can be clearly defined. One function that is of importance, at least in the developing animal, is elastogenesis in the vessel wall (Ross and Burnstein, 1971).

Other cells were seen in early baboon lesions by Geer *et al.* They included things that resembled monocytes and lymphocytes and were morphologically distinguishable from smooth-muscle cells. The overlying endothelial cells had abundant pinocytotic vesicles so that the overall picture of the lesion was one of great phagocytic and metabolic activity. Abundant interstitial material and the presence of degenerating cells support the view that cells are taking up material like lipoid, some dying and others proliferating as a regenerative phenomenon. Whether the presence of lipid is the cause of the whole process or is secondary to cell damage is one of the hotly debated subjects by atherologists. Attempts to solve this problem by tissue culture of vascular cells are too few (Pollak, 1969); one of the principal difficulties is the precise identification of the cells that grow in culture, but there can be little doubt that, for the future, tissue-culture methods may help considerably in our understanding of the mechanisms of atherogenesis.

A number of dietary experiments have been done in baboons in order to reproduce atherosclerosis. Geer *et al.* (1968) and Gresham and Howard (1965b) found this animal to be sensitive to an increase of dietary cholesterol, in that moderate rises of serum cholesterol readily occurred in most animals fed cholesterol. There is, however, a conspicuous variability in the response which represents a considerable problem in the interpretation of experimental results.

Howard *et al.* (1967) described experiments in which baboons fed on cholesterol and other lipids developed a β-lipoproteinemia, the β-lipoproteins preponderating over the α-lipoproteins. This situation is like that in man and emphasizes the need to study the whole question of lipid transport by these substances not only in the plasma but also through the vessel wall. Another factor that was discovered in these experiments was the high level of plasma insulin in this species of primate. Perhaps such elevated levels of insulin may predispose to excessive synthesis of fatty acids by the arterial wall, thus aggravating atherosclerosis. The association of high plasma insulin levels in some diabetic people may also explain the well-known association of atherosclerosis and diabetes mellitus.

The view that an excessive intake of carbohydrate, especially sucrose, in the diet may be a factor in causing arterial disease has been explored in the baboon. Feeding sucrose in excess will cause a rise of both phospholipids and triglycerides in the plasma (Coltart, 1969), but there was no evidence, from these experiments, that atherosclerosis was enhanced. I have also observed considerable obesity in baboons fed sucrose for a year but was unable to detect more than the usual slight degree of atherosclerosis in the aortas and coronary arteries of these animals. The elevation of serum triglycerides that follows carbohydrate-feeding varies with the kind of substance that is administered. Macdonald and Roberts (1967) fed a variety of carbohydrates labeled with ^{14}C and were able to show that the amount of ^{14}C incorporated into glycerides was greatest following sucrose-feeding. They concluded that the explanation was preferential incorporation of fructose by the liver.

The development of methods such as thin layer and gas–liquid chromatography for the more precise separation of lipids in blood and tissues has enabled the probable association between lipids and arterial disease to be pursued extensively. Though various lipids have from time to time been suspected, the only certain culprit seems to be cholesterol and even here the precise role is uncertain. Whether it is a primary cause of atherosclerosis or a secondary factor contributing to arterial occlusion remains unsolved. Attempts to lower cholesterol levels are nevertheless worthy endeavors and various devices have been tried in baboons and rabbits, such as the use of choleretic and other drugs (Howard et al., 1965).

Problems that arise in handling primates in general and baboons in particular have been reduced by the use of phencyclidine anesthesia (Vondruska, 1965). It is necessary to appreciate the different responses of various primate species and also to consider the effects that repeated anesthesia might have on such parameters as blood pressure. It is known that the drug has a hypertensive effect in squirrel monkeys and this, clearly, can be important in studies on atherogenesis.

B. A. Lapin (1966) has studied the effects of stressful situations in *Papio hamadryas;* these animals were restrained normally and without any anesthesia. He was able to show that repeated frustration of the male led to chronic persistent hypertension and to myocardial infarction in some of the animals. It is difficult to determine the effects of physical restraint itself in this situation, though control animals that were captured and handled similarly did not develop hypertensive vascular disease. Studies on stress are always difficult for no one can define it precisely and there is a great need to reach agreement on the exact meaning and measurement of it. So many changes occur that the picture is confused. When one considers how hypothalamic influences can produce such widespread effects, including a change in the level of circulating blood lipids (Gunn *et al.*, 1963), the task seems to be almost impossible.

VI. The Squirrel Monkey

Clarkson and his colleagues at Winston Salem, N. C. (Middleton *et al.*, 1964, 1967) have been largely responsible for the popular use of this genus in vascular research and the establishment of a field station in Columbia has ensured a good supply of animals. From here the animals are kept on a farm before final transfer to the medical school. As the animals become habituated they have been able to study the effects of captivity and experimentation on basic vascular and lipid parameters.

Rosenblum and Cooper (1968) have published a comprehensive book about the squirrel monkey; they deal largely with various normal aspects of the animal. A study of this sort is an essential prerequisite for the investigation of pathologic processes in any primate and it is to be hoped that more volumes of this kind will appear, such as the study of *Macaca mulatta* by Valerio *et al.* (1969).

Several primates have been used for studies on lipid metabolism; of all the squirrel monkey seems to be eminently suitable, for it shares, with man, a limited ability to absorb cholesterol from the diet. Thus massive hypercholesterolemia does not occur as in rhesus monkeys fed cholesterol. A study of cholesterol-feeding (Lofland *et al.*, 1970), together with various fats and oils, revealed an important point. The amount of cholesterol in the whole carcass was the same at the end of 2 years' feeding. However, the distribution of cholesterol was different. Plasma cholesterol levels were lower in monkeys given cholesterol and safflower oil and higher in those fed butter and cholesterol. So it seems clear that the degree of saturation of the fat that is fed with the cholesterol may greatly influence the blood level of the sterol. Though some indecision remains about the relative importance of blood and tissue levels of cholesterol in atherogenesis the observation emphasizes an often forgotten point, namely, that blood levels of any component do not necessarily reflect tissue levels. We observed this phenomenon in dietary experiments in rats some years ago (Gresham and Howard, 1963).

There are so many aspects of cholesterol metabolism that may affect the development of atheroma that it is often difficult to know which to select for study. Cholesterol deposition is in part related to the balance of endogenous synthesis and exogenous ingestion and is also determined by other things, such as the level of cholesterol ester hydrolases and other enzymes in the tissues under consideration. Young primates suppress endogenous synthesis more readily when fed cholesterol than do the older animals, and the mechanism of this remains to be elucidated. The remarkable way in which the newly hatched chick clears its aorta of lipid within a week or so of hatching might provide clues to this phenomenon. It seems clear that not all forms of cholesterol are histotoxic and irretrievable when deposited in the tissues and that particular esters of cholesterol are more damaging than others (Adams *et al.*, 1963).

It is not surprising that recent interest has been directed to the Glomsett enzyme (LCAT) in this respect, Portman *et al.* (1970) have done this in the squirrel monkey. They have also shown that the concentration of lysolecithin increases in the aortic intima and inner media when atherosclerosis is induced by diet in the squirrel monkey and in the rabbit. This rise in tissue level may partly be related to the increased levels of lecithin in the plasma of animals on experimental diets and a similar situation occurs in man. Because the fatty acyl transferase pathway is active in arterial tissue and endogenous levels of lysolecithin determine the incorporation of fatty acyl radicals into the arterial wall, it is clear that the whole enzyme complex is important in determining the type of cholesterol derivative that appears in the vascular intima. As we have already seen, the kind of cholesterol ester may well determine its pathologic effects. There is conclusive evidence that, in certain situations at least, cholesterol can directly damage the vessel wall, though whether this artificial situation in the squirrel monkey compares to the situation in man is still debatable. Maruffo and Portman (1968) demonstrated this point by feeding squirrel monkeys on diets that contained cholesterol and induced coronary atherosclerosis. The changes were those mainly of intra- and extracellular lipid accumulation after various intervals on a diet containing butter and cholesterol. These lesions did not reveal much in the way of cellular proliferation. However, in another group that had been fed the atherogenic diet for 3 months and were then given a control diet for a further few months a different situation appertained.

The coronary artery lesions now were more of a proliferative and degenerative nature. Cells were present which did not resemble smooth-muscle cells that have been observed in a variety of experimentally induced atherosclerotic plaques (Eyestone, 1965; St. Clair *et al.*, 1967; Norman, 1968), though the authors did consider it likely that these cells may at some stage have arisen from smooth-muscle cells. Another striking feature of the lesions was the presence of stainable mucopolysaccharide and other extracellular materials; in addition, there was calcification, fragmentation of elastic lamellae, and proliferation of collagen. These lesions produced by Maruffo and Portman resemble human lesions much more closely than those that have been produced in the past, and it looks as though intermittent, rather than continuous, lipoid onslaught on the vessel wall might be a mechanism in human atherogenesis. A similar situation using interrupted cholesterol-feeding in the rabbit produces a similar effect.

Much of the earlier work that recommended the squirrel monkey in this field of research was concerned with the demonstration of lipid-laden fatty streaks in the thoracic and abdominal aorta of animals trapped in the wild (Middleton, 1964). These workers found no lesions in the larger coronary arteries. Studies of this sort emphasize the problem that the degree of aortic disease is not necessarily an indication of factors that might be concerned in the induction of coronary atherosclerosis.

Unfortunately it is often true in this field of work that attempts are made to prevent the induction of arterial lesions in experimental animals long before the precise nature of the disease and its relationship to human atherosclerosis has been clearly defined. Morrison et al. (1966) studied the effects of chondroitin sulfate A given subcutaneously on the development of atherosclerosis in the squirrel monkey and claimed that this substance prevented the development of arterial disease in this animal. It is difficult to reconcile these observations with the delayed but potent hyperlipidemic effect of chondroitin sulfate when given parenterally to rabbits. It seems unlikely that this material will be of any value in man because oral administration will lead to its degradation to constituent hexose derivatives. Similar experiments by Malinow et al. (1968) using oral pyridinol carbonate, which is said to reduce aortic atherosclerosis in rabbits (Shimamoto et al., 1965), had no effect on the development of dietary-induced atherosclerosis in the squirrel monkey.

Much more needs to be known of basic mechanisms of atherogenesis and lipid accumulation before a therapeutic device to prevent the disease can be developed.

VII. The Chimpanzee

The chimpanzee has been used extensively for atherosclerosis studies by Sandler and Bourne (Sandler and Bourne, 1963; Bourne and Sandler, 1972). They showed that this animal on a nutritionally adequate diet produces aortic atherosis histogenetically similar to those of the human. Furthermore the developing atheromas in both chimpanzee and human show reduced 5' nucleotidase and ATPase activity.

VIII. Other Aspects of Cardiovascular Research in Primates

We have so far only considered the principal popular genera of primates that have been used in research on arterial disease. Many more have been and are being used but the reports are more sporadic and less extensive than those to do with the monkeys considered here. This last paragraph is concerned with a variety of aspects of cardiovascular research in primates other than atherosclerosis.

Primates often carry a wide variety of parasites and these can produce evidence of disease in many organs, including the cardiovascular system. It is clearly important not to confuse such parasitic lesions with those of spontaneous or induced degenerative disease. Soto *et al.* (1964) examined 20 recently imported rhesus monkeys from India; after a study of various biochemical and hematological parameters and electrocardiography, the animals were killed and further examined. They found that 18 of the animals showed varying degrees of myocarditis, which was usually focal but was occasionally confluent and extensive; lesions of the latter kind can readily be confused with areas of ischemic necrosis. These workers attempted to demonstrate viruses and bacteria but found none of significance, nor were antibody titers elevated. SGOT and SGPT levels could not be correlated with the degree of myocardial involvement, nor could electrocardiographic abnormalities be shown. A variety of organisms, including *Bordetella bronchiseptica,* cytomegalovirus, sarcosporidia, and helminths were present but did not seem to be responsible for the lesions. Clearly it is well to be aware of the possibility of parasitic causes for lesions that might be found in experimental animals.

Certain forms of myocarditis seem more common in New World rather than Old World monkeys, particularly toxoplasmosis, which commonly affects the myocardium (McKissick *et al.*, 1968). These workers reported an outbreak of the disease in 40 animals received at the Philadelphia Zoo; of these 9 died of toxoplasmosis and showed a variety of lesions in many organs. Often the organisms were readily visible as the free or encysted form but occasionally they were difficult to find; the heart is such an organ where necrosis was considerable but organisms few.

Thrombosis, when it occurs in animals, tends to be due to parasitic infestation. Very rarely indeed does it occur in association with atherosclerotic disease. Rarely do thrombi occur in primates for no apparent reason, though occasional examples are reported (Ulland, 1968). Thrombosis is an important occlusive event in human coronary arteries and it is desirable to try to replicate the process in the experimental animal. On the whole this is difficult, particularly in nonhuman primates. Studies of the basic coagulation parameters, such as platelets, coagulation factors, and fibrinolytic mechanisms, need to be done and many have already been started (Seaman and Malinow, 1968; Macfarlane, 1970). There is a general similarity between human and nonhuman primate coagulation mechanisms; if anything, there is a tendency to greater and more rapid coagulation response in nonhuman primates; this is illustrated by the increased tendency to develop contact activation to factors XI and XII and by a higher factors II and VII activity. Platelet counts tended to be higher in nonhuman primates, particularly the squirrel monkey, irrespective of the kind of diet that had been fed. All these results are rather surprising because of the rarity of detectable spontaneous thrombosis in these animals as compared to man. However, it is important to consider the whole coagulation spectrum in this regard. Thrombi and clots may form more easily in nonhuman primates but equally they may be removed more rapidly by fibrinolytic mechanisms. This view is supported by plasmin and plasminogen assays for a variety of mammals where it can be shown that monkeys, carnivores, and rats have higher plasminogen levels than has man, implying a greater fibrinolytic potential (Macfarlane, 1970).

Much work is now going on about the properties of platelets in several animal species. Phenomena of platelet adhesion, aggregation, and dissolution are all important aspects of thrombosis. Comparatively little work has been done on the platelets of nonhuman primates. First of all, it is difficult to decide which *in vitro* test most reliably indicates the *in vivo* position. Using the aggregometer Mills (1970) has shown a biphasic aggregation of baboon platelets when treated with ADP. That is to say two stages occur with an interval between before the platelets are irreversibly clumped. This aggregation is potentiated by epinephrine which is a feature peculiar to the primates and may be a factor of importance in thrombogenesis and atherogenesis. Studies on fibrinogen have revealed a tendency to lower levels in nonhuman primates as compared to man; this may again be bound up in some way with man's thrombotic propensity. Another interesting comparative study has been made by Doolittle and Mross (1970) on the analysis of fibrinopeptides. Although not of particular impact in the field of thrombogenesis, their work does show a remarkable identity of amino acids in three separate fibrinopeptides of the chimpanzee emphasizing the close relationship of man and the anthropoid at a molecular level and providing ground for a more extensive study of xenografts, particularly in the fields of cardiac and hepatic transplantation.

The problems of studying induced thrombosis in nonhuman primates remains. Several thrombogenic agents are available and have produced interesting results but so far attempts to induce coronary thrombotic occlusion are few and difficult. Monocrotaline derived from the seeds of *Crotalaria spectabilis* has the potent effect of damaging vascular endothelium and vascular lesions in animals have been described either following accidental ingestion in horses, cattle, and men or after deliberate administration as to turkeys. The oral or intraperitoneal administration of monocrotaline to adult *Macaca* species gives rise to endothelial damage in small branches of the hepatic venous system followed by exudation of fluid into the vessel wall and subsequent narrowing and thrombosis of the veins. Larger veins are not affected, which may be due to the fact that they have thicker, more muscular walls, which precludes the exudation of fluid into them. The kind of vessel that is affected by monocrotaline varies with the species of animal employed. Monkeys also develop a pulmonary arteritis, as do rats, and yet the latter animal does not develop lesions in the hepatic vasculature (Allen *et al.*, 1967, 1969). *Crotalaria* alkaloids are useful tools for the study of endothelial cell injury and deserve more attention by experimental pathologists in the field of vascular research.

Attempts to define normal parameters are essential for the proper understanding of work with animals. One such aspect of study has been electrocardiography which is an important adjunct in the assessment of myocardial disease. Generally speaking, the smaller the animal and the faster the rate of the heart the greater is the degree of ST elevation. This is a normal feature that must not be construed as providing evidence of ischemia (Hill *et al.*, 1960). In the squirrel monkey considerable variations in the P wave have been observed in normal animals; this may be an effect of anesthesia (Wolf *et al.*, 1969). Variations in the T wave, indicating repolarization differences, have been observed in a wide variety of nonhuman primates; these should be recognized as normal and not associated with any evidence of myocardial damage at necropsy. Malinow (1966) has observed similar phenomena in the *Macaca mulatta*.

Bristow and Malinow (1965) have also reported varieties of bundle branch block in otherwise normal rhesus monkeys. These workers used vector and intracardiac records in the interpretation of their results, emphasizing the need for thorough and extensive studies of such electrocardiographic variations in primates before the ECG can be used as an adjunct to the study of vascular disease. Despite minor electrocardiographic variations between normal primates, however, the basic anatomy

244   G. A. GRESHAM

of the conducting system is more similar to that of man than is the arrangement in such creatures as dog and calf. Experiments which involve laceration of the left side of the septum in baboons and other animals have shown electrocardiographic effects in the baboon that resemble septal damage in man (Watt *et al.*, 1965).

Studies of blood flow in and adjacent to ischemic myocardium have been done by several workers. Grayson and Irvine (1968) studied the effect of ligating a principal coronary artery in African green monkeys and baboons. They measured blood flow in the myocardium by means of heated probes and showed, as might be expected, that there was no blood flow in the ischemic area, though it took some 5 hours after ligation for the flow to cease altogether. Blood flow was also reduced in the areas adjacent to the infarct but the important point was that this fall in flow could be prevented by the prior administration of the α blocker bethanidine. This work emphasizes the important role of vascular spasm in increasing areas of myocardial infarction. It is clear that the answer to problems of myocardial ischemia is not to be found in the study of anatomic lesions alone but must involve studies of vascular reactivity as an important adjunct.

Work on valvar disease, its induction and treatment, has so far received little attention in primates, nor is congenital heart disease a feature of subhuman primates, so that little has been written about this. Detweiler (1964), in his excellent review of genetic aspects of cardiovascular disease in animals, says nothing of primates; the most extensive study was made in dogs. Valvar replacements, cardiac transplantations, and other kinds of constructive vascular surgery are now being widely done in primates, however, and because of similarities of size, blood groups, and so on certain members of the order such as baboons have been used for such work (Van Zyl *et al.*, 1968).

In 1969, 17,600 New World primates and 50,000 Old World primates were used for various kinds of research purposes. It is significant that only 300 chimpanzees were used for obvious reasons of expense and rarity and fewer are likely to be available in years to come without the efforts of primate centers. There can be little doubt of the desirability of non-human primates for cardiovascular research and it is to be hoped that the pioneering efforts to establish centers in the United States will be rapidly extended to Europe and other parts of the world.

REFERENCES

Adams, C. W. M., Bayliss, O. B., Ibrahim, M. Z. M., and Webster, N. W. (1963). *J. Pathol. Bacteriol.* **86**, 431.

Allen, J. R., Corstens, L. A., and Olson, B. E. (1967). *Amer. J. Pathol.* **50**, 653.

Allen, J. R., Corsters, L. A., and Katagiri, G. J. (1969). *Arch. Pathol.* **87**, 279.

Anitschkow, N. (1933). *In* "Atherosclerosis" (E. V. Cowdry, ed.), Chapter 10. 248–264. Macmillan, New York.

Armstrong, M. L., Connor, W. E., and Warner, E. D. (1967). *Arch. Pathol.* **84**, 227.

Bourne, G. H., and Sandler, M. (1972). "Atherosclerosis in Chimpanzees" *in* "The Chimpanzee" Vol. 6. Karger, Basel.

Bowden, D. (1966). *Folia Primatol.* **4**, 346.

Bowyer, D. E. (1967). Ph.D. Thesis, University of Cambridge.

Bowyer, D. E., and Gresham, G. A. (1971). Unpublished results.

Bowyer, D. E., Howard, A. N., Gresham, G. A., Bates, D., and Palmer, B. V. .(1968). *Progr. Biochem. Pharmacol.* **4**, 235.

Bristow, J. D., and Malinow, M. R. (1965). *Circ. Res.* **16**, 210.

Coltart, T. M. (1969). *Nature (London)* **222**, 575.

Cooper, R. W. (1964). *Lab. Anim. Care* **14**, 474.

Dahme, E. (1968). *Z. Gesamte Exp. Med.* **145**, 305.

Detweiler, D. K. (1964). *Circulation* **30**, 114.

Doolittle, R. F., and Mross, G. A. (1970). *Nature (London)* **225**, 643.

Duguid, J. B. (1949). *Lancet* **2**, 925.

Eyestone, W. H. (1965). *J. Amer. Vet. Med. Ass.* **147**, 1482.

Geer, J. C., Catsulis, C., McGill, H. C., and Strong, J. P. (1968). *Amer. J. Pathol.* **52**, 265.

Gillman, J., and Gilbert, C. (1957). *Exp. Med. Surg.* 181.

Goldblatt, H. (1937). *J. Exp. Med.* **193**, 671.

Goldblatt, H., Lynch, J., Hanzall, R. F., and Summerville, W. W. (1934). *J. Exp. Med.* **59**, 347.

Goldsmith, I. E., and Moor-Jankowski, J. (1969). *Ann. N. Y. Acad. Sci.* **162**, 80.

Goodall, J. (1965). *In* "Primate Behaviour: Field Study of Monkeys and Apes" (I. De Vore, ed.), p. 425. Holt, New York.

Grayson, J., and Irvine, M. (1968). *Cardiovasc. Res.* **2**, 170.

Gresham, G. A., and Howard, A. N. (1963). *Fed. Proc., Fed. Amer. Soc. Exp. Biol.* **22**, 1371.

Gresham, G. A., and Howard, A. N. (1965a). *Acta Cardiol., Suppl.* **11**, 189.

Gresham, G. A., and Howard, A. N. (1965b). *Ann. N. Y. Acad. Sci.* **127**, 694.

Gresham, G. A., and Howard, A. N. (1969). Unpublished results.

Gunn, C. G., Friedman, M., and Byers, S. O. (1963). *J. Clin. Invest.* **39**.

Haigh, M. V., and Scott, A. (1965). *Lab. Anim. Care* **15**, 57.

Hill, R., Howard, A. N., and Gresham, G. A. (1960). *Brit. J. Exp. Pathol.* **41**, 633.

Howard, A. N., Gresham, G. A., Jones, D., and Jennings, I. W. (1965). *Life Sci.* **4**, 639.

Howard, A. N., Gresham, G. A., Hales, C. N., Lindgren, F. T., and Katzberg, A. A. (1967). *In* "The Baboon in Medical Research" (H. Vugtborg, ed.), Vol. II. Univ. of Texas Press, Austin.

Hummer, R. L. (1965). *J. Amer. Vet. Med. Ass.* **147**, 1063.

Hunt, R. D., Garcia, F. G., and Hegsted, D. M. (1969). *Amer. J. Clin. Nutr.* **22**, 358.

Katzberg, A. A. (1966). *Anat. Rec.* **154**, 213.

Kawamura, R. (1927). "Neue Beiträge zur Morphologie und Physiologie der Cholinesterinsteatose," p. 267. Fischer, Jena.

Kritchevsky, D., Shapiro, I. L., and Werthessen, N. T. (1962). *Biochim. Biophys. Acta* 65, 556.

Lapin, B. A. (1966). Quoted by Bowden (1966).

Lapin, B. A., and Yakovleva, L. A. (1963). *In* "Comparative Pathology in Monkeys" (W. F. Windle, ed.), p. 132. Thomas, Springfield, Illinois.

Lindsay, S., and Chaikoff, I. L. (1957). *Arch. Pathol.* 63, 460.

Lindsay, S., and Chaikoff, I. L. (1966). *J. Atheroscler. Res.* 6, 38.

Lofland, H. B., and Clarkson, T. B. (1965). *Arch. Pathol.* 80, 291.

Lofland, H. B., Clarkson, T. B., and Bullock, B. C. (1970). *Exp. Mol. Pathol.* 13, 1.

McAndrew, G. M., and Ogston, D. (1965). *Amer. Heart J.* 70, 838.

Macdonald, I., and Roberts, J. B. (1967). *Metab. Clin. Exp.* 16, 572.

Macfarlane, R. G., ed. (1970). "The Haemostatic Mechanism in Man and Other Animals." Academic Press, New York.

McGill, H. C., Strong, J. P., Holman, R. L., and Werthessen, N. T. (1960). *Circ. Res.* 8, 670.

McGill, H. C., Frank, M. H., and Geer, J. C. (1961). *Arch. Pathol.* 71, 96.

McKissick, G. E., Ratcliffe, H. L., and Koestner, A. (1968). *Pathol. Vet.* 5, 538.

Malinow, M. R. (1966). *Amer. Heart J.* 71, 140.

Malinow, M. R., and Maruffo, C. A. (1966). *J. Atheroscler. Res.* 6, 368.

Malinow, M. R., and Storvick, C. A. (1968). *J. Atheroscler. Res.* 8, 421.

Malinow, M. R., Perley, A., and McLaughlin, P. (1968). *J. Atheroscler. Res.* 8, 455.

Mann, G. V., Andrews, S. B., McNally, A., and Stare, F. J. (1953a). *J. Exp. Med.* 98, 195.

Mann, G. V., Watson, P. L., McNally, A., Goddard, J., and Stare, F. J. (1953b). *J. Exp. Med.* 98, 196.

Maruffo, C. A., and Portman, O. W. (1968). *J. Atheroscler. Res.* 8, 237.

Meier, R. M., Greenhoot, J. H., Shorley, I., Goodman, J. R., and Porter, R. W. (1963). *Nature (London)* 199, 812.

Middleton C. C. (1964). *Arch. Pathol.* 78, 16.

Middleton, C. C., Clarkson, T. B., Lofland, H. B., and Prichard, R. W. (1964). *Arch. Pathol.* 78, 16.

Middleton, C. C., Clarkson, T. B., Lofland, H. B., and Prichard, R. W. (1967). *Arch. Pathol.* 83, 145.

Mills, D. C. B. (1970). *In* "The Haemostatic Mechanism in Man and Other Animals" (R. G. Macfarlane, ed.). Academic Press, New York.

Morris, J. N. (1951). *Lancet* i, 1 and 69.

Morrison, L. M., Murata, K., Quillagan, J. J., Schjeide, D. A., and Freeman, L. (1966). *Circ. Res.* 19, 358.

Norman, A. W. (1968). *Biol. Rev.* 43, 97.

Pollak, O. J. (1969). "Monographs in Atherosclerosis," Vol. 1. Karger, Basel.

Portman, O. W., and Alexander, M. (1966). *Arch. Biochem. Biophys.* 117, 357.

Portman, O. W., Soltys, P., Alexander, M., and Osuga, T. (1970). *J. Lipid Res.* 11, 596.

Rinehart, J. F., and Greenberg, L. D. (1949). *Amer. J. Pathol.* 25, 481.

Roberts, J. C., and Straus, R. (1965). "Comparative Atherosclerosis." Harper, New York.

Rosenblum, L. A., and Cooper, R. W., eds. (1968). "The Squirrel Monkey." Academic Press, New York.

Ross, R., and Burnstein, P. (1971). *Sci. Amer.* 224, 44.

St. Clair, R. W., MacNitch, J. E., Middleton, C. C., Clarkson, T. B., and Lofland, H. B. (1967). *Lab. Invest.* **16**, 828.

Sandler, M., and Bourne, G. H. (1963). "Atherosclerosis and its Origins." Academic Press, New York.

Scott, R. F., Jones, R., Daoud, A. S., Jumbo, O., Coulston, F., and Thomas, W. A. (1967). *Exp. Mol. Pathol.* **7**, 34.

Seaman, A. J., and Malinow, M. R. (1968). *Lab. Anim. Care* **18**, 80.

Shimamoto, T. F. (1969). *Acta Pathol. Jap.* **1**, 15.

Shimamoto, T. F., Numano, T., Fujita, T., Ishioka, T., and Atsumi, T. (1965). *Asian Med. J.* **8**, 825.

Smith, E. (1965). *J. Atheroscler. Res.* **5**, 224 and 241.

Soto, P. J., Beall, F. A., Nakamura, R. M., and Kupferberg, L. L. (1964). *Arch. Pathol.* **78**, 681.

Sperry, W. M., Jailer, J. W., and Engle, E. T. (1944). *Endocrinology* **35**, 38.

Taylor, C. B., Cox, C. E., Counts, M., and Fogi, M. (1959). *Pathol. Bacteriol.* **35**, 674.

Taylor, C. B., Cox, G. E., Manalo-Estrella, P., and Southworth, J. (1962). *Arch. Pathol.* **74**, 16.

Ulland, B. M. (1968). *Brit. Vet. J.* **124**, 245.

Valerio, D. A., Miller, R. L., Innes, J. R. M., Courtney, K. D., Pallota, A. J., and Guttmacher, R. M. (1969). "Macaca Mulatta (Management of a Laboratory Breeding Colony)." Academic Press, New York.

Van Riper, D. C., Fineg, J., and Day, P. W. (1967). *Lab. Anim. Care* **17**, 472.

Van Zyl, J. J. W., Murphy, G. P., and de Klerk, J. M. (1968). *Lab. Primate Newslett.* **7**, 17.

Vlodaver, Z., Medalie, J., and Neufeld, H. M. (1968). *J. Atheroscler. Res.* **8**, 923.

Vondruska, J. F. (1965). *J. Amer. Vet. Med. Ass.* **147**, 1073.

von Rokitansky, K. (1846). "Lehrbuch der Páthologischen Anatomie." Braunmüller & Seidel, Vienna.

Washburn, S. L., Jay, P. C., and Lancaster, J. B. (1965). *Science* **180**, 1541–1547.

Watt, T. B., Murao, S., and Pruitt, R. D. (1965). *Amer. Heart J.* **70**, 381.

Wilson, C., and Byrom, F. B. (1939). *Lancet* i, 136.

Wolf, R. H., Lehner, N. D. M., Miller, E. C., and Clarkson, T. B. (1969). *J. Appl. Physiol.* **26**, 346.

Younger, R. K., Scott, H. W., Butts, W. H., and Stephenson, S. E. (1969). *J. Surg. Res.* **9**, 263.

CHAPTER 10

Humanlike Diseases in Anthropoid Apes*

CLARKE STOUT

I. Introduction

In 1964, the Institute for Comparative Pathology was established by the University of Oklahoma Medical Center in conjunction with the Oklahoma City Zoo and the Oklahoma Zoological Society. The laboratory was located on the grounds of the zoo, and its primary purpose was the study of cardiovascular disease in captive wild mammals and birds. The project was supported by a grant from the National Heart Institute. A number of arterial lesions were encountered in zoo animals during the course of the work, and, in addition, several disease entities were found in anthropoid apes which were very similar morphologically to their human

* Supported in part by Grant HE 08725 from the National Heart Institute, United States Public Health Service.

249

counterparts. These latter observations form the basis for this report, since they may be important indicators of areas for future research with nonhuman primates. The findings are listed under the headings of the human diseases which they resembled morphologically. The majority of the cases have been previously published (Stout and Lemmon, 1969a,b, 1971; Stout and Snyder, 1969).

II. Materials and Methods

The animals to be described were culled from over 500 autopsies on mammals and birds dying in the Oklahoma City Zoo between 1965 and 1969. A few additional cases came from Dr. W. B. Lemmon's psychological research laboratory at the University of Oklahoma. For the most part, the autopsies were consecutive. The amount of ancillary clinical material available on the animals was extremely variable, but was fortunately relatively complete in the cases to be described here. Standard methods were used in the gross and microscopic examination of tissues, and in the grading of arterial lesions. Detailed accounts of methodology may be found in previous publications (Stout and Lemmon, 1969a,b, 1971; Stout and Snyder, 1969).

III. Atherosclerosis

Arterial disease in mammals and birds has been extensively scrutinized during the last few decades. Although most studies have been experimental in nature, the extent and distribution of arterial lesions have been determined in many species of captive wild animals, domestic and laboratory animals, and a few species of free-ranging wild animals. The majority of spontaneously occurring lesions have been found in the aortas, with the extramural coronary and large cerebral arteries being only rarely involved (Sandler and Bourne, 1963; Roberts and Straus, 1965). This same distribution of lesions, with an occasional exception (Kramsch and Hollander, 1968), has held true for most experimental studies involving the feeding of high-fat, high-cholesterol diets. The converse is frequently true in younger human victims of myocardial infarction. In these patients, the extramural coronary artery atherosclerosis is often equal to or more severe than that in the aorta, and there is a tendency for lesions to appear in the circle of Willis prematurely. Therefore, it is interesting that atherosclerotic lesions have been found in the extramural coronary and large cerebral arteries of the chimpanzee with considerable frequency. In the Oklahoma material, the coronary and cerebral arterial lesions occupied slightly less area, but were more advanced histologically (i.e., thicker and with more central intimal necrosis) than their aortic counterparts (Stout and Lemmon, 1969b). Coronary and cerebral arterial lesions were found in two of four chimpanzees examined. They were not related to age or sex, and this was also true in the larger series reported by Andrus et al. (1968). This important parallel between the degree of coronary and cerebral atherosclerosis in chimpanzees and young human ischemic heart disease patients has not attracted much notice. It is significant that in the one such experiment conducted thus far in chimpanzees, the ingestion of a high-fat, high-cholesterol diet did not potentiate the predilection of atherosclerosis for the larger coronary and cerebral arteries (Andrus et al., 1968). Although only limited numbers of free-ranging chimpanzees have been examined, it would appear from the information available, that the predominant coronary and cerebral arterial lesions are related to the captive state. The exact way in which captivity affects the vasculature remains to be proven; it is obviously an important area for further inquiry.

IV. Preeclampsia

Preeclampsia, or toxemia of pregnancy, is a disorder in humans characterized by albuminuria, edema, hypertension, and glomerular capillary endothelial swelling. It usually develops during the last trimester of pregnancy and is relatively common, occurring in some 5% of all pregnancies in the United States. Primigravidas are particularly vulnerable, and 15–20% of such females may be affected if they are single, teenaged and Negro. Although the changes of preeclampsia usually revert to normal within a few weeks after delivery, the danger of eclampsia with convulsions or death always exists. Moreover, it cannot be said with certainty that preeclampsia has not residual effects in terms of renal disease or hypertension.

A number of animal diseases resemble preeclampsia in humans in some respects (Hammond et al., 1950), but, at present, none have been found which replicate the complete syndrome (Craig, 1969). Glomerular capillary endothelial swelling was found in a 9-year-old chimpanzee from Dr. Lemmon's colony (Stout and Lemmon, 1969a). The animal died in the 8th month of her first pregnancy. Death occurred during anesthesia, which was given in order to treat bleeding ringworm lesions. The animal had appeared lethargic and edematous for 2 or 3 weeks prior to death, but hypertension and albuminuria were not documented. Because glomerular capillary endothelial swelling is thought to be pathognomic for preeclampsia in humans, its presence in a pregnant chimpanzee must be given considerable weight.

V. Essential Hypertension

Nearly all instances of hypertension in animals have been produced by experimental manipulations, such as constricting the renal artery (Goldblatt *et al.,* 1934), feeding excess salt (Dahl, 1961), and so forth. Selective breeding has yielded strains of animals which are more susceptible to the induction of hypertension, and occasional investigators have produced elevated blood pressure by putting the animal into various situations thought to be emotionally stressful (Henry *et al.,* 1967). The majority of these animals have been rats, mice, or rabbits, and only the Russians have reported hypertension in nonhuman primates, this being produced by social and sexual isolation or various conditioning procedures (Miminoshivili *et al.,* 1960). Investigators in the United States have thus far been unable to confirm the latter findings.

One chimpanzee from Dr. Lemmon's colony was suspected to have essential hypertension. Although the blood pressure was not measured during life, the circumstances of death and the anatomic findings in the cardiovascular system were compelling. The animal, a 20-year-old female, died suddenly and unexpectedly with acute pulmonary edema. The lungs showed chronic passive congestion in addition to the edema, and the left ventricle of the heart was markedly hypertrophied. Sections of the kidneys revealed focal pseudoelastic hyperplastic arteriosclerosis and hyaline arteriolar sclerosis, and in many arterioles, proliferation and hypertrophy of intimal and medial cells. Similar vascular changes were present in the adrenal capsule. In a human, this constellation of morphologic alterations would lead most pathologists to suspect hypertension very strongly (Stoddard and Puchtler, 1969), and some would insist that an error had been made if such a patient's blood pressure was recorded as normal during life. There would seem to be no logical reason to interpret these findings differently, just because they occurred in an anthropoid ape (Stout and Lemmon, 1971).

Isolated blood pressure measurements have been recorded on several chimpanzees at the Yerkes Regional Primate Center of Emory University. Some of these values have been quite high, but also quite variable. The determination of valid average blood pressure levels in chimpanzees presents real problems. The great strength of these creatures makes blood pressure measurement hazardous in the nonsedated animal, even on the isolated arm. Sedation, on the other hand, produces many artifacts. It would be useful to identify hypertensive chimpanzees during life, so that appropriate tissues could be serially biopsied, appropriate chemical analyses made repeatedly, and so forth.

VI. Ulcerative Colitis

An animal disease closely resembling idiopathic ulcerative colitis in the human has yet to be described in the literature. Furthermore, all attempts to reproduce this syndrome experimentally have failed (Kirsner, 1970). An 8-year-old siamang gibbon from the Oklahoma City Zoo became lethargic and died 6 weeks after the death of its long-time cagemate. Findings at autopsy were limited to the colon, the entire mucosa of which was focally ulcerated. Morphologically, the process resembled shigellosis in some respects, but *Shigella* organisms could not be cultured from the colon at autopsy, the clinical course was unlike that of typical nonhuman primate shigellosis, and none of the animals in adjacent cages displayed evidence of diarrheal disease. This observation was strengthened by the fact that similar colitis-like lesions were found on three separate occasions in siamang gibbons at the Philadelphia Zoo. Colitis developed in each of these animals when they were introduced into the cage of a more aggressive and dominant gibbon. In each instance, the more dominant animal was abusive, took food away from the introducee, and so forth. No overt injuries were observed. One of the animals died with an acute diarrheal illness very much like typical nonhuman primate shigellosis. In the other two animals, the diarrheas were less explosive. No stool cultures were obtained, but there were no parasites or amoeba found, and no adjacent animals developed diarrhea (Stout and Snyder, 1969).

VII. Etiology of Humanlike Disease in Anthropoid Apes

Dr. Lemmon and I have been impressed by the fact that each of the chimpanzees described in this report displayed abnormal behavior during life. This abnormal behavior was characterized by an increased number of stereotyped movements such as rocking, hoarding of food and other objects, poor socialization with peers, and so on, and tended to distinguish these animals from their cagemates. Our data are not sound enough to confirm the existence of a link between abnormal behavior and disease in anthropoid apes, but further research in this area would seem to be indicated. All animals are influenced by confinement, and this is particularly true of the anthropoid apes. The fact that these animals rarely reproduce in captivity is adequate proof that profound psychological and physiologic aberrations have occurred. Some of these aberrations are due to the fact that animals are often captured during infancy, and thus prevented from developing normal "personalities" because of isolation from their peers. Other aberrations are probably due to limited cage space, improper diet, conflicts with adjacent humans and animals, and a host of other poorly understood factors. In the past, most investigators have tended to ignore these forces, despite their obvious importance. The main reason for this is that these factors are hard to quantitate. However, advances in technology, and the recent development of the regional primate research centers with their large complements of anthropoid apes and other nonhuman primates, and their diversified faculties, which include veterinarians, behaviorists, neurophysiologists, pathologists, physiologists, and biochemists should make possible a sophisticated attack on problems of this nature.

VIII. Significance of These Findings

The very high prevalence of "humanlike" diseases in captive anthropoid apes in Oklahoma suggests that these diseases are not rare among such animals. It should be appreciated that some of the changes described were rather subtle, and would have been missed by me in the absence of expert consultative advice. This refers specifically to the renal lesions of preeclampsia, and also to the renal vascular changes in the animal thought to be hypertensive. For this reason, it is possible that the incidence of humanlike diseases has been underestimated in captive wild animals.

Because of the likelihood that humanlike diseases are not rare in captive anthropoid apes, these animals should be studied during life, so that diseases, when present, could be accurately identified and characterized. Morphology leaves much to be desired in terms of diagnosis of disease. Furthermore, the study of a live chimpanzee with essential hypertension could yield much valuable information. For example, using techniques currently available, the blood flow to different organs could be monitored and correlated with the physical and emotional state of the animal. The actions of various hypotensive agents could be determined in terms of vascular resistance in various organs, thereby permitting a better estimation of the long-term effects of the drugs. Certain regions of the brain or the peripheral autonomic nervous system could be stimulated electrically by telemetry or ablated in an attempt to alter the hypertensive state. These same manuevers could be conducted in a chimpanzee with preeclampsia, or some other disease process in which detailed physiologic monitoring would be rewarding.

Tremendous differences exist between humans and animals, and the diseases which affect them. These differences are often directly proportional to the phylogenetic "distances" between the species being considered. Because of this, one would anticipate that the study of diseases of anthropoid apes might yield information which could be extrapolated to the treatment of human disease more readily than the study of diseases of lower animals. Rous discovered that a virus could cause a sarcoma in chickens in 1910, but viruses have yet to be proven to cause human tumors, despite the fact that the evidence for such a conclusion appears stronger each year. What if Rous had discovered a naturally occurring virus-induced tumor in a chimpanzee?

IX. Conclusions

It appears likely that humanlike diseases may be expected to occur with appreciable frequency in captive anthropoid apes. Whether or not these diseases are related to the various stresses associated with captivity remains to be determined. Available captive anthropoid apes should be examined carefully, so that those animals with humanlike diseases can be identified and studied during life. It is anticipated that research using anthropoid apes will yield information which may be more readily extrapolated to human medicine than information obtained from research with lower animals.

REFERENCES

Andrus, S. B., Portman, O. W., Riopelle, A. J. (1968). *Progr. Biochem. Pharmacol.* **4**, 393–419.
Craig, J. M. (1969). *Fed. Proc., Fed. Amer. Soc. Exp. Biol.* **28**, 206–210.
Dahl, L. K. (1961). *J. Exp. Med.* **114**, 231–235.
Goldblatt, H., Lynch, J., Hanzel, R. F., and Summerville, W. W. (1934). *J. Exp. Med.* **59**, 347–379.
Hammond, J., Brown, F. J., and Wolstenholme, G. E. W., eds. (1950). "Toxemias of Pregnancy, Human and Veterinary." McGraw-Hill (Blakiston), New York.
Henry, J. P., Meehan, J. P., and Stephens, P. M. (1967). *Psychosom. Med.* **29**, 408–432.
Kirsner, J. B. (1970). *Scand. J. Gastroenterol.* **6**, Suppl., 63–91.
Kramsch, D. M., and Hollander, W. (1968). *Expt. Mol. Pathol.* **9**, 1–22.
Miminoshivili, D. I., Magakian, G. O., and Kokaia, G. I. (1960). *In* "Problems of Medicine and Biology in Experiments on Monkeys" (I. A. Utkin, ed.), pp. 103–123. Pergamon, Oxford.
Roberts, J. C., and Straus, R., eds. (1965). "Comparative Atherosclerosis." Harper, New York.
Sandler, M., and Bourne, G., eds. (1963). "Atherosclerosis and its Origin." Academic Press, New York.

Stoddard, L., and Puchtler, H. (1969). *In* "Pathology Annual" (S. C. Sommers, ed.), Vol. 4, pp. 253–268. Appleton, New York.

Stout, C., and Lemmon, W. B. (1969a). *Amer. J. Obstet. Gynecol.* **105**, 212–215.

Stout, C., and Lemmon, W. B. (1969b). *Exp. Mol. Pathol.* **10**, 312–322.

Stout, C., and Lemmon, W. B. (1971). *Exp. Mol. Pathol.* **14**, 151–157.

Stout, C., and Snyder, R. L. (1969). *Gastroenterology* **57**, 256–261.

CHAPTER 11

Cross-Circulation between Humans in Hepatic Coma and Chimpanzees

JOSEPH H. PATTERSON, ROBERT C. MACDONELL, JR., GERALD T. ZWIREN, H. F. SEIGLER, RICHARD METZGAR, and MICHALE KEELING

I. Introduction

The feasibility of treating individuals in hepatic coma by cross-circulation with primates was demonstrated by Bosman *et al.* (1968) and Saunders *et al.* (1968). Using a complicated technique, they described improvement following cross-circulation between a baboon (*Papio ursinus ursinus*) and a 20-year-old female in hepatic coma due to acute hepatitis. Subsequently, Hume and his colleagues (1969) used baboons (*Papio anubis*) with four adult patients who had chronic progressive hepatic failure unresponsive to medical management, with definite improvement in two and slight improvement in one.

Recently, Hollander *et al.* (1971) have described treatment of a patient with fulminant hepatorenal failure due to alcoholic hepatitis by cross-circulation with a baboon for 5½ hr. Though the baboon had developed pulmonary edema, the patient could be aroused by spoken commands and the procedure was terminated. The patient was able to follow simple spoken instructions 6 hr after completing cross-circulation; 39 days later was ambulating with a walker, appeared neurologically and behaviorally normal, and planned to return to full-time employment.

Cross-circulation presumably allows removal of toxic metabolites from the blood of an affected person by the primate liver, and may permit time for regeneration of the patient's damaged liver. In the original cross-circulation procedures carried out by Bosman and Saunders and their colleagues, no attempt to cross-match the human and baboon blood was made. Subsequently, Hume et al. (1969), Fortner et al. (1972), May et al. (1972), and Hollander et al. (1971), have stressed the need for ABO compatibility between patient and primate.

II. Cross-Circulation Using Yerkes Chimpanzees

The present authors have applied the cross-circulation technique to two patients, using chimpanzees (*Pan troglodytes*), each having a blood group (type O) identical with that of the patient. These were 3- and 5-year-old girls, patients in the Henrietta Egleston Hospital for Children, who developed hepatic coma secondary to presumptive infectious hepatitis. The experience of employing chimpanzees, rather than baboons, for the purposes of cross-circulation is described.

The first patient had developed malaise, nausea, and jaundice and was treated conservatively with apparent clinical improvement over the initial 10-week period. However, there was recurrence of these symptoms and signs which persisted during the 2 weeks prior to admission. On admission she was deeply icteric without other abnormal physical findings. The urine gave a strongly positive test for bile; there was marked increase in serum bilirubin, glutamic oxalacetic and glutamic pyruvic transaminase, and alkaline phosphatase levels. In addition, prothrombin and partial thromboplastin times were greatly prolonged. Despite treatment with bed rest, a low fat–high carbohydrate diet, and oral vitamin K, there was no clinical improvement. During the seventh hospital day, increased irritability and infrequent episodes of irrational speech and behavior occurred. Though the elevated transaminase levels had decreased somewhat, the coagulation values were more prolonged than those on admission. Despite dietary exclusion of fat and protein, therapy with intravenous glucose, intravenous corticosteroids, and oral and rectal neomycin sulfate, there was progressive deterioration in the child's level of awareness over the next 3 days.

In the morning of the 11th day, the child was semicomatose, but she had normal respirations and exhibited spontaneous movements of the extremities with purposeful withdrawal from painful stimuli. Later in the day she became more stuporous and spontaneous movement of the extremities disappeared. At this point, daily double blood volume exchange transfusions with heparinized fresh whole blood were begun, utilizing indwelling brachial artery and antecubital vein catheters. Decreases in bilirubin, alkaline phosphatase, prothrombin time, SGOT, and SGPT occurred following each replacement, with return to preexchange levels in 18–24 hr.

Though she did not regain full consciousness, initially there was transient improvement in neurologic status after each exchange, but after six daily replacement transfusions she was decorticate and demonstrated periodic respirations and hyperactive deep tendon reflexes. She also had gastrointestinal bleeding and abnormal coagulation values with decreases in factors V, II, VII, and X, with marked prolongation of prothrombin and partial thromboplastin times.

After assessment of the patient's unsatisfactory response, preparations were made for cross-circulation with a chimpanzee. The patient weighed 35 lb; a 5-year-old female chimpanzee, named Cassandra, weighing 65 lb, was selected. The animal had been housed in the Yerkes Primate Research Center since the age of 1 year, and had received regular evaluations for the presence of tuberculosis, malaria, and microfilariae with negative results. Single screenings for histoplasmosis and blastomycosis, were also negative. In addition, she had been screened for viral and fungal pathogens on numerous occasions. The donor animal's blood type was O positive. Standard blood cross-matching between the patient and the donor animal prior to cross-circulation demonstrated compatibility of the chimpanzee serum and the patient's erythrocytes, indicating the absence of naturally occurring heteroagglutinins for human ABO-compatible cells in the chimpanzee. Strong macroscopic hemagglutination reaction between the patient's serum and the chimpanzee's erythrocytes was observed. Because of the latter finding it was decided to perform a partial exchange transfusion with human blood on the donor animal prior to cross-circulation to dilute the chimpanzee erythrocytes.

Under sterile conditions in the operating room, the patient's right femoral artery and vein were cannulated with plastic catheters. No anesthesia was required. The chimpanzee was anesthesized and the left groin surgically prepared, and left femoral artery and vein cut-downs were performed. Both vessels were cannulated with plastic catheters. A rapid 20-min exchange transfusion was performed on the chimpanzee with 2 liters of fresh heparinized O-positive human blood compatible with that of the patient.

Cross-circulation was begun by connecting the catheters from the right femoral artery of the patient to the left femoral vein of the chimpanzee and the left femoral artery of the chimpanzee to the right femoral vein of the patient (Fig. 1). Blood pressure and pulse rate of each partner and central venous pressure of the patient were monitored. C clamps on both sets of catheters were used to adjust the flow but were not manipulated after the first 10 min, since the flow appeared to be balanced in both circulations. The actual cross-circulation was carried out for a period of 5 hr and 50 min without complication or significant change in pulse rate, blood pressure, or central venous pressure. Because of the poor urinary output during the first 2 hr of cross-circulation, the patient was given 25 gm of mannitol, intravenously, without response. One hour later, 50 mg of ethacrynic acid was administered, intravenously, with a diuresis of 800 ml and subsequent satisfactory urinary output by the patient. The chimpanzee urinated copiously.

On completion of the procedure, antibiotics, corticosteroids, and vitamin

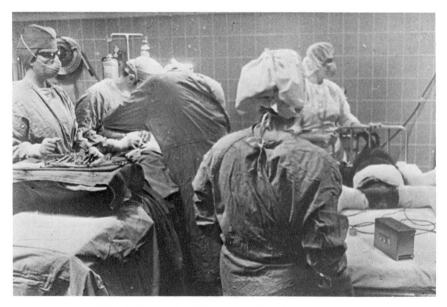

Fig. 1. Cross-circulation procedure in progress.

K were continued and prophylactic antituberculous drugs were added to the patient's regimen. There was no return to full consciousness, but respirations and deep tendon reflexes became normal, and the decorticate positioning disappeared. Bilateral plantar extensor responses persisted, and corneal and gag reflexes remained hypoactive. During the next 24 hr vital signs, hematocrit, and urinary output remained stable. The previous coagulation abnormalities of factor V, the vitamin K-dependent factors, and prothrombin and partial thromboplastin times were corrected. The abnormal bilirubin, alkaline phosphatase, SGOT, and SGPT levels transiently decreased toward normal during the cross-circulation but showed rebound to pre-cross circulation levels within 18–24 hr. On the 18th hospital day, 24 hr after completion of cross-circulation, the patient's condition again deteriorated. Despite performance of the seventh and eighth exchange transfusions progressive neurologic dysfunction resulted in death on the 20th hospital day.

Postmortem examination demonstrated a small shrunken liver with marked distortion of the architecture, an increase in fibrous tissue, and evidence of disorderly regeneration of liver cells. Examination of the brain was not permitted.

Serum obtained from the patient, 48 hr after the cross-circulation, revealed persistent agglutination with the chimpanzee's undiluted pre-cross circulation erythrocytes. This serum did not agglutinate the animal's post-cross circulation cells. Serum taken from the chimpanzee 48 hr after the procedure agglutinated with its own undiluted pre-cross circulation erythrocytes but did not react with the patient's cells from before or after the cross-circulation. Immune serum from another person, who had received and rejected a chimpanzee renal xenograft, strongly agglutinated the chimpanzee's undiluted pre-cross circulation erythrocytes and gave a positive, but weaker, reaction with the post-cross circulation cells. This immune serum did not agglutinate the patient's pre- or post-cross circulation erythrocytes.

III. Effect of the Cross-Circulation Procedure on the Chimpanzee

The chimpanzee had no immediate problems with the initial, partial exchange transfusion of compatible human blood or the cross-circulation (see Keeling and Moore, 1970). The animal was observed closely for the next 2 days and appeared normal; and 3 days after the cross-circulation, it was anesthetized with phencyclidine* and a blood sample was

---

* Sernylan, Biocenties, St. Louis, Mo.

taken. A routine physical examination showed no abnormality. At this time, major and minor cross-matches with blood obtained from the child showed that another cross-circulation would have been possible. Although a repeat cross-circulation was considered, it was not performed because such a procedure appeared inappropriate at that time.

There was a fall in red blood cell count, hematocrit, and hemoglobin compared with values obtained before the cross-circulation. The following day, the fourth, the animal showed some anorexia, inactivity, and depression. It was again anesthetized, and its blood sampled, demonstrating a slight further decrease in red blood cell count, but the hemoglobin and hematocrit were not remarkably altered. The animal became progressively inactive and developed pale vaginal and oral mucous membranes the next afternoon. On the sixth post-cross circulation day, the red blood cell count was $1.27 \times 10^6/mm^3$, the hematocrit 12%, and the hemoglobin 2.8 gm/100 ml. This represented massive hemolysis of human erythrocytes present in the animal's circulation. After major and minor cross-matching, the chimpanzee was transfused with 230 ml of packed red cells combined from other O blood-type chimpanzees in the colony. She displayed no hematuria or hemoglobinuria. On the following day, the animal showed remarkable clinical improvement. She was active, alert, and showed improvement in mucous-membrane color. Her total red cell count had increased to $3.12 \times 10^6/mm^3$. The hematocrit was 25% and the hemoglobin 7.2 g/100 ml. She was given 25 gm of dextrose in 500 ml of Ringer's solution to compensate for anorexia and fluid loss. The clinical improvement during the next few days was obvious, and the animal showed no other abnormalities for the duration of her 60-day quarantine period. The final CBC values closely approached the normal values shown for this animal before cross-circulation. The blood chemistry showed an insignificant elevation of SGPT, SGOT, and bilirubin in the 4-day postcirculation sample, but five subsequent determinations were normal during the next 2 months. The initial elevation may have been associated with blood products from the child not yet eliminated by the chimpanzee.

Bearcroft (1968) has described a similar elevation in the baboon. Before the animal's final release, two needle biopsies of the liver were taken at 3-week intervals, and, although an occasional hepatocyte showed acidophilic necrosis on the first biopsy, there was no other histologic evidence of hepatitis. The second biopsy showed no hepatic changes, and the negative clinical histologic and serum enzyme results indicated that the animal did not contract hepatitis. The animal was removed from quarantine 65 days after cross-circulation and has had no clinical problems subsequently.

IV. The Second Cross-Circulation

In the case of the second cross-circulation, the procedure was carried out for 14 hr and 45 min with Iyck, a 7-year-old male chimpanzee weighing 63 lb. A determination of bilirubin, transaminase, alkaline phosphatase, and prothrombin time values in the human patient, before and after each procedure, yielded results similar to those performed. At 5 hr after cross-circulation, dramatic improvement in the neurologic status of the subject was noted. She was awake, fully oriented, and asked and answered questions appropriately. Pupillary dilatation and blindness, believed to be of cortical origin, were present and persisted for 10 days, at which time spontaneous resolution occurred. Despite her considerable clinical improvement, there was no biochemical evidence of regenerating hepatocellular function. Coagulation studies were consistently abnormal and terminally, cutaneous and gastrointestinal bleeding occurred 24 days following cross-circulation. Despite another double volume exchange transfusion, coma recurred, and respiratory insufficiency, secondary to intrapulmonary hemorrhage, developed on the 40th hospital day. Death occurred on the 41st hospital day, 25 days after the cross-circulation. Autopsy revealed extensive hepatocellular autolysis of gastric mucosa. Examination of the brain demonstrated moderate edema, flattened convolutions, narrowed sulci, a narrowed fourth ventricle, and a few scattered petechiae. Similar aftercare was used, and clinical and laboratory results experienced with the chimpanzee used for this cross-circulation resembled those observed with the first chimpanzee.

V. Discussion

The species choice of the donor animal employed for cross-circulation with patients in hepatic coma is quite important. Obvious ethical considerations interdict human-to-human cross-circulation in most situations. Eiseman and Hume demonstrated that pigs and dogs tolerate exchange transfusion with human blood very poorly (Hume et al., 1969). Brief periods of improvement have been achieved utilizing heterologous perfusion with extracorporeal pig livers (Eiseman et al., 1965). Improvement in patients in hepatic coma following cross-circulation with baboons has been reported (Hume et al., 1969; Bosman et al., 1968; Saunders et al., 1968; May et al., 1972; Fortner et al., 1972; Hollander et al., 1971).

Problems due to human–baboon ABO blood groups incompatibility can be at least partially alleviated by replacement transfusion of the donor animal with human blood prior to cross-circulation (Hume *et al.,* 1969; May *et al.,* 1972; Fortner *et al.,* 1972; Hollander *et al.,* 1971). Blood-typing of 124 baboons revealed type B in 43%, type AB in 30%, and type A in 27% (Moor-Jankowski *et al.,* 1964). In contrast, 89% of chimpanzees have been shown to have type A blood, and 11%, type O, which more closely parallels the human distribution of 45% type O and 41% type A (Wiener *et al.,* 1963; Wiener and Moor-Jankowski, 1963; Smith, 1966).

Experience in tissue transplantation has demonstrated the desirability for compatibility in tissue antigens as well as ABO blood groups (Starzl *et al.,* 1965). Renal heterografts from both baboons and chimpanzees have functioned for limited periods, most having failed within 2 months. However, Reemtsma reported one patient with a chimpanzee renal heterograft who survived 9 months (Hitchcock *et al.,* 1964; Starzl, 1964; Reemtsma, 1966; Porter, 1964). Postmortem histologic findings, with some exception, have suggested more severe rejection phenomena in baboon-to-man than in chimpanzee-to-man heterotransplants (Porter, 1964; Metzgar and Zmijewski, 1966). Several investigators have demonstrated the presence of human histocompatibility antigens in the HL-A system in chimpanzees (Balner *et al.,* 1967; Shulman *et al.,* 1965; Dorf and Metzgar, 1972; Metzgar and Zmijewski, 1966). The above evidence suggests that the chimpanzee is the superior choice for the partner in human–primate cross-circulation.

The serologic data reported here is in agreement with previously published findings on the serum–erythrocyte interactions between humans and chimpanzees. A natural heteroagglutinin to chimpanzee erythrocytes is present in varying amounts in the serum of most humans and has posed a problem in chimpanzee-to-human xenografting (Metzgar and Seigler, 1972; Reemstma *et al.,* 1964a,b). Chimpanzee serum does not possess a natural heteroagglutinin for ABO-compatible human erythrocytes. The naturally occurring heteroagglutinin present in the patient's serum after the cross-circulation and a subsequent exchange transfusion was not able to detect the persisting chimpanzee erythrocytes present in the donor animal 48 hr after the partial exchange transfusion and cross-circulation. However, the immune agglutinin present in the serum of a human who had rejected a chimpanzee renal heterograft was capable of detecting these chimpanzee erythrocytes. Serum taken from the chimpanzee 48 hr after the partial exchange transfusion and cross-circulation produced hemagglutination with her own undiluted pre-cross circulation erythrocytes. This agglutination was attributed to residual "unbound" antibody present in the human plasma transferred to the animal during the partial exchange transfusion and cross-circulation. One cannot be certain of the presence of "coating" antibody because an antiglobulin test was not performed on the chimpanzee donor's post-cross circulation erythrocytes. Since there was unbound human heterohemagglutinating antibody in the chimpanzee's serum at this time, either the cells were coated with this antibody or the naturally occurring human antibody had a low avidity *in vivo* for chimpanzee cells. The animal tolerated the partial exchange transfusion and cross-circulation well, indicating that this human heteroagglutinin does not cause an immediate hemolytic transfusion reaction. The progressive, moderately severe, hemolytic anemia observed in the chimpanzee, which required transfusion with chimpanzee erythrocytes on the sixth day, was probably the result of the immune response to persisting human erythrocytes.

Pre-cross circulation studies of coagulation in the patient were interpreted as representing a combination of the effects of liver dysfunction and heparin. The prothrombin and partial thromboplastin times were prolonged and the levels of factor V and vitamin K-dependent factors were decreased. These values returned to normal following cross-circulation. However, 48 hr later, coagulation studies were again abnormal and were not corrected to the same magnitude following an exchange transfusion with fresh whole blood. The results observed suggest that cross-circulation may have been more efficacious than exchange transfusion in correcting abnormalities of coagulation.

The association between viral hepatitis and the presence of Australia antigen in the serum of affected patients has been described (Prince, 1968; Blumberg *et al.,* 1969; London *et al.,* 1969; Giles *et al.,* 1969; Wright *et al.,* 1969; Hirschman *et al.,* 1969). The antigen has been found in chimpanzee sera, and stated to be immunologically identical with that from humans (Hirschman *et al.,* 1969; Editorial, 1969). Outbreaks of infectious hepatitis have occurred among human handlers of chimpanzees, suggesting direct transmission of hepatitis virus from primate to human (Hillis, 1961; Held, 1962; Riopelle, 1963). Retrospective studies on stored serum performed 8 months after the cross-circulation revealed the antigen to be absent in the pre-cross circulation serum of the patient and present in that of the chimpanzee. Its presence was detected in the post-cross circulation serum for each partner. A positive Australia antigen reaction was also found in stored reference sera taken from the chimpanzee before cross-circulation. There is no explanation for this. The animal had shown no history of hepatic disease during the 4 years she had been in the Yerkes colony and had received no previous transfusion. No abnormalities were reported in the blood chemistries of personnel involved in the care and handling of either of the chimpanzees during the quarantine period. The

Australia antigen was detected in the sera of only 2 of 64 chimpanzees from the same colony. If the test for Australia antigen proves sufficiently sensitive in detecting primate carriers of hepatitis virus, as it has for humans, it will be a valuable tool in the selection of partners for experimental procedures with humans.

VI. Summary and Conclusions

These cross-circulation studies have demonstrated that this procedure has therapeutic possibilities and that chimpanzees are suitable primates to use with humans, although limitations are imposed by the extreme scarcity of these animals. Previously, the baboon had been chosen for this purpose, mainly for its size, availability, and management. However, not only are the chimpanzee's tissue antigens more compatible with man than those of any other primate, but there is an added advantage in that there is a 5–10% occurrence of O blood group in the chimpanzee population, whereas this blood group is not known to exist among baboons. It would appear advantageous to reproduce and raise more O blood-type chimpanzees for organ transplantation research and possible therapeutic cross-circulation with humans.

Consideration of carefully controlled investigative studies of cross-circulation with chimpanzees in comparison with other modes of therapy for hepatic coma is warranted.

ACKNOWLEDGMENTS

We are indebted to Geoffrey H. Bourne, Professor of Anatomy, Emory University School of Medicine; Director, Yerkes Regional Primate Research Center, Emory University for making the chimpanzees available for use in this procedure. Direct and invaluable contributions to the cross-circulation procedures were made by James W. Bland, James J. Corrigan, Daniel Caplan, Lynn Behrens, and Peter A. Ahmann, Department of Pediatrics, Emory University School of Medicine. Valuable assistance was offered by the following persons: James F. Schwartz, Department of Pediatrics; Thomas L. Tidmore, Department of Anesthesiology; Julius Wenger, Department of Internal Medicine; Randi V. Rosvoll, Department of Pathology, Emory University School of Medicine; Shirley L. Rivers, Atlanta Regional Red Cross Blood Center; Milford H. Hatch, National Communicable Disease Center, United States Public Health Service; Harold M. McClure, Yerkes Regional Primate Research Center, Emory University; and Mrs. Ann M. Dewart and the laboratory staff, Henrietta Egleston Hospital for Children. Acknowledgment is also made of the financial support given to the Yerkes Primate Research Center by grant RR00165 of the Division of Research Resources, National Institutes of Health; to USPH grant HE 11246 from the National Heart Institute; grant Am 08054 from the National Institute for Arthritis and Metabolic Diseases.

REFERENCES

Balner, H., van Leewun, A., Dersjant, H., and van Rood, J. J. (1967). *Transplantation* **5**, 624.

Bearcroft, W. G. C. (1969). *J. Med. Microbiol.* **1**, 1.

Bosman, S. C. W., Terblanche, J., Saunders, S. J., Harrison, G. G., and Barnard, C. N. (1968). *Lancet* **2**, 583.

Blumberg, B. S., Sutnick, A. I., and London, W. T. (1969). *J. Amer. Med. Ass.* **207**, 1895.

Dorf, M. E., and Metzgar, R. S. (1972). *Proc. Conf. Exp. Med. Surg. Primates, 2nd,* New York, 1969, Medical Primatology, 1970. Basel, S. Karger, p. 12.

Editorial. (1969). *J. Amer. Med. Ass.* **208**, 1694.

Eiseman, B., Liem, D. S., and Raffucci, F. (1965). *Ann. Surg.* **162**, 329.

Fortner, J. G., Beattie, E. J., Jr., Shui, M. H., Howland, W. S., Chaudry, K., Martini, N., Sherlock, P., Wright, W., Moor-Jankowski, J., and Weiner, A. S. (1972). *Proc. Conf. Exp. Med. Surg. Primates, 2nd, 1969* (in press).

Giles, J. S., McCollum, R. S., Berndtson, L. W., Jr., and Krugman, S. (1969). *N. Engl. J. Med.* **281**, 119.

Held, J. R. (1962). *CDC Vet. Pub. Health Notes* p. 8.

Hillis, W. D. (1961). *Amer. J. Hyg.* **73**, 316.

Hirschman, R. J., Shulman, N. R., Barker, L. F., and Smith, K. O. (1969). *J. Amer. Med. Ass.* **208**, 1167.

Hitchcock, C. R., Kiser, J. C., Telander, R. L., and Seljeskog, E. L. (1964). *J. Amer. Med. Ass.* **189**, 934.

Hollander, D., Klebanoff, G., and Osteen, R. T. (1971). *J. Amer. Med. Ass.* **218**, 67.

Hume, D. M., Gayle, W. E., Jr., and Williams, G. M. (1969). *Surg., Gynecol. Obstet.* **128**, 495.

Keeling, M. E., and Moore, G. T. (1970). *Lab. Anim. Care* **23**, 703.

London, W. T., Sutnick, A. I., and Blumberg, B. S. (1969). *Ann. Intern. Med.* **70**, 55.

May, A. G., Estero, R., Satran, R., and Turner, M. (1972). *Proc. Conf. Exp. Med. Surg. Primates, 2nd,* New York, 1969. Medical Primatology, 1970. Basel, S. Karger, p. 52.

Metzgar, R. S., and Seigler, H. F. (1972). *Proc. Conf. Exp. Med. Surg. Primates, 2nd,* New York, 1969. Medical Primatology, 1970. Basel, S. Karger, p. 2.

Metzgar, R. S., and Zmijewski, C. M. (1966). *Transplantation* **4**, 84.

Moor-Jankowski, J., Wiener, A. S., and Gordon, E. B. (1964). *Transfusion* **4**, 92.

Porter, K. A. (1964). *In* "Experience in Renal Transplantation" (T. E. Starzl, ed.), Chapter 25, p. 299. Saunders, Philadelphia, Pennsylvania.

Prince, A. N. (1968). *Proc. Nat. Acad. Sci. U. S.* **60**, 814.

Reemtsma, K. (1966). *Advan. Surg.* **2**, 285.

Reemtsma, K., McCracken, B. H., Schlegel, J. U., Pearl, M. A., DeWitt, C. W., and Creech, O., Jr. (1964a). *J. Amer. Med. Ass.* **187**, 691.

Reemstma, K., McCracken, B. H., Schlegel, J. U., Pearl, M. A., Pearce, C. W., DeWitt, C. W., Smith, P. E., Hewitt, R. L., Flinner, R. L., and Creech, O. (1964b). *Ann. Surg.* **160**, 384.

Riopelle, A. J. (1963). *In* "Proceedings of a Conference on Research with Primates," p. 19. Tektronix Found., Beaverton, Oregon.

Saunders, S. J., Terblanche, J., Bosman, S. C. W., Harrison, G. G., Walls, R.,

Hickman, R., Biebuyck, J., Dent, D., Pearce, S., and Barnard, C. N. (1968). *Lancet* **2**, 585.

Shulman, N. R., Moor-Jankowski, J., and Miller, M. C. (1965). *Ser. Haematol.* **2**, 113–122.

Smith, C. H. (1966). "Blood Diseases of Infancy and Childhood," 2nd ed., p. 70. Mosby, St. Louis.

Starzl, T. E., ed. (1964). "Experience in Renal Transplantation." Saunders, Philadelphia, Pennsylvania.

Starzl, T. E., Marchioiro, T. L., Terasaki, P. I., Porter, K. A., Paris, T. D., Herrmann, T. J., Vredevoe, D. C., Hutt, M. P., Ogden, D. A., and Waddell, W. R. (1965). *Ann. Surg.* **162**, 749.

Wiener, A. S., and Moor-Jankowski, J. (1963). *Science* **142**, 67.

Wiener, A. S., Moor-Jankowski, J., and Gordon, E. B. (1963). *Amer. J. Phys. Anthropol.* **21**, 271.

Wright, R., McCollum, R. W., and Klatskin, G. (1969). *Lancet* **2**, 117.

CHAPTER 12

The Cape Chacma Baboon in Surgical Research

J. H. GROENEWALD and J. J. W. VAN ZYL

The Cape Chacma baboon, which roams wild over a wide area of the Cape Province and other parts of South Africa, has for many years been the natural enemy of every farmer wherever encountered. Thousands of these humanlike animals, classified as vermin in South Africa, are killed annually by farmers because of the damage done to grain and fruit crops.

Since the establishment of the University of Stellenbosch Primate

Colony in 1967, these baboons, because of their similarity to man both anatomically and physiologically, were suddenly reflected in a different light. They now became a valuable asset in research directed toward the future well-being of the human race. Where previously the government of South Africa experimented with poisons to help the farmers in their struggle against these pests, the same government is now proposing the introduction of legislation to safeguard these animals for medical research purposes.

The University of Stellenbosch Primate Colony is possibly unique in the world in supplying large primates to a variety of acute and relatively acute surgical experimental projects. Although by far the largest number of animals is used in the various surgical projects at the Karl Bremer Hospital, University of Stellenbosch, many projects related to anatomy, physiology, biochemistry, nuclear medicine, bacteriology, microbiology, serology, immunology, and cardiac and pulmonary research are also supported. Projects related to immunology, especially with regard to tissue transplantation, form the second largest group.

From July 1967 until December 1969, the colony received and handled over 3000 baboons. No other primates are handled or supplied by the colony.

Some of the protocols supported by the colony, such as the search for improved immunosuppressive agents, are long-term projects utilizing hundreds of animals. Others are short-term and minor projects using only a few animals to perfect a technique or solve a relatively small problem before employing a particular method of treatment in humans. In both types of experimentation, however, the baboon has proved ideal as an experimental animal.

Of the various research projects that have been supplied with animals from the colony, the following represent some of the major advantages and encouragements in various areas of research that have been achieved to date or are under continuous investigation.

I. Mitral-Valve Replacement

As part of the development of the University of Stellenbosch mitral-valve prosthesis, and in the study of the problem of thromboembolism that occurs following the use of prosthetic valves, 20 mitral-valve replacements were carried out on Cape Chacma baboons (Barnard, 1969). Because baboons tolerate cardiopulmonary bypass well (even lengthy perfusions of up to 3 hr, using high flow rates of 100–125 ml min kg body weight, being followed by prompt and complete recovery) the baboon is an ideal experimental animal for this type of research.

In these replacement studies massive thrombus formation was never encountered, although small fibrin thrombi, a potential source of emboli, became loosely attached to the Dacron velour covering the prosthetic ring during the first eight postoperative weeks. Thereafter the smooth ingrowth of living tissue onto the prosthesis appears capable of preventing thrombo-embolism (Barnard, 1969). The valve is now routinely being used in human mitral-valve replacement at this institution.

A further advantage of using the baboon in this and many other types of experimentation where blood transfusion is required is that the baboon can be accurately typed with human ABO typing sera. As baboons are secretors it is possible to determine their blood groups by the ability of saliva to inhibit the activity of human erythrocyte typing antisera, and the sera of these animals can also be examined for agglutinins to human erythrocytes (van Zyl *et al.,* 1968a).

II. Cardiac Transplantation

Because of the above-mentioned advantages, namely, the tolerance of cardiopulmonary bypass and the possibility of ABO blood group-matching, baboons were also used at the Karl Bremer Hospital laboratories to study the techniques of cardiac transplantation with satisfactory results (Barnard and Heydenrych, 1972).

III. Corneal Transplantation

Corneal and vitreous human transplants were studied both as homo- and as heterotransplants; the latter utilized both baboon-to-human as well as human-to-baboon transplants (Ferreira and Bosch, 1972). Good results in all categories were obtained despite the fact that no immunosuppressive treatment whatever was used. Successful baboon-to-human transplants were obtained with return of vision in patients who, prior to operation, had no vision at all. The baboon has also been investigated as a means of storing human corneas for later transplantation to another human patient. It seems as if the baboon might well be suitable for this type of preservation.

IV. Lung Transplantation

This study, which included a variety of surgical procedures, was controlled with physiological studies using ^{133}Xe to determine the functional effect of the various pulmonary surgical procedures on lung perfusion and ventilation. The baboon is eminently suitable for this purpose for a number of reasons. The relatively wide chest is easy to scan; clotting of venous and arterial anastomoses are less frequent than in the dog; the baboon has a well-developed mediastinum compared with the relatively rudimentary mediastinum in the dog. Chronic survival is easily obtained after contralateral pneumonectomy, in contrast to the dog which usually dies after contralateral pneumonectomy. This study is still under investigation and procedures performed include complete hilar stripping, division of bronchial arteries and lymphatics, division of vagal branches to one lung, reimplantation, and homotransplantation (Lubbe *et al.*, 1972).

V. Auxiliary *ex Vivo* Extracorporeal Liver Perfusion and Hepatic Assist

A system for the *ex vivo* extracorporeal perfusion of the liver and cross-circulation with patients with hepatocellular damage has been perfected. The system has been extensively investigated in a baboon model with experimentally created hepatocellular damage, and more than pleasing results have been achieved. Because human blood can be successfully perfused through the baboon liver with complete maintenance of function the animal provides a valuable method for hepatic-assist treatment of patients with hepatocellular damage in liver coma. In this instance the baboon is not only a valuable experimental animal but is of even greater value in the actual treatment of patients in hepatic failure.

VI. Immunology and Immunosuppressive Therapy after Tissue Transplantation

In this study alone over 1800 baboons have been used. These studies, using kidney transplantation as a model, had the prime purpose of (1) studying the serologic rejection patterns especially to try and find earlier indications of the rejection crisis; (2) searching for better and improved immunosuppressive agents; (3) studying the morphologic pattern of rejection. All three studies were directed solely toward application to human transplantation.

A. SEROLOGIC REJECTION PATTERNS

Investigation into the microbiologic pattern showed that three serologic tests are of value in the diagnostic judgment of renal allotransplant reactions. These are heteroagglutinins, heterohemolysis, and the complement system. While each of them alone is insufficient, combined they allow insight into the transplant immunologic status. Serologic heterohemagglutinin determinations are of diagnostic value, especially if other sero reactions are also considered. A dilution rise of up to 1:30 from a normal 1:2 is often recorded at the time of rejection. Heterohemolysins appear earlier and increase sooner after transplantation. According to observations in baboons, a rise in titer of heterohemolysins must be regarded as an indicator of the preparatory stage of a rejection reaction, and a sudden loss of free heterohemolysins is indicative of an acute rejection episode, especially if the complement titer drops at the same time (Brede and Murphy, 1968). These valuable indications are now used in human transplantation subjects.

B. IMMUNOSUPPRESSIVE THERAPY

In the search for improved immunosuppressive therapy both chemical and biologic agents were evaluated. Of the many drugs tried only a few of the more stimulating results will be briefly discussed. Subcellular kidney cell fraction (SKCF), prepared in our laboratories, prolonged renal allograft survival on a dose-related basis from a mean survival of 9.6 days for untreated controls to a mean survival of 20.0 days in treated animals (Table I) (Groenewald et al., 1972a). These animals all had kidney grafts transplanted from ABO-compatible donors after bilateral nephrectomy (Groenewald et al., 1968).

By mechanically converting the electric charge on these subcellular kidney cell fractions we created what we termed Contra-antigens (Loiseleur,

TABLE I

SURVIVAL TIMES OF UNTREATED CONTROLS AND BIOLOGICALLY TREATED ALLOGRAFTED BABOONS[a]

Treatment	Dosage schedule	Survival in days
Untreated controls		9.7 ± 5.6
SKCF (6)	0.02 ml single dose im	12.0 ± 4.0
SKCF (13)	2.0 ml im 7 days; 2 ml every 2nd day	20.0 ± 21.0^{b}
SKCF (5)	20.0 ml im at operation and 7th postoperative day	7.0 ± 4.0
SKCF (6)	0.2 ml at operation and every 2nd day im	16.0 ± 9.0^{b}
SKCF (6)	0.02 ml pre- and postoperatively every 2nd day	7.0 ± 5.8
SKCF (6)	0.02 ml pre- and postoperatively every 2nd day im	9.0 ± 6.3
SKCF (4)	2.0 ml pre- and postoperatively every 2nd day im	12.6 ± 5.8

[a] All values are mean SD. Numbers in parentheses indicate number of animals in each experimental subgroup.
[b] Statistically significant $P < 0.05$.

1950). These so-called Contra-antigens proved to be even better than the SKCF in prolonging renal allograft survival (Groenewald et al., 1970b). As indicated in Table II, survival in the groups treated with Contra-antigens, as in the SKCF treatment groups, was dose-related. In treated animals a maximum mean survival time of 16.5 days was achieved with these Contra-antigens. This was, however, further increased to a very significant mean of 26.5 days in animals which were pretreated, using similar dosages of Contra-antigens (Table II). Apart from prolonged survival good histologic evidence of suppression of the rejection phenomenon was also seen in these animals treated with Contra-antigens (Weber et al., 1969a).

Baboons proved of value in the evaluation of many other immunosuppressive agents. These included chloroquine, which gave histologic and biochemical evidence of immunosuppression, but did not prolong allograft survival beyond a mean survival time of 11.8 days (Groenewald et al., 1970d). Thalidomide, on the other hand, prolonged allograft survival to a mean of 21.2 days, but showed no histologic evidence of immunosuppression (Murphy et al., 1972).

These are but a few of over 40 conventional and newer immunosuppressive agents that were evaluated in the baboon. Because of the similarity of the rejection patterns, both serologically and histologically, to the human and because of the similarity of the hematologic and biochemical changes, the baboon is considered to be ideal for this type of research work. In contrast, for instance, to the dog it was easy not only to ABO-type and even tissue-type and match baboons for transplantation, but follow-up studies were much more easily performed. The baboon is also very easy to catheterize and therefore renal clearance studies can be performed with ease and renal function after transplantation can be easily assessed. As baboons drink well from a nipple on a graduated bottle, intake and output

TABLE II

Survival Times of Untreated Controls and Biologically Treated Allografted Baboons[a]

Treatment	Dosage schedule	Survival in days
Untreated controls		9.7 ± 5.6
Contraantigens (6)	0.2 ml im every 2nd day	11.8 ± 6.9
Contraantigens (6)	5.0 ml im every 2nd day	16.5 ± 7.1^b
Contraantigens (6)	20.0 ml im every 2nd day	12.4 ± 7.3
Contraantigens (6)	2.0 ml im pre- and postoperatively every 2nd day	26.5 ± 19.5^b

[a] All values are mean SD. Numbers in parentheses indicate number of animals in each experimental subgroup.
[b] Statistically significant $P < 0.05$.

over periods of 24 hr or otherwise can be measured relatively accurately. Blood pressure can be measured in the standard clinical manner obviating the problems encountered in most other animal species.

An added advantage found was the availability of an apparently inbred colony of baboons in the Cape Point nature reserve in which the effects of transplantation between such inbred animals could be studied (van Zyl *et al.*, 1968a).

C. PATHOLOGY

The histologic pattern of rejection in the baboon has been extensively studied. In the untreated renal transplanted animals the rejected kidney showed destruction by two mechanisms. These included first, interstitial peritubular infiltration destroying the tubules, and second, vascular changes causing vascular obstruction by endothelial proliferation and even thrombosis at an early stage. As a whole the picture was very similar to that observed in kidney transplants undergoing rejection. Depending on the immunosuppressive agents employed, modifications of the usual rejection pattern encountered in the baboon were clearly defined. These treated animals often showed a milder inflammatory vascular reaction and no fibrinoid necrosis, a constant finding in untreated control animals (Weber *et al.*, 1969b).

VII. Antilymphocyte Serum

The production of antilymphocyte serum (ALS) and antilymphocyte globulin (ALG) and its evaluation in the baboon is discussed separately from other immunosuppressive agents, because of the special value of using subhuman primates to evaluate these preparations prior to clinical use in humans (Balner *et al.*, 1968). The lack of suitable *in vitro* tests makes evaluation in the subhuman primate all the more important at the present time (Balner *et al.*, 1968). We found the Cape Chacma baboon, although lower on the evolutionary scale than the chimpanzee, a suitable animal for this type of testing, giving constant results (Brede *et al.*, 1972b). In the first instance we used baboons to evaluate goat-antibaboon lymphocyte serum as a control in the development of a parallel program to develop a potent goat-antihuman, and later horse-antihuman lymphocyte serum (Brede *et al.*, 1972a). More extensively the baboon skin-graft model was used to evaluate ALS and ALG from local as well as from various centers throughout the world, prior to human application of these preparations. The ability of these antihuman lymphocyte preparations to prolong both first and second set skin-graft survival in the baboon served as a valuable indication of the predicted potency of the preparations in human applications (Brede *et al.,* 1972a). Furthermore, by following these baboons with regular hematologic studies, especially thrombocyte counts and serology, both the potency of the preparation and its toxicity could be evaluated in the test animal for safety prior to human application (Brede *et al.,* 1972b).

VIII. Drug Evaluation

Drug evaluation and toxicity studies must be regarded as more applicable to man when performed in subhuman primates, such as the baboon, than when carried out in less closely related species, where the pattern of reactions are often quite different from those found in man. Much valuable data in this regard can be collected from drug studies in intact animals and in isolated organ perfusions, especially of kidneys (Groenewald *et al.,* 1970c; Murphy *et al.,* 1968, 1969a,b). Other centers in South Africa, such as the South African Council for Industrial and Scientific research, have used the baboon even more extensively in nutritional and drug-evaluation studies.

IX. Liver Transplantation

The baboon is probably a more suitable animal for this type of research than the pig (which is at present most commonly used for liver transplantation research because of the anatomic difficulties encountered in the dog). For some unknown reason the pig does not readily reject a transplanted liver and therefore has certain disadvantages. An additional advantage of the baboon is that, because of an excellent collateral circulation, bypass is not necessary during liver transplantation, making the operation easier.

X. Organ Preservation

Successful preservation of organs prior to transplantation has become all-important. We have found the baboon an excellent experimental model for 24-hr kidney-preservation studies (Groenewald *et al.,* 1969b, 1970e,f, 1972b). Here again, more accurate functional evaluation of the preserved transplanted organ could be carried out in the baboon. Correlation with human results was again possible and because parameters of functional clearance values were available, the kidneys, postpreservation, could also be evaluated in the isolated perfusion system in which renal clearance as well as metabolic studies could be performed (van Zyl *et al.*, 1968b).

XI. Cardiodynamic Studies

Various protocols in this category have been, or are being, investigated in the baboon. These include the renal and systemic effects of experimentally created aortic coarctation (Groenewald *et al.*, 1970a). At present the baboon is being used as a model in shock experiments. Both flow and pressure studies can be readily measured, and, because of the anatomic advantages over the dog, the implantation of catheters for pressure recordings in the larger vessels are readily carried out. The baboon is also found suitable for experimentation involving heart catheterization.

XII. Normal Anatomic and Physiologic Studies and Values

Valuable information has been obtained in studies on the anatomy and physiology of the baboon prostate (Schoones *et al.*, 1970) preparation of a stereotactic brain atlas and erythropoietin production and release (Groenewald *et al.*, 1969a; Mirand *et al.*, 1968, 1969). The baboon in itself has also been studied intensively and normal values recorded in over 3000 animals (Weber *et al.*, 1972). These values are useful for correlation of results.

XIII. Handling

An added factor in favor of the Cape Chacma baboon in preference to other primates and some other animal species is the ease with which these animals can be handled. Because they are naturally timid, the Cape Chacma baboon will not attack unless threatened in a position from which he cannot escape. Even then he will not attack if you turn your back on him. This lack of aggressiveness makes it almost pleasant to handle the animals in contrast to most other types of subhuman primates.

Tranquilization can easily be obtained with Sernylan [Compound CI-395 or (1-phenylcyclohexyl)piperidine hydrochloride]. This agent is a potent calming and hallucinating drug that apparently acts by interfering with sensory inflow into the cortex at midbrain or thalamic level (Chen and Weston, 1960). This pharmacologic block of sensory pathways results in indifference to pain and a catatonic or cataleptic state with loss of consciousness at higher doses. The biting reflex is depressed at all doses. The absence of this reflex, together with the catatonic state, allows safe handling even when the animal is apparently conscious.

Postoperative care has few problems in the baboon. They do not interfere in any way with sutures or dressings and appear to do their best to safeguard such areas against possible injury.

Animals needing special attention postoperatively can readily be kept in restraining chairs for prolonged periods of time. In such a chair, while the animal feeds and waters himself, intravenous drips and endothoracic tubes or drains can be left *in situ* without problems. Furthermore, the animal can be placed in an oxygen tent, or receive such treatment or examination as required while fully conscious. It is easy to draw blood, to collect urine, and measure the blood pressure of baboons. These are all factors which make these animals obviously ideal for experimentation.

XIV. Conclusions

The foregoing remarks emphasize the value of having the Cape Chacma baboon available for medical research. The obvious advantages over the dog are also clearly indicated. Any surgeon who has had the opportunity of working with both types of animals will have no doubt in his mind as to the numerous advantages that the baboon offers.

REFERENCES

Balner, H., Eysvogel, V. P., and Cleton, F. J. (1968). *Lancet* 1, 19.
Barnard, P. M. (1969). M.D. Thesis, University of Stellenbosch.
Barnard, P. M., and Heydenrych, J. J. (1972). *Primates Med.* 8 (in press).
Brede, H. D., and Murphy, G. P. (1968). *S. Afr. Med. J.* 42, Suppl. 17 Aug.
Brede, H. D., Groenewald, J. H., and Murphy, G. P. (1972a). *Symp. Transplant. Biol., 1970* (in press).
Brede, H. D., Murphy, G. P., Weber, H. W., van Zyl, J. J. W., de Klerk, J. N., and Rudman, E. R. (1972b). *Proc. Symp. Primates Transplant. Res., 2nd, 1969* (in press).
Chen, G. M., and Weston, J. K. (1960). *Anesth. Analg. (Cleveland)* 39, 132.
Ferreira, A. B. W., and Bosch, A. H. (1972). *S. Afr. Med. J.* (in press).
Groenewald, J. H., van Zyl, J. A., Schoonees, R., van Zyl, J. J. W., and Murphy, G. P. (1968). *S. Afr. Med. J.* 42, Suppl. 17 Aug. 1968.
Groenewald, J. H., Mirand, E. A., Mostert, J. W., Takita, H., and Murphy, G. P. (1969a). *Circulation* 40, 97 (abstr.).
Groenewald, J. H., van Zyl, J. J. W., Weber, H. W., and Murphy, G. P. (1969b). *Trans. Amer. Soc. Artif. Intern. Organs* 15, 219–224.
Groenewald, J. H., van Zyl, J. J. W., and Murphy, G. P. (1970a). *Invest. Urol.* 7, 299.

Groenewald, J. H., Murphy, G. P., Weber, H. W., and Brede, H. D. (1970b). *In* "Infections and Immunosuppression in Subhuman Primates" (Balner and Beveridge, eds.), pp. 211–215. Munksgaard, Copenhagen.

Groenewald, J. H., Retief, C. P., and Murphy, G. P. (1970c). *S. Afr. Med. J.* 44, 1002.

Groenewald, J. H., Weber, H. W., de Klerk, J. N., and Murphy, G. P. (1970d). *Primates Med. Surg.* 2.

Groenewald, J. H., Weber, H. W., van Zyl, J. J. W., and Murphy, G. P. (1970e). *Rev. Surg.* 70, 170.

Groenewald, J. H., van Zyl, J. J. W., and Murphy, G. P. (1970f). *Cryobiology* 6, 500.

Groenewald, J. H., Brede, H. D., Weber, H. W., and Murphy, G. P. (1972a). Medical Primatology (Goldsmith, Moor, and Jankowski, eds.). Basel, S. Karger.

Groenewald, J. H., Weber, H. W., van der Walt, J. J., Greeff, M. J., and van Zyl, J. J. W. (1972b). *Transplant. Proc.* 3, 634.

Loiseleur, J. (1950). *Ann. Inst. Pasteur, Paris* 78, 151.

Lubbe, J. J. de W., Barnard, P. M., White, J. J., and Lötter, M. G. (1972). *S. Afr. Med. J.* (in press).

Mirand, E. A., Murphy, G. P., Steeves, R. A., Groenewald, J. H., van Zyl, J. J. W., and Retief, F. P. (1968). *Proc. Soc. Exp. Biol. Med.* 128, 785.

Mirand, E. A., Groenewald, J. H., Kenny, G. P., and Murphy, G. P. (1969). *Experientia* 25, 1104.

Murphy, G. P., Groenewald, J. H., Schoonees, R., van Zyl, J. J. W., Retief, C. P., and de Klerk, J. N. (1968). *Invest. Urol.* 6, 294.

Murphy, G. P., Groenewald, J. H., Kenny, G. M., and Reynoso, G. F. (1969a). *J. Surg. Oncol.* 2, 1169.

Murphy, G. P., Schoonees, R., Groenewald, J. H., Retief, C. P., van Zyl, J. J. W., and de Klerk, J. N. (1969b). *Invest. Urol.* 6, 466.

Murphy, G. P., Brede, H. D., Weber, H. W., Groenewald, J. H., and Williams, P. D. (1972). *J. Lab. Clin. Med.* (in press).

Schoonees, R., de Klerk, J. N., and Mirand, E. A. (1970). *Invest. Urol.* 8, 103.

van Zyl, J. A., Murphy, G. P., Weber, H. W., Brede, H. D., van Zyl, J. J. W., van Heerden, P. D. R., Retief, F. P., Retief, C. P., and Botha, M. C. (1968a). *S. Afr. Med. J.* 42, Suppl. 17 Aug. 1958.

van Zyl, J. J. W., van Zyl, J. A., Groenewald, J. H., Schoonees, R., Lochner, A., and Murphy, G. P. (1968b). *S. Afr. Med. J.* 42, Suppl. 17 Aug.

Weber, H. W., Brede, H. D., Murphy, G. P., and Groenewald, J. H. (1969a). *S. Afr. Med. J.* 43, 541 (abstr.).

Weber, H. W., Brede, H. D., Retief, F. P., Retief, C. P., van Zyl, J. A., Groenewald, J. H., van Zyl, J. J. W., and Murphy, G. P. (1969b). *J. Urol.* 101, 465.

Weber, H. W., Brede, H. D., Retief, C. P., Retief, F. P., and Melby, E. C., Jr. (1972). *Nat. Acad. Sci.—Nat. Res. Counc., Publ.* (in press).

CHAPTER 13

Degenerative Diseases

OSCAR FELSENFELD

I. Introduction

The purpose of this chapter is to present a brief discussion of a selected set of degenerative diseases chosen from three large groups of such disorders. One of these groups consists of degenerative diseases described in man as well as in free-living nonhuman primates. The second contains diseases of human beings that can also be produced in other primates by experimental means. The third consists of conditions not yet reported in apes and monkeys but which may soon be studied also in these animals because the results of recent research indicate that nonhuman primates may serve as models for their investigation.

Degenerative diseases have been treated from the point of view of the pathologist who is primarily interested in the ills of man and who uses nonhuman primates only as an experimental model. The emphasis has been on phenomena easily observed in the gross specimen and with the aid of the ordinary optical microscope. This does not negate the primary mission of the pathologist as the interpreter of laboratory findings, including microscopic studies, in understanding clinically observed phenomena. Nor should the emphasis on histology be construed as a disbelief in the possibility of biochemical changes preceding morphologic disturbances. As a matter of fact, primary anatomic changes induce diminished cell function with consequent physiologic disturbances more often than not.

Letterer (1959) related the history of attempts to reconcile the views separating "heteroplastic" and "metabolic" disorders. In spite of the expanding functional concepts of pathology, it was felt that primary metabolic alterations with prevailing irregularities of enzymic, hormonal, nutritional, and related biochemical activities should be considered in the respective chapters of this book. The same holds true concerning genetic and other "inherited" disorders. Most of the latter are not easily reproduced in primates.

Considerable space has been devoted to the description of the histologic alterations related to degenerative diseases but not to their symptoms and course, because those who are interested in using nonhuman primates as an experimental model for the investigation of such diseases are well acquainted with them. The same applies to the choice of references. The literature on nonhuman primates is large, even though "normal" physiologic values and histologic details of several primate species in general and some organs in particular are still lacking or incomplete. Most of the currently used monographs on primates have not been included among the references because they are readily available and known to the majority of the readers. The choice of quotations in this chapter reflects the personal preference of this contributor for imaginative literature, because creative imagination is the strongest motivation for productive research.

II. General Considerations

The cause of degeneration may be summarized as involution, including that observed postpartum and in old age; genetic; chemical and physical factors such as infections, drug effects, trauma, pressure, tension, radiation, and inactivity; metabolic disturbances, including loss of fluid and ions; endocrine disorders; and lack of nervous stimulation. Degenerative phenomena may have more than one cause in the same individual or organ.

When the primate does not receive sufficient food or the intestinal absorption is impaired, first the adipose tissue loses fat, then the muscles are reduced. The lipids of the brain appear to be quite resistant to starvation. Antimetabolites and disturbances of metabolism affecting a single essential nutrient or a group of these have consequences often characterized by a known clinical and pathologic picture. The lack of neural stimulus may reveal itself not only by muscular wasting but also by osteoporosis, as in syringomyelia. The manifold manifestations of pressure are well known, as those resulting from circulatory stasis in the liver, pressure of the enlarged thyroid and the aorta on adjacent bones and obstruction of blood and lymphatic vessels and of the flow of cerebrospinal fluid. Many of these and related phenomena can be reproduced in nonhuman primates by surgical means, including electrosurgery, X-ray irradiation, laser beams, deficient diets, administration of antimetabolites, incorporation of toxic materials, large amounts of hormones, and other means which need not be repeated in this chapter.

The morphologic changes appearing in the degenerating cells have been studied also with the electron microscope. They have been succinctly summarized by Anderson and Scotti (1968) as alterations beginning in and around the mitochondria which appear shrunken and dense, vesicular distention of the protoplasm, depletion of ribosomes, followed by swelling of the mitochondria and loss of their structure, then formation of lipid droplets.

The normal intracellular mechanism (G. H. Bourne, 1962) is modified as a result of changes of the membrane permeability and the active transport mechanism, and by alterations in the enzyme cycles with which the energy-producing mechanism is intimately allied. The damage to intracellular membranes permits the penetration of enzymes into intracellular structures and allows them access to their substrates in other parts of the cell from which they are separated under normal conditions. This may result in cell deterioration (Green, 1967).

Degenerative changes demonstrable with the aid of the optical microscope usually consist of changes apparent in morphologic studies that may, however, require special histologic staining methods that permit the visualization and differentiation of chemical metamorphoses. It may be stated that the exact composition of most materials appearing in degenerating cells and tissues is not yet fully known in spite of the considerable progress in modern histochemical methodology.

The nomenclature and classification of degenerative changes vary from one textbook and medical dictionary to another because the underlying intracellular processes have not yet been fully elucidated by electron microscopy and microchemistry.

The most common type of degenerative change is cloudy swelling, called also parenchymatous degeneration. It is a result of mitochondrial damage. This type of degeneration is frequently transient and, as a rule, reversible. The designation "parenchymatous degeneration" originated from the observation of gross changes in the liver, at autopsy, in infectious diseases. It is characterized by intra- and extracellular water and ion transport shifts, disturbances of the permeability of intracellular structures, including the intracellular limiting membranes. Protein-like material(s) is seen in the cells that may become granular, sometimes vesiculated and strongly acidophilic. The nuclei remain unchanged. Parenchymatous degeneration represents mild regressive changes (Letterer, 1959). It is a common phenomenon, seen principally in infectious diseases in the liver and renal tubules. The evaluation of minimal parenchymatous degenerative changes may cause difficulties in postmortem material. Hydropic and vacuolar degeneration are related to it and are due to swelling of the mitochondria and intracellular edema with congestive enlargement of the cell lysosomes (Dianzani et al., 1969).

Hyaline or vitreous degeneration consists of the appearance of a structureless, glassy, protein-like substance that is not easily resorbed. Hyaline degeneration of collagen fibers is not uncommon. Hyaline degeneration develops in the female breast with increasing age. Reversible hyaline degeneration is observed in the mammary glands during the first part of the intermenstruum. Calcification or the deposit of other insoluble materials is observed frequently in connective and fibrous tissue undergoing degeneration, e.g., in arteriosclerosis and in pulmonary silicosis.

Fatty metamorphosis, formerly called fatty degeneration, is the result of accumulation of lipids within the cells due to hypoxia and toxins. Lipids may appear also in fatty infiltration of the interstitial spaces, for instance, when liberated by disintegrating myelin. Histiocytes, perhaps also fibrocytes, may take up excess fat and carry lipids in the interstitial spaces. Fatty metamorphosis is frequent in the heart, liver, and kidneys. In the liver diffuse, peripheral, and centrolobular fatty metamorphosis are distinguished but lipids may appear also in perivascular and peribiliary spaces. It has been stated that mostly neutral fats and cholesterol esters are found in hepatic fatty degeneration of that organ. Fatty metamorphosis of the kidneys seldom reveals itself by the formation of large droplets as in the liver. Nevertheless, patchy or more diffuse desquamation is a frequent sequel. Fatty degeneration and infiltration may follow the course of myocardial fibers in the heart, principally in the large papillary muscles of the ventricles and in the septum, giving it a "tigroid" appearance. A combination of hyaline degeneration and fatty changes is often seen in the renal tubules, and in other localizations.

Mucoid degeneration is accompanied by the appearance of mucopolysaccharides or mucoproteins, particularly in the heart. This type of degeneration is seen also in neoplasms.

Fibrinoid or fibrinogenic degeneration may not be a single process. It is of considerable importance in degenerative vascular changes.

Primary amyloidosis is uncommon but can be produced in some nonhuman primates by allergens (Felsenfeld and Greer, 1972). Secondary amyloidosis has been reported in a squirrel monkey (Banks and Bullock, 1967).

Wallerian degeneration will be discussed in Section VIII.

Endogenous and exogenous pigments may appear in degenerative processes. They may be diffuse or localized. Lipofuscin is often observed in "brown atrophy." Lipofuscin appears in the liver cells near the nuclei in the central and intermediary zones. Its presence in hepatic structures, muscles, ganglion cells, and other locations is a sign of insufficient circulation. Lipofuscin may disappear after recovery. Melanin is not a homologous matter. At least three types are known. It is produced by intracellular enzymes, while the chromogen is a metabolic product. Hematogenic pigments originate from hemoglobin and its components and derivatives, as hemosiderin, bilirubin, porphyrin, and hematoidin.

We relegate "glycogenic degeneration" to the group of metabolic disorders, as well as gout, ochronosis, porphyria, and related disorders.

When the degenerative processes of individual organs are considered, a few general statements may be of value.

Involution, as the fatty metamorphosis of the thymus during maturation, actually should not be considered a degenerative process. The brain seldom shows homogeneous degenerative changes. Such alterations are usually localized. The frontal lobes are more frequently involved. The gray matter reveals degenerative changes more frequently than does the white. The Nissl substance of the nerve cells is a *locus minoris resistentiae*. Malnutrition first influences the lymphopoietic apparatus, whereas the reticuloendoethelial system remains intact for a long time. The bone marrow is frequently replaced by adipose or fibrotic tissue during degenerative processes. Splenic and lymph node follicles decrease in size and numbers, hemosiderosis becomes apparent, and adipose and connective tissues increase. Loss of fibrils and striation, brown atrophy, and increase in fatty tissue are common in the degenerating heart. Connective tissue proliferates and lipofuscin is found in degenerating livers. The cortex is often involved in the kidney. The glomeruli degenerate less easily than do the tubules.

The balance between bone formation and resorption is disturbed in degenerative processes of the skeletal system. There may be an increase of the spongiosa, and hyperplasia of adipose tissue.

These changes are common to man and all other primates. However, the borderline between degeneration and necrosis is often not defined precisely.

III. Aging

Zeman (1962) and McKeown (1965) reviewed the pathologic changes in senescence. The general concept of aging is of a degeneration caused by wear and tear, stress, past infectious diseases, repeated exposure to toxic agents, and lack of replacement of oligobiotic cells and tissues, e.g., of neurons. It has been suggested that hereditary factors, including some that influence vascular function and elastic tissue quality, play a role in aging. On the cellular level, cell permeability undergoes alterations. Cell division and cell growth are reduced. The rate of metabolism, especially of O_2, is reduced. Repair diminishes. As a consequence of these qualitative and quantitative changes organ atrophy sets in, often accompanied by an increase in the stroma. Lipochrome deposit may alter the color of the tissues.

Shock (1970) commented on the probability that death increases logarithmically with age. He pointed out the dilemma arising from lack of demonstrable significant differences in dietary requirements and food absorption on one hand, inadequacies in cellular nutrition on the other hand, during aging.

Strehler (1969) believed that selective gene activation and depression are the cornerstone of molecular gerontology, probably at the level of the control of translation of genetic messages through the DNA–RNA system.

Comfort (1970) was in agreement with Strehler's ideas and developed a logic diagram of the questions raised and studied by the latter. He listed substances that may modify the rate of aging as those preventing an attack on DNA, radioprotectants, breaking the vicious cycle if faulty protein synthesis develops, lysosome stabilizers, suppressants of the aging effect of immune divergents, agents preventing crosslinks in molecules with a long life span, hormonal factors influencing protein synthesis and storage, and antimetabolites.

Walford (1969) reviewed the immunologic aspects of aging, by studying the distribution of serum globulins, blood anti-A and anti-B antibodies, precipitins against horse serum, and autoantibodies in man. Walford studied kidney lysozyme activity and autoimmunity in mice of different ages. He could not decide if these phenomena cause senescence or are the result of aging. The problem is the same in primates.

Further notes on changes related to senile degeneration will be found in the chapters dealing with specific organs and organ systems because investigation of senile degeneration appears to be in need of much additional study. Monkeys and apes are valuable experimental animals for such studies, principally those which have been reared in primate centers and have a reliable and well-recorded history which is a prerequisite for longitudinal studies.

IV. Diseases of the Circulatory System

A. ARTERIOSCLEROSIS

Anderson and Scotti (1968) classified arteriosclerosis as: (1) atherosclerosis with principal changes in the intima appearing in large vessels of the elastic types; (2) Mönckeberg's medial sclerosis with involvement of the media in arteries; (3) and diffuse arteriosclerosis proper with pathologic findings in both layers of small vessels. The last may be differentiated into two forms. Thickening of the vessel wall and narrowing of the lumen are typical for both forms. In hyaline arteriosclerosis hyaline deposits prevail; in hyperplastic arteriosclerosis cellular proliferation and fibrosis predominate.

Speaking of arteriosclerosis as a term including all three forms, loss of elasticity of the blood vessels due to pathologic changes in the intima or of the media (Mönckeberg's medial sclerosis) are observed, with narrowing of the lumen and in many instances with subsequent thrombotic phenomena. The elastic laminae deteriorate. Connective tissue increases. Calcium is deposited. Hyaline degeneration with lipid deposits form "fibrous plaques." The subendothelial layer of the capillaries, the muscularis of the arterioles, and the intima and the media of the arteries are often involved.

It has been suggested by Whereat (1970) that arteriosclerosis is basically a disease of the mitochondrial respiratory assembly.

King and Iacono (1970) believed that calcium metabolism and serum lipid disturbances are the principal factors influencing the development of arteriosclerosis. The role of biosynthetic cholesterol, its endogenous sources, and the feedback mechanism protecting against hypocholesterolemia was studied by Wilson and Dietschy (1966) in monkeys and summarized in man and animals by Ho and Taylor (1970).

Partial or total occlusion of blood vessels by thrombosis is a much-feared consequence of arteriosclerosis because it leads to ischemic heart and cerebrovascular disease, one of the major health problems of developed countries today. In addition, obliteration of the vessels of the extremities (endarteritis obliterans) may result in gangrene. Masironi (1970) showed in a worldwide survey that the development and severity of arteriosclerosis cannot be confidently attributed to any single dietary factor, nor to elevated blood cholesterol. Stress, physical inactivity, diabetes, and hereditary factors may influence the development of the disease.

Lipids accumulate in or under the intima forming "fatty streaks" in atherosclerosis with thrombotic deposits. Atherosclerosis presents with yellow lipid and gelatinous waxy patches of the intima, which is edematous, thrombotic deposits, later ulcerations, fibrosis, and calcifications. Not only the aorta but also the coronary vessels of the heart and those of the Willis circle of the base of the brain undergo these changes.

The immunologic theory of atherosclerosis has been tested by D. F. Davies (1969). He included in his concept the action of antigen–antibody complexes aggregating with blood platelets, the opening of the intracellular gaps between endothelial cells (the "adhesion plates") by histamine and related substances liberated by mast cells activated by the antigen–antibody complexes, and consequent excessive filtration of lipoproteins through the arterial wall.

The "thrombogenic" theory of atherosclerosis states that mural thrombi of blood platelets and fibrin on the unaltered intima constitute the primary event. These thrombi become organized and incorporated into the intima, then lipid is deposited. It has been also proposed that increased acid mucopolysaccharide and connective-tissue alterations due to endocrine changes altering the lipid metabolism are the basis of atherosclerosis. These concepts require further elucidation.

Nonhuman primates are known to develop arteriosclerosis in nature. Among other observers, Lindsay and Chaikoff (1963) noticed atherosclerosis in chimpanzees as early as at the age of 6 years. Strong et al. (1969) found chimpanzees, baboons, and spider monkeys often, marmosets less frequently, suffering from this disease. Lindsay and Chaikoff (1966) published a survey of atherosclerosis in different species. Middleton et al. (1964) described atherosclerosis in squirrel monkeys in nature; Malinow and Maruffo (1965) in free-ranging howler monkeys.

Chawla et al. (1967) reported on arteriosclerosis and subsequent thrombotic phenomena in wild rhesus monkeys; Stout and Lemmon (1969) on coronary and cerebral atherosclerosis in nonhuman primates.

Sandler and Bourne (1968) and Eggen et al. (1969) compared selected species in experimentally induced disease. Taylor (1965) wrote an extensive review on similar studies. Clarkson *et al.* (1969) investigated New World monkeys. Strong *et al.* (1969) and Gresham and Howard (1969) had favorable experiences with studies on baboons. Middleton *et al.* (1964) preferred squirrel monkeys. Buck (1963) worked with capuchin and rhesus monkeys. Kritchewsky (1969) compared nonhuman primates and other species as experimental animals. Stout and Groover (1969) found numerous parallels between natural and induced atherosclerosis in monkeys.

Buck (1963) carried out considerable work on the histogenesis and morphology of arteriosclerosis in nonhuman primates. He found that multipotential mesenchymal cells of the media accumulate low-density protein and may be stimulated to produce more collagen and less elastin. Buck considered whether fats, as those present in coconut oil, may play a role in the development of the disease in free-living monkeys. Cholesterol metabolism was related to the age of the primates in the experiments of MacNintch *et al.* (1967). Phospholipid composition and metabolism in nonhuman primates were studied in detail by Portman (1969) with results similar to those found in man.

Lofland *et al.* (1967) and Sandler and Bourne (1968) carried out extensive biochemical and histochemical studies in nonhuman primates with atherosclerosis. The latter used chimpanzees, baboons, rhesus, squirrel monkeys, marmosets, and woolly monkeys. The results of the studies of Sandler and Bourne demonstrated that deficiency of essential fatty acids may lead to atheroma formation, secondary to ATPase deficiency that is linked to cell-membrane activities. Diet-induced atherosclerosis could be prevented by the administration of essential fatty acids.

The disease may show regression in the rhesus (Armstrong *et al.,* 1970) as it is often similarly observed in atherosclerosis with a cyclic course in man.

Arteriosclerosis and, less often, Mönckeberg's sclerosis have been observed under natural conditions, but atherosclerosis is more easily produced in nonhuman primates with the aid of specific diets, and even by pyridoxine deficiency without special lipid administration. Atherosclerosis proper is more difficult to induce. Nevertheless, much can be learned about this disease from studies in nonhuman primates.

B. HYPERTENSIVE DISEASE

Koster (1970) listed the forms of hypertension of known origin as endocrine disturbances (Conn's syndrome, Cushing's disease, pheochromocytoma), renal diseases (stenotic arteries and parenchymatous disorders), some collagen diseases (lupus erythematosus, polyarteritis nodosa), anatomic (coarctation of aorta), drugs (e.g., oral contraceptives), and psychosomatic (stress-induced).

Goldblatt (1948) used monkeys in his studies of the renal origin of hypertension by clamping the renal artery. It is believed that this manipulation causes an alteration in the adrenal function of its cortex that responds with hormonal excretion of the glomerular zone resulting in increased blood pressure probably due to the increased renin influx from the ischemic kidney.

Weiner (1970) reviewed the literature on the influence of complex social stimuli that may raise the blood pressure of animals, as well as sodium chloride–steroid interaction that causes hypertension. He was in agreement with the multifactorial and mosaic theory of essential hypertension. Attention was called to the involvement of the vasomotor center as well as to its cortical control.

It is believed that all forms of hypertension, including the transient and intermittent varieties, and especially primary hypertension of unknown origin, require investigation in nonhuman primates. The mode of action and generation of the renin–angiotensin system participating in the process of blood pressure regulation are particularly timely problems to which experiments on apes and monkeys may yield proper answers.

V. Gastrointestinal System

There is no histologically proven degenerative disorder that could be attributed to senile degeneration alone. Starvation, however, may induce degenerative changes of the intestinal wall and variable changes in the ileal acid phosphatase activity. Malabsorption of vitamin B_{12} and ceroid deposits appeared in the rhesus kept on protein-deficient diets (reviewed by Pfeiffer, 1970). In our experiments (Wolf et al., 1970) reduction of cell division and suppression of the renewal of the intestinal epithelial and lymphoid-plasma cells were observed in patas monkeys fed with a low-protein diet. This phenomenon resembles reversible degeneration in starving people and should be investigated further.

The importance of the hyaline degeneration of the pancreatic islets in diabetes of the adult has not yet been elucidated. It may be the cause or the sequel of the disease. Subendothelial hyalinization is found as a rule, followed by fibrosis which may also develop, however, without hyaline degeneration.

Maruffo et al. (1966) described pigmentary liver disease in free-ranging howler monkeys. The intensity of the degenerative changes increased with age. Lipofuscin deposits, multinucleated hepatocytes, diffuse fatty meta-morphosis, and focal necrosis accompanied this condition. Of other ante-cedent factors of liver cirrhosis in nonhuman primates, infectious hepatitis has been studied in chimpanzees and patas monkeys (reviewed by Smetana and Felsenfeld, 1969), as well as nutritional deficiencies (Smetana and Greer, 1967; de la Iglesia *et al.*, 1967, in squirrel monkeys; Rutherford *et al.*, 1969, in the rhesus). Of the many toxic materials causing liver cirrhosis or degenerative changes in the liver, the recent literature contains a report by Allen and Carstens (1970) who succeeded in producing a syndrome similar to that of Budd-Chiari in monocrotaline-intoxicated monkeys. Transient autoimmune phenomena have been produced with heterogeneous material in patas (Felsenfeld *et al.*, 1967). It may be ex-pected that the increasing emphasis on nutritional studies and on exogenous toxic materials (including pesticides), and the hitherto accumulated knowl-edge of the histology and physiology of the nonhuman primate liver will be utilized in further experimentation on this model.

VI. Urinary Tract

Degenerative nephrosis and calcinosis, degeneration of the tubular epi-thelium, and lipid droplets are not infrequently seen in free-living monkeys (Labunowa, 1965; Kaur *et al.*, 1968). In advanced age loss of nephrons, perhaps due to vascular changes, are seen. X-Ray irradiation results in interstitial fibrosis and hypertension, as well as in "radiation cystitis." Acute tubular necrosis can be introduced with chemicals (mercuric chloride, potassium bichromate); lower nephron nephrosis with myoglobinuria by causing the crush syndrome; and osmotic nephrosis-like kidney with coarse vacuoles of the proximal convoluted tubules by potassium deficiency.

The relationship between kidney and hypertension was discussed previously.

The variety of pathologic phenomena observed in nature as well as those inducible in the laboratory indicate the usefulness of nonhuman primates in the study of renal disorders.

VII. Respiratory System

Senile emphysema without frank bullous formation, reduction of the ciliated epithelium and the mucous glands, rigidity of the bronchial tree, and the relationship of these phenomena to cardiopulmonary disease have been studied in wild rhesus monkeys by Vohra (1967). It would be of interest to investigate also the pathophysiologic relationship between the upper and lower respiratory tract in nonhuman primates as Drettner (1970) did in man. Moreover, the respiratory system of nonhuman primates lends itself to longitudinal studies of degenerative changes due to air pollution. However, some components of the upper respiratory tract, as the tonsils, vary from one species to another (Felsenfeld *et al.*, 1972) and may require a careful selection of the experimental species.

VIII. Nervous System

The complicated anatomic and physiologic relations of this system require a more detailed discussion.

There is a wide choice of reviews and books on the central nervous system (CNS) and peripheral nerves of nonhuman primates. The papers of Holloway (1967), Schneider (1968), and Sanides (1969) on comparative neuroanatomy, Bogolepova (1969a,b) on the hypothalamus, Verhaart (1966) on the pyramidal tract, and Manocha *et al.* (1970) on CNS enzymes contain many recently acquired and valuable data.

When degenerative changes of the nervous system are considered, it is important to realize that the neurons constitute the anatomic and functional basis of its activity. Neurons are composed of a cell body that has one or more processes. The neurofibrils cross the cells and enter the processes. They can be visualized only with the aid of special staining methods. Routine histologic staining with hematoxylin and eosin demonstrates cell nuclei and their membranes. The Nissl substance varies from one cell to another but is dispersed and readily studied after staining with thionine or related dyes. The size and shape of neuron cells vary greatly, and the axons may or may not have a myelin sheath that is rich in lipids. The cell is usually elongated, rhomboid, quadrangular with concave outlines, or elliptical, pear-shaped, or round. The nucleoli are centrally located.

The fibers of the oligodendroglia support the myelin sheaths in the CNS, whereas Schwann cell sheaths envelop the myelin layer in peripheral nerves. They are interrupted by the so-called Ranvier nodes. The axons end with a terminal "bouton" in peripheral nerve endings or at synapses.

Neuroglia are commonly differentiated as oligodendroglia and astrocytes, the former having few processes that appear as supporting and as "satellite" cells in the vicinity of neuron cells. Astrocytes are numerous in the gray matter and rich on fibrils. Some of them appear to be attached to blood vessels. Microglia or mesoglia are fixed mesodermal cells that become mobilized by myelin degeneration and acquire the function of phagocytes. They frequently take up fat granules.

Clasmatodendrosis is a degenerative process with intracellular edema, pycnotic nuclei, and loss of fibrils. Gemistocytes are edematous astrocytes with displaced, often multiple, nuclei. Vacuolation is not unusual. Gliosis is a term designating an increase in the number of neuroglia that may form glial nodes. Satellitosis is an increase of the supporting cells normally present around neuron cells.

Whereas the hyaline-like corpora amylacea, consisting of concentric lamellae, appear in old age, glial plaques or Alzheimer's sclerotic nodules are seen also in other degenerative processes.

Shrinkage, nuclear pycnotic, and clumping of the Nissl substance are signs of nerve-cell degeneration. Vacuolation and the appearance of abnormal or excess pigment may ensue. At that time, the Nissl substance may be pushed to the periphery, then disappear. Neuronophagia by macrophages is usually a phenomenon related to inflammation.

Demyelination and neuroma formation, degeneration of the boutons and other end organs, fragmentation of the fibrils, and synaptic changes form part of the wallerian degeneration (*vide infra*).

Fatty metamorphosis of glia and ganglion cells with small granules of neutral fat occurs in the CNS. Gangliosides accumulate in Tay-Sachs amaurotic idiocy, cerebrosides in Gaucher's disease. The lipids in the white matter are not easily liberated but degeneration of the neural fibrils may lead to their separation, principally of lecithin. Disintegration of myelin in the presence of lecithin. Disintegration of myelin in the presence of excess fluid produces a mixture of partially decomposed lipids that may be difficult to differentiate with the aid of the commonly used fat stains. Hemosiderin may be deposited and increase the damage further. Lipids are often seen in astrocytes during senescence.

In the old, shrinking of the neurons, but not necessarily a diminution of the Purkinje cells (Schenk and Enters, 1970), increase of lipochrome, appearance of lipid droplets also in astrocytes, sometimes amyloid, senile plaques, narrow gyri, and excess subarachnoid fluid are found. Senile plaques are less densely disseminated when dementia is absent. They are found mostly in the cortex of the frontal lobes and the cornu ammonis. The origin of senile plaques has not yet been fully elucidated. They consist probably of lipid and so-called amyloid materials. Alzheimer's neurofibrillary change may be apparent. Circulatory disturbances are frequent. The CNS vessels are thin-walled and have no external lamina. Intimal and medial fibrosis, superimposed by atheromatous degeneration, are not rare. Fibrosis of the media of arterioles is seen also in the young, capillary fibrosis in the old.

Nerve-cell loss may result from hypoxia. In hypoxidosis or hypoxia, swelling of the cytoplasm, disintegration of the nuclei, gliosis, and disappearance of the Nissl substance are seen. As a rule of thumb it may be said that if the Nissl substance had disappeared and the nucleus is contracted, the changes are irreversible. Several variants of this process, with dark, triangular nuclei, or contraction of the entire nerve cells with pycnotic nuclei, have been described. The globus pallidus and subthalamic nerve cells are usually involved in carbon monoxide, nitrous oxide, cyanide, and carbon disulfide poisoning.

Hypoxia may lead also to destruction of the myelin fibers if their supporting oligodendroglia became affected. Since the blood supply of the gray matter is more abundant, ischemic changes may be confined to the white matter, as is sometimes seen in Biswanger's vascular encephalopathy. Ischemic hypoxidosis is observed also in blood dyscrasias, as an effect of high altitude, and as a result of a number of poisons.

Ischemia of the globus pallidus has been said to be one of the possible causes of parkinsonism. The great number of other causative agents, from infection to manganese poisoning, and the sometimes variable morphologic findings complicate the experimental production of this disease in primates, even though Lewy bodies are seldom absent from the substantia nigra and the substantia ferruginosa in induced parkinsonism.

In long-standing cerebrovascular disease the convolutions are reduced, principally of the frontal lobes. The loss of volume of the white matter is less. The ventricles, especially the lateral pair, are dilated. This may be due to atrophy of the brain substance or to arachnoidal degeneration. The loss of nerve cells is most evident in the frontal lobes, basal ganglia, and brainstem. Satellitosis is frequent. Neuronophagia may be observed. The nuclei become more basophilic but the Nissl substance decreases, while lipofuscin deposits appear. The pyramidal cells degenerate also in amyotropic lateral sclerosis and in postencephalitic states. Gliosis is apparent but in the spinal cord principally the fasciculus gracilis is involved. Corpora amylacea occur in areas of gliosis. The results of thrombosis and infarction in the cranial cavity depend on the localization and extent of the accident. Due to the increase of fibrous materials in the intima and adventitia, the basilar artery is often tortuous. Atheromas are frequent in such vessels. Deposits of iron and hyalinization may follow. Arterioles often show only hyalinization at the same time but amyloid may be deposited. Cerebral arteriosclerosis is frequent in the young but in the old ischemic disease of the heart often appears before intensive cerebral atheromatosis develops. These changes may antedate aortic atheromas.

Arteriosclerotic dementia is a degenerative disease, as is Pick's temporal and frontal lobe syndrome without senile plaques and neurofibrillar changes.

Cerebral angiography is applicable also to nonhuman primates (Ryan et al., 1969), which may be helpful in such investigations and related studies.

The causes and effects of increased intracranial pressure and spinal cord injury appear to be the same in all primates (White *et al.,* 1969; Bourke *et al.,* 1970; Shenkin and Bouzarth, 1970) and can be produced as well as alleviated experimentally.

A large group of diseases is connected with demyelination. The myelin sheath disintegrates while the axon remains intact or degenerates. Primary demyelination is seen in multiple sclerosis. Other causes are vascular disturbances, infection, trauma, nutritional deficiency, and allergy. Demyelination can be introduced in experimental animals, e.g., in cats. Popova (1967) observed it in nonhuman primates.

Cortical changes and alterations of other parts of the CNS can be introduced by surgical methods. Jacob-Creutzfeld corticopallidospinal degeneration was produced in a chimpanzee by a slow virus (Lampert *et al.,* 1970). Postencephalitic, arteriosclerotic, drug-induced, and hypoxic parkinsonism characterized by degenerative lesions of the cortex and basal ganglia can be evoked in nonhuman primates. The Klüver-Terzian syndrome, originally described in monkeys after temporal lobotomy, has been observed in man as a result of atrophy of the cerebral cortex. Experimental epilepsy has been described by Killiam (1969). Chusid and Kopeloff (1969) prefer alumina cream implantation to elicit the condition in monkeys.

The study of wallerian degeneration in nonhuman primates has been a favorite subject of neuropathologists in the past. This phenomenon appears after severing a peripheral nerve, and resembles autolysis. Axon and myelin disintegrate. There is edema, then breakdown of lipids. The nodes of Ranvier become wider. Motor nerves and unmyelinated fibers disintegrate more slowly. The nuclei are pushed toward the cell membranes, under which the Nissl material appears only as a narrow sheath. The perikaryon shows reactive changes, followed by building of new axonic protoplasm, but retrograde degeneration follows when connection with the end organ of the nerve cannot be established. Myelin-sheath regeneration takes place as a result of the activity of the Schwann cells. The role of ischemia, principally in concurrent arteriosclerosis, deserves attention because motor nerves are very susceptible to lack of oxygen.

Retrograde changes of central nervous synapses and subsequent wallerian degeneration after peripheral as well as cerebral nerve selions have been studied in nonhuman primates by Beresford (1965) and Guillery (1965). These authors reviewed the literature and the results of their own experiments which proved the presence of limited retrograde degenerative changes in CNS nervous synapses as well as chromatolysis of the neuron cells. Further experiments seem to be indicated, principally with the aid of electron microscopy because silver impregnation methods used to study such phenomena are not quite informative.

Many important degenerative diseases, such as syringomyelia and amyotropic lateral sclerosis, still await the development of a feasible technical approach. The influence of alcohol, tobacco, lack of vitamin B_{12} on the CNS, and other questions require further study. One of the additional problems is the role of allergy in CNS lesions. Diederichsen and Pyndt (1970) reported antibodies against neurons in epilepsy and in cerebral palsy. The autoimmune mechanism was studied in schizophrenia by Tkach and Hokama (1970); antimyelin antibodies by Edgington and Dalessio (1970). Perhaps we will acquire a better understanding of CNS diseases by exploring this field also. The contributions to neuropathology achieved by the use of nonhuman primates in experimental work has been gratifying. It may be anticipated that further interesting and valuable investigations will be undertaken in this area.

IX. Endocrine System

The endocrine system is under humoral–neural control. An oversimplified schema of its main flow is the following: cerebral cortex → mesencephalon → pituitary   stalk → anterior   hypophysis → ACTH → adrenal cortex → corticosteroids → peripheral circulation → feedback to hypophysis. The products of the pineal gland are also delivered into the systemic circulation and with it to the cerebrum, brainstem, and hypothalamus (Quay, 1970a,b).

G. H. Bourne (1961) reviewed the role of several endocrine glands in aging.

It appears that most of the experiments in studying degenerative changes of the hypophysis and their consequences have been based on surgical or radiation damage of that organ (Zernes and Pickren, 1969). However, other factors, such as the exhaustion of adrenocortical hormones after stress or long administration of corticosteroids may also lead to regressive changes.

Similarly degenerative phenomena were noted in the adrenals of monkeys after total body irradiation (Goncharov, 1969).

Degenerative changes in the thyroid may be caused by lack of pituitary stimulus, trauma, surgery, and drugs. In old age, the size of the follicles decreases and the epithelial cells become tall cuboidal. Flattened epithelium is observed in induced hypothyroidism.

Primary degeneration of the parathyroid is very rare and not easily induced.

Senile ovaries are atrophic. Endometrial dysplasia is frequent with or without cystic degeneration. Lipofuscin in the endometrium has been found also in squirrel monkeys (Graham, 1970).

Spermatogenesis on the testicular–hypophyseal feedback mechanism (Johnsen, 1970). Perhaps the only constant finding in the testes of the old is the thickening of the basement membrane and of the lamina propria of the seminiferous tubules. The Leydig cells show increased pigmentation. Degeneration of the seminiferous epithelium can be induced by immobilization of monkeys (Zemjanis *et al.*, 1970).

The reader is referred for further studies on endocrine disorders to Bourne (1961).

X. The Skeleton

Wright and Bell (1964) and Reed and Driewaldt (1967) described bone dyscrasias that occur in monkeys living in their natural habitat. Fractures and dislocations are often observed in captive monkeys.

Pathologic changes of the bones are frequently based on the disturbance of the Ca:P ration. Lack of P, endocrine disturbances, lack of vitamins, increased deposits of Ca in bradytrophic tissues, lack of phosphatase, pH changes, perhaps also neural influences as in generalized calcinosis, loss of Ca in demineralization, and other chemical and physical factors may influence the normal process of bone formation. Inactivity and ischemia are additional causes of such disturbances. The influence of Ca intake and excretion, sex, and age on the skeletal system was discussed by R. H. Davies *et al.* (1970) and by Catt (1970). The latter emphasized the role of calcitonin in such conditions.

Ossification may be enchondral, consisting of cartilage formation at the epiphyseal end that is replaced by osteoblasts forming bone; subperiostal; and in the skull, intramembranaceous. Osteoclastic activity assures balance between bone formation and resorption, principally near marrow cavities. Circulating Ca and P are quite stable except when Ca is mobilized in excess amounts from the bones by parathyroid hormone activity. The hypophysis and the thyroid gland appear to have an effect on the equilibrium in the bones.

With progressing age bone tissue is reduced, principally the cortical type in the long bones, which becomes more porous. The trabeculae decrease in diameter, and the lacunae widen, especially in interstitial bone. On the other hand, Haversian canals and the lacunae may become overcharged with Ca. Ischemia develops as a result of the degenerative changes in blood vessels and their consequences. Circulatory disturbances are frequent after trauma. They may lead to disturbances of the physiology of the bones, e.g., when the ligamentum teres of the femoral head and its artery are damaged. Patchy ischemic changes are frequent in arteriosclerosis. Osteochondritis is usually an idiopathic condition. The term "osteochondritis dissecans" is reserved for loose osteocartilaginous bodies in the joints, usually appearing as a result of trauma. The amount of bone tissue is reduced in osteoporosis but bone calcification is undisturbed while the total amount of Ca is reduced. Inactivity or disuse osteoporosis is most evident in the metaphyses of long bones and near cartilages. After extensive inactivity, osteoporosis may become diffuse and involve also the cortex. When the whole body is immobilized, Ca will be lost. It is believed that Sudeck's acute atrophy developing after trauma, especially after fractures, is of neurovascular origin.

Bone absorption may be exaggerated in senescence. This, together with endocrine disturbances, principally the lack of estrogens and the subsequent disturbances of the mineral metabolism, increase the chances for the development of spinal and pelvic degenerative changes in postmenopausal women. At that time a narrowing of the trabeculae in the vertebrate may appear, with reduction of the osteoblasts, resulting in a decrease of the volume and compression of part or all of one or more vertebrae. Kyphosis develops.

Similar changes can be produced in experimental animals by cortisone overdosage; epiphyseal disturbances by thyroid damage.

Excess renal excretion of P leads to chronic osteodystrophy. Osteoclastic hyperactivity with bone resorption is not unusual in this condition. In so-called fibrous osteitis (von Recklinghausen's disease) the osteoclastic hyperactivity results in dystrophy of the cortical bone, particularly in the phalanges. Bone and marrow are replaced by fibrous tissue and a substance without a lamellar structure is produced. Fibrous dysplasia may be observed in one or more bones, may appear in the skeleton of the face and in long bones where fibrous tissue and nonlammelar trabeculae are first seen in the center, then on the periphery of the afflicted bone. Cyst formation, fractures, and deformities may result. If the cysts are subchondral, an osteoarthritic condition develops. However, the joints do not become involved (Woods, 1961).

Such alterations have been produced in experimental animals but little research has yet been performed on nonhuman primates.

Osteoarthritis is very common in the old, both in man and in other primates. It is usually a slowly progressing chronic disorder. First, the articular surface of the cartilages begins to show degenerative changes. The destructive process is initially superficial. The surface of the cartilage becomes rough. Later, the degenerative changes reach deeper. Fissures and crevices are formed, sometimes accompanied by reactive hyperplasia. Finally, the entire cartilage or its major part is destroyed. Then eburneated bone will form the articular surface. It does not seem that true osseal ankylosis develops in this disease. There are, however, fibrosis of the joint capsule, proliferation of the synovial membrane with formation of villi, Heberden nodes, osteophytes in the tendons, and fragmentation of the bone into the joint. The joints that carry weight are often affected. The hip joints are more often afflicted in males than in females, whereas Heberden nodes are less common in men. The Heberden nodes may result from hypertrophic bone changes or represent osteophytes of the tendons. Stress, trauma, especially repeated, and other injuries, are supposed to be the underlying factors of osteoarthritis. Even in old persons without osteoarthritis the cartilage is less elastic, appears opaque and often yellowish as a result of pigment deposit.

Degenerative arthritis may be accompanied by calcification of the proliferating villi (Margo and Owens, 1970).

Aseptic idiopathic vascular necrosis of the head of the femur and the similar osteochondritis juvenilis deformans Calvé-Legg-Perthes appear in young people as a distortion of normal growth (Katz, 1970), whereas osseous ankylotic spondylitis or osteochondritis Bekhterev-Strümpel-Marie affects principally the posterior intervertebral, costovertebral, and sacroiliac joints. The latter is usually classified with rheumatoid arthritis of chronic proliferative nature.

Stevens (1970) distinguished osteoarthritis as (1) primary generalized, characterized by more common appearance in women, with symmetrical involvement of the distal finger joints, carpometacarpal joints of the thumbs, cervical and lumbar areas of the spine, the knees, the first metatarsophalangeal joints, and less often the hip joints; (2) primary localized osteoarthritis that is more common in men, particularly in heavy workers, affecting most often the hips, sometimes the knees; (3) secondary osteoarthritis which appears usually in the younger, without sex predominance in its distribution, and may be congenital, developmental, traumatic, inflammatory, or ischemic.

Trauma has been considered one of the leading causes of bone and joint diseases of chronic-degenerative nature. Beeler (1970) produced further evidence that spondylolysis and spondylolisthesis are of acquired nature and that their relationship to trauma cannot be denied.

Osteoporosis, osteoarthritis, acquired osteochondritis, synovitis, and related conditions merit study in experimental animals not only because of our scientific interest in them but also for their function as causative agents of deformities, subsequent invalidity, and the socioeconomic consequences closely allied to such disabilities. There is considerable spadework to be performed to develop feasible experimental animal models for their study. Trauma can be inflicted under laboratory conditions, with all parameters (pressure, direction, time, frequency, and other elements) controlled with ease. The application of modern anesthetic methods and post-traumatic administration of pain relievers keeps the animal more comfortable than a human person is with chronic intraosseous tension due to excess fluid extravasation, long-lasting increased intraarticular hydrostatic pressure, and pain at night as well as after rest (Stevens, 1970). Cortisone-induced arthropathy in rabbits and sulfanilamidoinazole-arthritis in rats are examples of drug-induced arthritis that need further evaluation and elaboration also in nonhuman primates in which nonmicrobial arthritis is difficult to induce but acute and chronic traumatic osteoarthritis can be elicited. The importance of such studies for the prevention and treatment of orthopedic crippling diseases is apparent.

XI. Muscles

Diaphragmatic hernias are often congenital but frequently cause considerable diagnostic difficulties in adults. They can be produced in experimental animals by surgical means, and to a certain extent in various shapes and sizes. Although it is appreciated that nonhuman primates can be used in such experiments, the available literature shows that little experimentation in this field has in fact been performed.

Angiosclerotic myasthenia is not accepted as a disease *sui generis* by most pathologists. It is considered the result of vascular changes that provoke excess fatigue in muscles with consequent insufficient circulation. Inertia and sedentary habits may aggravate the condition.

Myasthenia gravis is a disease of unknown etiology. It is characterized by progressively decreasing function of the myoneural junctions either for lack of acetylcholine or excess of cholinesterase. Disturbances of the thymus and autoallergic phenomena have been considered as the underlying factors. It may be a multiple metabolic failure, perhaps due to intracellular degenerative changes, but proof for this theory is not yet available (Kozhnikov, 1970).

When muscle fibers are injured, the sarcoplasm becomes swollen, striation is lost, and granular or hyaline changes appear. Zenker's degeneration is an example of far-reaching degenerative muscular tissue changes, often accompanied by amyloidosis. In extensive damage of the muscles the so-called crush syndrome may develop, with nephrotic alterations and myoglobinuria. After loss of muscle tissue, the excretion of creatine increases and that of creatinine decreases. Creatine kinase and aldolase attain higher levels in the muscles, but not when nerve damage is present simultaneously. The nuclei of the muscle cells are displaced during degenerative processes, calcification may set in (traumatic myositis ossificans), and fatty metamorphosis and infiltration with or without fibrosis may be seen. Denervation reveals itself by clumping of the nuclei, rounding up or irregular outline of the fibers, and replacement of the muscle tissue by fat.

Traumatic myopathia either as a result of one single injury, or subsequent to repeated traumas, is of great interest in industrial medicine, as well as in the forensic sciences. Infarction after arterial or venous occlusion and focal necrosis appear not only in the heart, but they may also be seen in practically any muscles in the aged. However, anatomic changes are not always discernible under the ordinary optical microscope, as in gastric myasthenia or senile atony of the intestinal tract. Investigation of such changes in primates presents a challenge.

XII. Skin

The physiology and pathology of the skin of the nonhuman primate differ from those of man in several respects, although there are numerous points of similarity in the histology of the skin of all primates.

The study of the skin of nonhuman primates is the subject of a different chapter.

Degeneration of collagen, hyaline changes, and loss of elasticity are signs of senescence. Loss of fat from the tegument is no rarity. Keratohyalinization also occurs. Senile keratosis in man resembles solar hyperkeratosis. The prickle cell layer becomes atrophic or acanthotic, with rifts between the basal and the prickle cell layers. The entire superficial part of the dermis may appear permeated by amorphous or fibrous material. Degeneration of the glandular structures is progressive with age. The timing, sequence, and extent of these events merit further investigation in all primates, including man.

XIII. Sensory Organs

A. THE EYE

Fatty degeneration of the cornea produces arcus senilis, with appearance of intra- and extracellular lipid deposits in and between the peripheral lamellae of the cornea. Hyaline degeneration of the cornea, principally in its lower half (Kozlowski's syndrome), deep corneal opacities that may be senile or presenile (cornea farinata Vogt), and endothelial corneal focal degeneration (cornea guttata) can be seen in all primates.

Cataract of the lens, resulting from increased pH and proteolytic enzyme activity with subsequent disintegration of the structure of the lens, increasing osmotic pressure followed by extrusion of water, and formation of detritus that reduces transparency, is seen in old primates.

Vascular degeneration of the choroid, usually with secondary degeneration of the pigment layer of the retina, is often a primary disease. Sclerosis of the vascular and choroid vessels, rather patchy than diffuse, is observed in the old. The pigment layer of the retina may show degeneration of its epithelial cells and deposit of hyaline material, particularly in the area of the macula. Hemorrhages between the basal membrane and the retinal pigment layer, sometimes also lacerations of Bruch's membrane, may follow.

Glaucoma, with obstruction of the drainage of the aqueous humor, is not infrequent in senescence. It is observed in man with or without arteriosclerosis or diabetes. It may be toxic or posttraumatic.

Many of these conditions can be produced experimentally. Our society is becoming sight-conscious. Perhaps the time has arrived when we will not be restricted to merely observing these disorders in primates but will expand our experimental efforts to acquire much needed additional knowledge of these diseases.

B. The Ear

Kelemen (1968) described nonexperimental aural pathologic phenomena observed in nonhuman primates.

In addition to posttraumatic degenerative changes, otosclerosis deserves mentioning. It may appear in any age group but is more frequent in females. The lamellar bony structure is replaced by cancellous bone containing many vessels and intercommunicating spaces filled with marrow. It appears that both osteoblasts and osteoclasts are hyperactive. Ankylosis of the hearing bones in the middle ear may ensue. Degenerative processes are more apparent in Menière's syndrome in which the endolymphatic system is distended, and the sensory elements of the labyrinth degenerate. They are seen in infectious diseases, blood dyscrasias, tumors, drug-poisoning, and other conditions.

Noise-induced hearing loss is becoming an increasing problem in our society, principally in certain industries, and in large cities. Nonhuman primates, responding to voice and noise, are feasible models for its study, as well as for the testing of ototoxic drugs.

XIV. Conclusions

A number of degenerative diseases common to man and nonhuman primates have been discussed. Those which are analyzed in other chapters and several disorders that have been extensively studied, e.g., those of the liver, have been mentioned only briefly. The intention of this contributor was not to present a review of all degenerative disorders that could and should be studied in nonhuman primates but rather to point out degenerative diseases that are feasible for study in such animals from the point of view of their histopathology.

The most common degenerative disease is senescence, whatever theory we accept as its cause. It can be observed in all animals. Rodents and fowl, with their shorter lifetime, have been attracting the attention of experimentors. Nonhuman primates live much longer than those animals and birds. Therefore the study of aging in nonhuman primates is a long-range project. However, the results should be more rewarding because apes and monkeys represent experimental models where the investigation is more apt to lead to results that are applicable to human beings than those observed by experimentation on rats and mice with histophysiologic attributes frequently quite different from those of primates.

The demands of the ever-expanding industrial development and the use of machinery in agriculture, also, in developing countries require more attention to be focused on occupational diseases, whether of traumatic or toxic origin. Oehme (1970) in his review of animal models used in toxicologic studies pointed out the importance of using the proper species for such investigations. Here we find again the nonhuman primate is the animal of choice. While healing of wounds is usually faster in nonhuman primates than in man and fibrous responses, particularly of serous membranes, are more intensive (Greer, 1970), nonhuman primates appear to be excellent subjects for surgical experimental studies.

Induced degenerative changes do not threaten the life of nonhuman primates to the same extent as in man because regeneration and repair are faster and more intensive. Nevertheless, surviving animals yield results applicable in principle to man.

It appears that before pathology can be studied competently, a number of physiologic factors in nonhuman primates, principally of the CNS and the endocrine system, have to be further elucidated. "Normal" or "average" values according to species, sex, age, feeding, and environmental conditions are not well known. This might be due to the shortage of tissue and clinical pathologists studying nonhuman primates. It is believed that if every nonhuman primate that dies for any reason whatsoever would be autopsied and all organs and a reasonable number of skin, bone, and joint samples examined thoroughly in all of them, our knowledge of "normal" and "borderline" conditions would broaden. Unfortunately, in addition to the shortage of full-time pathologists, there is the problem that nonhuman primates are often presented for autopsy many hours after death, without proper refrigeration of the carcasses, in many zoologic gardens and animal farms.

Pollution of the environment used to be a matter of prevention rather than a pathogenic problem. We may be inclined to overlook the importance of some pollutants or become overzealous and tend to condemn agents appearing in connection with pollution as pathogenic just because they are present in the atmosphere, water, or food. Experiments on nonhuman primates may yield data that could prove applicable to man also in this field. Perhaps the "sentinel monkey," much used in arboviral studies, may assist us to evaluate properly the effect of many pollutants.

While nonhuman primates cannot furnish us with answers to all medical problems in man, and caution is necessary especially in the study of degenerative diseases, frequently more facts applicable to man can be derived from an experiment on a few chimpanzees than on dozens of rats. We must hope that our cost-conscious society will realize this fact and think of nonhuman primates more as experimental animals than as entertaining expensive pets. The more extensive use of apes and monkeys in applied medical research is necessitated by the growing need for that type of study which, unfortunately, has revealed a declining tendency during recent years.

REFERENCES

Allen, J. R., and Carstens, L. A. (1970). *Amer. J. Pathol.* **59**, 81.
Anderson, W. A. D., and Scotti, T. M. (1968). "Synopsis of Pathology," 7th ed. Mosby, St. Louis, Missouri.
Armstrong, M. L., Warner, E. D., and Connor, W. E. (1970). *Circ. Res.* **27**, 59.
Banks, K. L., and Bullock, B. C. (1967). *J. Amer. Vet. Med. Ass.* **151**, 839.
Beeler, J. W. (1970). *Amer. J. Roentgenol., Radium Ther. Nucl. Med.* **108**, 796.
Beresford, W. A. (1956). *Prog. Brain Res.* **14**, 33.
Bogolepova, I. N. (1969a). *Zh. Nevropatol. Psikiat. im S. S. Korsakova* **69**, 67.
Bogolepova, I. N. (1969b). *Arkh. Anat., Gistol. Embriol.* **57**, 61.
Bourke, R. S., Nelson, K. M., and Naumann, R. A. (1970). *Exp. Brain Res.* **10**, 427.
Bourne, G. H. (1961). "Structural Aspects of Aging." Pitman, London.
Bourne, G. H. (1962). "Division of Labor in Cells." Academic Press, New York.
Buck, R. C. (1963). *In* "Atherosclerosis and Its Origin" (M. Sandler and G. H. Bourne, eds.), p. 28. Academic Press, New York.
Catt, K. J. (1970). *Lancet* **2**, 255.
Chawla, K. K., Murthy, C. D., and Chakravarti, R. N. (1967). *Amer. Heart J.* **73**, 85.
Chusid, J. G., and Kopeloff, L. M. (1969). *Epilepsia* **10**, 239.
Clarkson, T. B., Lofland, H. B., and Bullock, B. C. (1969). *Ann. N. Y. Acad. Sci.* **162**, 103.
Comfort, A. (1970). *Hum. Deveolp.* **13**, 127.
Davies, D. F. (1969). *J. Atheroscler. Res.* **10**, 253.

Davies, R. H., Morgan, D. B., and Rivlin, R. S. (1970). *Clin. Sci.* 39, 1.
de la Iglesia, F. A., Porta, E. A., and Hartroft, W. S. (1967). *Exp. Mol., Pathol.* 7, 182.
Dianzani, M. V., Baccino, F. M., and Rita, G. A. (1969). *Acta Anat.* 73, Suppl., 152.
Diederichsen, H., and Pyndt, I. C. (1970). *Brain* 93, 407.
Drettner, B. (1970). *Ann. Otol., Rhinol., & Laryngol.* 79, 499.
Edgington, T. S., and Dalessio, D. J. (1970). *J. Immunol.* 105, 248.
Eggen, D. A., Strong, J. P., and Newman, W. P., III. (1969). *Ann. N. Y. Acad. Sci.* 162, 110.
Felsenfeld, O., Greer, W. E., and Felsenfeld, A. D. (1967). *Ann. Allergy* 25, 6.
Felsenfeld, O., and Greer, W. E. (1972). Unpublished data.
Felsenfeld, O., Greer, W. E., and Cvjetanović, B. (1972). Unpublished data.
Goldblatt, H. (1948). "The Renal Origin of Hypertension." Thomas, Springfield, Illinois.
Goncharov, N. P. (1969). *Radiobiol. Radiother.* 10, 497.
Graham, C. E. (1970). *Amer. J. Obstet. Gynecol.* 107, 837.
Green, D. D. (1967). *Arch. Biochem. Biophys.* 119, 312.
Greer, W. E. (1970). Personal communication.
Gresham, G. A., and Howard, A. N. (1969). *Ann. N. Y. Acad. Sci.* 162, 99.
Guillery, R. W. (1965). *Prog. Brain Res.* 14, 57.
Ho, K.-J., and Taylor, C. B. (1970). *Arch. Pathol.* 90, 83.
Holloway, R. L., Jr. (1967). *Brain Res.* 7, 121.
Johnsen, S. G. (1970). *Acta Endocrinol. (Copenhagen)* 64, 193.
Katz, J. F. (1970). *Clin. Orthop. Related Res.* 71, 193.
Kaur, J., Chakravarti, R. N., and Chugh, K. S. (1968). *J. Pathol. Bacteriol.* 95, 31.
Kelemen, G. (1968). *Acta Oto-Laryngol.* 66, 399.
Killiam, K. F., Jr. (1969). *Ann. N. Y. Acad. Sci.* 162, 610.
King, L. R., and Iacono, J. M. (1970). *Int. Nutr. Reps.* 2, 43.
Koster, M. (1970). *In* "Psychosomatics in Essential Hypertension" (M. Koster, H. Musaph, and P. Visser, eds.), p. 1. Karger, Basel.
Kozhnikov, M. T. (1970). *Ukr. Zh. Ortoped.* 7, 18.
Kritchewsky, D. (1969). *Ann. N. Y. Acad. Sci.* 162, 80.
Labunowa, T. (1965). *Acta Biol. Med. Gdansk* 9, 273.
Lampert, P. W., Gibbs, C. J., and Gajdusek, D. C. (1970). *Amer. J. Pathol.* 59, 4.
Letterer, E. (1959). "Allgemeine Pathologie," pp. 126–335. Thieme, Stuttgart.
Lindsay, S., and Chaikoff, I. L. (1963). *In* "Atherosclerosis and Its Origin" (M. Sandler and G. H. Bourne, eds.), p. 350. Academic Press, New York.
Lindsay, S., and Chaikoff, I. L. (1966). *J. Atheroscler. Res.* 6, 36.
Lofland, H. B., St. Clair, R. M., and MacNintch, D. E. (1967). *Arch. Pathol.* 83, 211.
McKeown, F. (1965). "Pathology of the Aged." Butterworth, London.
MacNintch, J. E., St. Clair, R. M., and Lehner, N. D. (1967). *Lab. Invest.* 16, 444.
Malinow, M. R., and Maruffo, C. A. (1965). *Nature (London)* 206, 948.
Manocha, S. L., Shantha, T. R., and Bourne, G. H. (1970). "Macaca Mulatta: Enzyme Histochemistry of the Nervous System." Academic Press, New York.
Margo, M. K., and Owens, J. N., Jr. (1970). *Clin. Orthop. Related Res.* 71, 202.
Maruffo, C. A., Malinow, W. R., Depaoli, J. R., and Katz, S. (1966). *Amer. J. Pathol.* 49, 1966.

Masironi, R. (1970). *Bull. WHO* **42**, 103.
Middleton, C. C., Clarkson, T. B., Lofland, H. B., and Pritchard, R. W. (1964). *Arch. Pathol.* **78**, 16.
Oehme, F. W. (1970). *Clin. Toxicol.* **3**, 5.
Pfeiffer, C. J. (1970). *Postgrad. Med.* **47**, 110.
Popova, L. M. (1967). *Zh. Nevropatol. Psikhiat. im S.S. Korsakova* **6**, 997.
Portman, O. W. (1969). *Ann. N. Y. Acad. Sci.* **162**, 120.
Quay, W. B. (1970a). *Amer. Zool.* **10**, 237.
Quay, W. B. (1970b). *Anat. Rec.* **168**, 93.
Reed, O. M., and Driewaldt, F. H. (1967). *J. Amer. Vet. Med. Ass.* **151**, 923.
Rutherford, R. B., Boitnott, J. K., Donohoo, J. S., Margolis, S., Sebor, J., and Zuidema, G. D. (1969). *Arch. Surg.* **98**, 720.
Ryan, K. G., Symeone, F. A., and Cortese, D. A. (1969). *Invest. Radiol.* **4**, 34.
Sandler, M., and Bourne, G. H. (1968). *Ann. N. Y. Acad. Sci.* **149**, 666.
Sanides, F. (1969). *Ann. N. Y. Acad. Sci.* **167**, 404.
Schenk, V. W. D., and Enters, J. H. (1970). *Psychiat. Neurol. Neurochir.* **73**, 77.
Schneider, H. R. (1968). *J. Comp. Neurol.* **133**, 411.
Shenkin, H. A., and Bouzarth, W. F. (1970). *N. Engl. J. Med.* **282**, 1465.
Shock, N. W. (1970). *J. Amer. Diet. Ass.* **56**, 491.
Smetana, H. F., and Felsenfeld, A. D. (1969). *Virchows Arch., A* **348**, 309.
Smetana, H. F., and Greer, W. E. (1967). *Recent Advan. Gastroenterol.* **3**, 25.
Stevens, J. (1970). *Clin. Orthop. Related Res.* **71**, 152.
Stout, C., and Groover, M. E., Jr. (1969). *Ann. N. Y. Acad. Sci.* **162**, 89.
Stout, C., and Lemmon, W. B. (1969). *Exp. Mol. Pathol.* **10**, 312.
Strehler, B. L. (1969). *Naturwissenschaften* **56**, 57.
Strong, J. P., Eggen, D. A., Newman, W. P., III, and Martinez, R. D. (1969). *Ann. N. Y. Acad. Sci.* **162**, 882.
Taylor, C. B. (1965). *Acta Cardiol., Suppl.* **11**, 1.
Tkach, J. R., and Hokama, Y. (1970). *Arch. Gen. Psychiat.* **23**, 61.
Verhaart, W. J. (1966). *J. Comp. Neurol.* **126**, 43.
Vohra, R. K. (1967). *Indian J. Pub. Health* **11**, 117.
Walford, R. L. (1969). *Klin. Wochenschr.* **47**, 599.
Weiner, H. (1970). *In* "Psychosomatics in Essential Hypertension" (M. Koster, H. Musaph, and P. Visser, eds.), p. 58. Karger, Basel.
Whereat, A. F. (1970). *Ann. Intern. Med.* **73**, 109.
White, R. J., Albin, M. S., Harris, L. S., and Yashon, D. (1969). *Surg. Forum* **20**, 103.
Wilson, J. D., and Dietschy, J. M. (1966). *J. Clin. Invest.* **45**, 1083.
Wolf, R. H., Felsenfeld, O., Brannon, R. B., and Greer, W. E. (1970). *Amer. J. Dig. Dis.* [N.S.] **15**, 819.
Woods, G. S. (1961). *J. Bone Joint Surg.* **43**, 758.
Wright, D. H., and Bell, T. M. (1964). *Lancet* **2**, 970.
Zeman, A. D. (1962). *Arch. Pathol.* **73**, 126.
Zemjanis, R., Gondos, B., Adey, W. R., and Cockett, A. T. K. (1970). *Fert. Steril.* **21**, 335.
Zernes, N. T., and Pickren, K. S. (1969). *Endocrinology* **85**, 949.

CHAPTER 14

Modeling of Neurogenic Diseases in Monkeys

G. M. CHERKOVICH and B. A. LAPIN

Various animals have been used as models for various human diseases for a long time. Different animals can be used as models according to the requirements of the experiment. Thus such lowly organized animals as frogs, in which the main principles of neuromuscular physiology are established, can also be used for the modeling of certain processes common for all animals in general. The modeling of highly complicated phenomena involving the functioning of the brain demands the use of animals with a higher neural organization, e.g., monkeys and dogs.

When modeling different pathologic processes of infectious origin it is essential for the experimenter to be sure that an identical picture with that of the human disease is produced, for example, measles, dysentery, etc. in monkeys (Dzhikidze a. oth.).

We can arrive at the same conclusion in the study of noninfectious human diseases. A number of diseases which will be described in this section are related to disturbances of the regulation of physiologic processes by the normal brain. This so-called neurogenic somatic pathology can best be produced in monkeys (*vide infra*).

Before passing on to the description of the pathogenetic similarities of some human and simian diseases, developed as a result of neurosis, we shall discuss some important questions concerning the peculiarities of experimental development of neurosis in monkeys. As in this area the peculiarities supporting the closer similarity of monkeys to man than to other animals, including dogs, become obvious. Thus, studies of monkeys give more significant data for the determination of the pathophysiology of neurosis in man than that obtained from other animals. This appeared in the attempts to develop experimental brain disturbances or, more appropriately described, "higher nervous activity disturbances in the category of neuroses." All the kinds of influences used for producing experimental neurosis in different animals are virtually similar. They were elaborated by Pavlov in dogs and all produce a psychological overstrain due to inhibition and excitation processes. It is possible to produce these effects by different methods and this is where the differences between primates and animals on a lower evolutionary level are most pronounced.

It is not difficult to produce the neurotic state in different animals, including dogs, since it is not necessary to use biologically natural signals to produce the condition of psychological overstrain.

It is possible to use any indifferent signal (light, bell, siren, metronome, etc.) which in the course of the experiment gradually acquires either the significance of the positive conditioned signal, due to constant reinforcement with food, or that of a defense stimulus. The significance of inhibitory signals (differentiations) may also be acquired if they are never supported by unconditioned food or defense reward.

Experimental neurosis usually develops as a result of the following effects: (1) a very long differentiation signal action leading to overstrain of the brain cortical inhibition process; (2) development of other kinds of internal inhibition, producing the same action; (3) effects making great demands on the mobility of the nervous process; (4) alteration of the signal significance of conditioned stimuli (when the positive signal is no longer reinforced and the differentiation is reinforced); (5) a quick change of signals when the positive conditioned stimulus is applied immediately, i.e., without any interval, after the action of the conditioned inhibition (the differentiation stimulus) or vice versa. The last effect demands a great mobility of nervous processes, i.e., the ability to rapidly replace the excitation with the inhibitory process. The development of neurosis as a result of the effects mentioned above can be estimated by the disturbances of the conditioned reflex activity and behavioral changes. However, the first such experiments in monkeys (Kaminski, 1948; Kriazhev, 1955) showed that both the force of nervous processes and their mobility surpassed those in other animals considerably, and this enabled them to resist such a strain and cope with the complicated tasks without the development of the neurotic state. Thus, neither the prolongation of the inhibitory signal to 20 min. nor the use of the inhibitory stimulus just after the positive signal showed any significant changes of higher nervous activity. Even more complicated tasks, such as the alteration of two pairs of stimuli, i.e., four signals in one experiment (the former inhibitory stimuli M_{60} (metronome) and red light were reinforced with food, and the former positive stimuli M_{120} and white light were left without food reinforcement), also did not result in the development of neurosis in monkeys with highly mobile nervous processes. We agree with the view that the higher the evolutionary level of the animal the more difficult it is to produce neurosis and the more resistant search carried out in monkeys by other experimenters (Norkina, 1949) the nervous system is to harmful effects (Birukov, 1955). Further research carried out in monkeys by other experimenters (Norkina, 1949; Miminosh-vili, 1960; Miminoshvili et al., 1960) confirmed the fact that experimental neurosis in nonprimate animals can be produced by less complicated means than in primates. The extent of adaptation to complicated tasks is higher in monkeys than in dogs and closer to that in man. The same conclusion was drawn in the study of electrophysiologic peculiarities occurring in experimental neurosis in monkeys (Lagutina and Sysoeva, 1965). The study by these authors on the direct changes in the main physiologic parameters (excitation thresholds, the character of electrical activity, etc.) in different cortical regions (orbital, temporal, frontal, occipital, motor) and in subcortical formations (hippocampus, striopallidum, nonspecific hypothalamus nuclei) permitted them to arrive at the conclusion that the key difference between monkeys and other animals is in their greater resistance to the production of pathologic-slow-wave-components of EEG. According to the authors' view this phenomenon and also the whole picture of electrical activity changes and monkey brain reactivity in experimental neurosis is similar to those in human neurotic patients.

Such important factors as a gregarious organization and community life in simians produce a similarity in the physiology and pathology of higher nervous activity in monkeys and man.

We do not intend to identify the monkey herd with the human society as do Zuckerman (1932) and Yerkes and Yerkes (1935). However, the prehistory of communal organization undoubtedly has its foundation in the monkey social group, which left its mark on the higher nervous activity and this basically, with certain modifications, has extended to human communities. Social relations are actually much more complicated in monkeys than in many other animals. Gregarious intercourse and intragroup signaling by means of a variety of sounds, facial expressions, and gestures occurs in monkeys. There also exists a strict hierarchy in the monkey herd, especially in baboons. Many of the reactions of monkeys are highly emotional. Because of these reasons Miminishvili considered gregarious conflict situations in searching for the most biologically adequate stimuli, which could be used to induce neurosis in monkeys. This proved more fruitful than the less complicated methods which have been successfully used in other animals.

It should be mentioned that in man, as a social creature, neurosis also appears as a result of social conflict situations. It is quite natural that the new quality of the higher nervous activity in man—higher consciousness—differentiates man from monkeys and does not permit the transfer of all the data from monkeys to man. However, there is no doubt that there is considerable similarity between human and monkey and the latter can rightfully be considered as intermediate link between man and the rest of the animals.

Emotional strain, especially due to emotions having a negative connotation, plays an important role in the development of neurosis in man as well as in monkeys. No other animals below monkeys have such vivid emotional reactions expressed in an obvious behavioral fashion. Emotions in man and monkeys as well as in other animals not only produce external behavioral effects but have also serious internal manifestations which involve not only the central nervous system but also the endocrine and other systems. These internal components also affect the external reaction and cannot be voluntarily suppressed. This was noted by Darwin in his "third principle" of emotions, according to which "the excited nervous system produces direct effects upon the body irrespective of will" in emotional excitement.

Neurosis is characterized not only by higher nervous activity disturbances but also by quite a number of somatic disturbances, among which cardiovascular diseases occupy an important place.

In a human clinic such diseases as hypertension and coronary insufficiency were connected with different nervous stress including negative emotions from a long time previously. However the etiologic and pathogenic significance of neuroemotional strain with these diseases can only be studied and identified in special experiments.

At the Institute of Experimental Pathology and Therapy (Sukhumi, USSR) monkeys were chosen as subjects for developing models of neurogenic diseases. Disturbances of the different norms of interrelations in a human community can cause violent emotions. By disturbing such interrelations in a social group of monkeys it should be possible to produce similar reactions resulting in neurosis. This neurosis and the somatic pathology which seems to accompany it should reflect the consequences of neuroemotional stress in man.

Kaminski (1948) was the first to express the opinion that biologically adequate stimuli should be used to produce experimental neurosis in monkeys. He made such a conclusion after unsuccessful attempts to produce neurosis in monkeys by the methods which proved successful in dogs. Miminoshvili (1960) put this idea into practice. He produced negative emotions in a male hamadryad baboon leader by disturbance of the social hierarchy. The males were first of all extracted from the group in which they were leaders and put into an isolated cage around which cages with the females and juveniles of his group were placed. The very fact of isolation caused restlessness, shouts, attempts to open the door and escape. Through wire walls of the cages the isolated dominant male saw that the junior monkeys were fed before him, whereas according to the law of social relations he was always the first to receive food and he strictly punished junior monkeys for disobeying the law. In this situation the dominant male experienced negative emotions, made attempts to tear up the net separating him from the animals infringing the rules, shouted threateningly and made menacing gestures and facial expressions. The situation was changed for the worse when another male was put into the cage with females formerly subordinate to him. In this case the aggression was so strong that it was necessary to separate the cages with a double net to avoid traumas. There were cases in which males broke their teeth trying to bite their rivals through the double net and their hair became very tousled. Negative emotions were thus enormously pronounced but all their efforts failed to resolve the situation and restore the previous order and hierarchy so the stress remained active.

The dominant males seem to experience similar negative emotions in the natural habitat when other males become mature and try to force out the leader. But in such a situation a conflict finds its natural solution, i.e., these negative emotions help to restore the normal conditions. The males fight, and one of them either dies or forms a separate herd with a few females. In the experimental situation these negative emotions traumatize the monkey daily without any respite. Overexcitement caused by negative emotions acquires the character of a "dominanta" and according to the law of the dominanta (Ukhtomski, 1966) stimuli which formerly did not have any relation to this situation became the cause of emotional disturbance. A stimulus which did not cause aggression earlier (the appearance of man approaching the cage, attempts to take pictures, etc.) now causes the full negative emotion reaction. Later on these negative emotions assume the character of neuroemotional trauma and lead to the development of neurosis in the course of the next few months.

We do not think it right to attribute everything described in the situation above to sexual excitement alone as certain authors have done (Markov, 1967). The collision of two different excitements, sexual and defensive (aggression), was very important in this situation. In the intact group the aggressive reaction of two rival males toward each other and the fights between them ceased to be sexual excitement. Sexual excitement was replaced by aggression, which might be described as active-defensive behavior. The feeding of monkeys in neighboring cages before the dominant male produced this active-defensive reaction. Miminoshvili discussed the reasons for the development of neurosis in monkeys according to the technique suggested by him. This technique provided the collision of competitive excitements in the brain cortex and subcortical regions which was caused by biologically heterogeneous reflexes—sexual, food, and defense.

Another method of developing experimental neurosis was also based on the use of natural, i.e., biologically adequate stimuli. It was carried out by diurnal rhythm disturbances of light and feeding (Cherkovich, 1959a). A number of stimuli were repeated daily in a dynamic stereotype in which the animals developed conditioned reflexes to the time when these stimuli (light, food, sounds, etc.) were active. Pavlov has shown that time can be a stimulus and it is possible to make a condition stimulus of it. These stimuli (light, food, etc.) repeated periodically and alternated caused alternating periods of excitement and inhibition in the brain together with an increase and decrease of a number of physiologic functions which were the basis for elaboration of a conditioned reflex to time. The diurnal rhythm is not a simple unconditioned reflex to time. The diurnal rhythm is not a simple unconditioned reflex response of certain organism systems to the change of various stimuli in the diurnal stereotype. This is proved by the fact that a monkey placed in artificial light and feeding at night and kept in the dark during the day did not at first show an immediate change in the diurnal periodicity of physiologic functions, but after a certain period of time the physiologic functions were gradually reconstructed in accordance with the new regimen. If the animal was again placed in natural conditions after a consolidation of this new artificially created diurnal rhythm, again the changes would not be immediate. During several days the diurnal rhythm would be the same in spite of the new conditions. The absence of immediate change following alteration of the stimulus and the presence of a pronounced aftereffect was characteristic of each conditioned reflex stereotype; time was needed for the extinction of former relations and the formation and consolidation of new ones. We therefore prefer to call diurnal rhythm the diurnal stereotype, because the latter makes it easier to understand the basic mechanisms.

Changing the place and time of the action of diurnal stereotype components, i.e., using the artificial light and feeding at night or repeating them twice during 24 hr for a long period of time, caused the elaboration of a new reflex to time in the form of an inverted or biphasic diurnal rhythm (Bykov and Slonim, 1949; Shcherbakova, 1949; Cherkovich, 1959a). If the diurnal stereotype was replaced by one absolutely opposite before the reorganization of physiologic functions according to the new regimen was fully completed induced brain cortex disturbances resulting in neurosis.

The production of neurosis using diurnal stereotype disturbances in monkeys proved possible since monkeys have the same diurnal rhythm (day monophase) as man. This technique might not prove fruitful in dogs as a polyphase diurnal rhythm without a well-expressed day activity concentration period, and night rest is typical of them.

All kinds of stereotype disturbances and a certain rhythmical pace in work were endured with difficulty by both animals and people.

Special studies (Kupalov, 1956; Livanov and Korol'kova, 1949; Danilov, 1958) showed that heterorhythmical stimulations were endured with difficulty by the nervous system and resulted in neurosis in dogs and rabbits. According to Tsuker and Hvan and Pirah (1955) people working in different shifts (day and night) also developed neurosis and somatic pathology. The neurosis caused by diurnal stereotype disturbances and the manifestations of somatic pathology accompanying neurosis in monkeys are similar in nature to certain disturbances which can be seen in a human clinic.

Startsev (1967) offered a technique of experimental production of neurosis in monkeys based on Ukhtomsky's learning on the dominanta as an excitement focus. This appeared in any central nervous system part and changed the current work of the nervous centers by means of deviation of the majority of impulses to that focus which caused another reflex reaction in the absence of the dominanta.

This technique is based on defensive dominanta formation and the transition of some vital factor into the signal of this defensive dominanta.

Thus, according to the author, a certain natural stimulus exciting a certain physiologic system brought this system to a pathologic state, becoming a conditioned signal of the defensive dominanta.

Operating with this or that natural stimulus from a certain physiologic system and changing it into a signal of the defensive dominanta it would probably be possible to produce purposeful functional systemic disturbances of this system especially (Startsev, 1967).

The above-mentioned methods of developing experimental neurosis in monkeys suggested by Miminoshvili (the disturbances of normal social relations) and Cherkovich (diurnal stereotype disturbances) were used by other experimenters. Thus, Dzhalagonia (1967) successfully used the technique of diurnal stereotype disturbances and the disturbances of social interrelations in baboons. The latter method was employed by Dzhalagonia with a certain modification—the traumatizing of monkeys was not carried out discontinuously but in two periods between which there was a short period of rest.

Sysoeva (1967) also used the technique of diurnal stereotype disturbances for developing neurosis in macaques. This technique was employed by the author in combination with rigid board or chair restraint and the use of an extremely powerful sound stimulus and a weak electric skin stimulus.

Markov (1967) combined these two methods in experiments in baboons. The diurnal stereotype was disturbed and when, according to experimental condition, the monkeys were subjected to artificial light and fed, social relations were also disturbed. Thus, summing up everything mentioned above, it is possible to say that the nervous processes in monkeys in their strength and mobility bring monkeys closer to man and place them higher than dogs for they can cope with the tasks causing neurosis in these latter animals. The similarity of stresses causing neurosis in monkeys and man should not, however, be too literally interpreted but in some cases the resemblance is very strong; for example, the similarity of neuroses caused in monkeys by violation of the diurnal stereotype and those observed in man in cases of diurnal stereotype disturbances is even greater. The experiments on monkeys in the field of experimental neuroses are important, because with them it is possible to induce most of the somatic diseases which are known as diseases of neurotic genesis in man. This has to be more extensively studied in monkeys.

To evaluate the state of the higher nervous activity during neurosis, simple food–motor conditioned reflexes had been elaborated in monkeys which had been trained with them for a long time till they became stable. Most often the stimuli were light signals (white and red light) one of which was always reinforced by food after the monkey had pressed the lever and served as a positive conditioned reflex signal, while the other was never followed by food reinforcement, sound signals (bells, sirens, metronomes) one of which was also a positive conditioned signal and the other a differentiation signal were also used.

Neuroses caused stable disturbances in the clear-cut conditioned reflex activity. Some monkeys developed general excitation expressed in intensified and sometimes multiple pressures on the lever, as well as in pressures during intersignal intervals usually absent in normal conditions. They also expressed excitation in disinhibition of differentiations, and in the appearance of and ultraparadoxical phase when, in answer to a differential signal, the monkey pressed the lever, while there was no pressure on the lever as a result of a positive signal. Such changes were noted in experimental monkeys by Miminoshvili (1960), Cherkovich (1959a), Markov (1967), and Lagutina and Sysoeva (1965). However, these were not unique disturbances of conditioned reflex activity. Some monkeys developed changes which were exactly opposite. The monkeys became inhibited and left some of the signals unanswered. Disturbances of diurnal stereotype were often first accompanied by disappearance of a certain part of the responses to light stimuli. Then with the intensification of inhibition, partial disappearance of responses to other signals became noticeable. In such cases the monkey either stereotypically walked in a circle, without paying any attention to the signals, or sat in the corner crying monotonously. Sometimes that state of inhibition developed as a subsequent stage of the period of overexcitation.

Among conditioned reflex disturbances, various incomplete reactions were very frequent. They found their expression in the inhibition of certain links of the conditioned reflex chain which was observed both in the initial and in the final links, that is; in response to a signal the monkey ran half the way to the lever which it had to press and then came back for food reinforcement. Sometimes the monkey made a peculiar hand movement in the direction of the lever and then returned for the reinforcement, or it would reach the lever, but instead of pressing it only touched it and then went for the reinforcement. Another category of changes concerns the second half of the links. In response to a signal the monkey pressed the lever and remained in this position, which sometimes was very strange and awkward—immobile, cataleptoid—and did not go for the reinforcement. Or the monkey would run in the cage, press the lever in response to a signal, and continue running without taking the food and even not looking at it. Many of the above-described conditioned reflex disturbances are accompanied by intensified vocal reactions. Various mistakes in the monkey's activity are frequently followed by piercing shrieks and so-called reactions to itself, i.e., the monkey bites its hands and arms and pulls out its hairs.

In addition to this, characteristic behavioral changes are observed in the neurotic state. These changes can be seen not only in the chamber where the conditioned reflex activity is studied, but in the living cage as well. Obsessional and stereotypically repeated movements are noticeable. Thus, neurotic monkeys can sit quickly rocking to and fro in a pendulum way for a long time. In some monkeys stereotype movements are expressed by running in a circle or from corner to corner (diagonally) with a typical throwing back of the head when turning.

Similar stereotype movements were noted by all the authors who worked with monkeys during the period of neurosis (Norkina, 1949; Miminoshvili, 1960; Cherkovich, 1959a; Lagutina and Sysoeva, 1965; Lagutina et al., 1970). Some monkeys developed manifestations that looked like hallucinations. For instance, when running about the cage the monkey avoided one of the corners, looking at it in fright each time it passed by, sometimes uttering groaning sounds. As a rule, the monkey quickly left the living cage for a portable one in which it was carried to the conditioned reflex chamber, and at the end of the experiment it quickly entered the portable cage again and was carried back into the living cage. The monkeys got used to that cage and the experimental procedure in general, as they had been trained to them for a long period of time, while the conditioned reflexes were elaborated and strengthened. When neurosis developed we noticed that some monkeys were afraid to enter the cage. The monkey quickly made its way to the cage, but at the very threshold it suddenly stopped, stepped aside, bent, looked into the cage and went away. We had to try more than once to make the monkey enter the cage and only then did it do so. Sometimes the monkey suddenly began to make movements as if trying to fight something back or seize something, the latter movements being accompanied by slow opening of the first and examining something presumably imagined lying on the palm.

The manifestations mentioned above occur in human beings in reactive states and psychogenics. But they are more than neuroses. They are elements of reactive psychoses which are usually associated not with acute emotional shocks, but with chronic psychogenic factors. The existence of such states renders the use of monkeys for modeling different reactive states especially important. Naturally we do not intend to model the whole complex syndrome of human reactive states and psychopathology, yet there are many symptoms of central nervous system functional disturbance that can be induced in a well-pronounced fashion in monkeys.

It has been considered that hallucinations develop as a result of different degrees of excitation of cortical sensory centers, as is observed in the state of hypnotic trance. Phase states (the transitory phases between the waking and sleeping state) occur widely in monkeys when neuroses develop; they can be experimentally induced and serve as the basis for the development of hallucinations. Thorough observations of neurotic monkeys, as well as the possibility of conducting large-scale experiments on them seem to help in better understanding some pathophysiologic causative mechanisms of hallucination which still remain not quite clear.

One of the most important moments in the manifestation of experimental neurosis in monkeys is that they produce some somatic pathology as a result of the activity disturbances of certain brain systems regulating definite bodily functions. In the studies of some authors, as was noted by Lagutina *et al.* (1966), the affect on brain functions was of a systemic character. It was impossible to determine in advance which system would be disturbed first in the production of experimental neurosis, as it depended on a number of unknown factors. The methods by which purposeful influence and production of pathology localized in this or that system can be interpreted are difficult to ascertain, but some attempts in this direction are being made at present. They will be discussed below.

Some of the types of somatic pathology produced in monkeys as a result of experimental neurosis have been recorded. Of 59 baboons and macaques involved in experiments on developing neurosis by the methods mentioned above, hypertension developed in 16 cases, coronary insufficiency in 19 cases, gastric achylia in 15 cases, amenorrhea in 6 cases, sexual weakness in males in 2 cases, and a pathologic increase of vagal tone in 3 cases.

The most important new addition to the concept of hypertension pathogenesis was made by Soviet authors. They determined the principal factor, the central nervous factor (Lang, 1950; Miasnikov, 1965), which interacted with the humoral factor, playing an important role in the prolongation and continuous maintenance and consolidation of hypertension. Humoral factors, of which adrenal and hypophysis hormones are the most important, produce a pressor action by direct influence on both peripheral mechanisms and the central nervous system, on the structures forming different emotions and causing an increase in the secretion of these hormones, and also the structures regulating blood pressure (Markov, 1967).

The opinion about the leading central nervous factor in the development of hypertension was originally formed in the main on the basis of thorough clinical observations and confrontation of the anamnesis of patients. However, in our opinion the most important experimental confirmation of the etiopathogenic role of emotional overstrain in the appearance and development of hypertension was seen only in monkeys. Monkeys have pronounced emotions and a high organization of social relations. It was only because of this that it was possible to prove convincingly that the changes in the surroundings associated with stereotyped disturbances, and social conflict situations (causing overstrain of negative emotions, when even the most emotional reactions are unable to restore status quo) lead in a great percentage of cases to the development of hypertension and coronary pathology in monkeys as in man. Various authors using different methods described above have noted a stable increase in blood pressure. Four of five monkeys (three baboons and two rhesus macaques) employed in Miminoshvili's experiment, developed hypertension. The dynamics of this hypertension was characterized by the fact that an increase of blood pressure in three monkeys followed a hypotension stage when the blood pressure reduced from the initial 120/90 to 78/45. This decrease was noted 4 months after the beginning of neurotizing effects. After several months the blood pressure increased and reached the level of 170–180/120–110 mm Hg. The fourth monkey did not show the stage of hypotension. The blood pressure began to increase immediately and reached the same high values. The authors considered that the diurnal rhythm of pulse rate and blood pressure was disturbed before a significant increase of blood pressure and in the stage of hypotension in monkeys. Diurnal variations of blood pressure in the period of the development of neurosis have a considerably larger range than in the norm.

These experiments also showed distortions of the diurnal blood pressure curve. In the period of great blood pressure variations maximum blood pressure values were often noted early in the morning or in the evening, while in the day they were normal. Similar changes of diurnal blood pressure variations occur in man in the initial stages of hypertension (Lang, 1950).

In the initial period of developing neurosis Miminoshvili *et al.* noticed prolonged or even inverted reactions on cold and certain pharmacologic stimuli—nitroglycerine, atropine, and pilocarpine. Similar disturbances in initial stages of hypertension took place in man (Miasnikov, 1965).

When the experiment was interrupted in the early stages of developing hypertension and the animal went back to the herd the blood pressure was normalized. If the experiment was continued further and the animal was not brought back to the herd the blood pressure was increased and its high values were stable and caused brain hemorrhages in some cases.

In experiments with the disturbances of the diurnal rhythm one monkey developed hypertension which disappeared soon after the end of the experiment when the monkey was returned to its group. Later, in Markov's experiments on two baboons (adult males), combined effects were used to produce experimental neurosis when the method of the diurnal rhythm disturbances (Cherkovich, 1959a) was used simultaneously with the method of the disturbance of social relations (Miminoshvili, 1960). As a result of this both monkeys developed hypertension. The other two males were subjected to the same treatment, but in addition were sensitized by three inoculations of 2 ml/kg of horse serum, and hypertension developed earlier. The author confirms the data of Miminoshvili *et al.* (1960) that pressure reactions in neurosis can be enhanced or prolonged in response to one and the same substance dose in confrontation with the control and the reaction of these monkeys before neurosis.

Eight monkeys 15 years of age developed hypertension in Magakian's experiments in which neurosis was induced by the collision of electric defensive reflexes and food reflexes. Four baboons in Sysoeva's experiments (1967) developed hypertension as a result of neuroses. The hypertension reached maximum values and stabilized by the end of the second year when the neurotic traumatizing effects were over. The author also confirmed the conclusion made earlier on the absence of reaction or the presence of an inverted reaction on hypotensive substances in monkeys in neurosis. Other authors produced hypertension in two rhesus monkeys in neurosis-induced experimentally. After a long rest the blood pressure normalized, but when the experiment was repeated the blood pressure increased again. In some cases neurosis and hypertension developed in captive monkeys without any special experiment involving the conflict situation. As a matter of fact, the conditions of life in captivity do not always permit a group and an open enclosure to be available for each sexually mature male. Very often males employed in other experiments according to experimental conditions have to sit in separate cages, while nearby in other cages a group of females with a male or a group of young monkeys are kept. Here again, according to experimental conditions, the monkeys of neighboring cages are fed earlier. Thus a conflict situation is created which is also used in a special experiment on the production of neurosis. We therefore do not exclude the possibility of developing neurosis and resulting hypertension outside a special experimental situation in monkeys living in captivity, kept in separate cages but not in an open enclosure, subjected to different procedures associated with hand fixation or different primate chairs, and also stress connected with the disturbance of normal social relations.

In fact among monkeys living in captivity in the surroundings described above so-called spontaneous hypertension has been encountered. This "spontaneous" hypertension had undoubtedly the same genesis as the experimental one because hypertension had never been detected in newly arrived monkeys imported from the native habitat.

Fourteen monkeys (6 macaques and 8 baboons) showing increased blood pressure when alive were examined on autopsy later. Thirteen of these monkeys died of different reasons or experiments. One of the monkeys died of a typical complication of hypertension—brain hemorrhages. The blood pressure in 9 monkeys by the time of death showed a stable increase and in 5 monkeys fluctuated widely (170/95–100/50 mm Hg) as it takes place in the transitory phase of hypertension in man. Myocardial hypertrophy was pronounced in 7 monkeys with persistent high blood pressure and 1 monkey with the transitory form of hypertension. There was an apparent correlation between the duration and degree of blood pressure increase and the extent of myocardial hypertrophy. Besides this the myocardium showed small and larger scars substituting atrophied and dead muscular fibers. Changes typical of hypertension were detected in the kidney of 6 monkeys. Sclerotic changes were detected in 3 of these 6 monkeys. Formation of solid homogenous masses of protein could be observed in the glomerular lumen and along the glomerular arterioles. A transformation of glomeruli into hyaline nodules was also noted. In one case protein exudation and sclerosis were noted in the glomeruli. As a result the glomeruli were converted into a connective tissue nodules. The scars in the kidneys often were wedge-shaped, the base facing out toward the kidney surface. Superficial contractions were noted in the same place. Such a kidney had a granular form. The process of sclerosing in the kidney had a diffused character in 2 monkeys with a stable blood pressure increase. The rest of the animals did not show changes in the kidneys.

Thus, hypertension in monkeys involves kidneys, the heart, the brain, and small vessels, which brings monkey hypertension close to human hypertension.

Hypertension was not the only form of pathology of the cardiovascular system which developed as a consequence of experimental neurosis. Besides a blood pressure increase in monkeys coronary insufficiency and myocardial infarction were developed in a number of cases. Sometimes coronary pathology and hypertension were concomitant, sometimes they developed separately.

The experiments in monkeys in this field are of a special interest and importance due to the fact that the role of spasm in the functional coronary vessel in the development of coronary pathology in the absence of atherosclerosis is revealed. In the wild, atherosclerosis in monkeys occurs only in old monkeys and is not of the stenosed character. In our expedition to Vietnam one of the purposes of which was to determine which diseases occurred in monkeys in the wild, and in which 44 monkeys, shot in the forest, were autopsied, atherosclerosis was not detected in any of the cases (Lapin *et al.*, 1963). Lipoidosis of aortic intima has been clearly observed in a group of rhesus monkeys; however, no signs of sclerosing along the internal surface of the intima in places of lipid accumulation could be seen. Half of the monkeys investigated were 6–8 years old, the rest of them were 2–5 years old. A similar study of monkeys caught during the next expedition to Africa directly in the forest revealed atherosclerosis only in one old male (*Cercopithecus aethiops*) which showed atherosclerotic plaques in the aorta. According to Savinov and Tufanov (1956) and also Porubel *et al.* (1958) who investigated on autopsy some thousands of newly arrived monkeys imported from the wild and killed for the preparation of vaccine against poliomyelitis, atherosclerosis was not detected in any animal.

Data on animals kept in colonies and zoos for a number of years or born there, demonstrate that atherosclerosis can occur in these animals, but only in a small percentage of cases and in old monkeys. In autopsy findings of 1179 monkeys of different species which died in the Sukhumi colony, atherosclerosis was detected in 30 cases mostly in macaques, baboons, and green monkeys from 5½ to 27 years. The plaques were small, located in the aorta, very seldom in the coronary vessels, and never in brain vessels. Thus everything discussed above makes it possible to involve monkeys with healthy vessels in experiments which involve the development of hypertension and coronary pathology and to study in the isolated form the functional component (spasm) of coronary vessels which can itself be the basis of these pathologic states or it may be combined with atherosclerosis. Clinical and experimental data confirm the essential importance of the reflex coronary spasm vessels in the clinical picture of angina pectoris and myocardial infarction. These functional changes are likely to play an even more important role in the pathogenesis of coronary insufficiency in cases of pronounced atherosclerosis, as it is known that atherosclerosis causes an increased tendency for coronary vessels to produce long spasmodic contractions (Davidovski, 1938). Such purely functional changes of coronary blood flow can play an important role in the pathogenesis of not only coronary insufficiency but even myocardial infarction in man (Lapin, 1956; Lukomski and Tareev, 1958; Miterev, 1959). The experiments in monkeys show that apart from the influence causing the reflex spasm of coronary arteries from certain interoceptive zones of other organs, such a spasm can also take place in neurosis as a consequence of persistent exitation focus formation in the vasomotor centers or the formation of the vicious circle supporting overexcitement in these centers. As a result of experimental neurosis produced by the same methods which were described above and which can produce hypertension, the conditions of coronary insufficiency and myocardial infarction (as mentioned above), may also occur in the absence of atherosclerotic vessel lesions and without thrombosis. The first data in this field belong to Miminoshvili *et al.* Of 5 monkeys employed in experiments by these authors, 4 developed coronary insufficiency. One of these 4 animals developed myocardial infarction causing its death. Coronary insufficiency and myocardial infarction were diagnosed electrocardiographically and then confirmed in the autopsy after the death of the monkeys.

Myocardial infarction detected by electrocardiography developed in a baboon (*Papio hamadryas*) in Cherkovich's (1959b) experiments as a result of neurosis caused by the disturbance of the diurnal stereotype. The autopsy revealed microinfarctions in the left ventricle. In the tests by Sysoeva (1967) coronary insufficiency developed after neurosis in four baboons. It led to myocardial infarction in one of the animals causing its death. In Magakian's experiments (1956), coronary insufficiency developed in 14 of 15 monkeys being accompanied by hypertension in 8 cases. Four monkeys died of myocardial infarction. A thorough pathohistologic study revealed neither atherosclerotic manifestations in these animals nor thrombosis of coronary vessels. All the monkeys were young.

Usually the ECG of newly arrived monkeys or in captive monkeys living in open enclosures in a large group was normal; we did not observe cases of coronary insufficiency or death from infarction. If the monkeys were kept in small cages and were often subjected to neuroemotional traumas as mentioned earlier, they could develop coronary pathology "spontaneously." Thus in monkeys living in cages and never released into open enclosures being often employed in different experiments, even without the use of unpleasant painful procedures, coronary insufficiency developed in a certain percentage of cases. A *Macaca irus* living in isolation from the group in a cage and used for a long time only in experiments on the study of reflex conditioned activity, as it became clear in the autopsy, suffered two infarctions when alive as two aneurisms of the left ventricle were revealed. The vessels of this monkey were not affected by atherosclerosis.

A baboon (*Papio anubis*) died of myocardial infarction also without the manifestations of atherosclerosis or thrombosis in the Sukhumi colony in a cage, after living there for a few years. This baboon was caught in Africa probably just after fighting and being chased out from the herd following it but never approaching and showed traces of fighting, e.g., a hole in the cheek pouch.

According to the data of Groover *et al.* (1963), available in the literature, the ECG did not reveal coronary pathology during the capture of wild baboons in Africa. However, after some time, during which these monkeys were often investigated, subjected to different procedures, restrained, etc., seven animals showed ECG changes indicating the appearance of angina pectoris in them. In one of the monkeys myocardial infarction was diagnosed by electrocardiography. This diagnosis was confirmed by autopsy findings. In a thorough vessel study the authors observed neither atherosclerotic changes nor those of thrombosis. The authors consider that coronary insufficiency and myocardial infarction in this case were caused by functional spasms of coronary vessels due to stress caused by the mode of surroundings in captivity. Thus, coronary insufficiency and myocardial infarction can often develop in monkeys as a consequence of experimental neurosis. The experiments on monkeys show the great role of the reflex functional spasm in the development of coronary pathology, even in the case of absence of atherosclerosis.

Monkeys, especially baboons and macaques, are the most suitable subjects for the study of sexual ovariomenstrual cycles as only monkeys and chimpanzees have a menstrual cycle like humans differing from estrus in other animals, including dogs. This physiologic area as well as other systems are regulated by the nervous system and the humoral mechanisms. The disturbance of sexual cycles can be observed in various disturbances of the higher nervous activity. It is known that such disturbances as amenorrhea are associated with nervous stress. The experiments in monkeys helped to prove the association between the functional higher nervous activity disturbances—neuroses and various menstrual cycle disorders.

Alekseeva (1959) produced experimental neurosis in three females (*Papio hamadryas*) by means of conflict situations in the herd. As a result they developed different disturbances of sexual cycles. The prefollicular phase reduced considerably resulting in bringing two consequent cycles very close to each other. These disturbances became even more profound later, the prefollicular phase being reduced to zero and the next two consequent cycles fused.

The second monkey showed disturbances in the follicular phase. The two subsequent cycles finished without any menstruation. It is possible that these cycles passed without ovulation. The third monkey showed little or no change in its sexual cycle. These sexual cycle disturbances were present in the two monkeys mentioned above during a few months. It took 11 months for the higher nervous activity and sexual cycles to become normal. The experiment was then repeated. The creation of conflict situations was employed again to induce neurosis and disturbances of the sexual cycles reappeared but in a more pronounced form in the three monkeys. Two monkeys developed amenorrhea and in one of them it was observed to extend for 5 months. The third monkey showed disturbances of the rhythm of sexual cycles with some of the cycles failing to produce menstruation. Neurogenous amenorrhea also developed in a female sitting in a cage with a male that developed impotence (Startsev, 1967). These disorders developed because of the unfavorable surroundings of this couple. Their cage was surrounded by cages with adult males threatening the couple constantly and rushing toward the end of the cage each time the male attempted to mount the female, thus disturbing the normal procedure of the sexual act. The male and the female were constantly in a defensive tense state. Incomplete sexual acts also caused aggression of the male against the female. He began to bite the female. These two monkeys developed motor disturbances (adynamy, shivering of the body and limbs). The male died after some time and the female was transferred to another male with neurogenous impotence. The state of amenorrhea lasted about 11 months in this female. At the end of these 11 months, the female was separated from the male.

The disorders of the sexual cycles mentioned above were the consequence of developing neurosis. This was supported by Alekseeva's data on reflex activity disorders in these monkeys studied before and after neurotizing effects. The monkeys ceased to react upon positive conditioned signals.

The behavior of these animals changed, they became alert and they lost weight. Besides this, long-term disturbances of sexual cycles appeared. Short-term stress situations, for example, surgical operations, also caused sexual cycle disturbances in monkeys. These disturbances did not last long and were restricted only by the cycle which followed next after the stress. Subsequent sexual cycles remained normal (Startsev, 1967). In the case of a developing neurosis the changes are long-term and involve many cycles. These data obtained from monkeys experimentally support the conception that psychogenic sexual disturbances occur in man (Kvater, 1961; Ivanov, 1966; and others).

The possibility of having a model of such disturbances in monkeys opens perspectives for a detailed understanding of their pathogenesis and of finding out methods of rational therapy.

Disturbances of the alimentary system could also be produced as a result of experimental neurosis. Such disturbances, studied by Startsev, were expressed in the form of gastric achylia. The detailed study of gastric achylia in monkeys showed that this state could serve as a good model of a similar disease in man. Two kinds of achylias are known for man: organic and functional. The reason for organic achylia is the atrophy of the mucosa, and that for functional achylia is a psychonervous factor. However, this has not been finally proved and accepted universally. Dogs did not appear suitable subjects for the creation of gastric achylia. They developed only short-term achylia, being far from the chronic form observed in man and from the type of chronic achylia produced by Startsev in monkeys. Gastric juice secretion in monkeys is in many of its features more similar to that in man than that in dogs is to the human disease. Thus, gastric juice on an empty stomach is secreted discontinuously in monkeys and in man. The phenomena did not always take place in dogs (Startsev, 1962, 1963, 1967). The acidity of the gastric juice in monkeys is similar to that in man. The gastric juice in monkeys surpasses that in dogs in its digestive ability and comes closer to the gastric juice in man. Gastric juice secretion is inhibited in monkeys in the same way as in man during night hours. However, this is characteristic of the gastric juice in free-ranging monkeys which were not restrained on an animal board. Such a juice could be obtained from monkeys having a gastric fistula when they were taken by hand directly from living cages only for a few minutes necessary for juice extraction via a fistula. After this procedure the animals were immediately returned to their cages before the taking of next gastric juice portion. Restraint in a board for a long time caused inhibition of the juice secretion immediately. Only the first sample has an acid reaction, that is the juice secreted before restraint. All subsequent portions would have an alkaline reaction. The mechanism of this gastric juice anacidity in monkeys in conditions of restraint are connected with the influence of negative emotions caused by restraint on gastric glands via nervous pathways. According to Startsev's data, the alkaline gastric juice was also secreted from Pavlov pouches when the animals were restrained on the board. In the case of the Heidenheim pouch in which the branches of the vagus were cut, acid gastric juice would be secreted even under conditions of restraint. Thus, negative emotions led to inhibition only when the pathways of the stomach with the central nervous system were preserved. These factors were used to produce experimental chronic neurogenic achylia in monkeys. It was produced by the combination of feeding the monkeys and consequent immobilization on the board. According to the authors (Startsev, 1967) the act of feeding turned into the signal for the defensive dominanta.

Of extreme interest was the fact that all the investigators producing experimental neurosis in monkeys by certain methods noted the appearance of a number of motor disturbances characteristic of hysteria in man. These were all kinds of hyperkinesis, tremor, seizures, paresis, and obsessive motions (Miminoshvili, 1960; Norkina, 1949; Cherkovich, 1959b; Startsev, 1967; Lagutina et al., 1970); all these phenomena are typical of hysteric symptoms in man. These motor disturbances were studied by Startsev and Kuraev (1967) in detail. The authors noted that all the motor disturbances mentioned above were of a functional character and in most of the cases were reversible. Tonic or tonicoclonic spasms were never observed during sleep. They were always associated with neuroemotional overstrain. Hyperkinetic reactions more often overlapped different muscular groups and first of all those which received and increased muscular efforts. Thus, if a monkey in reflex-conditioned experiments pressed the lever with its right hand, in the elaboration of differentiation the seizures were noted in the right hand. When the right hand was accidentally wounded, and the monkey began to use its left hand, the seizures were observed in the left hand. After the strengthening of differentiation the seizures ceased. Similar manifestations are known for man. It is known that in hysteric motor disturbances professional motions associated with these muscular groups, which experience the greatest overstrain daily, suffer first of all. These are hand and finger seizures in pianists, typists, "writing seizures," etc. (Dzerjinsky, 1921; Davidenkov, 1956).

Experimental material obtained by different experimenters at the Sukhumi Institute of Experimental Pathology and Therapy of the USSR Academy of Medical Sciences and presented in this chapter showed that the neurogenic models of various human diseases could be most fruitfully reproduced in monkeys. In spite of the fact that the experiments carried out on other animals (rats, rabbits, cats, and dogs) gave valuable data improving our understanding of the etiology and pathogenesis of a number of human diseases, the experiments in monkeys contributed much new data to these problems and made it possible to receive models of diseases brought close to pathology appearing in man to the maximum. It is natural that there cannot be a complete sign of equality between the models of simian and human diseases. There exists a barrier consisting of such phenomena as consciousness, suggestibility, speech, social relations, etc. It will never be possible to overcome this barrier and put a sign of equality between these manifestations in monkeys or any other animals and man.

However, it is necessary to approximate the picture of the pathology produced in animals to that in man to the maximum extent. In this respect it is difficult to overestimate the role which monkeys play as experimental subjects. The monkeys reached the steps where such phenomena as natural stimuli, social relations, and conflict situations play the leading role in functional pathology. All this leaves no doubt of the advantages of using monkeys in experiments on modeling human diseases.

REFERENCES

Alekseeva, L. V. (1959). *Probl. Endocrinol. Horm. Ther.* **5**, 55–62.

Birukov, D. A. (1955). *In* "Voprosy sravnitel'noi fiziologii i patologii vysshei nervnoi deiatel'nosti" (Problem of Comparative Physiology and Pathology of the Higher Nervous Activity), p. 5. Leningrad.

Bykov, K. M., and Slonim, A. D. (1949). *In* "Opit isuchenia peridicheskih izmenenij fiziologich. funktsiy v organizme" (Experience in Studying Periodical Changes of Physiological Functions in the Organism), pp. 3–15. Moscow.

Cherkovich, G. M. (1959a). *Bull. Exp. Biol. Med. (USSR).* **8**, 21–24.

Cherkovich, G. M. (1959b). *Patol. Fiziol. Eksp. Ter.* **6**, 22–26.

Cherkovich, G. M., and Kokaya, G. Ya. (1962). *Int. Symp. Zoo Anim. Pathol. 2nd, 1962,* pp. 185–188.

Danilov, I. V. (1958). *Zh. Vyssh. Nerv. Deyatel I. P. Pavlova* **8**, 537–545.

Davidenkov, S. N. (1956). "Clinical Lectures on Nervous Diseases." Leningrad.

Davidovski, I. V. (1938). "Pathological Anatomy and Pathogenesis of Human Diseases." Medgiz.

Dzhalagonia, Sh. L. (1967). *Med. Primatol. Tbilisi,* pp. 97–106.

Groover, M. E., Seljeskog, E. L., Haglin, J. J., and Hitchcock, C. R. (1963). *Angiology* **14**, 409–416.

Ivanov, N. V. (1966). "Questions of Psychotherapy of Functional Sexual Disorders." Moscow.

Kaminski, S. D. (1948). "Dinamic Disturbances in the Activity of Brain Cortex" (K. M. Bykov, ed.). Moscow.

Kriazhev, V. Ya. (1955). "Higher Nervous Activity in Intercourse Conditions." Medgiz.

Kupalov, P. S. (1956). "Trudy konferentsii posviaschennoi probleme nevrozv" (Proceedings of the Conference Dedicated to the Problem of Neuroses), p. 5–10. Petrozavodsk.

Kvater, E. I. (1961). "Hormonal Diagnostic and Therapy in Obstetrics and Gynecology." Moscow.

Lagutina, N. I., and Sysoeva, A. F. (1965). *Vestn. Akad. Med. Nauk SSSR* **11**, 20–30.

Lagutina, N. I., Norkina, L. N., and Sysoeva, A. F. (1970). *Probl. Physiol. Patol. Vyssh. Nerv. Deiatel.*, pp. 94–110.

Lang, G. F. (1950). "Hypertensive Diseases." Medgiz.

Lapin, B. A. (1956). *Arkh. Patol.* **18**, 92–95.

Lapin, B. A., and Yakovleva, L. A. (1963). "Comparative Pathology in Monkeys." Thomas, Springfield, Illinois.

Lapin, B. A., Norkina, L. N., Cherkovich, G. M., Yakovleva, L. A., Kuksova, M. I., Alekseeva, L. V., Fufacheva, A. A., and Startsev, V. G. (1963). "The Monkey as an Object of Biomedical Experiments." Sukhumi.

Lapin, B. A., Dzhikidze, E. K., and Yakovleva, L. A. (1965). *Meditsina.*

Livanov, M. N., and Korol'kova, T. A. (1949). *Gagra Discuss.* **1**, 301–311.

Lukomski, P. E., and Tareev, E. M. (1958). *Tr. Vses. S'ezda Ter.* **14**, 207 and 269.

Lunev, D. K. (1956). *Vestn. Akad. Med. Nank SSSR* **4**, 80–83.

Magakian, G. O. (1956). Thesis, Sukhumi.

Markov, Kh. M. (1967). "Allergic Sensibilization and Neurogenic Hypertension." Sofia.

Miasnikov, A. L. (1965). "Hypertensive Disease and Atherosclerosis." Moscow.

Miminoshvili, D. I. (1960). *In* "Theoretical and Practical Problems of Medicine and Biology in Experiments on Monkeys" (I. A. Utkin, ed.), pp. 53–67. Pergamon, Oxford.

Miminoshvili, D. I., Magakian, G. O., and Kokaia, G. Ya. (1960). *In* "Theoretical and Practical Problems of Medicine and Biology in Experiments on Monkeys" (I. A. Utkin, ed.), pp. 103–121. Pergamon, Oxford.

Miterev, L. G. (1959). *Sov. Med.* **1**, 57–62.

Norkina, L. N. (1949). *Tr. Sukhum. Biol. Stantsii, Moscow,* pp. 147–161.

Porubel, L. A., Prokhorova, I. A., Sergeev, A. N., and Slavin, Ya. M. (1958). "1st Scientific Conference for the Hearing of Reports," Theses and summary reports, p. 41. Moscow Sci. Res. Inst. Poliomyelitis Prep., Moscow.

Savinov, A. P., and Tufanov, A. V. (1956). "1st Scientific Session of the Institute for the Study of Poliomyelitis," Theses of the reports, p. 36. Moscow.

Shcherbakova, O. P. (1949). *In* "Opyt izuchenia periodicheskih izmenenij fiziologicheskih funktsij v organizme" (Experience of Studying Periodical Changes of Physiological Functions in the Organism), Theses and summary reports, pp. 42–64. Moscow.

Startsev, V. G. (1967). *Med. Primatol., Tbilisi* pp. 148–164.

Startsev, V. G., and Kuraev, G. A. (1967). *Zh. Nervopat. Psikhiat. S. S. Korsakova* **67**, 880–886.

Sysoeva, A. F. (1967). *Med. Primatol., Tbilisi,* pp. 74–87.

Ukhtomski, A. A. (1966). "Dominanta." Nauka.

Yerkes, R. M., and Yerkes, A. M. (1935). *In* "A Handbook of Social Psychology" (C. Murchison, ed.), pp. 973–1033. Clark Univ. Press, Worcester, Massachusetts.

Zucker, M. B., and Khwan, L. M. (1956). Cited by Lunev (1956).

Zuckerman, S. (1932). "The Social Life of Monkeys and Apes." Harcourt, Brace, New York.

CHAPTER 15

Development of a Brain Prosthesis*

LAWRENCE R. PINNEO

* About half of this research was carried out while the author was on the faculty of the Delta Regional Primate Research Center, Tulane University, and was supported by Contract Nonr-475(11) from the Office of Naval Research, and by National Institutes of Health Grant FR-00164. The remainder of the work was completed at the Stanford Research Institute, and was supported by Contract N00014-68-C-0184 from the Office of Naval Research. Design and development of the computer-controlled stimulators was by Mr. E. A. Elpel of SRI, the software by Mr. P. C. Reynolds of Stanford University, and the interfacing hardware and software by Mr. J. H. Glick also of Stanford University. Dr. J. N. Kaplan, Mrs. P. A. Johnson, Mr. E. Davis assisted with surgery and testing of animals. I wish to thank all the many students, graduate students, colleagues, and visiting fellows who contributed to these studies.

I. Introduction

This chapter reports details of development of an experimental brain prosthesis using primates. By "brain prosthesis" is meant "artificial brain," a term used for the brain just as "artificial heart," or "artificial lung," or "artificial kidney" are used to denote devices and techniques for replacing functions of those organs lost by injury or disease. This is in contradistinction to replacing a peripheral function controlled by the brain, such as movement of an arm, by devising an artificial arm when the brain has been injured. Our approach, instead, is to replace the lost peripheral function due to brain injury by altering the function of the brain itself. "Experimental" means that so far our studies have been limited to animals, primarily monkeys, and no attempt has been made to extend the work to the clinic. Of course, we do hope and expect that with improvements in technique and instrumentation our results will be applicable to humans. The detailed description in this report is of a successful prosthesis for simulated stroke (Pinneo *et al.*, 1972), though our attempts to apply the method to retinal blindness, problems of consciousness (as in coma), and mental retardation are also briefly described.

The major problem in developing a brain prosthesis was to convince ourselves that it could be done. The fact that the brain is so complex and that we really know so little about its function, especially with regard to the highly developed human brain, represents a very difficult barrier indeed. Certainly if one begins with the attitude that it is impossible, then it becomes so. Only by maintaining a strong belief that it *is* possible, and then searching for ways to proceed incrementally, were we able to cope with this problem. An even more difficult problem, which at times was almost insurmountable, was to convince our scientific colleagues of the possibilities of developing a brain prosthesis, especially those reviewing our research proposals for granting agencies. This problem is still not solved, but we hope that this report will at least convince them that it is feasible and more research should be done.

II. Programmed Stimulation of the Brain

The basis of our method is programmed electrical stimulation of the brain, whereby brain tissue *not* damaged, but somehow still involved in the lost function, is stimulated to function in time as it might normally. This approach originated from several years of study of brain mechanisms controlling purposive movement, details of which are published elsewhere (Pinneo, 1966a,b, 1968, 1972).

In order to use the principal of programmed stimulation for experimental stroke, we had to determine just where in the brain, other than the cerebral cortex and/or subcortical structures affected by the stroke, we could stimulate to produce purposeful movements. Several studies (e.g., Hess, 1949, 1954; Dow and Moruzzi, 1958; Pompeiano, 1959; Brodal *et al.*, 1962) had shown that electrical stimulation of subcortical structures of the brain can produce skeletal motor activity, such as movements of the head, foreleg, hindquarters, and muscles of the face. Even "higher" levels of behavior can be elicited by stimulation of brainstem nuclei (von Holst and von St. Paul, 1960, 1962; Delgado, 1964, 1965), including normally appearing progressions in time and space such as attack, withdrawal, sitting, standing, preening, seeking and eating of food, and the like. These findings suggested to us that electrodes could be placed in various areas of the brain and each locus electrically stimulated to produce some elementary movement much as the brain produces the movement normally. Then, if many electrodes could be stimulated simultaneously and sequentially according to some pattern, or program, related to normal behavior, we should be able to produce coordinated limb movements and thus any "goal-directed behavior" which might be eliminated by a stroke.

Our procedure for locating areas in the brain for producing the elementary movements was in three steps:

1. Motion pictures were made of simple motor acts of monkeys living in a compound. Such behaviors as eating, grooming, locomotion, play, aggressive behavior, responses to novel stimuli, and the like were analyzed to determine which elementary movements entered into which behaviors as a function of time.

2. Extensive brain-mapping studies were conducted to determine where in the brainstem one must stimulate to produce the various elementary movements recorded on film in step 1. Using stereotaxic procedures with several species of monkeys and apes, but primarily the rhesus monkey (*Macaca mulatta*), electrodes were placed into the nuclei of the cerebellum and the vestibular nuclei of the brainstem, as well as the reticular formation and many other brainstem structures. Almost countless locations were found in which elementary movements could be produced by electrical stimulation (Pinneo, 1966a). These elementary movements included flexion and extension of all four limbs at the wrist, elbow, shoulder, ankle, knee, or hip; clenching and spreading of the fingers and fine digital movements; opening and closing of the mouth; movements of the tongue in and out; curling or sideways movement of the tail; movements of the eyes, singly and together, and dilatation of the pupils; and many autonomic responses, such as modification of heart or respiration rate. In *every* case,

the precision and control of movement were far superior to those produced by cortical stimulation, and in fact could not be discriminated by observers from similar normal movements.

In all cases it was found that the extent and complexity of movement was directly related to current strength. The optimum electrical parameters for motor production were 200 to 300 rectangular, monophasic, negative pulses per second, 0.1–1.0 msec pulse duration, with base-to-peak currents varying from 50 to 200 μA for a complete limb movement. A stereotaxic atlas and brain-stimulation manual describing these results, which can be used by others for brain prosthesis in the monkey, is being published separately (Pinneo, 1972).

3. The final step was to try to combine the first two steps to produce, or directly control, the behaviors seen in an awake animal. Up to as many as 60 electrodes were chronically implanted in each of several monkeys under aseptic surgery. When an animal had fully recovered from the implant procedure (usually 2 weeks to a month), programmed stimulation of the brain was then carried out in the awake animal. This was done on a moderate scale by using separate biological stimulators, such as the Grass Model S-4, for each of up to six electrodes, and to program the operation of the stimulators by electronic "gates" produced by waveform generators such as the Tektronix 160 series. Of course, with such a relatively few electrodes under programmed control only the simplest of behaviors could be reproduced.

III. Experimental Prosthesis for Stroke

A. SIMULATED STROKE PRODUCTION

From these preliminary experiments it was evident that programmed stimulation of the brain at sites other than the cortex could indeed produce a "purposive" sequence of motor behaviors, and we could therefore proceed with an experimental prosthesis for stroke. Selected monkeys were made monoplegic or hemiplegic by removing cortical and subcortical regions in one hemisphere of the brain involved with control or limb movements on the opposite side of the body. It is recognized that this procedure is not an exact equivalent to stroke, since stroke is due to lack of blood in the affected cortical and/or subcortical region (such as the putamen) by a cerebrovascular accident of one sort or another. On the other hand, selective cortical ablation is much more precise in producing paralysis than is occlusion of portions of the anterior or middle cerebral arteries, especially when only one limb or parts of a limb are to be paralyzed.

The general procedure for producing lesions is fairly standard, and includes mapping the cortex with electrical stimulation to locate areas producing movements in the desired limb and then aspiration or coagulation of those areas. Adult monkeys are preferred to younger animals because skull growth is complete, an important consideration when electrodes and connectors are cemented to the skull for long periods of time.

In the case of the monkey Bruno, whose brain is illustrated in Fig. 1, electrical stimulation on both sides of the central fissure and near the top of the brain (the upper bone edge is the midline in this monkey) produced movements in the right arm, the limb to be paralyzed in this case. Labeled numbers were placed on the cortex, as shown in (A) delimiting the portions sensitive to stimulation, and the portion of the brain between the numbers was removed (the inner dotted line in B). This particular result was surprising, since the area of the brain we excised is usually the portion involved in controlling the hindlimbs, while the forelimb is controlled by an area more lateral. Nevertheless, when this monkey was sufficiently recovered from surgery, it was indeed the right arm that was paralyzed to voluntary movement, though there also was evident weakness in the right hindlimb.

Following surgery, monkeys are observed for at least 1 month in order to determine the nature and extent of their paralyses. During this time they are observed alone in a standard primate cage and later in a larger 12 ft × 10 ft × 12 ft compound. Tests are administered to determine their ability to pick up and manipulate small objects and otherwise use the disabled limb. We have found that recovery in the paralyzed limbs is much more rapid after the animals are put into the larger compound and have the opportunity to move around and use their limbs. Often within a few hours after being moved into these larger living quarters a monkey will display an appreciable amount of function in a limb that was never observed to be used while the monkey was housed in a smaller cage.

The animals that show a substantial degree of recovery in their paralyzed limb(s) are operated on a second time during which additional cortical and/or subcortical tissue is removed. For example, in Bruno much of the function of the right arm had returned after 2 months. Electrical stimulation of the exposed cortex during the second operation revealed that the tissue adjacent to that which had been removed earlier responded to electrical stimulation by producing right-arm movement, whereas during the first operation it had not so responded. Such a finding also has been reported by others (Glees and Cole, 1950). This second portion of cortex was also removed electrosurgically and by suction (the outer dashed line of Fig. 1B), resulting in a new paralysis of the right arm with little or no recovery over an extended period of time.

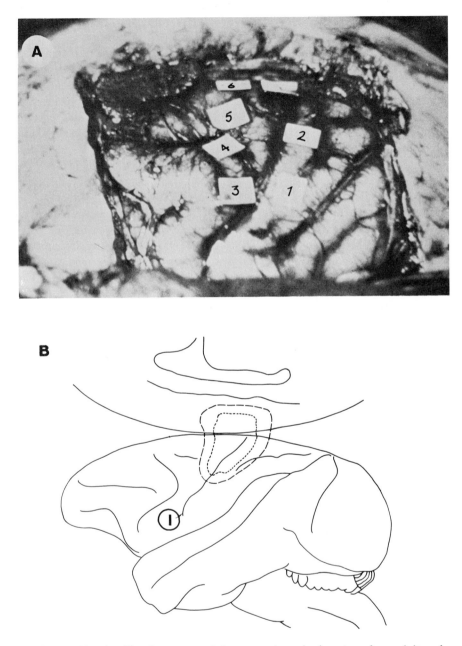

Fig. 1. (A) A still enlargement of the exposed cerebral cortex of an adult male
rhesus monkey named Bruno; from a 16-mm film of the surgery producing paralysis

B. ELECTRODE IMPLANTATION

The next step was to implant electrodes in the brainstem for eliciting movements in the paralyzed limbs with electrical stimulation, using a technique for inserting electrodes into the brain of awake animals restrained in a chair (MacLean, 1967; Robinson, 1967). This technique is particularly suited to our purposes since stimulating electrodes can be inserted into the brain under exactly the same conditions as when an animal is tested at a later date, and electrodes can be permanently fixed at only those locations that elicit discrete and distinctly different movements in the paralyzed limbs. Also, since many of the regions in the brainstem that elicit elementary motor responses with stimulation are located near areas that control such vital functions as respiration, heart rate, and blood pressure, problems of surgical trauma that might arise from inserting electrodes into these areas when the animal is anesthetized are obviated by using awake subjects.

This technique involves permanently attaching an electrode guidance platform to the paralyzed monkey's skull while the monkey is anesthetized and in a stereotaxic instrument. The platform can be made cheaply of dental acrylic; it contains an array of holes through which electrodes can be inserted into the brain with stereotaxic accuracy (Fig. 2). It is aligned above the monkey's head with the stereotaxic apparatus and is attached to the skull with screws and acrylic dental plastic. The skin below the platform is removed so that electrodes can be inserted directly through the skull without piercing the skin first.

The electrodes we use are multiple-chain stainless steel that are commercially available.* These are constructed of No. 316 stainless steel wire, 0.075 mm diameter, with quad Teflon-coated leads composed of six contacts of 1-mm exposed wire each and separated by 2 mm.

In a given implantation session, a chain of electrodes is lowered into the brain of a tranquilized monkey (Innovar-Vet, McNeil Laboratories) in 1.0-mm steps through a hole predrilled in the skull. Motor responses

* Mr. Henry Schryver, Developmental Design, 110 West Packard Avenue, Fort Wayne, Indiana.

---

in Bruno's right arm much as an actual stroke might do. Numbers are bits of paper placed on the cortex to identify the margins of cortex which, upon direct electrical stimulation, are identified with right-arm or hand movement. (B) A drawing of a side view of the rhesus monkey brain; the upside-down portion represents the medial surface of the hemisphere. The circled "1" identifies the central fissure, which in (A) is the blood vessel running between numbers 2 and 3 and beyond. The inner dotted line of (B) delineates the margin of cortex identified by the numbers in (A). All cortex within this region was removed as shown by the dashed line. (After Pinneo *et al.*, 1972.)

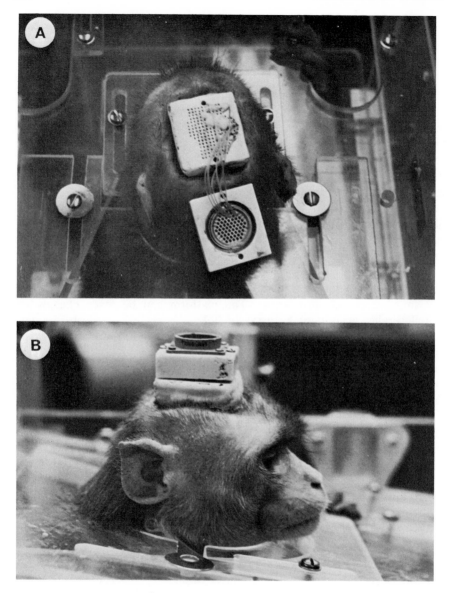

Fig. 2. Monkey Mildred, awake but tranquilized, with an electrode guidance platform and Amphenol plug attachment for implanting electrodes into the brain. (A) The guidance platform contains an array of holes for inserting electrodes with stereotaxic accuracy; cemented-over holes contain electrodes already implanted. (B) Connector screwed to baseplate for stimulation and testing of electrodes. (After Pinneo *et al.*, 1972.)

to stimulation are initially tested at each step as the electrode is lowered. After a location is found that produces a distinct elementary motor response, electrodes are permanently fixed by attaching it to the platform with acrylic (see Fig. 2A). The electrodes are then connected to an Amphenol plug that is housed in a box made of acrylic or aluminum. The box is attached to the guidance platform with screws (Fig. 2B), and is removed from the platform only when electrodes are inserted into the brain.

In Bruno, 13 locations were found that upon stimulation produced movement in his paralyzed right arm. Normally appearing movements in this limb produced by brainstem stimulation included rotation of the wrist, arm turning in toward the body from the shoulder, arm up from shoulder, arm up from elbow, arm *back* from shoulder, arm straight out from the body, rotation of forearm out from the body at the elbow, flexion of the thumb, and several other elementary movements. Exact location of each electrode can only be determined by histologic examination, so we are not too certain of the structures in which the electrodes presently reside. This is because more than one location can cause the same movement, and because there is not yet a suitable stereotaxic atlas for an adult rhesus monkey (but see Pinneo, 1972). For Bruno, our placements ranged from A1, L3, H-2, arm up from elbow, to P13, L3, H-5, arm up from shoulder.

IV. Multiple-Electrode Programmable Brain Stimulators

A. THE SYSTEM

The stimulating and programming equipment referred to above for our preliminary work was too limited to successfully control the full operation of a paralyzed limb, let alone the two or more limbs that might be affected in an actual stroke. We therefore designed and constructed a system for multiple-electrode brain stimulation that could be programmed and operated by a laboratory-sized computer, in our case a LINC-8 (manufactured by the Digital Equipment Corporation, Maynard, Mass.). Only 10 channels of stimulation were developed at first, with the possibility for future expansion to any number of electrodes.

The system consists of four parts (Fig. 3): (1) A 10-channel programmable brain stimulator; (2) the LINC-8 computer with appropriate interfacing equipment; (3) the software programs controlling the system; and (4) a monkey whose movements are to be produced by programmed stimulation of the brain. The 10-channel programmed brain stimulator

Fig. 3. The complete brain prosthetic system for stroke with programmed stimulation of up to 10 electrodes. Monkey Louie in the foreground illustrates how a paralyzed right arm can be made to extend up and out from the body, while the right hand is made to be open. In the left background is the LINC-8 computer, while in the right background is the programmed brain stimulator. A teletype, used for calling forth a particular program of movements, is not shown. (After Pinneo *et al.*, 1972.)

(PBS) actually consists of 10 separate stimulators, each containing its own control circuitry.

The basic stimulator consists of a current generator that provides negative going pulses, from 0 to 1 mA, into any low impedance load. The remainder of each stimulator circuit consists of pulse generators, ramp generators, time generators, and solid-state multiplex switches. These devices, all constructed on a 5 in. × 5 in. standard DEC computer card

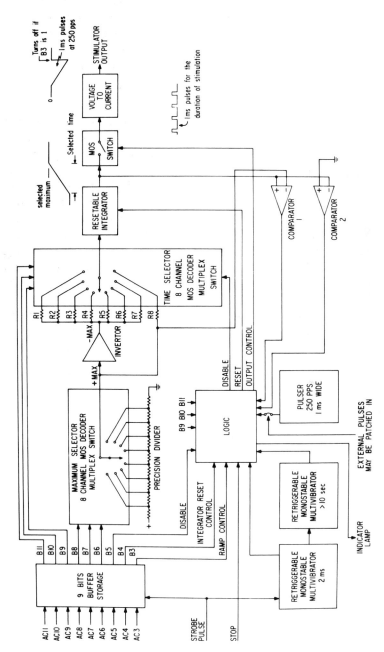
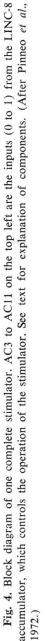

Fig. 4. Block diagram of one complete stimulator. AC3 to AC11 on the top left are the inputs (0 to 1) from the LINC-8 accumulator, which controls the operation of the stimulator. See text for explanation of components. (After Pinneo *et al.*, 1972.)

of mostly integrated circuits, control the output of the current generator so that pulses of 1-msec width are delivered to the animal's implanted electrodes at the usual repetition rate of 250 pps. Current intensity, controlled by the ramp generator, increases as a function of time to make a given limb move a given distance in a finite amount of time.

B. STIMULATOR DESIGN AND OPERATION

Figure 4 shows the block diagram of one such stimulator. Parameters describing the desired waveshape of the ramp integrator which controls current intensity to the animal, are passed to a given stimulator by strobing bits 3–11 of the LINC-8 accumulator into a separate buffer register (top left in Fig. 4) for each accumulator. That is, though the accumulator bits are common to all stimulators the strobe pulse is unique to a given stimulator. From then on the stimulator operates independently of the computer. This has the advantage of leaving the computer free to address another stimulator or for other calculations. But it also means that the maximum amount of current the animal will receive, and the amount of time for the current to reach that maximum, must be specified in the 9-bits sent the stimulator from the LINC-8 accumulator.

To illustrate the design and operation of a stimulator, Table I gives the significance of the accumulator bits, while Table II gives the maximum current and the time to reach maximum current (supplied the stimulator by bits 6–11). Bits 3, 4, and 5 when set or not set, control the ramp-generating integrator. The output of the integrator is applied to a MOS

TABLE I

SIGNIFICANCE OF BITS OF LINC-8 COMPUTER FOR CONTROLLING INTEGRATOR AND OUTPUT OF PROGRAMMED BRAIN STIMULATOR

Bit no.	Not set = 0	Set = 1
0		
1	Not used	
2		
3	Hold integrator at maximum current	Provide ramp function only
4	Do not reset integrator	Reset integrator
5	Disable channel	Enable channel
6		
7	Maximum current[a]	
8		
9		
10	Time (in seconds) to reach maximum current[a]	
11		

[a] See Table II for octal code specifying value. (After Pinneo et al., 1972.)

TABLE II

OCTAL COMPUTER CODES FOR SPECIFYING MAXIMUM STIMULATING CURRENT (BITS 6, 7, 8 TABLE I) AND THE TIME TO REACH MAXIMUM (BITS 9, 10, 11 TABLE I)

Octal code	Maximum current (μA)	Time to reach maximum current (sec)
0	50	7
1	100	5
2	200	3
3	300	2
4	400	1
5	500	0.5
6	750	0.2
7[a]	1000	−0.5[a]

[a] The last time is for a *negative* integrator charging rate and is actually the time to reach zero from any maximum. (After Pinneo et al., 1972.)

switch (top right components of Fig. 4), which either switches at a rate of 250 pps, or is turned off depending on the control logic. The output of the switch is applied to a voltage-to-current circuit, which is the current generator supplying pulses to the animal. Thus, *when* the integrator is reset, *and* the integrator is allowed to charge to a maximum in a certain amount of time, *and* the output control is enabled, *then* 1.0 msec pulses of increasing current intensity are delivered to the animal at 250 pps. The logic conditions imposed by these three bits are:

1. The integrator is reset when B4 is 1 *and* the 2-ms multivibrator is 1.
2. The integrator charging is enabled when:
   (a) A positive slope is selected (Time 0–6 Table 2), *and* the integrator output is less than the maximum setting, *or*
   (b) A negative slope is selected (Time 7 Table 2), *and* the integrator output is greater than zero.
3. The output control is disabled (no output) when:
   (a) The integrator output is less than zero, *or*
   (b) The disable bit is set (B5 = 0), *or*
   (c) Bit 3 is 1 (ramp only) *and* the maximum is reached, *or*
   (d) The maximum time (≈ 12 sec) is exceeded without receiving a new command.

In order to control the maximum amount of current and the time to reach maximum, separate MOS multiplex switches are used on the stimulator board. As shown in Fig. 4, bits 6, 7, and 8 are applied to an 8-channel MOS decoder, multiplex switch. This in effect connects a tap on a precision voltage divider to one terminal of a comparator (comparator

1). Another comparator (comparator 2) terminal is connected to the integrator's output, and hence this comparator senses when the integrator output reaches the desired maximum; that is, when it reaches the voltage on the selected divider tap. The output of the maximum selector multiplex switch is applied through an inverter and a series of resistors to the second multiplex switch. This second switch is controlled by bits 9, 10, and 11, which select the resistor to be connected to the integrator and hence control the charging rate of the integrator. Since the integrator inverts the signal, the negative input voltage causes a positive integrator output. A single positive input can be selected (R8) for the rare case when a negative ramp is desired (the last "time" in Table II with octal code 7). Because the desired maximum is applied as an input, the time to reach maximum is independent of the maximum value and hence completely controlled by the value of the selected resistor. There is a disable input on the time-selector multiplex switch which is, in effect, a ninth open-circuit position for the switch. When disabled, integrator charging is stopped and the integrator output remains at a constant level.

The strobe pulse which loads the LINC accumulator bits also triggers a 2-ms monostable multivibrator (MMV), which in turn triggers a 12-sec MMV. The first MMV is used in conjunction with the logic circuits described earlier, to reset the integrator and to disable the output control during resetting. The 12-sec MMV is a safety feature which turns off the stimulator after 12 sec if no new commands are received. Both MMV's are completely resettable; that is, they measure time from the last pulse received. Hence, new commands may be given at any time and the stimulator will correctly accept the new states.

V. Computer Programming and Control

A. PROGRAMMING LOGIC

The time of occurrence of each stimulator relative to other stimulators, and the rate at which current intensity of the output pulses increases with time, are determined by the LINC-8 computer and the software to operate the system. The LINC-8 actually consists of two computers that work in conjunction: a PDP/8 computer that operates either independently or as a control unit, and a LINC computer that operates as a peripheral device of the PDP/8. The LINC is dependent upon the PDP/8 for software interpretations of many of its console functions and operations. The stimulators are under the control of a LINC-8 program called PCB (programmed control of the brain). Due to the system configuration just explained, some portions of PCB are written in LINC machine language and others in PDP/8 machine language. However, PCB is logically a single program and can be described as such.

From a computational point of view, the program control of stimulators for inducing coordinated movement is simply the transmission of information to multiple peripheral devices as a function of real-time. The user must specify which stimulators are to be turned on, at what maximum current level, and at what time relative to a real-time clock. This information is encoded in a table format, called TIMETABLE, and all the running program must do is monitor the real-time clock and strobe information into the stimulator buffers at the time specified by TIMETABLE. The control of the stimulators is completely determined by the values encoded in the TIMETABLE: PCB has no on-line feedback capability, and any modification of output must be done by writing a more precise TIME-TABLE. The advantage of this system is that it is compatible with various configurations of stimulators, various movements, and various brain implantation sites. Moreover, once a particular configuration of hardware and TIMETABLE values is shown to be effective for a certain evoked movement, the TIMETABLE can be stored on magnetic tape for future use. This permits PCB to develop a repertoire of control programs for various movements.

Because PCB derives its versatility from its TIMETABLE repertoire, TIMETABLES have been made as easy to write as possible. PCB possesses teletype input routines that accept information for a TIME-TABLE one line at a time, where a "line" is defined as the real-time interval that is to pass before the next stimulator is to be addressed, the particular stimulator channel, and the 9-bit data word specifying what the stimulator is to do. Lines of data can be added, inserted, or removed from the table independently of each other, and the whole TIMETABLE can be coded more efficiently to save space in core storage. PCB is integrated with the LINC systems programs to permit the easy storage of TIMETABLES on magnetic tape and the reading of tape files into core for modification or execution. In addition to provisions for both teletype and magnetic tape input of TIMETABLES, PCB has a manual control option that enables the user to control the stimulators without a TIME-TABLE via the teletype and sense switches on the computer console. This is useful in testing the peripheral equipment and in establishing values for new TIMETABLES.

B. OPERATION OF PROGRAM

The PBS's are interfaced with the LINC-8 computer with DEC R and W series Flipchip Modules (Fig. 5). The PDP-8 accumulator (used to

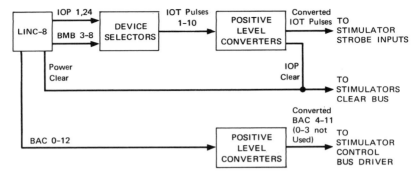

Fig. 5. Block diagram of the interfacing between the LINC-8 and the Programmed Brain Stimulators. (After Pinneo *et al.*, 1972.)

program the stimulators) is converted from DEC logic levels of -3 V and ground to the ground and $+5$ V levels used in the stimulator circuitry. Bits 3–11 of the buffered output from the accumulator are led in parallel into W601 positive output converters. The converter outputs are then fed by cables to inverting drivers on the stimulator rack, while the driver outputs are bussed to the input buffers of all the stimulators. These converted accumulator bits are then strobed, in parallel, into any given stimulator buffer by a control pulse decoded and level converted in the interface. Standard DEC IOT (input/output) pulses are used as the control pulses and are produced in W103 device selectors by IOT instructions. Each stimulator has a specific IOT instruction associated with it that upon execution loads the accumulator bits into the input buffer of that stimulator. A separate IOT instruction is combined with the power clear line from the computer, to disable all stimulators in an emergency. This "safety stop" can be instituted from the computer or from a "panic button" on the stimulator rack.

C. OPERATION FOR SIMULATED STROKE PROSTHESIS

In practice, the system works like this (see Figs. 3, 4, and 5). A monkey with multiple implanted electrodes, each of which, when electrically stimulated, produces a consistent motor movement of a limb paralyzed to voluntary control, is connected by cable to the PBS. Previously, each electrode will have been tested by a single stimulator to determine the frequency of stimulation and the maximum current required for production of a given amount of movement. When it is known *which* of any 10 or fewer electrodes are to be used to produce a given sequence of movements, *when* each electrode is to be stimulated relative to all other electrodes, the *maximum* current to be supplied each electrode, and *how long* it should take to reach the maximum current, then all of this information is supplied TIME-

TABLE via the teletype of the computer. The control portion of the PCB program running the stimulators is called forth, and when the "start" button is pressed the animal will be stimulated and the types and amount of movement programmed will occur. Note that with a little training the animal can be given a set of switches that tell the computer what set of movements to produce; thus, he can enter into the control of his own behavior.

VI. Programmed Brain Stimulation for Visual Prosthesis

If our experimental brain prosthesis is to be successful, it must be capable of restoring lost functions other than just skeletal motor activity. Our attempts at prosthesis for brain deficits other than those due to stroke have only recently begun, but enough progress has been made to suggest that the programmed stimulation method will be successful. The approach is the same as that used in the stroke experiments: stimulation in time of critical brain areas *not* damaged, yet involved in the lost function, such that the nervous tissue acts as it would "normally."

For example, our approach to visual prosthesis, published in detail elsewhere (Pinneo, 1971), is an attempt to mimic normal vision and is based on our research for the past 9 years on the neurophysiology of brightness vision in primates (Arduini and Pinneo, 1961, 1962a,b, 1963a,b; Brooks, 1966; Pinneo *et al.,* 1967). Ours is by no means the first attempt at visual prosthesis, or even with stimulation of the brain to pro-duce "vision." Recently, in England, Brindley and Lewin (1968) have implanted 80 platinum electrodes over the calcarine and neighboring cortex of the right hemisphere of a 52-year-old nurse suffering from glaucoma and retinal detachment. Stimulation of these electrodes produced an experience of "seeing" a very small spot of light. These investigators hope that by programming the stimulation of each electrode using objects in the visual field, blind patients will be able to avoid obstacles when walking and possibly read print or handwriting. The major difference be-tween our approach and theirs is that they are simply stimulating the cortex to produce any sort of visual "experience" in space and time, while we are attempting to reproduce "normal" vision by making the visual system act as it would physiologically with light stimulation of the retina (see Pinneo, 1971, and other reports in the same volume).

A. NEUROLOGY OF BRIGHTNESS VISION

So far we have only been concerned with brightness vision, the simplest of visual processes, but the fundamental one from which all other vision is derived. To find out how the brain acts "normally" in brightness vision, we have recorded the electrical activity of the primary visual pathway, from the retina to the visual cortex, associated with light and dark adaptation, brightness discrimination, flicker and fusion discrimination, brightness enhancement, and the "Talbot brightness" of high-frequency flicker. Most of the experiments were carried out with chronically implanted rhesus and squirrel monkeys, though we also have had an opportunity to work with humans (Pinneo and Heath, 1967) in which verbal report was available as well as electrical activity. Also, in most of our experiments macroelectrodes were used, and the multiple units thus obtained were integrated to reflect the average amount of activity in the visual pathway. Nevertheless, our basic results were confirmed by microelectrode controls, and by others (Jung, 1964; Straschill, 1964, 1966) using microelectrodes exclusively.

The visual system activity shown in Fig. 6A represents the kind of activity seen in the optic nerve, optic chiasma, optic tract, the lateral geniculate nucleus, and the optic radiations. It has been recorded in our laboratory from the cat, the squirrel and rhesus monkeys, and from man, and has been correlated with psychophysical data in both monkey and man. The activity shown on the top line of Fig. 6A is an oscilloscope recording of instantaneous activity following 15–30 min dark adaptation. At the arrow a light is turned on and the retinas of fixated eyes are illuminated. Note the initial "on" discharge, typical of a *changing* light stimulus, followed by a dramatic depression of activity when *steady* illumination persists (as seen by the activity 5 min later, shown on the second line, with the retina still constantly illuminated). In the third line of Fig. 6A, the light is turned off producing a large "off" discharge, again typical of a *changing* light stimulus. The fourth line is 5 min later, illustrating that the dark discharge after prolonged illumination of the retina remains rather high. The last two lines are control records showing that steady illumination definitely has a depressing effect (compare second line with the last two).

Figure 6B shows what happens to visual system activity as the *steady* illumination, representing background light in the normal visual field, is varied over a wide range. Of great significance is the fact that the activity of the visual system decreases as a power function of increased retinal illumination in the same way that brightness vision in humans increases. Furthermore, if a flash of light is superimposed upon the steady light, as is done when obtaining brightness-discrimination functions, then we find that the greater the level of background illumination the lower the electrical response of the visual system to the flash of light. If the background light is sufficiently intense, then no response occurs in the visual system and, indeed, no flash is seen (Brooks, 1966). The relationship between visual pathway activity and light stimulation is in fact so close, that we have been able to predict with 100% accuracy the perceptual capacity of humans

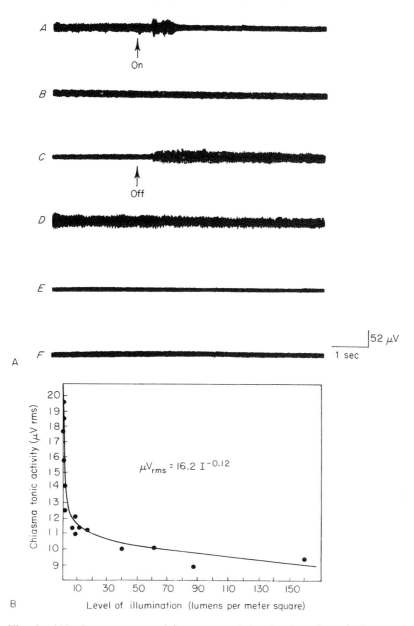

A

B

Fig. 6. (A) Spontaneous activity measured in the lateral geniculate nucleus of the cat during dark and light adaptation. *A:* In dark adaptation. At arrow, a diffuse steady light is turned on, illuminating the retina. *B:* Five minutes after *A,* with light still on. *C:* Ten minutes after *B.* At the arrow, light is turned off. *D:* Five minutes after *C.* *E:* Noise level of the dead animal. *F:* Record obtained from deafferented retina during retinal ischemia. Compare with *B* and *E* to see inhibiting effect of steady light. (After Arduini and Pinneo, 1963a.) (B) Plot of the absolute level of activity in the optic chiasma of a cat as a function of level of illumination. (After Arduini and Pinneo, 1962b.)

347

(Pinneo and Heath, 1967) to see a threshold flash of light against any background illumination from the electrophysiologic activity alone.

B. DESIGN OF A VISUAL PROSTHESIS

From these results and others obtained over the last few years we have formulated a theory of how "brightness" is perceived in terms of the activity of the primary visual pathway. Needless to say, our particular interpretation has met with criticism from other investigators who interpret the data differently. However, their criticisms are not cogent as far as visual prosthesis is concerned, so they will not be discussed here. The facts still remain, and it was upon these that we have devised our visual prosthesis experiments. The facts may be summarized as follows:

1. A relatively prolonged depression of the overall activity of the primary visual pathway represents an increase in perceived brightness of a background illumination. Therefore, if we could somehow depress the random dark discharge of a blind animal, or an intact one in the dark, then he should "perceive" a steady background light whose intensity is related to the amount of depression by a power function.

2. A brief, synchronous, burst of activity in the primary visual pathway represents a transient illumination, or flash. Repetitive synchronous discharges represent flicker. Therefore, if the primary visual pathway could be made to fire in synchronous bursts, then flashes and/or flickering lights would be "perceived" by a blind animal, or an intact one in the dark. Furthermore, by combining the bursts with depression of activity as described in 1, then the combination of "flashes" upon a "steady" background illumination should be perceived.

To date we have been able to accomplish these changes in the activity of the primary visual pathway, but are just now testing the "perceptual" validity of these changes. Without describing our methods in detail here (see Pinneo, 1971) we found that direct current polarization across the laminar layers of the lateral geniculate nucleus will produce a significant decrease in the random dark discharge as measured in the radiation fibers. By appropriate adjustment of the amount of polarization, the activity level in the radiation fibers could be made to span the range produced by 4 logarithms of light illumination (maximum about 150 ft-L). Polarization across the optic tract or chiasma was effective in some instances, but not nearly as effective as in the geniculate and therefore was discontinued. By placing other electrodes in the geniculate and stimulating there with brief volleys, we were also able to produce synchronous bursts in radiation fibers comparable in *size* to the phasic response to light flashes; these were also equal to a 4-log unit change of intensity.

Our next step is to train implanted rhesus monkeys in the dark to make "brightness discriminations" to these two forms of visual pathway electrical stimulation. When they reach a suitable criterion of correct responses, we will retest them in an actual brightness discrimination situation using light. If on these tests their ability to make correct discriminations is significantly better than controls, then we may assume we have replicated some form of "brightness vision," at least comparable to the retest situation.

The use of the brain prosthesis system for visual prosthesis would be much like that for stroke. Chains of vertical electrodes would be implanted in the lateral geniculate nucleus. Using an expanded version of the PBS, each electrode would be connected to a separate stimulator which in turn, via the LINC-8, would be controlled by an operational amplifier responsive to a region of light in the visual field. In this way we should be able to present a three-dimensional mosaic of stimulated points representing the entire visual field, that is, "pattern vision." For the present we will only consider black-and-white representation. The level of background illumination would be determined by the computer by averaging the outputs of all the operational amplifiers. This averaged signal would in turn modulate a polarizing signal across the geniculate layers to control the average amount of activity of the geniculate and its responsiveness to flashes of light. Thus, "perception" of the patterns of flashes would be related to the level of background illumination as in normal vision.

VII. Conclusions and Future Possibilities

In this chapter details have been given of a successful brain prosthesis for simulated stroke in the monkey, which may also be applied to certain types of blindness with a high probability of success. We are confident that with sufficient future research the methods employed may be extended to many other forms of deficit due to brain injury or disease. For example, two other deficits in which we are just now beginning to research as likely candidates are problems of consciousness, as in coma, and mental retardation. Our approach to both of these deficits is based on the concept of the reticular activating system of the brainstem, by which efficient normal behavior is a function of the level of consciousness, or "arousal" (Lindsley, 1951; Hebb, 1955; Malmo, 1959). By electrical stimulation of various areas of the brain (Hess, 1954; Clemente and Sterman, 1963; Hernández-Peón and Ibarra, 1963; Parmeggiani, 1964), such as the nucleus reticularis, the inferior thalamus, and the caudate nucleus, sleep-like states may be produced, which in many ways mimic the level of alert-ness of mentally retarded children or animals. Conversely, electrical stimulation of the mesencephalic reticular formation (Morruzzi and Magoun, 1949) produces arousal, even in an anesthetized animal, while destruction of this area produces comalike behavior. To date our experiments using programmed brain stimulation have been reasonably successful in controlling level of arousal (Pinneo, 1966a, 1968) both upward and downward, and to return to consciousness an animal suffering from experimental coma. It still remains to be seen whether this type of stimulation will effect learning rate in a retarded animal.

For many critics, the greatest objections to application of our experimental brain prosthetic system to human use will undoubtedly be the relatively high cost, the impracticality of large computers for individual use, and the necessity for implanted electrodes. Costs, like anything else, can only be reduced for a user by subsidization or mass production. If our system should, in time, prove practical for human use, subsidization is a likely prospect from public clinics and national health organizations. Meanwhile, advances in computer hardware miniaturization (such as for NASA), and other electronic techniques, should ultimately lead to practical, relatively low-cost, general-purpose computers small enough to be worn or carried by a human being as part of his clothing. Finally, the use of implanted electrodes in humans, while not desirable, is nevertheless becoming more and more acceptable as this becomes a more widely used practice in the treatment of several neurologic and mental diseases (Ramey and O'Doherty, 1960; Sheer, 1961). Certainly for the stroke or blind victim, implanted electrodes may well be the lesser evil if it means restoration of his lost function. However, even the necessity of implanted electrodes will probably pass, for experiments in our laboratory (Pinneo et al., 1971) and those of others show that it may soon be possible to stimulate selective locations deeply in the brain using external electrodes without the intervention of any surgery. Should this become possible, brain prosthesis may well be used for other types of deficits than those produced by injury or disease, including disorders of behavior.

REFERENCES

Arduini, A., and Pinneo, L. R. (1961). *Boll. Soc. Ital. Biol. Sper.* **37**, 430–432.
Arduini, A., and Pinneo, L. R. (1962a). *Arch. Ital. Biol.* **100**, 415–424.
Arduini, A., and Pinneo, L. R. (1962b). *Arch. Ital. Biol.* **100**, 425–448.
Arduini, A., and Pinneo, L. R. (1963a). *Arch. Ital. Biol.* **101**, 493–507.
Arduini, A., and Pinneo, L. R. (1963b). *Arch. Ital. Biol.* **101**, 508–529.
Brindley, G. S., and Lewin, W. S. (1968). *J. Physiol.* (*London*) **196**, 479–493.
Brodal, A., Pompeiano, O., and Walberg, F. (1962). "The Vestibular Nuclei and

their Connections, Anatomy and Functional Correlation." Oliver & Boyd, Edinburgh.

Brooks, B. A. (1966). *Exp. Brain Res.* **2**, 1–17.

Clemente, C. D., and Sterman, M. B. (1963). *Electroencephalogr. Clin. Neurophysiol., Suppl.* **24**, 172–187.

Delgado, J. M. R. (1964). *Int. Rev. Neurobiol.* **6**, 349–449.

Delgado, J. M. R. (1965). *Science* **148**, 1361–1363.

Dow, R. S., and Moruzzi, G. (1958). "The Physiology and Pathology of the Cerebellum." Univ. of Minnesota Press, Minneapolis.

Glees, P., and Cole, J. (1950). *J. Neurophysiol.* **13**, 137–148.

Hebb, D. O. (1955). *Psychol. Rev.* **62**, 243–254.

Hernández-Peón, R., and Ibarra, G. C. (1963). *Electroencephalogr. Clin. Neurophysiol., Suppl.* **24**, 188–198.

Hess, W. R. (1949). "Das Zwischenhirn." Schwabe, Basel.

Hess, W. R. (1954). *In* "Brain Mechanisms and Consciousness" (E. D. Adrian, F. Bremer, and H. H. Jasper, eds.), pp. 117–136. Thomas, Springfield, Illinois.

Jung, R. (1964). *Ber. Deut. Ophthal. Ges.* **66**, 69–111.

Lindsley, D. D. (1951). *In* "Handbook of Experimental Psychology" (S. S. Stevens, ed.), pp. 473–516. Wiley, New York.

MacLean, P. D. (1967). *Electroencephalogr. Clin. Neurophysiol.* **22**, 180–182.

Malmo, R. B. (1959). *Psychol. Rev.* **66**, 367–386.

Moruzzi, G., and Magoun, H. W. (1949). *Electroencephalogr. Clin. Neurophysiol.* **1**, 455–473.

Parmeggiani, P. L. (1964). *Progr. Brain Res.* **6**, 180–190.

Pinneo, L. R. (1966a). *Nature (London)* **211**, 705–708.

Pinneo, L. R. (1966b). *Psychol. Rev.* **73**, 242–247.

Pinneo, L. R. (1968). *In* "The Squirrel Monkey" (L. A. Rosenblum and R. W. Cooper, eds.), pp. 319–346. Academic Press, New York.

Pinneo, L. R. (1971). *In* "Visual Prosthesis—The Interdisciplinary Dialogue" (T. Sterling, E. Bering, S. Pollack, H. Vaughan, Jr., ed.), pp. 109–123. Academic Press, New York.

Pinneo, L. R. (1972). "Stereotaxic Atlas and Manual for the Electrical Production of Motor Behavior in the Adult Rhesus Monkey (*Macaca mulatta*)." Williams & Wilkins, Baltimore, Maryland (in preparation).

Pinneo, L. R., and Heath, R. G. (1967). *J. Neurol. Sci.* **5**, 303–314.

Pinneo, L. R., McEwen, B., and Hansteen, R. (1967). *Fed. Proc., Fed. Amer. Soc. Exp. Biol.* **26**, 655.

Pinneo, L. R., Erickson, E. E., and Kinney, R. A. (1971). *In* "Neuroelectric Research" (D. Reynolds and A. Sjoberg, eds.), pp. 405–425. Thomas, Springfield, Illinois.

Pinneo, L. R., Kaplan, J. N., Elpel, E., Reynolds, P., and Glick, J. (1972). *Stroke* **3**, 16–26.

Pompeiano, O. (1959). *Arch. Sci. Biol. (Bologna)* **43**, 163–176.

Ramey, E. R., and O'Doherty, D. S., eds. (1960). "Electrical Studies on the Unanesthetized Brain." Harper (Hoeber), New York.

Robinson, B. W. (1967). *Physiol. Behav.* **2**, 455–457.

Sheer, D. F., ed. (1961). "Electrical Stimulation of the Brain." Univ. of Texas Press, Austin.

Straschill, M. (1964). *Pfluegers Arch. Gesamte Physiol. Menschen Tiere* **281**, 84–85.
Straschill, M. (1966). *Kybernetik* **3**, 1–8.
von Holst, E., and von St. Paul, U. (1960). *Naturwissenschaften* **18**, 409–422.
von Holst, E., and von St. Paul, U. (1962). *Sci. Amer.* **206**, 50–59.

CHAPTER 16

Visual Refractive Characteristics and the Subhuman Primate*

FRANCIS A. YOUNG

* This investigation was supported in part by U. S. Public Health Service research grant EY 00284 from the National Eye Institute; U. S. Public Health Service Institutional Grant to Washington State University; the 6571st Aeromedical Research Laboratory, Holloman Air Force Base, New Mexico; and U. S. Public Health Service research grant NB 05459 from the National Institute of Neurological Diseases and Blindness.

353

I. Introduction

A. NATURE OF THE PROBLEM

Research on the solution of most human health problems is expedited if the investigator can find a suitable animal substitute for the human. In cancer research, for example, a wide variety of animal substitutes is readily available. However, the number of different animals capable of developing atherosclerosis is greatly limited; thus only a few animals (e.g., the baboon) may be used for this type of research.

1. Control of Variables

When an investigator attempts to determine the variables which may contribute to the development of visual refractive characteristics in humans, he faces a situation in which he has little control. For example, many investigators have argued that near work has a significant effect upon the visual refractive characteristics of the eye such that an individual who engages in a large amount of near work is likely to develop a refractive anomaly known as nearsightedness or myopia. It is, at present, impossible to determine accurately the amount of reading or other near work to which human subjects may be exposed. The effect of a near visual space on the development of myopia in humans can be studied much more efficiently with a suitable animal substitute.

2. Substitute Animals

A suitable substitute must share most of the important visual characteristics of the human. Further, the substitute's adaptation and response to human environmental conditions must be humanlike. For reasons outlined in Section II, the subhuman primate meets these criteria better than any other animal.

3. Limitations of Substitute Animals

One of the problems of primary importance with respect to the development of the refractive characteristics of the human or subhuman primate eye is the role played by heredity and environment. The two general approaches to the study of this relationship, which have traditionally been used with humans, are the foster parent approach and the co-twin control approach. The first of these may be used quite effectively with subhuman primates, but the second cannot be used in any meaningful way because twinning is extremely rare among subhuman primates, and it is not known whether the few twins born are identical or fraternal.

B. EXPERIMENTAL APPROACHES

1. *Same Environment—Different Heredity*

In the foster parent approach with humans, the investigator observes a child who has been transferred from his natural parents to foster parents with a different type of environment, and then evaluates the effect of the environment on the child by determining the similarity between the characteristics of the child and the foster parents and the child and the true parents. If the child appears to be more similar to the foster parents than to the true parents (e.g., in language ability), environment is considered to be the determinant. In order to make this approach efficient, one should transfer the child at a very early age, and both the foster parents and the environment they provide should be considerably different from that which would have been experienced with the true parents. Obviously, it is extremely difficult to maintain these conditions with humans, since modern communication systems, particularly television, tend to level environments. Further, the investigator's ability to shift young children from true parents to foster parents and to provide thereby a complete difference in environment is very limited.

2. *Different Environments—Randomized Heredity*

With subhuman primates, however, it is much easier to set up two entirely different environments and to evaluate the effects of these environments on subjects randomly assigned to each. For example, if a group of very young subhuman primates are randomly assigned to two experimental conditions, one involving a near visual environment, the other a far visual environment, and, after exposure to these environments for a period of time, there is a difference in the refractive characteristics of the two experimental groups, then it is clear that the environment has a sufficient effect. Since most of the other aspects of the two environments can be controlled with subhuman primates, it is possible to determine the effects of differences between the environments upon groups with random hereditary characteristics. If, on the other hand, after a period of time there is no difference between the two experimental groups, one could conclude that the differences between the two visual environments have no effect upon the visual characteristics of the subhuman primate subjects.

3. *Different Environments—Same Hereditary*

It is necessary in any design (such as that described in Section I,B,2) that all other sources of variation be either controlled or determined and measured if one is to draw any conclusions about the relationship between the visual environment and changes in refractive characteristics. While it is not possible to utilize co-twin controls with subhuman primates, it is possible to use them with human subjects. Studies have been made of identical twins who have been separated from their true parents and placed in considerably different environmental situations. If these environmental situations are different enough, if the twins are exposed to them for a sufficient length of time, and if they are effective in changing the characteristics of the twins, such effects should show up over time. If they are not effective in changing the characteristics of the twins, the twins should retain their virtual identity of characteristics. Where such environmental differences have no effect, one could conclude that the environment plays little or no role in developing the characteristics of the subjects whereas, if the different environments have a measurable effect then one could conclude that they do have an effect on developing the characteristics of the subjects.

Obviously, an investigator using a co-twin control study must make certain that there is a true difference in the environments to which the subjects are exposed. Thus, ideally, if one were interested in investigating the effect of a near visual environment on the visual refractive characteristics of identical twins, one member of each set of twins should be exposed to a great deal of near work while the other member of the set should be exposed to virtually no near work. Since it is not possible to learn to read under ordinary conditions without engaging in near work, it should be clear that one twin would be prevented from learning to read while the other twin would be encouraged strongly to learn to read. Under our present social system, such a differential treatment of identical twins is virtually unthinkable. For these reasons, if one is to make use of an experimental approach to the determination of the effects of heredity or environment on the visual refractive characteristics of human and subhuman primates, one is virtually forced to use subhuman primates since only these can be controlled in terms of environmental exposure.

In Section II the visual characteristics of human and subhuman primates are compared in terms of anatomic, optical, and binocular characteristics as well as in terms of which subhuman primates could most effectively be employed for the study of the development of visual refractive characteristics.

II. Comparative Visual Refractive Characteristics of Human and Subhuman Primates

A. DIMENSIONAL CHARACTERISTICS OF THE EYE

1. *Shape of the Eye*

Figure 1 shows a cross-sectional slide of an adult chimpanzee eye placed over a similar slide of an adult human eye, with the two lenses aligned. The anterior chambers and corneal curvatures of the eyes are very similar. The vitreous chambers are similar in shape but not in size, although the two cross-sectional slides are somewhat misleading since the human eye shows its full normal curvature, whereas the chimpanzee eye has lost much of its fullness. In both eyes the retina has been detached (chimpanzee) or partly detached (human) from the choroid and sclera

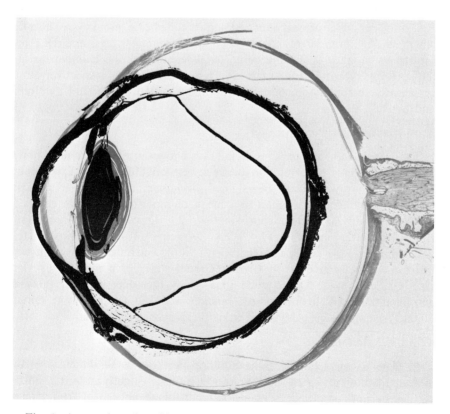

Fig. 1. An overlay of a chimpanzee eye (black) and a human eye (gray). Both eyes 3.75×.

as a result of the fixation and cutting processes involved in making such slides.

The great similarity of shape and basic relationship between the retina (inner), choroid (middle), and sclera (outer) layers of the tunic of the eye is apparent in these slides. All human and subhuman primate eyes are characterized by a very large vitreous chamber, approximately equal depth of the lens and anterior chamber, and distention of the tunic of the eye by the vitreous gel and the aqueous humor in the anterior chamber in front of the lens.

2. Size of the Eye

Studies by Young and Leary (1967) provide the following dimensions for human, chimpanzee, and rhesus monkey eyes; all of these measurements represent means based on five or more eyes. The depth of the anterior chamber is 3.42 mm for humans, 3.89 mm for chimpanzees, and 3.08 mm for rhesus monkeys. The lens thickness is 3.87 mm for humans, 3.62 mm for chimpanzees, and 3.04 mm for rhesus monkeys. The axial length is 25.1 mm for humans, 22.7 mm for chimpanzees, and 18.7 mm for rhesus monkeys. The measured axial lengths for the chimpanzee and human eyes shown in Fig. 1 are 19.26 and 23.10 mm, respectively; thus both eyes are slightly smaller than the mean axial length found by Young and Leary.

3. Refractive Error and Size

The refractive characteristics of the eyes shown in Fig. 1 are not known. However, there is a correlation between refractive error and axial length so that the more myopic or nearsighted the individual, the longer the axial length. Both the mean human and mean chimpanzee refractive errors in the Young and Leary data approximate −2.5 diopters. If the human and chimpanzee eyes shown in Fig. 1 had refractive errors of approximately +2.5 diopters, the obtained axial length would have been consistent with this type of refractive error. Cross-sectional slides of a monkey eye and an infant eye, made at the same time, using the same process, show an axial length of 17.30 mm for the monkey and 15.04 mm for the human infant. Again, the refractive errors of these eyes are not known.

4. Axial Relationships

In general, for adult eyes the anterior–posterior axial dimension from in front of the cornea to the rear of the retina is slightly greater than the vertical or horizontal axial dimension from top to bottom or from side to side of the globe. In more hypermetropic individuals all three dimensions are virtually identical (Deller *et al.*, 1947). This suggests that the hypermetropic eye shows no changes following early anatomic development, probably determined by heredity, whereas the emmetropic and myopic eyes show increased anterior–posterior axial length as a result of the action of some variable(s) other than heredity.

B. BASIC STRUCTURE OF THE EYE

1. *Characteristics of the Globe*

All human and subhuman primate eyes are characterized by three layers of the membrane or tunic: the outermost white sclera; a middle vascular layer, the choroid; and the innermost neurosensory layer, the retina. These layers surround and are distended by the large vitreous gel and the smaller aqueous humor which maintain an intraocular pressure of $\sim$ 17 mmHg. A smooth-muscle system, the ciliary body, controls the shape of a refractive element, the lens, located between the anterior chamber and the vitreous chamber. The eyes of human and subhuman primates are also characterized by a clear cornea which is similar in structure for all primates, but increases in curvature and optical refractive power from humans through the larger apes to the smaller monkeys. The cornea serves as the major refractive element in the primate eye. Immediately in front of the lens is the iris with an opening in the center, the pupil, which serves to control the amount of light entering the eye and to improve the optical quality of the eye under higher levels of illumination by restricting entering light to the center of the lens area.

2. *Gross Retinal Characteristics*

All primate eyes are characterized by a small, approximately circular area which consists only of cone receptor cells which mediate maximum visual acuity or resolving power of the eye. This small area, the fovea centralis, encompasses a visual angle of approximately 1.5° and provides the equivalent of 1:1 projection of the retinal image on the visual cortex of the brain. The fovea contains only the color-receptor cones, and responds to all colors as well as to black and white. Since cones are less sensitive in low levels of illumination than are the rods which share surrounding peripheral parts of the retina, the fovea does not respond at all under levels of illumination which elicit rod responses.

An ophthalmoscopic examination, or fundus camera photograph, of various primate eyes reveals a basic similarity of distribution of the arteries within the eye. Other similarities are the shape, size, and characteristics of the optic disc through which the optic nerve enters along with the retinal arteries and veins, and the size, shape, and location of the fovea, which is seen as a small depression near the center of the retina (Wolin and Massopust, 1967).

3. *Microretinal Characteristics*

The microstructure of the retina of human and subhuman primate eyes has been studied by Polyak (1957), who demonstrated that the eyes of these subjects are extremely similar in distribution of rods and cones: a rod-free fovea and a layered structure of sensitive receptor cells (rods and cones) in the outermost layer of the retina, lying close to the choroid itself. Thus light must pass through all the other layers of the retina before reaching the sensitive elements which face toward the back of the eye rather than toward the lens of the eye as might reasonably be expected. The foveal type of retinal system found in the primates requires that the eye be capable of a wide degree of movement in order to bring visual stimuli onto the foveal area of the retina.

C. BINOCULAR CHARACTERISTICS

1. *Visual Fields*

The forward-placed eyes of the human and the larger subhuman primates provide for virtually complete overlap of the visual fields in the two eyes. Actually, the eye placement and the shape of the nose and head in the subhuman primates allow for an overlap which is more comparable to the Mongolian eye position than to the Caucasian eye position, since in the latter the high nose bridge restricts the nasal visual field. Cowey (1963) has developed a method of perimetry for use with monkeys and other subhuman primates and Cowey and Weiskrantz (1963) carried out a biometric study of visual-field defects in monkeys. They found that the monkey responds in the perimetry situation essentially the same as the human, although they were not able to obtain measurements as far out in the periphery of the eye in monkeys.

2. *Ocular Movements*

The subhuman primates show the characteristic humanoid ocular movements of convergence upon viewing an approaching object, divergence on viewing a receding object, conjunctive movements to either side or in the vertical plane, as well as rotational movements. Both the human and subhuman primates have similar extraocular muscles consisting of the medial, exterior, superior and inferior rectus muscles, and the superior and inferior oblique muscles.

3. *Disjunctive Movements*

When the eyes perform the disjunctive movements of convergence and divergence, accompanying changes occur in accommodation represented by a change in the shape and thickness of the lens of the eye. Under primary conditions when no convergence is exercised and the visual axes of the eyes are essentially parallel, there is a minimal amount of accommodation induced and the eye is adjusted for distance vision. When the eyes are converged upon a near object accommodation is induced, which clarifies the image of the near object on the retina. As the object continues to approach the eye, the amount of both accommodation and convergence must be increased in order to maintain a sharp single image of the near object. Ordinarily, these changes in accommodation and convergence are coordinated, but occasionally it is possible for an individual to maintain accommodation without convergence or convergence without accommodation. The breakdown in the coordination between accommodation and convergence is particularly noticeable under conditions of ocular fatigue, which may be induced by a great amount of reading or other intensive near work.

D. DEMOGRAPHIC CHARACTERISTICS

1. *Availability*

In consideration of the use of various subhuman primates as substitutes for the human in visual studies, the availability of such substitutes must be taken into account. At the present time monkeys, particularly the *Macaca mulatta* or rhesus monkey, are the most widely available (and widely used) of all possible human primate substitutes. *Macaca nemistrina* or pigtail monkey and *M. speciosa* or stumptail monkey are also fairly commonly used as laboratory primates, though less frequently than the rhesus. The larger subhuman primates, such as the gibbon, orangutan, chimpanzee, and gorilla, are much less available than the monkeys, although the eyes of the chimpanzee and gorilla are more similar to those of the human than the eyes of any other subhuman primates.

2. *Size, Cost, and Control Factors*

Although the chimpanzee and gorilla eyes most nearly approximate the human eye, the size and strength of these animals is so great that they can only be used effectively with exceptionally large, strong cages and properly designed control facilities. With the development of Sernylan,* control of all subhuman primates has been greatly simplified, and through the use of dart guns it is possible to anesthetize these larger animals without actual restraint, although most well-equipped laboratories still utilize some form of restraining cage to permit easy anesthetization of these larger primates. Chimpanzees are the most readily available of the larger primates and may be purchased for approximately $700 each. Gorillas are much rarer and correspondingly more expensive, approximating $1500 each. Gibbons and orangutans are not as good visual substitutes as are chimpanzees and gorillas. The orangutan especially is very rare and the cost of such an animal is in the neighborhood of $2000; the gibbon is more available with cost approximating those of the chimpanzees.

Most of the laboratory monkeys can be purchased easily for costs ranging between $50 for stumptail and $85 for the pigtail. Fully adult females and young adult males are about the size of a medium-size dog, weighing around 15–20 lb. Large males weigh up to 35–40 lb. The large males are difficult to handle and do not make practical substitutes for the human in visual research; however, the female and younger males are relatively easy to handle by one person utilizing proper techniques.

III. The Development of Visual Refractive Characteristics

A. NORMAL OCULAR DEVELOPMENT IN HUMAN AND SUBHUMAN PRIMATES

1. *The Human Eye at Birth and in Early Years*

Those investigators, e.g., Deller *et al.* (1947), Sorsby and Sheridan (1960), and van Alphen (1961), who have attempted to measure the axial length of eyes with X-ray or direct measurement, show that the infant eye is approximately three-quarters the size of the adult eye. Infant eyes grow rapidly so that by the end of the second year of life, they have virtually reached the normal adult size in all dimensions, including the curvature of the cornea—except for the anterior–posterior axial length, which increases as the child grows older (Sorsby *et al.*, 1961) and finally stops between 15 and 30 years of age, depending upon the environmental conditions to which the person is subjected (see Section III,A,3). Human and subhuman primate infant eyes generally show a high level of hypermetropia or farsightedness, as much as 5–7 diopters (Mehra *et al.*, 1965). As the eye increases in size, it decreases in farsightedness and moves toward normal-sightedness or emmetropia.

---

* Trade name (Parke, Davis & Co.) for phencyclidine hydrochloride.

2. *The Subhuman Primate Eye at Birth and in Early Years*

The same sequence of development is found in infant monkey and chimpanzee eyes: approximately three-quarters of the size of the adult eye, and hypermetropic refractive errors of up to 8–9 diopters, decreasing as the animal grows older and reaches the age of 9 months to 1 year. If the monkey is put in an especially restricted visual environment, he will pass beyond the state of emmetropia or zero refractive error into a minus refractive error or myopia (Young, 1963b). Again, if the monkey is not subjected to any type of near visual environment, he does not show changes toward myopia (Young, 1965a). Some investigators, e.g., Brown (1938) and Hirsch (1952), have found a slight increase in hypermetropia as children progress from ages 1 and 2 through 7 to 9. This slight increase is followed by a decrease into emmetropia in most cases. Young et al. (1971), in their investigations of the development of refractive characteristics of young monkeys and chimpanzees, have not found this apparent increase in hypermetropia at the comparable human age of 7 to 9. The latter investigators believe that this increase in apparent hypermetropia in intermittently followed subjects is essentially an artifact resulting from the inability to achieve a good level of cycloplegia in very young human and subhuman primate subjects. In general, the younger the subject the more difficult it is to bring about a complete relaxation of accommodation, without which it is impossible to determine the total amount of hypermetropia present since any accommodation reduces the amount of measured hypermetropia. As the subject grows older it becomes easier to achieve a level of cycloplegia, and consequently the investigator tends to find more hypermetropia as the subjects age up to the point where the movement toward emmetropia begins. Thus, it is possible that the reported increase in hypermetropia from 2 to 6 or 7 is a result of an improved level of cycloplegia within this age span.

3. *The Adult Human and Subhuman Primate Eye*

Early studies (Cohn, 1867; Kempf et al., 1928) indicated that the change toward myopia seems to stop around age 16 to 17. Later studies (Young et al., 1954, 1969, 1970) seem to show that the slowing down and stopping of progression toward myopia may not occur until 24 or more years of age. Recent clinical records (Riffenburgh, 1965) and studies (Weitzman et al., 1966) suggest that some individuals (both human and subhuman primates) may not even start to develop myopia until they are 20 or more years of human age (Young, 1961). These studies also suggest that the obtained results are dependent upon the particular environmental conditions which existed at the time. Thus, the early studies performed before the 1930's found relatively little change toward myopia after age 16 because, for practical purposes, near-work stimulation ceased around that age since few individuals finished high school. Studies in the 1950's show that there are myopic changes occurring as late as the college years, and are more prevalent in the high-school-age range through 18 years. Today even more children are going through high school and entering college. Studies of normal males with no myopic refractive errors, placed in atomic submarines, indicate that even these older individuals are capable of developing myopic refractive errors when in a near visual situation for a long period of time (Weitzman *et al.,* 1966).

The studies of Sato (1957) and Otsuka (1967) in Japan also demonstrate a close relationship between the amount of accommodation in near visual work done and the development of myopic refractive errors in high school and college age subjects.

B. REFRACTIVE ERRORS

Astigmatism is essentially a lack of sphericity of the cornea of the eye. Human populations will show as much as 25–40% astigmatism (with the principle axis of corneal curvature in the vertical direction), whereas monkey populations may show as little as 5–15% (Young, 1964).

Hypermetropia is the condition which results when the eye is too short and the optical system brings the image of a distant point source to a focus behind the retina. In young children hypermetropia can be reduced by use of accommodation so that in most cases they can see clearly both at far and near. In some cases the hypermetropic child does not have a sufficient amount of accommodative ability to see clearly at near, in which case he may overconverge in order to boost his accommodative ability and becomes crosseyed or esotropic. As the hypermetropic child grows older, he loses his ability to induce a sufficient amount of accommodation to compensate for his hypermetropia and then he must wear plus lenses in order to see clearly at far and near.

In the myopic condition, the eye is too long, and the optical system brings the image of a point source to a focus in front of the retina. The myopic individual will need minus lenses in order to see clearly at far, although he can usually see clearly at near without any additional lenses.

In emmetropia, the relaxed eye brings the image of the distant point source to a focus on the retina because the eye is essentially the correct length for the optical system.

While no subjective studies have been made on the refractive characteristics of subhuman primates, studies of visual acuity in these primates suggest that they function in much the same way as human subjects and that the monkey or chimpanzee with a myopic refractive error cannot see clearly at a distance but can see clearly at near point (Young, 1970).

C. ABNORMAL DEVELOPMENT OF REFRACTIVE CHARACTERISTICS

1. *The Role of Heredity*

Most young children move away from hypermetropia toward emmetropia and myopia. This process of moving toward emmetropia appears to involve the optical components of the eye and has suggested to Sorsby and his associates (1957) that there must be a process of "emmetropization" working in the eye to bring about the conditions of emmetropia. They suggest that this process is essentially a biologic coordination of the separate optical components each of which are normally distributed so as to bring about a normal distribution of the total refractive characteristics of the eyes of a large human population.

These conclusions are based upon the fact that little or no correlation exists between any of the optical components, such as the curvature of the cornea, the depth of the anterior chamber, the curvature of the front lens surface, the thickness of the lens, the curvature of the rear lens surface, the depth of the vitreous chamber, or the axial length. Generally, the only optical component which shows a high correlation with a refractive error is axial length, as would be expected from the description given in Section III,B for hypermetropia, emmetropia, and myopia.

2. *The Role of Environment*

Some investigators, e.g., Cohn (1867) and Stenstrom (1948), have suggested the possibility that the change from hypermetropia toward emmetropia and myopia results from a change in the axial length of the eye. Donders (1864) and Cohn (1867) have suggested that these changes result from the accommodation and convergence exerted in near work. Studies by Young (1961, 1963b, 1965a) in monkeys have demonstrated that randomly selected animals placed in a near visual environment show changes toward myopia, which is accompanied by an increase in axial length, while equally randomly selected monkeys kept in an open environment show no such changes, either of refractive error or of axial length. Young (1965a), Bedrossian (1966), and Gostin (1963) have demonstrated that monkeys and children who are kept under atropine do not show changes toward increased myopia or increased axial length, even though they are continued in a near visual environment or a reading situation. Thus, these studies suggest that accommodation may play a significant role in the changes of refractive errors which occur among children and young adults. Recent studies by Young and his co-workers (1969, 1970) on Eskimo populations demonstrate that Eskimos who have not learned to read show a hypermetropia of 2 or more diopters on the average, while their children who have learned to read show emmetropia or myopia.

Since both heredity and environment have been implicated in the process of "emmetropization," it is desirable to determine the contribution of each to this process as well as to the development of myopia among both human and subhuman primates.

IV. The Role of Heredity in the Development of Visual Refractive Characteristics

A. NATURE OF THE HEREDITARY CONTRIBUTION

1. *General Considerations*

Much of the difficulty involved in the attempt to separate the contribution of heredity from that of environment results from a failure to understand the nature of the interaction between heredity and environment. Any particular visual characteristic could be evaluated in terms of the relative contribution of three aspects: (1) the actual genetic constitution received as a result of fertilization of the egg, (2) the effect of the intrauterine environment, and (3) the effect of the general environment after birth. Since the neonate must pass through the intrauterine environment before it reaches the outside environment, there is a possibility that visual conditions previously attributed to the effects of heredity may possibly result from the effects of the intrauterine environment, which is not well understood at the present time. For a number of years it was assumed that there was little or no interaction between the conditions of the mother and the conditions of the fetus, but the effects of thalidomide, heroin, and similar substances taken by the mother have been demonstrated to have serious effects upon the conditions of the fetus and may result in major alterations of the visual characteristics of the fetus. Since these are relatively extreme cases, it may even be possible that much less dramatic aspects of the mother's condition have major effects upon the development of the fetus and that therefore, conditions which were thought to be due to heredity may actually be due to the intrauterine environment.

The considerations pertaining to the relationship between heredity and the intrauterine environment also pertain to the relationship between heredity and the outside environment. In order to be able to say that something is due to the effects of heredity and not to the effects of the intrauterine or outside environment, it will be necessary to hold the variation in the intrauterine or outside environment constant while varying the effects of heredity. Clearly, this is extremely difficult to accomplish and yet it must be accomplished before one can make any statements about the contributions of either heredity or environment to the development of visual or other characteristics of the fetus, infant, and child. This position does not deny that the individual grows in all respects over time and will not reach the growth determined by heredity until some time after he is born, but whether he actually achieves the growth determined by heredity will depend upon environmental variables, so that even in this situation there is an interaction between heredity and environment which is difficult to separate.

2. Approaches to Study of Heredity

a. *The Pedigree or Family Tree.* One of the oldest and most commonly used approaches to the study of the effects of heredity on any characteristic is the development of a pedigree or a family tree in which the individuals within the family demonstrating the particular characteristics are plotted. Since this approach basically utilizes measurements which indicate that the characteristic is present or not present, there is no control whatsoever for environmental conditions which might also contribute to the presence or absence of the particular characteristic. Thus, for example, one might demonstrate that one parent has a myopic refractive error while the other parent has a hypermetropic refractive error. These parents have three children; all three children have myopic refractive errors. One might thus conclude that the myopia is genetically dominant and is responsible for the myopic refractive errors in the children. On the other hand, since a considerable amount of evidence has been accumulated in recent times to demonstrate that exposure to reading and other near-work situations has an effect upon the development of myopia, it might only indicate that the parents are encouraging the children to do a lot of reading in order to achieve a better academic standing and success in intellectual functions and that the reading results in the development of myopia. Since no control has been exerted in developing their pedigree, it is impossible to say whether the myopia demonstrated in the children is related to the near-work environment or the effects of heredity.

b. *The Intersibling Correlation Approach.* Karl Pearson (1904) developed an approach to the study of heredity in which one assumes that the measurement and correlation of characteristics between identical twins, fraternal twins, and siblings would show a higher relationship in decreasing order from identical twins to siblings than would be found in the correlation of the same characteristics in randomly paired individuals. In the intersibling correlation approach, instead of using two different measures, such as height and weight, on the same individual and correlating these measures, one uses identical measures, e.g., height only, on pairs of individuals and correlates these measures. Thus, the correlated heights of identical twins yield very high correlations of $+0.90$. Fraternal twins show a much lower correlation, not much different from the 0.50 shown for the siblings, whereas the intercorrelation between randomly selected pairs of individuals will approximate 0.00. On this basis, one could conclude that those characteristics which show correlations of 0.50 or higher represent a strong influence of heredity.

Again, this approach does not take into account the operation of various environmental influences upon the characteristics under study. If it can be demonstrated that environment does have an effect upon the characteristics being studied, it is essential that control be exerted. Thus, if one were intercorrelating height in family groups and some of the families had excellent diets and some had very poor diets, there would tend to be a reasonable correlation between siblings in terms of height, which is partially due to the differences in diet of the groups. In order to use this approach effectively, one must control for the environment in some way, either by equating it or randomizing it across all pairs of individuals.

c. *The Co-twin Control Approach.* The approach to the control of heredity which has traditionally been considered the most effective approach is the co-twin control approach utilizing identical twins. Since identical twins develop from a single zygote, they have identical hereditary characteristics. If these individuals are separated, with one individual placed in one type of environment and the other individual in a different type of environment, we could rightly assume that any differences found in the measured characteristics are related to the differences in environment since the individuals have identical hereditary characteristics. Unfortunately, the co-twin control approach is difficult to achieve since the investigator has little or no influence over the type of environment or exposure presented to these twins. Again, it is important to control the imitative behavior which is so prevalent in identical twins since, if the characteristics under study are influenced by the environment, similarity of obtained results may be due to a similarity of environment rather than to a similarity of heredity. Even in the co-twin control approach it is essential to determine the effects of the environment on these identical twins and to insure that the environment really does differ in order to determine the effects of differences in environment.

d. The Foster Home Approach. It is also possible to take children with different hereditary backgrounds and place them in "the same environment." If the investigator could actually achieve a similarity of environment with individuals with different hereditary constitutions, he could then determine the effects of heredity under a constant environment with respect to the development of particular characteristics. This method offers the greatest promise as far as human substitutes are concerned since it is more nearly possible to achieve identical environments with animals than it is to achieve it with human beings. The major difficulty in this approach is the requirement that the environment to which any particular group is submitted is identical for all members in the group. The further requirement that hereditary characteristics be randomized over the group can be achieved much more effectively.

B. Results Obtained in Studies Designed to Evaluate the Role of Heredity

1. *Theoretical Considerations*

Relatively little work has been done in the attempt to evaluate the role of heredity in the development of visual characteristics in human or subhuman primates. This lack of evaluative studies is dependent primarily upon the pervasive influence of Steiger (1913), who argued for the random distribution of the ocular components within individuals which would result in a normal curve of refractive errors in adult human populations. If the seven optical components discussed in Section III,C,1 are each distributed randomly in the human population, each optical component should fit a normal distribution and the combination of all optical components within a large group of individuals should also fit a normal distribution. The algebraic sum of the optical components randomly received by the given individual would determine the refractive error of that individual. Since these optical components are randomly distributed within the human populations, there should be no intercorrelation between any one of the optical components and refractive errors.

Steiger assumed that the random distribution of individual optical components resulted from the biologic variability which is assumed to occur as a result of heredity, and thus the distribution of refractive errors is entirely dependent upon heredity and has nothing whatsoever to do with environmental effects. Steiger based this reasoning upon a gross exaggeration of the obtained distribution of refractive errors in a large population of adult humans. Since the distribution of refractive errors was not normally distributed, but rather was extremely leptokurtic and skewed, the forcing of this obtained distribution into a normal curve vitiates his reasoning and puts in question his whole concept of the random distribution of optical components leading to a random distribution of refractive errors (Young, 1955a). Studies of wild and laboratory monkeys by Young (1965b) indicate that one can develop any shape distribution of refractive errors desired by manipulating the environment to which subhuman primates are exposed.

2. Intersibling and Parent–Sibling Correlations

Young (1958, 1965a) has attempted to evaluate the effect of heredity on visual refractive errors in both human and subhuman primates using the intersibling correlation and parent–sibling correlation approach. Using human siblings ranging in age between 5 and 18 and correcting for age differences through the use of partial correlation technique, 207 pairs of siblings were correlated on measures of height, weight, interpupillary distance, and refractive errors, while 148 pairs were correlated for IQ. These correlations represent a mixture of male, female, and mixed-sex sibs, with the latter totaling 108 pairs and the like-sex sibs, 99 pairs. The correlations are somewhat reduced as compared with the like-sex sibs; however, the relationships between the correlations are constant for all groups. Thus, for the like-sex sibs, the correlation between interpupillary distance was 0.48; height, 0.43; weight, 0.23; refractive errors of the right eye, 0.15; and IQ, 0.35. These results suggest that there is a higher contribution of heredity to the interpupillary distance than to any of the other measures and, in descending order, height, IQ, weight, and refractive error, the least affected by heredity.

When the intercorrelation approach is applied to parents and offspring, such as was possible with monkey subjects, the same basic results are obtained since the corrected correlation coefficients with age removed by partial correlations were +0.12 for the right eye and +0.11 for the left eye, as compared with the correlation between human siblings of +0.14 for all sibling pairs combined. Both the human and monkey studies suggest that heredity plays a relatively small role in the development of refractive characteristics.

3. Twin Studies

Very few studies of the effects of heredity on development of refractive characteristics utilizing the co-twin control approach have been reported. One of the early studies by de Roetth (1937) indicates that it is possible for identical twins to show as much as a 3-diopter difference in refractive characteristics when subjected to considerably different visual environments. However, he also found that most identical twins show very similar refractive characteristics under similar environments. The rather extensive twin study by Sorsby and his co-workers (1962) unfortunately made no attempt to control the effect of the environment and therefore no statement can be made of the relative contribution of either to the development of refractive characteristics of the eye.

4. Similar Environment—Different Heredity Approach

There were no studies reported on human subjects until very recently by Young and his co-workers (1969, 1970) and Cass (1966). Both these studies indicate that when Eskimo parents are refracted, they show primarily a fairly high level of hypermetropia, approximating +2 diopters as a mean, while their children show a fairly high level of myopia, approximating −2 diopters as a mean among the older children. The primary difference between these two groups is the fact that the parents engage in virtually no reading since they have no written language and never learned to read any language, whereas the children have been taught to speak and read English in regular schools; therefore, the parents have a relatively distant visual environment, while the children have a relatively near visual environment. The obtained differences are so striking that it is virtually inconceivable that the differences could be explained solely on the basis of heredity and must represent a considerable influence of the environment on the development of refractive characteristics of the eyes. In these Eskimo studies, the possibility that a change in diet has occurred, which may affect the development of refractive characteristics, cannot be disregarded.

However, studies carried out by Young using monkeys rule out the possibility of a difference in diet affecting the visual refractive characteristics of these subjects. In these studies (Young, 1961, 1962, 1963b, 1965a) an average of three-quarters of all animals placed in a near visual environment have shown changes in refractive characteristics toward increased myopia. Since both experimental groups and control groups have been maintained on exactly the same monkey chow, the differences between the experimental and control group are most probably due to the effect of restriction of visual space so as to create essentially a near-point visual environment which requires the maintenance of accommodation and convergence in order to be able to see clearly. If the environment were the sole determinant of the changes in refractive characteristics, then all of the monkeys in the near visual situation should have shown the myopic changes. Since approximately one-quarter did not show such changes, there are undoubtedly other variables operating to bring about these differences.

V. The Role of Environment in the Development of Visual Refractive Characteristics

A. THE EFFECTS OF DIET

1. *Human Subjects*

Diet has frequently been cited as a cause of the development of abnormal refractive characteristics (e.g., Cass, 1966). Unfortunately, most of these citations are based upon opinion and not upon controlled studies. Several studies have been performed in an attempt to relate different diet characteristics to refractive characteristics in humans. A series of studies of English children by Gardiner (1956a,b, 1958, 1960; Gardiner and MacDonald, 1957) has shown that severe deficiency in nutrition apparently does have an effect upon development of normal visual refractive characteristics, but normal diet cannot be shown to have any effect. Similar results were obtained by Young (1955b) when he compared the diets of myopic and nonmyopic children: (1) 100 male myopes ate significantly more cake than 230 nonmyopes; (2) 37 males with >1 diopter myopia ate significantly fewer eggs and cookies but more fresh meats other than beef or pork than 49 males with >1 diopter hypermetropia; (3) 106 female myopes ate significantly more butter, oleomargarine, and bread and significantly fewer eggs than did the 197 nonmyopes; (4) 39 female myopes of 1 diopter or more ate significantly less potato than 55 comparable hypermetropes; and (5) both male and female myopes tend to eat fewer eggs than nonmyopes (apparently the most consistent difference in all the comparisons).* This study indicated that there were no consistent differences in the number of times per week that myopes and nonmyopes ate a particular food. Both groups ate a fairly well-balanced diet and the lack of differences suggests that there is no close relationship between diet and myopia, at least insofar as diet can be determined in terms of the frequency of eating various foods. In this particular population, the high socioeconomic level was virtually identical for both the myopes and nonmyopes; consequently, there was no nutritional deficiency of any magnitude in either group of subjects.

2. *Subhuman Primates*

No studies have been carried out systematically varying the diet in subhuman primates to determine the effects of diet on visual refractive

---

* One unexpected finding was that myopes sleep a significantly fewer number of hours per night than the nonmyopes.

characteristics; all such subjects are provided virtually the same diet, although it is far easier to control the diet of subhuman primates than human subjects. In any diet study, of course, it is not possible to say how effectively different subjects utilize the food they eat. Thus there is the real possibility that even though the subjects consume comparable amounts per weight of the same food, there may be a difference in dietary deficiency among individual subjects.

B. The Effects of Diseases

1. *Human Subjects*

Hirsch (1957) made a study of the relationship between different types of disease conditions and the development of visual refractive characteristics. His study demonstrated that measles detrimentally effect the visual refractive characteristics of the neonate. Gardiner and James (1960) found a positive relationship between "toxemia," including hypertension, preeclampsia, and renal disease, in the mother and myopia in the infant. Other conditions, such as venereal diseases, may lead to eye problems but usually do not directly effect the visual refractive characteristics per se.

2. *Subhuman Primates*

Subhuman primates have not been systematically studied in an effort to determine the relationship between disease and visual refractive characteristics. Many of the diseases found in humans apparently do not occur in subhuman primates, and those diseases which do occur, such as tuberculosis, pneumonia, and intestinal infections, appear to be unrelated to the development of visual refractive problems in the offspring, primarily because infected female monkeys usually are disposed of immediately or die rapidly as a result of these conditions. For these reasons, it is impossible to make any statement concerning the relationship between diseases and visual refractive characteristics in subhuman primates.

C. The Effects of a Near Visual Environment

1. *Near-Work Effects in Humans*

A large number of studies reviewed by Baldwin (1964) and Young *et al.* (1954) have considered the relationship between reading and other forms of near visual work and the development of myopia. While the authors of these investigations have frequently concluded that such visual near work leads to the development of a myopic refractive error, the studies were not designed to rule out the influence of heredity, and thus the conclusion that near work leads to the development of myopia becomes a matter of belief rather than a demonstrated experimental result.

2. *Reading, Intelligence, and Myopia*

A common statement made in the older visual refraction literature is that individuals who are nearsighted tend to seek occupations which they can do efficiently and to avoid occupations which are difficult for them. Thus such individuals move into the intellectual occupations and away from nonintellectual occupations. The acceptance of this interpretation leads to some very peculiar results when one considers the relationship between years of schooling, intelligence, and refractive error. Both the level of intelligence and the proportion of individuals showing myopia increase with the level of schooling. Individuals in high school are more intelligent than individuals in grade school; they also have more myopia than individuals in grade school. Individuals in college show a higher average intelligence level than individuals in high school and also show more myopia than individuals in high school, while individuals in graduate school show again a higher general level of intelligence plus a higher percentage of myopia. It would appear that there is a relationship between intelligence and myopia, especially in light of the hypothesis that both myopia and intelligence are determined by heredity. This would mean that those poor individuals who have no myopia are necessarily less intelligent than those individuals who have myopia.

Several studies (Hirsch, 1959; Nadell and Hirsch, 1958; Young, 1955b, 1963a) have investigated the relationship between refractive characteristics and intelligence tests scores and find no basic relationship between these two sets of data. If there is no relationship, how does one explain the finding that both intelligence level and percentage of individuals with myopia increase with years of schooling. Clearly, if years of schooling were responsible for the development of myopia, there would be no problem in handling both an increase in myopia with years of schooling and the lack of relationship between intelligence tests scores and refractive characteristics. While there is no relationship between intelligence tests scores and refractive characteristics, there is a relationship between refractive characteristics and reading ability as well as ability to make grades. In both of these areas, the myopic individual exceeds the nonmyopic individual. It is, of course, a possibility that since the myopic individual spends a great deal more time reading than does the nonmyopic individual in general, this greater amount of reading leads to the development of myopia as well as to better performance in the academic situation.

3. *Near-Work Effects in Subhuman Primates*

Young (1961, 1962, 1963b, 1965a) has studied the relationship between visual refractive characteristics and a restricted near visual space in subhuman primates. These studies have consistently demonstrated that it is possible to create myopia in subhuman primates solely by exposing them to a near visual environment. The earlier the subhuman primate is exposed to the near visual environment, the greater the amount of myopia developed, even though it requires a longer period after exposure to the near visual environment to start the development of myopia than is required in older subhuman primates. Through the use of ophthalmophakometry and ultrasonography, it is possible to demonstrate that the development of myopia in subhuman primates parallels that in human subjects since there is an increase in axial length in both groups and the correlation between the axial length and refractive error is very high in both chimpanzee and human subjects (Young *et al.*, 1966).

The development of myopia in both human and subhuman primates apparently occurs in two steps. The first step is the development of what is usually called a "spasm of accommodation" which results in an inability on the part of the subject to relax the ciliary muscle so that the lens remains consistently accommodated for near point. Once this spasm of accommodation has set in and the lens remains accommodated over a period of weeks or months, it appears to be followed by a second step which is the increase in axial length resulting from the stretching in the posterior part of the primate eye. The exact mechanism which causes the increase in axial length to occur following the development of the spasm of accommodation is not clear at the present time, although Coleman (1970) has shown that there is an increase in the intraocular pressure in the vitreous chamber of the eye when a human accommodates. It is conceivable that if accommodation is maintained over a period of time this increased pressure may also be maintained. The part of the rear of the eye which is least supported is the posterior part, and, thus, the effects of increased pressure is more likely to be demonstrated in the posterior part than in any other part of the eye surrounding the vitreous chamber.

4. Ocular Changes in the Near-Work Situation

When an individual views an object at near point, he must increase his accommodation as well as his convergence. While Coleman (1970) has shown that both accommodation and convergence bring about an increase in pressure in the vitreous chamber of the eye, it is possible to demonstrate that if monkeys (Young, 1965a) or humans (Bedrossian, 1966; Gostin, 1963) are kept under atropine, the changes in refractive characteristics toward myopia do not occur. This is in spite of the fact that the individual may have converged when exposed to a near visual situation but cannot accommodate. Thus, it would appear that accommodation plays a more important role in the process of the development of myopia than does convergence. Further support for the role of accommodation in the development of myopia comes from a study performed by Young (1962) in which monkeys were placed in a near visual environment under different levels of illumination. Under a high level (25 ft-c) and under a very low level (0.25 ft-c) the animals showed less change toward myopia than animals kept under a medium level (4 ft-c) of illumination. Under very low levels of illumination, the animals could not see well enough to accommodate, whereas under higher levels of illumination, the animals had a constriction of the pupil which reduced the amount of accommodation required. Under the medium level of illumination the animals necessarily exercise more accommodation than under either the high or the low levels. These animals developed significantly more myopia than the animals under the high and low levels. Clearly, the animals under the medium and high levels should have exerted approximately the same amount of convergence, although they exerted different amounts of accommodation.

Further, when animals do not develop the spasm of accommodation, they apparently do not develop the increase in axial length. Again, it is not known at the present time why some animals develop a spasm of accommodation and some do not. This difference in the ability to develop a spasm of accommodation could possibly be related to differences in innervation or in the blood supply of the ciliary muscle, which in turn would be related to the hereditary characteristics of the subject.

VI. Conclusions

The subhuman primate has visual characteristics which are very similar to those of the human primate (Young and Farrer, 1972). These basic similarities make the subhuman primate an ideal substitute for the human primate in studies designed to determine the effects of heredity and environment on the development of visual refractive characteristics.

Since the subhuman primate's behavior and environmental exposure can be controlled much more accurately than can the human primate's environmental conditions, it is possible to utilize an experimental approach in which different environmental conditions are provided for groups with randomly selected hereditary characteristics. Thus all individuals in a particular group are exposed to virtually the same environmental conditions, whereas individuals in a different group are exposed to different environmental conditions, but the same for all individuals in the particular group. This approach enables one to literally control heredity through a randomization procedure while systematically varying the environment. Thus it is possible to utilize one of the most efficient and effective techniques for studying the relationships between environment and heredity. If randomly selected groups are placed in different environments and no differences in visual characteristics result from exposure to these different environments, one may conclude that the environments have no effect upon the development of visual refractive characteristics. If, on the other hand, subjects in one environment develop one type of visual refractive characteristic, while subjects in a different environment show no changes or develop a different type of visual refractive characteristic, one may conclude that the environment does have an effect upon the development of visual refractive characteristics. Further, if the environment can create differences in visual refractive characteristics which cover all of the differences found in randomly selected individuals, then it is possible to conclude that the environment plays the dominant role in the development of these visual refractive characteristics.

The studies utilizing subhuman primates to determine the effects of heredity and environment on visual refractive characteristics show quite clearly that the environment plays a major role in the development of emmetropia and myopia in subhuman primates. Because of the essential similarity of the visual characteristics of the human and subhuman primates, one may generalize that the environment also plays a major role in the development of emmetropia and myopia among human primates.

REFERENCES

Baldwin, W. R. (1964). Unpublished Ph.D. Thesis, Indiana University, Bloomington.
Bedrossian, R. (1966). *Proc. Int. Congr. Ophthalmol., 20th,* Int. Congr. Ser. No. 146, pp. 612–617.
Brown, E. V. L. (1938). *Arch. Ophthalmol.* 19, 719–734.
Cass, E. (1966). *Proc. Int. Congr. Ophthalmol., 20th,* Int. Congr. Ser. No. 146, pp. 1041–1053.
Cohn, H. (1867). "Untersüchungen der Augen von 10,060 Schulkindern, nebst Vorschlägen zur Verbesserung der den Augen nachtheiligen Schuleinrightugen." Verlag von Friedrich Fleischer, Leipzig.
Coleman, D. J. (1970). *Amer. J. Ophthalmol.* 69, 1963–1979.
Cowey, A. (1963). *Quart. J. Exp. Psychol.* 15, 81–90.
Cowey, A., and Weiskrantz, L. (1963). *Quart. J. Exp. Psychol.* 15, 91–115.
de Roetth, A. (1937). *Klin. Monatsbl. Augenheilk.* 98, 636–652.
Deller, J. F. P., O'Connor, A. D., and Sorsby, A. (1947). *Proc. Roy. Soc. Ser. B* 134, 456–467.
Donders, F. C. (1864). "On the Anomalies of Accommodation and Refraction of the Eye" (transl. by W. D. Moore). New Sydenham Soc., London.
Gardiner, P. A. (1956a). *Brit. Med. J.* 2, 699–700.
Gardiner, P. A. (1956b). *Trans. Ophthalmol. Soc. U. K.* 76, 171–180.
Gardiner, P. A. (1958). *Lancet* 1, 1152–1155.
Gardiner, P. A. (1960). *Proc. Nutr. Soc.* 19, 96–100.

378   FRANCIS A. YOUNG

Gardiner, P. A., and James, G. (1960). *Brit. J. Ophthalmol.* **44**, 172–178.
Gardiner, P. A., and MacDonald, I. (1957). *Clin. Sci.* **16**, 435–442.
Gostin, S. B. (1963). *Guildcraft* **37**, 5–15.
Hirsch, M. J. (1952). *Amer. J. Optom. Arch. Amer. Acad. Optom.* **29**, 445–459.
Hirsch, M. J. (1957). *Amer. J. Optom. Arch. Amer. Acad. Optom.* **34**, 289–297.
Hirsch, M. J. (1959). *Amer. J. Optom. Arch. Amer. Acad. Optom.* **36**, 12–21.
Kempf, F. A., Collins, S. D., and Jarman, B. L. (1928). "Refractive Errors in the Eyes of Children as Determined by Retinoscopic Examination With a Cycloplegic," Pub. Health Bull. No. 182. U. S. Govt. Printing Office, Washington, D. C.
Mehra, K. S., Khare, B. B., and Vaithilingam, E. (1965). *Brit. J. Ophthalmol.* **49**, 276–277.
Nadell, M. C., and Hirsch, M. J. (1958). *Amer. J. Optom. Arch. Amer. Optom.* **35**, 321–326.
Otsuka, J. (1967). *Acta Soc. Ophthalmol. Jap.* **71**, Suppl., 1–212.
Pearson, K. (1904). *Phil. Trans. Roy. Soc. London, Ser. A* **203**, 53–86.
Polyak, S. (1957). *In* "The Vertebrate Visual System" (H. Kluver, ed.), pp. 207–281. Univ. of Chicago Press, Chicago, Illinois.
Riffenburgh, R. S. (1965). *Amer. J. Ophthalmol.* **59**, 925–926.
Sato, T. (1957). "The Causes and Prevention of Acquired Myopia." Kanehara Shuppan Co. Ltd., Tokyo.
Sorsby, A., and Sheridan, M. (1960). *J. Anat.* **94**, 192–197.
Sorsby, A., Benjamin, B., Davey, J. B., Sheridan, M., and Tanner, J. M. (1957). *Med. Res. Counc. (Gt. Brit.), Spec. Rep. Ser.* SRS-293.
Sorsby, A., Benjamin, M., Sheridan, M., Stone, J., and Leary, G. A. (1961). *Med. Res. Counc. (Gt. Brit.), Spec. Rep. Ser.* SRS-301.
Sorsby, A., Sheridan, M., and Leary, G. A. (1962). *Med. Res. Counc. (Gt. Brit.), Spec. Rep. Ser.* SRS-303.
Steiger, A. (1913). "Die Entstehung der sphärischen Refraktionen des menschlichen Auges." Karger, Basel.
Stenstrom, S. (1948). *Amer. J. Optom. Arch. Amer. Acad. Optom., Monogr.* **58**.
van Alphen, G. W. H. M. (1961). *Ophthalmologica* **142**, Suppl., 1–92.
Weitzman, D. O., Kinney, J. A., and Ryan, A. P. (1966). "A Longitudinal Study of Acuity and Phoria Among Submarines," Rep. No. 481. U. S. Nav. Submarine Med. Center, Groton, Conn.
Wolin, L. R., and Massopust, L. C., Jr. (1967). *J. Anat.* **101**, 693–699.
Young, F. A. (1955a). *Amer. J. Optom. Arch. Amer. Acad. Optom.* **32**, 354–366.
Young, F. A. (1955b). *Amer. J. Optom. Arch. Amer. Acad. Optom.* **32**, 180–191.
Young, F. A. (1958). *Amer. J. Optom. Arch. Amer. Acad. Optom.* **35**, 337–345.
Young, F. A. (1961). *Amer. J. Ophthalmol.* **52**, 799–806.
Young, F. A. (1962). *Amer. J. Optom. Arch. Amer. Acad. Optom.* **39**, 60–67.
Young, F. A. (1963a). *Amer. J. Optom. Arch. Amer. Acad. Optom.* **40**, 257–264.
Young, F. A. (1963b). *Invest. Ophthalmol.* **2**, 571–577.
Young, F. A. (1964). *Exp. Eye Res.* **3**, 230–238.
Young, F. A. (1965a). *Amer. J. Optom. Arch. Amer. Acad. Optom.* **42**, 439–449.
Young, F. A. (1965b). *Eye, Ear, Nose Throat Dig.* **27**, 55–71.
Young, F. A. (1966). *Optom. Weekly* **57**, 44–49.
Young, F. A. (1969). *Proc. 77th Annu. Conv., Amer. Psychol. Ass.* pp. 239–240.
Young, F. A., and Farrer, D. N. (1972). Visual Similarities of Nonhuman and

Human Primates. *In* Goldsmith, E. I. and Moor-Jankowski (eds.) "Medical Primatology 1970," pp. 316–328. Karger, Basel.

Young, F. A., and Leary, G. A. (1967). *Proc. 75th Annu. Conv., Amer. Psychol. Ass.* pp. 89–90.

Young, F. A., Beattie, R. J., Newby, F. J., and Swindal, M. T. (1954). *Amer. J. Optom. Arch. Amer. Acad. Optom.* 31, 111–121 and 192–203.

Young, F. A., Leary, G. A., Baldwin, W. R., West, D. C., Box, R. A., Harris, E., and Johnson, C. (1969). *Amer. J. Optom. Arch. Amer. Acad. Optom.* 46, No. 9, 676–685.

Young, F. A. (1970). *Amer. J. Optom. Arch. Amer. Acad. Optom.* 47, No. 3, 244–249.

Young, F. A., Farrer, D. N., and Leary, G. A. (1971). *Amer. J. Optom. Arch. Amer. Acad. Optom.* 48, No. 5, 407–416.

Young, F. A., Leary, G. A., and Farrer, D. N. (1966). *Amer. J. Optom. Arch. Amer. Acad. Optom.* 43, No. 6, 370–386.

CHAPTER 17

Contribution of Primate Research to Sensory Physiology

HIDEO SAKATA

As is true of most biomedical sciences, the ultimate purpose of sensory physiology is to study human function. Hence neurophysiologists in this field want to elucidate the neuronal mechanisms of human sensation and perception. As far as we are concerned with basic principles of receptor mechanisms, even invertebrate animals offer excellent material for experimentation because of their simplicity and large cell size, e.g., limulus eye (Hartline, 1934) and crustacean stretch receptor (Kuffler, 1954). Species differences become a problem when we try to correlate the neuronal activities of experimental animals with our own subjective experience of sensation. A straightforward analogy of the electrophysiologic findings in lower animals with human sensation is often unsatisfactory or even misleading. This is especially true when we wish to know the functional significance of the morphologic details of human sense organs, to find out some quantitative correlation between sensory events and neuronal activity, or to analyze the mode of neural coding in the central nervous system (CNS). We have to look for animals that have structural design of sense organs and CNS similar to man, that is the primate. The primates are characterized by prehensile

381

hands and well-developed eyes. Le Gros Clark (1959) pointed out these two evolutionary trends as follows: (1) the replacement of sharp, compressed claws by flattened nails associated with the development of highly sensitive tactile pads on the digits; (2) the elaboration and perfection of visual apparatus with development of varying degrees of binocular vision. Indeed, investigations of sensory physiology in primates have made outstanding contributions in these two systems, somatosensory and visual, and much more is expected in the future.

I. Studies in the Somatosensory System

The similarity of the monkey's somatosensory system to the human is twofold. First is the well-known feature of ridged glabrous skin of hands and feet with its characteristic sensory receptors. Second is the remarkable development of the dorsal column lemniscal system with an essential similarity of the postcentral sensory cortex to human cerebrum. The species difference of the somesthetic system is well demonstrated by Fig. 1. This is a schematic outline of body representation in the ventrobasal thalamic complex in rabbit, cat, and monkey (Rose and Mountcastle, 1959). There is a dominance of the trigeminal (facial) representation in the rabbit, and a relative increase of the representation of the limbs in cat and monkey. It seems to be related to the progressive abbreviation of the snout or "olfactory" muzzle (another evolutionary trend of Le Gros Glark, 1959). The relative increase of the representation of hand and fingers in monkey may also be noted in Fig. 1. This is of course related to the density of innervation and receptor population in the periphery. If you look at the picture of the monkey's hand (Fig. 2B), its similarity to the human hand is striking. As a matter of fact the forepaw of the rabbit and the cat (Fig. 2A) are not even called the hand, and its pad with glabrous skin is only vestigial. As the glabrous skin of hand and fingers has a central role in the tactile perception of man, using the monkey for the study of this sensory modality is indispensable. The similarity is not only in its appearance as symbolized

Fig. 1. Schematic outlines of body representation in the ventrobasal thalamic complex in rabbit, cat, and monkey. (Rose and Mountcastle, 1959.)

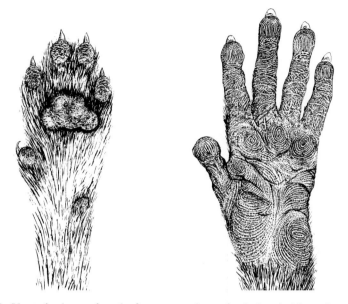

Fig. 2. Ventral views of cat's forepaw and monkey's hand. Note the glabrous skin of cat does not have any ridges like that of monkey.

by fingerprints but also in cutaneous end organs. The glabrous skin of the monkey's hand has three peculiar receptors such as that of man, called "triad"—Meissner corpuscle, Merkel's disc, and Pacinian corpuscle—distributed similar to the human hand. Recent psychophysical experiments with the rhesus monkey demonstrated a very high tactile sensitivity and acuity comparable to man (Mountcastle *et al.*, 1972).

One of the first neurophysiologic studies of the somatosensory system was the mapping of somatotopic representation in the cerebral cortex of the cat by Adrian (1941). A little later Woolsey *et al.* (1942) devised a topographical map of the monkey cortex, which was much more precise than that of the cat. The crucial difference between cat and monkey is the central sulcus separating the somatosensory area (SI) from the motor cortex. This important landmark of cerebral fissuration can be seen only in higher primates. Also the SI of the monkey is a longitudinal strip along the postcentral sulcus, so that the linear dermatomal sequence of somatotopical organization was clearly demonstrated by Woolsey, Marshall, and Bard, as shown in Fig. 3. The same principle of topographical organization was found recently in the second somatosensory area (SII) as shown by the insert diagram in Fig. 3 (Whitsel *et al.*, 1969). A continuous line from the head down to the tail is called "dermatomal trajectory" along the axis of SII. On the other hand, the boundary between motor and sensory areas

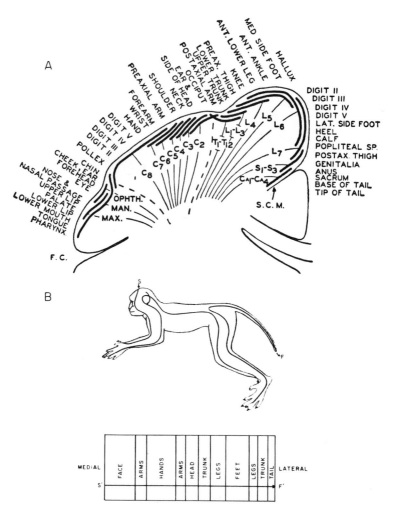

Fig. 3. (A) Schematic frontal section through postcentral gyrus of monkey illustrating somatotopic representation according to dermatomal order. Note that the order of cervical dermatomes is reversed. (Woolsey *et al.*, 1942.) (B) Dermatomal trajectory in SII of monkey. Continuous path on the body surface (S–F) corresponds to the mediolateral array of cortical neurons (S′–F′) in the dorsal view of an unfolded SII (lower portion of the figure). Note the dermatomal order is the same as in SI, though the receptive fields are bilateral in SII.

is a small obscure dimple (postcruciate dimple) in the cat. This seems to be one of the reasons why many neurophysiologists consider the two areas inseparable, and prefer the ambiguous term "sensorimotor" cortex (Terzuolo and Adey, 1960). Moreover, the comparison of the evoked potential study in animals with the stimulation experiments in human subjects (Foerster, 1936; Penfield and Boldrey, 1937) was essential to establish the sensory function of the postcentral gyrus. The sensory projection per se, as demonstrated by the evoked potential method, does not necessarily indicate the sensory function of a given area, but it may sometimes be related to the control of movement and other behaviors. In fact, a strong projection from the peripheral sensory nerve was demonstrated in the motor cortex as well as in SII (Woolsey, 1958), and was considered to be evidence that the "motor" cortex has some sort of sensory function. Although this stimulated a reconsideration of the classical theory of functional localization, recent experiments combining single-unit recording and microstimulation suggested sensory input to the motor cortex might be related mainly to the motor output (Sakata and Miyamoto, 1968; Asanuma *et al.*, 1968). In general sensory function of the cerebral cortex is demonstrated by three different approaches: (1) evoked response to sensory stimuli; (2) sensory effect of the stimulation; and (3) sensory loss by ablation. The

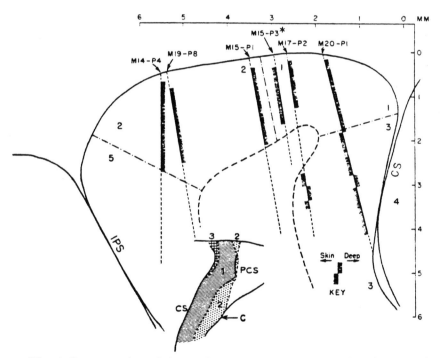

Fig. 4. Reconstruction of the tracks of six electrode penetrations in the sagittal section of postcentral gyrus of monkey at the mediolateral level marked C on the inset. Cross-hatching to the right indicates units driven from deep tissue and that to the left, skin units. (Powell and Mountcastle, 1959.)

latter two are very difficult to achieve in animals, and are possible only with elaborate experimentation involving animal psychophysics. Therefore, the significance of clinical observations in human subjects who provide verbal expression of subjective experiences can never be overemphasized.

The most important application of the microelectrode technique in sensory physiology was single-unit analysis in the central nervous system, which made it possible to study information-processing in CNS. Mountcastle (1957) was one of the pioneers in this field in his study of the cat's SI. He used subtle natural stimulation instead of electrical stimulation, and found that the neurons of the same submodality were clustered in a vertical column. In more extensive studies on monkeys Powell and Mountcastle (1959) constructed a scheme of columnar organization by elaborate histologic procedures showing the various penetrations (see Fig. 4). It is clear in this figure and many other reconstructions that the penetration which was vertical to the cortical surface contained neurons of the same submodality, either deep or skin, but those which were oblique to the surface (for example, the two penetrations in the deeper part of the left side) contained neurons of different submodalities appearing alternately. Moreover, a differential distribution of joint and skin submodalities according to the cytoarchitectural areas was found in monkey experiments (Fig. 5). Namely, the majority of area 3 neurons were skin units, whereas most of the area 2 neurons were joint units. The cytoarchitectural difference between these areas is so delicate that clear-cut demonstration of its functional significance might be possible only in monkey cortex. This positive evidence of Brodmann's theory of functional localization according

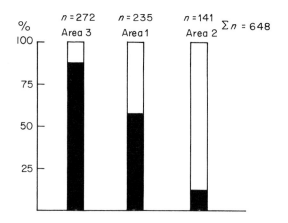

Fig. 5. Bar graph indicates the relative prevalence in each cytoarchitectural area of postcentral gyrus of neurons sensitive to cutaneous stimuli (black bar) and those activated by the movement of joints (white bar). (Mountcastle and Powell, 1959.)

to the cytoarchitectural criteria is very important because, in spite of the fact that the functions of many of Broadmann's subdivisions still remain unknown, it is the essential basis of the correlative investigations of animal and human cerebral cortex.

Mountcastle and his associates have been using monkeys as experimental animals, and we shall describe their recent contributions later in some detail.

II. Studies of the Visual System in Primates

In his famous book "The Vertebrate Visual System," Polyak emphasized the important role of vision in the evolution of primates and man (Polyak, 1957). He argued that the skilled manipulative use of hands, highly developed with bipedalism in human ancestors, was only possible under the control of vision with high acuity and stereopsis. Vision was also considered to have contributed to the development of intelligence by supplying the brain with an immense amount of information. Indeed, Polyak designated vision as "the leading sense in the development of Man from an infrahuman primate," being "a step ahead of hearing and speech during evolution." However, it also means that there is no essential difference in vision between human and higher primates.

There are four main features of primate vision: (1) obligatory binocular stereopsis, in which both eyes are coupled into a conjugate team (focused on a common fixation point); (2) high central acuity made possible by the formation of a specially differentiated central fovea; (3) trichromatic color vision related to the diurnal habit; and (4) a highly developed visual pathway with a six-layered geniculate body and conspicuously laminated striate cortex.

Yet, viewed individually, most of these features are not exceptional for the primates. For example, stereopsis can be observed in the cat and other carnivora, or even in eagles, swallows, and other birds and lizards. A central fovea is also found in various vertebrates with good vision, including reptiles and birds. Color vision is present in teleost fishes, whose retinas were used preferentially for basic experiments. (This might be the reason why monkeys were not used preferentially for the neurophysiologic studies of vision until recently.) The most important feature seems to be in the mode of organization of the whole system, especially the progressive dependence on the cortex. Alternatively, the synaptic organization of the primate retina is simpler than the frog and lower mammals (rabbit, etc.), as demonstrated in recent electron microscopic investigations (Dowling and Boycott, 1966).

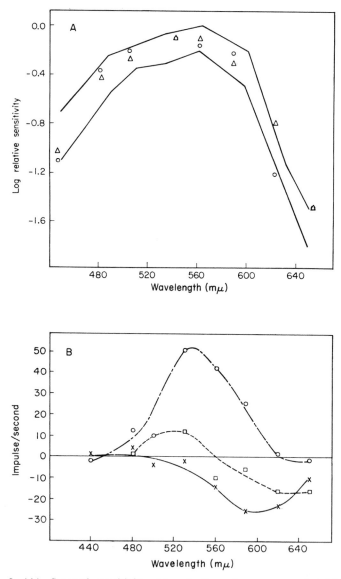

Fig. 6. (A) Spectral sensitivity determinations from averaged inhibitory (△) and excitatory (○) cells in squirrel monkey LGN. The lines delimit the range of the behaviorally determined photopic luminosity curves. (B) Plot of the responses of a spectrally opponent ($+ G - R$) cell to various monochromatic adaptation from different spectral regions. Impulse per second is with respect to spontaneous rate. (×) 510 mμ bleach; (□) no bleach; (○) 680 mμ bleach. (De Valois, 1965.)

One of the earliest electrophysiologic experiments in the visual system was the study of retinotopic representation by S. A. Talbot and Marshall (1941). Since their subjects were mainly cats, a detailed mapping of the primate visual cortex was done only recently (Daniel and Whitteridge, 1961). In fact, anatomic and clinical studies demonstrated the essential feature of the retinotopic representation of man and primates (Holmes, 1945). Likewise, most of the important research in the mammalian visual system has been done with cats and other infraprimate animals (Kuffler, 1953; Hübel and Wiesel, 1959; etc.). The necessity of primate experiments was first recognized in the problem of color vision, as the cat was found to be poor in hue discrimination, having only one type of cone. De Valois (1965) was one of the first to seriously evaluate this subject. He noticed the dangers inherent in cross-species comparison of human psychophysical studies and physiologic data of animal experiments. Therefore, he made some detailed psychophysical studies of color vision in various monkeys. According to him, there are considerable species differences even among the primates, and some of the New World monkeys, such as squirrel monkeys and *Cebus* monkeys are both protanomalous trichromatic, the former having severer color deficiency. However, macaque monkeys have red–green color vision system, identical with normal human observers, being more sensitive than man to wavelengths below 500 mμ (blue).

In the electrophysiologic studies of lateral geniculate neurons, using monochromatic diffuse illumination of the retina, De Valois (1965) found two classes of cells: broad-band cells and spectrally opponent cells. Broad-band cells were further subdivided into excitator and inhibitor. The photopic luminosity curve, as determined behaviorally for the squirrel monkey, and the curves from the averaged broad-band cells agree very well with each other, as shown in Fig. 6A. This indicates that broad-band cells carry information of brightness. On the other hand, spectrally opponent cells give excitatory response to a certain range of wavelength and inhibitory response to the other, as shown in the example of Fig. 6B. In this $+G-R$ cell, selective bleaching with red light eliminated the inhibitory component and greatly enhanced the excitatory response to green light, while bleaching with green light elicited only an inhibitory response.

Subsequently, Wiesel and Hübel (1966) studied the monkey lateral geniculate neurons with a small monochromatic spot and analyzed the spatial distribution of spectral sensitivity in the receptive fields. They found two types of color-coding neurons, one with center–surround arrangement and the other with uniform receptive field. The type I cells were excited with monochromatic light of a certain spectral range applied in the center and inhibited with another range of spectrum in the surround, as shown diagrammatically in Fig. 7 (red on–center, green off–surround). For type II

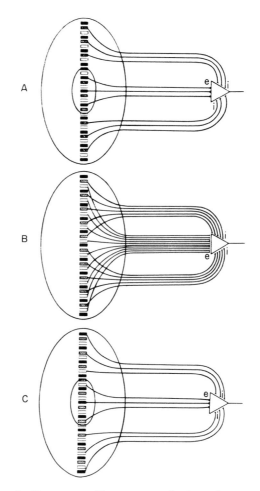

Fig. 7. Schematic diagrams to illustrate contribution of cones to types I, II, and III geniculate neurons. (A) Type I cell: excitatory input (e) from red-sensitive cones in center and inhibitory input (i) from green-sensitive cones in periphery. (B) Type II cell: excitatory input from green-sensitive cones and inhibitory input from blue-sensitive cones all over the receptive field. (C) Type III cell: input from all three types of cones, excitatory from center and inhibitory from periphery. (■) Red; (▨) green; (□) blue. (Redrawn from Wiesel and Hübel, 1966.)

neurons one color was excitatory and the other was inhibitory all over the receptive field. It was assumed that the two different color receptors having antagonistic connection to this type of cell are distributed homogeneously in the receptive field (Fig. 7). Type III neurons of Wiesel and Hübel are the broad-band cells with center–surround arrangement, either on–center or off–center.

These results of the studies of color coding cells in LGN reminds one of Hering's opponent color theory accounting for color contrast, either simultaneous = spatial or successive = temporal, fairly well (De Valois, 1960).

On the other hand, the Young-Helmholtz theory of the trichromatic system was nicely confirmed at the level of sensory cells of the retina by the microphotometric study of cone pigments. Marks *et al.* (1964) applied this method to the primate retina and found three different groups having peaks at 570 mμ (red), 535 mμ (green), and 445 mμ (blue), respectively. These values are exactly the same as human cones, whereas there were some discrepancies with the data of goldfish retina.

The study of color vision has also been initiated in the cortex. Motokawa *et al.* (1962) studied spectral sensitivity of single units in the visual cortex of the macaque monkey and found some opponent color units of narrower range of wavelength than in LGN (Fig. 8). In a recent report Gouras (1971) described trichromatic variety of color-sensitive neurons in monkey striate cortex. For example, cyan–on red–off cell was considered to receive excitatory input from both green and blue cone mechanisms and inhibitory input from red ones. Another excitatory combination was magenta–on (red + blue) and yellow–on (red + green). The trichromatic types are in contrast to the dichromatic (red–on, green–off, etc.) cells found in LGN, indicating a greater degree of color discrimination in the cortex.

Recently, Hübel and Wiesel (1968) extended their investigations of information-processing of vision to the monkey striate cortex. They found that the size of the receptive fields of the monkey near the fovea was much

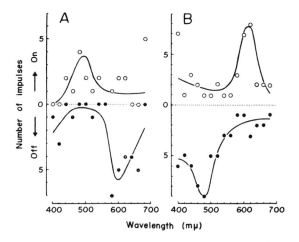

Fig. 8. Spectral curves for two opponent-type neurons of visual cortex of monkey. Open circle indicates on-discharge; filled circle, off-discharge. (A) is the blue-sensitive unit and (B) is the red-sensitive unit. (Motokawa *et al.*, 1962.)

smaller than in the cat, in relation to its high acuity. Another important finding in this study was the functional differences of the cortical layers. By detailed histologic examinations, simple cells were found to be localized in layer IVB, where the radiation fibers terminate, and complex cells were found both in layers III and V where the efferent pyramidal neurons are located (Fig. 9). It is interesting to note that cortical lamination is related to the hierarchy of neurons. A more recent study of monkey visual cortex by Hübel and Wiesel (1970) is concerned with stereoscopic vision. Neurons in area 18 were sensitive to certain specific disparity of the left and right eyes, corresponding to the spatial depth, as shown in Fig. 10. This type of cell was first discovered in the cat's visual cortex by Barlow's group (Barlow *et al.,* 1967) and Bishop's group (Nikara *et al.,* 1968) but their distribution in regard to the cytoarchitectural area was not clear.

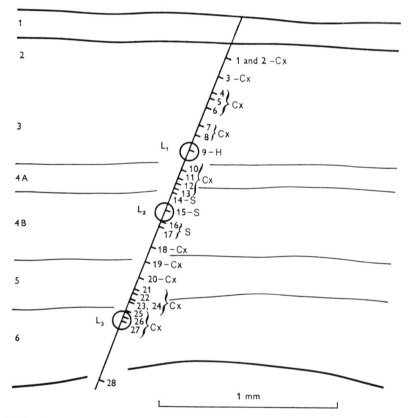

Fig. 9. Simple (S), complex (Cx), and hypercomplex (H) cells recorded in each layer of the striate cortex of monkey. Note simple cells area found only in layer 4B. (Redrawn from Hübel and Wiesel, 1968.)

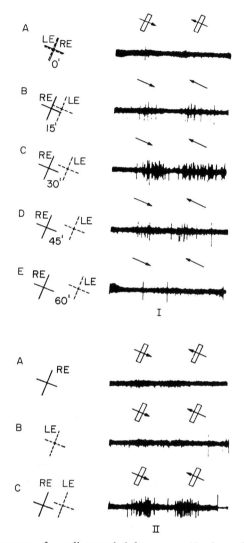

Fig. 10. Responses of a cell recorded from area 18 of monkey. It responded best to a slit in an orientation as shown on the inset. (I) Both eyes stimulated together in anatomically corresponding regions in A. In B–E, position of the stimulus to left eye was horizontally displaced to the right on the screen in steps of 15 min. Maximum responses were obtained with a shift of 30 min. (II) A: Left eye, stimulated alone; B: right eye, stimulated alone; C: both eyes, stimulus to left eye shifted 30 min to the right. (Hübel and Wiesel, 1970.)

Thus area 18 in the monkey may now be considered to have the function of stereovision, which was also demonstrated behaviorally with Julesz's random dot pattern (Bough, 1970).

III. Correlative Studies of Psychophysics and Neurophysiology

Until Adrian initiated the neurophysiologic study of sensation (Adrian, 1926), the history of modern sensory physiology had been the history of experimental psychology (Boring, 1950). The main concern of nineteenth century investigators was the mechanisms of "elementary" sensations and structure–function interrelations of sensory organs. They formulated many important theories of sensation, which offered good working hypotheses for the neurophysiologist, such as Young-Helmholtz' and Hering's theories of color vision mentioned above. However, it often led to endless disputes or erroneous conclusions, as their method of *introspection* could never demonstrate sensory events objectively. Psychophysics established by Fechner (1862) is not an exception.

Fechner's law of logarithmic intensity function was based on the assumption that a *relative* increment in the stimulus energy produces a *constant* increment in the apparent magnitude. This assumption was challenged by Stevens who believed that the *relative* increment of the stimulus should correspond to the *relative* increment of sensation. Hence, the power function was proposed as a general principle, and was demonstrated by the experiment of "subjective magnitude estimation" (Stevens, 1957).

Neurophysiologic evidence to support Stevens' power law was presented by Mountcastle *et al.* (1963) in their single-unit analysis of the ventrobasal complex of the monkey thalamus. They measured the number of impulses of slowly-adapting joint units at varying joint angles. A linear relation in a log-log scale was obtained between the two as shown in Fig. 11, indicating that response R is the power function of the stimulus S; i.e., $R = KS^n$, where R is the impulse number at given angle θ minus spontaneous discharge, S is θ minus threshold angle θ_T, K is a proportionality constant, and n is an exponent. Much precaution was taken in this study (1) to use the unanesthetized animal in which conscious sensory experiences were expected to be going on; (2) to precisely control the stimulus with a joint-rotator machine; and (3) to eliminate the effect of variability of central cell response by data-processing with the aid of computer. However, this was only to confirm the general principle of intensity function. In fact no psychophysical experiment of the estimation of joint angle had been performed.

In order to get more direct correlation between the neuronal activity

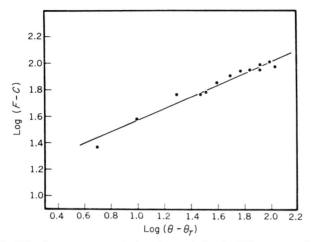

Fig. 11. Stimulus–response relation in a thalamic VB neuron driven by knee extension. Logarithm of impulse per second is plotted against logarithm of joint angle in degree. $F = 13.9 \, (\theta \cdot \theta)^{0.429} + 24$; $r = 0.979$. F, steady-state discharge at given angle θ; C, spontaneous discharge; and θ_T, threshold angle. (Mountcastle *et al.*, 1963.)

and the sensory experience it is necessary to combine the neurophysiologic and psychophysical experiments together in a similar situation. The main advantage of combining the two types of observation is as follows (W. H. Talbot *et al.*, 1968). Electrophysiologic studies can now provide precise measures of the neural coding from periphery to cerebral cortex. But they have so far provided little understanding of those cerebral mechanisms which may lead to subjective sensory experiences. On the other hand, psychophysical studies analyze those experiences and provide quantitative measures of the sensory performance of the intact, behaving organism. These two methods are complementary for the common aim of understanding sensory mechanisms. Mountcastle's group undertook a series of experiments on the glabrous skin of the hand, using the same mechanical stimulator for monkeys (neurophysiologic) and humans (psychophysical). The striking similarity of the hands of the two species (see Fig. 2) favored the assumption that "what monkeys and humans feel with their hands is in principle the same."

The intensity function of the steady light pressure was compared in two sets of experiments. One was the recording of single large myelinated fibers of the median nerve of the monkey innervating the glabrous skin of the hand (Mountcastle *et al.*, 1966). These fibers are readily divisible into two classes: One is quick-adapting and the other is slow-adapting. The latter respond to the steady indentation of the skin with continuous discharge. The stimulus–response relation of the slow-adapting units were

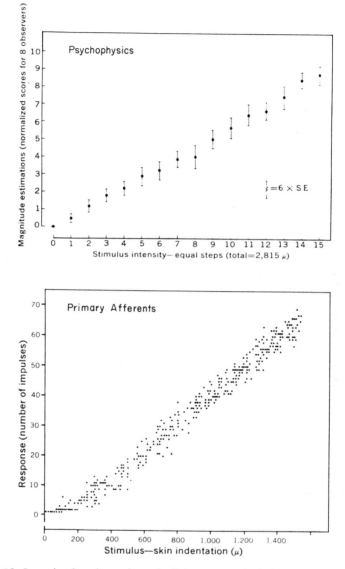

Fig. 12. Intensity functions of steady light pressure both in human psychophysics and afferent nerve of monkey, obtained with the same mechanical stimulator. Upper graph is the plotting of magnitude estimation in human subjects. Lower graph is the plot of number of impulses in single units of median nerve innervating monkey's hand. (Mountcastle *et al.*, 1966.)

linear as shown in Fig. 12B. On the other hand, the subjective magnitude estimation of the amplitude of the same pulse identation applied to the human hand was also linear to the stimulus intensity (Fig. 12A). This is good evidence that the sense of steady pressure is mediated by the slow-adapting unit. Moreover, it seems to support a general hypothesis that "the central nervous system follows in *linear* fidelity whatever image of the world is reflected to it over peripheral receptors" (Mountcastle, 1967). The physiologic property of this type of unit is quite similar to the Iggo corpuscle which was found as an isolated "touch spot" in the hairy skin of the cat and monkey (Iggo, 1963).

A morphologic study of this touch corpuscle showed that it is innervated by a single myelinated axon which branches profusely under the epithelial layer, and their terminals are encapsulated with a specific type of cell called Merkel's cell (Iggo and Muir, 1963). The whole arrangement is quite similar to the Merkel's disc of glabrous skin. It is generally believed that the latter is also a slowly-adapting mechanoreceptor. This was confirmed by a recent study of Munger *et al.* (1971) in the raccoon. Consequently, the Meissner corpuscle is now considered to be the quick-adapting skin receptor. The Pacinian corpuscle, the last of the glabrous skin triad, was found to be sensitive to the vibration of high frequency in isolated

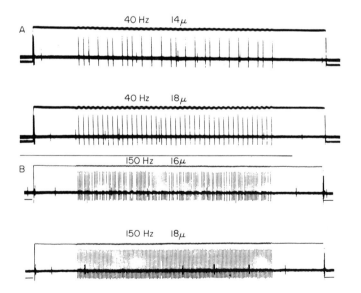

Fig. 13. Responses to sine wave oscillation in two types of quick-adapting units. (A) Glabrous skin unit (Meissner type). (B) Pacinian-type unit. For each set the upper trace is the movement of stimulus probe, and lower trace is nerve impulse. (W. H. Talbot *et al.*, 1968.)

preparation (Sato, 1961). In a recent investigation of vibratory sensibility W. H. Talbot *et al.* (1968) examined the response of median nerve units of monkey to sine wave oscillation, and found two distinct types. One was a glabrous skin quick-adapting unit (Meissner corpuscle) responding optimally at 30–40 Hz, and the other was a subcutaneous unit exquisitely sensitive at 200–300 Hz, which was considered to be Pacinian (Fig. 13). At the same time they studied the human threshold in the same area of the hand, using exactly the same pattern of stimulus. The threshold curve (what they call "tuning curve") of the cutaneous unit was parallel to the human threshold function in low-frequency range below 40 Hz (Fig. 14), while that of the Pacinian unit was parallel to the latter in high-frequency range above 60 Hz (Fig. 15). This evidence suggests that vibratory sensa-

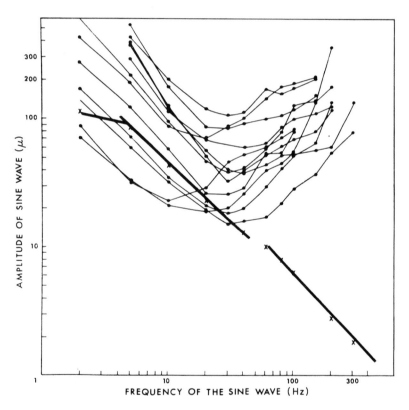

Fig. 14. Tuning curves of glabrous skin quick-adapting units of monkeys (light lines) superimposed on a human threshold curve measured in fingertip (thick line). Each point indicates the sine wave amplitude required to elicit 1 impulse/cycle. (W. H. Talbot *et al.*, 1968.)

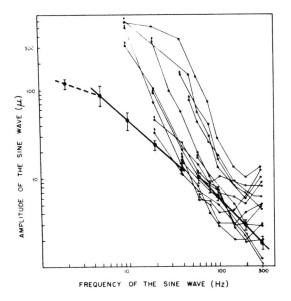

Fig. 15. Tuning curves for Pacinian units. (W. H. Talbot *et al.*, 1968.)

tion is mediated by two different sets of receptors, which are categorized as "flutter" (in the low-frequency range) and "vibration" (in the high-frequency range).

Subsequently, Mountcastle *et al.* (1969) studied the response of cortical neurons in SI to vibratory stimulus in unanesthetized monkeys. The majority of cortical neurons were found to be receiving afferent input from one particular type of receptor. An example of response pattern and data analysis of the cortical neurons of the cutaneous quick-adapting type is shown in Fig. 16. Periodicity of discharge in relation to the sine wave frequency was fairly well preserved in the cortical cells. The main difference from the periphery was the greater variability in the temporal pattern of impulses. On the other hand, the human capacity of frequency discrimination is considered to be related to the variability of "pitch" sensation. A recently developed application of statistical decision theory to psychophysics (Green and Swets, 1966) has made it possible to estimate the variability of "sensory events" independently from the motivational factor in a decision-making process. Therefore a modified method of "yes–no" procedure was performed in the task of discriminating two different frequencies in human subjects. The probability of correct "yes" (hit) was plotted against that of incorrect "yes" (false alarm) in Fig. 17. From this "receiver operation characteristic" (ROC) curve a standard deviation of

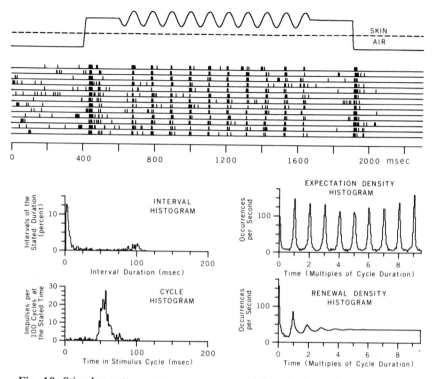

Fig. 16. Stimulus pattern, nerve responses, and four basic methods of analysis of SI neurons sensitive to sine wave oscillation. Oscillation was superimposed on a steady indentation after 200 msec delay for 1 sec. Note transient discharges at onset and end of step indentation. For the histogram analysis data were collected only during actual sine wave, excluding onset and offset transients. (Mountcastle *et al.*, 1969.)

underlying sensory events was calculated (Fig. 17). This value was approximately the same as the variability of cortical neuronal events measured in the cycle histogram (Fig. 16B). This direct comparison of cortical neuronal events with sensory events seems to be a step forward in the understanding of cerebral mechanisms underlying the sensory experience. More recently, Mountcastle and associates initiated the investigation combining the psychophysical and neurophysiologic experiments in the same animal (trained monkeys), using the chronic single-unit recording technique (Evarts, 1965; Mountcastle, 1972). This seems to be the most promising type of experimental setup without the danger of cross-species comparison. The monkey's intelligence is of a great help in the complicated situation of psychophysics.

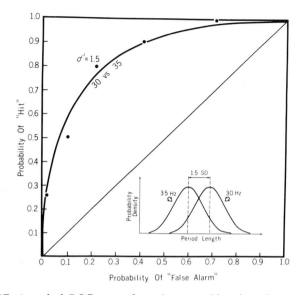

Fig. 17. A typical ROC curve for a human subject in a frequency-discrimination task. Standard frequency, 30 Hz; comparison frequency, 35 Hz. Inset: two probability distributions of the sensory events calculated from the curve, indicating the variation in pitch.

IV. Studies of Sensory Perception in Primates

Recent advances of the pattern recognition theory in the information sciences have shed new light on the problem of perception (Sayre, 1966). In this context neurophysiologic studies in the visual system by Hübel and Wiesel, Barlow's group, and P. O. Bishop's group have made a great contribution to the understanding of neural mechanisms of visual pattern perception, especially in regard to the process of feature extraction. However, for complete pattern recognition leading to the identification of objects, many more steps, including decision-making processes, seem to be necessary. In fact our present knowledge of the cerebral mechanisms of perception depends mainly on the clinical studies of agnosia and neuropsychological investigations in primates. The function of perception and recognition has been ascribed to the "association areas" in the parietal, temporal, and preoccipital cortex (Chow and Hutt, 1953).

The main difficulty for the neurophysiologist who wants to study association areas is that they are "silent" in anesthetized animals and no response is elicited with sensory stimuli. This can be overcome by the technique of single-unit recording in unanesthetized animals, as mentioned earlier. Another problem is the difficulty in finding the "best" stimulus for the neurons in these areas, probably because of its complexity. Now that we have a much better understanding of sensory information-processing up to the primary sensory and adjacent areas, it might be a good time to start neurophysiologic investigations in association areas. A recent single-unit study of monkey inferotemporal cortex by Gross et al. (1969) is one of the few examples of this type of investigation. Since Klüver and Bucy (1938) described the symptom of "psychic blindness" in bilateral temporal lobe ablation, many neuropsychological investigations have shown that the inferotemporal cortex is related to visual-discrimination learning (Mishkin and Pribram, 1954; etc.). The main findings by Gross et al. were that receptive fields of inferotemporal units were all extremely large and included the fovea, and they had highly specific response properties. One of the interesting units which responded to dark rectangles "responded much more strongly to a cut-out of a monkey hand."

In the somatosensory system also, a process of "feature extraction" was confirmed by the finding of directionally selective cutaneous units in SI of unanesthetized monkeys (Mountcastle et al., 1969; Werner and Whitsel, 1970; Schwarz and Fredrickson, 1971). Subsequently, single-unit analysis of the parietal association area was initiated in similar preparations (Sakata, 1972; Duffy and Burchfiel, 1971). Neurons in area 5 were found to be activated only with a complicated pattern of stimulus, such as certain combinations of multiple joints, or the combination of joint and skin stimulation. Figure 18 shows an example of joint and skin unit in area 5 (Sakata, 1972). The "best" stimulus for this unit was the combination of elbow flexion and the skin rubbing (trace 1, as shown in the inset diagram). The elbow flexion alone elicited only weak transient response (trace 2), and the cutaneous stimulus alone was almost ineffective (trace 3). The specific convergence of joint and cutaneous input seems to be related to the recognition of tactile objects by palpation, which was considered to be the function of the parietal association area (Denny-Brown and Chambers, 1958).

As these examples indicate, monkeys are uniquely suitable for the study of sensory information-processing at higher levels. This may be attributed to the large number of neurons in primates; as Barlow (1969) pointed out "the number of neurons seems to increase at higher levels," while "sensory messages are recoded into a less redundant form at relays in the sensory pathways, leading to economy of impulses." Moreover, the hierarchical organization of the corticosensory system seems to be much clearer in monkeys than infraprimate mammals, as demonstrated in a

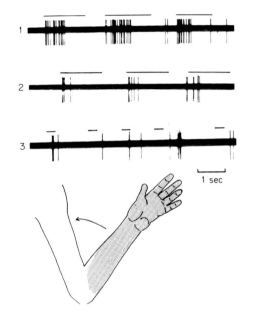

Fig. 18. An example of neurons in area 5 of monkey driven by a specific combination of joint and skin stimulations. Trace 1 is the response to combination of elbow flexion and skin rubbing; Trace 2, elbow flexion alone; Trace 3, skin rubbing alone. The inset indicates the cutaneous receptive fields and direction of joint movement. (Sakata, 1972.)

recent anatomic study of corticocortical connection (Jones and Powell, 1970). Indeed, the elaboration of association areas is the main factor of expansion of the cerebral cortex in higher primates. In conclusion, research in higher primates is likely to become indispensable for future investigations of the neural mechanisms of sensory perception.

REFERENCES

Adrian, E. D. (1926). *J. Physiol. (London)* **61**, 49–72.
Adrian, E. D. (1941). *J. Physiol. (London)* **100**, 159–191.
Asanuma, H., Stoney, S. D., Jr., and Abzug, C. (1968). *J. Neurophysiol.* **31**, 670–681.
Barlow, H. B. (1969). *In* "Information Processing in the Nervous System" (K. N. Leibovic, ed.), pp. 209–230. Springer-Verlag, Berlin and New York.
Barlow, H. B., Blakemore, C., and Pettigrew, J. D. (1967). *J. Physiol. (London)* **193**, 327–342.
Boring, E. G. (1950). "A History of Experimental Psychology," 2nd ed. Appleton, New York.

Bough, E. W. (1970). *Nature (London)* **225**, 42–44.

Chow, K. L., and Hutt, P. J. (1953). *Brain* **76**, 625–677.

Daniel, P. M., and Whitteridge, D. (1961). *J. Physiol. (London)* **159**, 203.

Denny-Brown, D., and Chambers, R. A. (1958). *Res. Publ., Ass. Res. Nerv. Ment. Dis.* **36**, 35–117.

De Valois, R. L. (1960). *J. Gen. Physiol.* **43**, Part 2, 115–128.

De Valois, R. L. (1965). *In* "Contributions to Sensory Physiology" (W. D. Neff, ed.), Vol. 1, pp. 137–178. Academic Press, New York.

Dowling, J. E., and Boycott, B. B. (1966). *Proc. Roy. Soc., Ser. B* **166**, 80–111.

Duffy, F. H., and Burchfiel, J. L. (1971). *Science* **172**, 273–275.

Evarts, E. V. (1965). *J. Neurophysiol.* **28**, 152–171.

Fechner, G. T. (1862). Elemente der Psychophysik. Breitkopf and Härtel, Leipzig.

Foerster, O. (1936). *Brain* **59**, 135–159.

Gouras, P. (1971). *Proc. Int. Union Physiol. Sci.* **8**, 46–47.

Green, D. M., and Swets, J. A. (1966). "Signal Detection Theory and Psychophysics." Wiley, New York.

Gross, C. G., Bender, D. B., and Rocha-Midanda, C. E. (1969). *Science* **116**, 1303–1306.

Hartline, H. K. (1934). *J. Cell. Comp. Physiol.* **5**, 229–247.

Holmes, G. (1945). *Proc. Roy. Soc., Ser. B* **132**, 349–361.

Hübel, D. H., and Wiesel, T. N. (1959). *J. Physiol. (London)* **148**, 574–591.

Hübel, D. H., and Wiesel, T. N. (1968). *J. Physiol. (London)* **195**, 215–248.

Hübel, D. H., and Wiesel, T. N. (1970). *Nature (London)* **225**, 41–42.

Iggo, A. (1963). *Acta Neuroveg.* **24**, 175–180.

Iggo, A., and Muir, A. R. (1963). *J. Anat.* **97**, 151.

Jones, E. G., and Powell, T. P. S. (1970). *Brain* **93**, 793–821.

Klüver, H., and Bucy, P. C. (1938). *J. Psychol.* **5**, 33–54.

Kuffler, S. W. (1953). *J. Neurophysiol.* **16**, 37–68.

Kuffler, S. W. (1954). *J. Neurophysiol.* **17**, 558–574.

Le Gros Clark, W. F. (1959). "The Antecedents of Man." Edinburgh Univ. Press, Edinburgh.

Marks, W. B., Dobelle, W. H., and MacNichol, Jr., E. F. (1964). *Science* **143**, 1181–1183.

Mishkin, M., and Pribram, K. H. (1954). *J. Comp. Physiol.* **47**, 14–20.

Motokawa, K., Taira, N., and Okuda, J. (1962). *Tonoku J. Exp. Med.* **78**, 320–337.

Mountcastle, V. B. (1957). *J. Neurophysiol.* **20**, 408–434.

Mountcastle, V. B. (1967). *In* "The Neurosciences" (G. C. Quarton, T. Melnechuk, and F. O. Schmitt, eds.), pp. 393–408. Rockefeller Univ. Press, New York.

Mountcastle, V. B., LaMotte, R. H., and Carli, G. (1972). *J. Neurophysiol.*, **35**, 122–136.

Mountcastle, V. B., Poggio, G. F., and Werner, G. (1963). *J. Neurophysiol.* **62**, 807–834.

Mountcastle, V. B., Talbot, W. H., and Kornhuber, H. H. (1966). *Touch, Heat Pain, Ciba Found. Symp. 1965* pp. 325–344.

Mountcastle, V. B., Talbot, W. H., Sakata, H., and Hyvarinen, J. (1969). *J. Neurophysiol.* **32**, 452–484.

Munger, B. L., Pubols, L. M., and Pubols, B. H. (1971). *Brain Res.* **29**, 47–62.

Nikara, T., Bishop, P. O., and Pettigrew, J. D. (1968). *Exp. Brain Res.* **6**, 353–372.

Penfield, W. G., and Boldrey, E. (1937). *Brain* **60**, 389–443.

Polyak, S. (1957). "The Vertebrate Visual System." Univ. of Chicago Press, Chicago.

Powell, T. P. S., and Mountcastle, V. B. (1959). *Bull. Johns Hopkins Hosp.* **105**, 133–162.

Rose, J. E., and Mountcastle, V. B. (1959). *In* "Handbook of Physiology" (Amer.

Physiol. Soc., J. Field, ed.), Sect. 1, Vol. I, pp. 387–430. Williams & Wilkins, Baltimore, Maryland.

Sakata, H. (1972). *In* "Somatosensory System" (H. H. Kornhuber, ed.). Thieme, Stuttgart (in publication).

Sakata, H., and Miyamoto, J. (1968). *Jap. J. Physiol.* **18**, 489–507.

Sato, M. (1961). *J. Physiol. (London)* **159**, 391–409.

Sayre, K. M. (1966). "Pattern Recognition, A Study in the Philosophy of Artificial Intelligence." Univ. of Notre Dame Press, Notre Dame, Indiana.

Schwarz, D. W. F., and Fredrickson, J. M. (1971). *Brain Res.* **27**, 397–401.

Stevens, S. S. (1957). *Psychol. Rev.* **64**, 153–181.

Talbot, S. A., and Marshall, W. H. (1941). *Amer. J. Ophthalmol.* **24**, 1255–1263.

Talbot, W. H., Darian-Smith, I., Kornhuber, H. H., and Mountcastle, V. B. (1968). *J. Neurophysiol.* **31**, 301–334.

Terzuolo, C. A., and Adey, W. R. (1960). *In* "Handbook of Physiology" (Amer. Physiol. Soc., J. Field, ed.), Sect. 1, Vol. II, 797–835. Williams & Wilkins, Baltimore, Maryland.

Werner, G., and Whitsel, B. L. (1970). *IEEE Trans. Man-Machine Syst.* **11**, 36–38.

Whitsel, B. L., Petrucelli, L. M., and Werner, G. (1969). *J. Neurophysiol.* **32**, 170–183.

Wiesel, T. N., and Hübel, D. H. (1966). *J. Neurophysiol.* **29**, 1115–1156.

Woolsey, C. N. (1958). *In* "Biological and Biochemical Basis of Behavior" (H. F. Harlow and C. N. Woolsey, eds.), pp. 63–81. Univ. of Wisconsin Press, Madison.

Woolsey, C. N., Marshall, W. H., and Bard, P. (1942). *Bull. Johns Hopkins Hosp.* **70**, 399–441.

CHAPTER 18

Performance Studies in Biomedical Research

DONALD N. FARRER

Applied research efforts designed to answer specific questions have a long historical record of recombining traditional academic disciplines. The current direction of applied biomedical research is not an exception to this rule. The wide variety of questions asked about "performance" of men has led to combining such disciplines as psychology, physiology, toxicology, and pathology for the answers. If the questions of job performance include parametric queries such as, "How much environmental change can be tolerated?", then health hazards demand utilization of nonhuman experimental subjects with extrapolation of the data to human beings. Studies that have a high probability of lethality must be explored with techniques different from the assessment of a subtle decrement effect.

Performance capabilities should not be assessed until the basic parameters of the independent variables are known: First, and usually most economically, determinations of the average lethal "dose" (LD_{50}) must be made. Second, this measurement should be followed with determinations of the variability of that mean value. Quartile values, LD_{75} and LD_{25}, as well as standard deviations are helpful preliminary calculations. Third, fiducial limits can be calculated at selected levels of type I and type II risks. All these preliminary experiments obviously can be conducted on untrained animals of research quality at minimal cost, and the results provide valuable data for the approximation of exposure levels calculated to produce sublethal behavioral perturbations. Because research costs increase rapidly with attempts to identify precisely the magnitude and duration of the effects on performance, it is important to make every effort to extract a maximum amount of information from these preliminary experiments. Hence, there is value in careful observations of the behavior of the subjects during the lethal-dose determinations. Utilization of independent observers with fixed rating categories has been helpful (e.g., recording time of onset and identification of first clinical signs; sensory involvement; motor involvement; alterations in activity patterns: excitation, depression, aggression, etc.; and incapacitation). These observations may lead to working hypotheses about the nature of the incapacitation as well as influencing the decisions regarding the selection of the appropriate performance tests.

Most commonly, performance tests are designed to assess specific functions, such as vision, audition, locomotion, or reaction time, as resulting from environmental insults. Methodology required to assess performance decrements involves a significant increase in time and effort. While the screening (LD determinations) studies require research grade and acclimated primates, the background training time required to obtain asymptotic performance often consumes several months before the subjects can be exposed to an environment likely to produce performance changes. This preexperiment training time is dependent largely upon the degree of complexity of the primate behavior that is required to answer the questions about the environment. (For example, changes in reaction time due to a tranquilizer requires less sophisticated techniques than measuring changes in functional vision with discrimination between 20/15 and 20/20 targets.) Each environmental hazard requires specific designs rather than "off-the-shelf" standardized tests of performance decrements. Due to the expense and requisite training time, comprehensive tests of all sense modalities and reactions are unwarranted unless the environmental hazard is so novel and poorly understood pharmacologically that intelligent guesses cannot be made about the sensory or motor impairment likely to result from exposure.

Based on equipment designed by Koestler (1965), Jeter and Hurst (1970) employed an omnibus test by training primates to perform a sequence of tasks designed to measure: (1) reaction time to visual stimuli; (2) reaction time to auditory stimuli; (3) visual discrimination; (4) continuous work activity; and (5) appetite. These tasks were combined to provide varying workloads over an 8-hr workday. Activity and concentration were divided into three levels: high, medium, and low. Peak workload consisted of high activity and high concentration, and was defined as follows: Simultaneous tasks were presented to the subject so that continuous lever-pressing was required on one lever while push-button reaction time responses were required at 2-sec intervals to both auditory and visual cues. Responding "correctly" resulted in successfully avoiding a punishing shock. Medium concentration was defined as a visual-discrimination task which required viewing a Landolt ring. The Landolt ring targets were presented in three sizes representing viewing angles of 20/400, 20/200, and 20/100 snellen, with the gap oriented up, down, left, or right. The orientation of the gap determined the "correct" lever. (Responding by pressing the "correct" lever resulted in a 0.3-gm food reward.) Medium activity was defined as the continuous lever-pressing task without any reaction-time tasks superimposed upon this basic work task. Low concentration and low activity were defined as a rest period interrupted occasionally with brief warning signals clearly signaling a simple reaction-time task. Following the simple reaction-time task the rest period resumed. Recycling high-, medium-, and low-activity tasks allowed sampling performance during a workday under "normal" circumstances for baseline purposes.

This technique was used to measure work decrements resulting from exposure to several toxic agents of unknown chemical composition. Additionally, the procedure was used to study behavioral decrement effects of staphycococcal enterotoxin B. This compound is frequently associated with food-poisoning, and is not thought to act on the CNS directly. However, work decrements, even short-term incapacitation, is expected because of the discomfort involved. As expected, performance on tasks requiring high activity and high concentration was significantly impaired, whereas performance on low concentration and low activity was not impaired as long as the motivation was kept very high. That it to say, shock-avoidance motivation was consistently aversive and decrements were produced with high workloads, and none were found with light workloads. Unfortunately, food-acquisition motivation was reduced with this compound as would be expected with any gastrointestinal disturbance.

Motivation is a serious consideration in any performance-decrement study due to "side effects." Shock-avoidance motivation has been used in many performance-decrement studies such as the above-mentioned study because of its stability. However, it should be noted that shock-avoidance motivation may mask a decrement effect. First, there is the reinforcement density issue. Asymptotically trained primates routinely perform at or near 100% correct levels. In terms of shock stimulation this efficiency percentage means that the animal may receive only one shock for each 200 or 300 trials during the baseline test phases. However, if a drug is administered which produces a drowsy reaction a slight performance decrement would occur causing the efficiency to drop, which means that the animal would be punished with shock much more frequently than the baseline rates. This significant increment in CNS stimulation can offset the drowsy reaction. Thus, the effect of the drug on performance was minimized due to the shift in reinforcement density. Conservative interpretation is obviously warranted in such situations. Second, shock-avoidance motivation is capable of eliciting emotional behaviors as an additional side effect. Tasks which require fine discrimination frequently are compromised by the urgency inherent in avoidance training. Threshold studies have effectively avoided these problems by utilizing new experimental methodologies which use the emotional response to advantage. Estes and Skinner (1941) described the conditioned emotional response (CER) as a learned response to any signal of imminent delivery of unavoidable shock. Basically this empirically derived notion is the description of highly reliable responses elicited after the animal learns to associate the signal with the shock. For example, a tone presented for 15 sec followed by a 0.5-sec shock soon elicits emotional behavior during the 15 sec prior to the delivery of the shock. Typically, overt behavior is temporarily suspended, and the animal is immobilized until the shock has been delivered. More recent investigators introduced an additional component to this response, and the result is described as conditioned suppression: First, the animal is trained to perform a simple task such as a variable interval food-reinforcement schedule for lever-pressing behavior. Such training procedures result in highly predictable and consistent lever-pressing behavior. Animals trained in this manner press the lever at relatively rapid rates for the food rewards which occur at the various time intervals. Second, after stable, asymptotic behavior is achieved on this simple task the CER procedure is superimposed on the task. That is to say, while the animal is working on the food-acquisition task, the tone is presented for 15 sec and the 0.5-sec shock is delivered as the tone terminates. Within a very few trials, the lever-pressing behavior terminates as soon as the tone is presented. No lever presses are made during this interval, and the lever-pressing behavior continues following the delivery of the unavoidable shock at the original rate. Once the conditioned suppression has been established, then the threshold study can be initiated. For example, the auditory threshold for the tone could be easily established by presenting a series of tones varying in intensity using a sequence determined by one of the established psychophysical techniques, and measuring the number of responses which occur during the 15-sec interval of the tone. Many investigators who use the conditioned-suppression technique analyze their data with a statistical treatment called the suppression ratio to quantify the degree of suppression as suggested by Hoffman et al. (1963). The ratio is computed as follows:

$$\text{Suppression ratio} = \frac{T_1 - T_2}{T_1}$$

where T_1 is the number of responses during the 15-sec period immediately preceding the tone-exposure trials, and T_2 is the number of responses during the 15-sec period of exposure to the tone. Thus, complete suppression results in a ratio of 1.00 and if responding is equal during T_1 and T_2 the ratio is 0.00. If responding should increase during T_2 a negative number would result. The conditioned-suppression method has been successfully employed in a variety of discrimination studies. As an example, Taylor *et al.* (1967) used this method to demonstrate that rhesus monkeys can detect the onset of X-ray stimulation. Detection of X-rays was evident in four monkeys after 20 trials in which X-rays and unavoidable shock were paired. The dose rate was reduced to 0.03 R/sec for the 15-sec exposure and a high level of response suppression was evident for all four subjects during the presence of the X-rays. The suppression ratios ranged from 0.70 to 0.90 during the experimental trials. These investigators used control trials in which all X-ray apparatus was operated, but the X-ray tube was shielded and directed away from the monkeys to ascertain that the suppression was not due to some artifact as an auditory cue of the apparatus. Control trials suppression ratios ranged from 0.11 to 0.26, and the authors conclude that monkeys can detect X-rays by using this conditioned-suppression technique. Smith (1970) has summarized the many psychophysical applications of this behavioral technique.

Recent technological advances and mushrooming applications for high-energy density lasers required assessment of the eye-damage potentials of these new systems. Ophthalmoscopically observable lesions had been noted, and threshold studies of minimum energy required to produce an observable lesion had been accomplished, but the significant functional question remained: i.e., is there a decrement in functional vision at energy densities too low to produce an ophthalmoscopically observable lesion. A series of experiments were executed to determine functional vision changes as a result of laser exposure. Due to the eye-hazard potential for permanent decrements in vision rhesus monkeys were selected as the research subject of choice. Farrer and Graham (1967) developed a monocular and binocular technique for determining subjective visual acuity in the rhesus monkey. Basically, the monkeys were taught to discriminate between a series of Landolt rings. A Landolt ring was projected onto a backlit projection screen in front of the subject, and the opportunity to press one of the four levers was provided. If the animal could discriminate the position of the break in the ring (up, down, left, or right) and press the lever associated with that position, a food reward was delivered. By varying the subtended visual angle of the Landolt ring a threshold for discrimination was obtained. The following Snellen fraction size targets were used: 20/200; 20/100; 20/50; 20/40; 20/30; 20/20; 20/10. Monocular acuity was obtained by occluding vision in the contralateral eye for one viewing condition. Seven size targets, four orientations of the break in the ring, and 10 trials for each of the above conditions resulted in 280 trials for a threshold determination for one viewing condition (i.e., OD, OS, and OU). Thus, 840 trials were required for the subjective visual acuity of each rhesus monkey.

The large number of trials required for statistical reliability required too much time to assess immediate or rapid decrements in vision. Thus, the procedure was revised (Graham et al., 1970) to incorporate the von Békésy (1947), Blough (1958), and Scheckel (1965) self-adjustment procedures. Basically, this method allows the subject to constantly adjust the size of the stimulus around the threshold level. This method has been demonstrated to be an efficient psychophysical procedure for scotopic and photopic sensitivity (Blough and Schrier, 1963; Schrier and Blough, 1966) and critical flicker frequencies (Symmes, 1962) in rhesus monkeys. For the study of visual decrements the positive-reward motivational procedures used by the above investigations were thought to be less effective than the shock-avoidance situation. The subjects were trained to press a lever on the right side of the performance panel when the Landolt ring gap was oriented to the right, and to press the left lever when the orientation of the gap was to the left. The orientation of the gap was randomized, and correct lever presses resulted in a tone and incorrect lever presses resulted in a shock. Correct responses automatically reduced the stimulus size of the Landolt ring, while incorrect responses increased the stimulus size. (Maximum target was 20/200 and minimum target was 20/10.) For the assessment of functional vision under circumstances where threshold acuity is not required (e.g., interest may center around 20/20 average reading ability), then additional stability can be introduced while punishing shock density is reduced by limiting the automatic stepper to the 20/20 targets. Preliminary experiments have been conducted with a variable shock intensity to reduce the punishment for errors in discrimination among small visual angle targets. The self-adjusting procedure with shock-avoidance motivation has proven effective in the assessment of visual functioning immediately following exposures to a variety of lasers and high-intensity light sources. Large laser-induced lesions in the macula produce blindness (less than 20/200 visual acuity) as first shown by Yarczower et al. (1966). Experiments in our laboratory with smaller foveal lesions which produce destruction of some foveal receptor cells have resulted in altering the 20/15 baseline visual acuity to 20/50 immediately after the exposure. Three-week follow-up testing of these subjects provided some evidence for partial recovery of functional vision (to 20/30 threshold). Both ophthalmoscopy and histopathology examinations provide ample evidence for destruction in photoreceptor cells of the fovea, but a sufficient number of cells remain intact to account for the performance at the 20/30 levels. Based on an extensive series of studies of high-energy density thermal eye damage, we have concluded that visual acuity does not change unless an ophthalmoscopically visible lesion is also present. However, the converse does not hold: that is to say, some very small ophthalmoscopically observable foveal lesions do not alter baseline visual acuity. It is assumed that the retina is sufficiently redundant to resolve 20/15 targets with some loss of receptor cells.

Killeen (1970) developed a deceptively simple performance task for the assessment of drug-induced changes in the behavior of rhesus monkeys. His limited interval avoidance procedure involved training the subject as follows: a cue light signaled the initiation of a 2-min trial in which the only requirement is a lever press during the 0.5 sec at the end of the 2-min time period to avoid shock. Obviously, time perception of such precision is not possible, and the rhesus monkey maximizes the probability of a response in the critical time period by developing a high rate of bar presses after approximately 1 min. Thus, typical behavior by a primate trained on this task includes no responses during the first minute of the trial followed by rapid responding for the second minute. Killeen has shown that this task is altered by barbiturates in the following ways: (1) the bar-pressing is not initiated at the baseline 1-min mark, but is delayed an additional fraction of the remaining 1 min; (2) the bar-pressing rate is reduced; and (3) more shocks are received. Additionally, he has data that amphetamines affect behavior in the opposite direction (e.g., bar-pressing is initiated earlier, few shocks received, and higher response rates result). Thus, this simple bar-pressing task can be used with unknown drugs to compare to known CNS-depressant and/or stimulant effects.

In summary, performance-decrement testing adds a dimension to classical physiological and medical determinants of environmental contaminants. Operant behavioral methodology applied to discrimination tasks yield subtle changes in performance which might remain unnoticed with usual examination procedures. These procedures yield data compatible with quantitative analysis which allow rather accurate determinations of the magnitude of the performance decrement (from lethal to barely detectable), as well as the duration (from permanent irreversible effects to brief transient effects) of the behavioral change. Automated equipment required to test experimental subjects without exposing humans (experimenters, assistants, and handlers) adds to the costs inherent in this methodology, and it is difficult to find substitutes for the long training time required for some discrimination tasks. New developments in behavioral methods are being reported in the literature which help offset the required leadtime. The new methods which capitalize on behaviors which previously inhibited research are encouraging (for example, Pavlov described "experimental neurosis" as emotional behavior which prevented the dogs from making fine discriminations between circles and ellipses, whereas the conditioned-suppression model uses this "experimental neurosis" to advantage).

REFERENCES

Blough, D. S. (1958). *J. Exp. Anal. Behav.* **1**, 31–43.
Blough, D. S., and Schrier, A. M. (1963). *Science* **139**, 493–494.
Estes, W. K., and Skinner, B. F. (1941). *J. Exp. Psychol.* **29**, 370–400.
Farrer, D. N., and Graham, E. S. (1967). *Vision Res.* **7**, 743–747.
Graham, E. S., Farrer, D. N., Crook, G. H., and Garcia, P. V. (1970). *Behav. Res. Methods Instrum.* **2**, No. 6.
Hoffman, H. S., Fleshler, M., and Jensen, P. (1963). *J. Exp. Anal. Behav.* **6**, 575–583.
Jeter, R. D., and Hurst, C. M. (1970). "Effects of Staphylogeal Enterotoxic B on Complex Operant Behavior in Monkeys," Tech. Rep. No. ARL-TR-70-14. 6571st Aeromed. Res. Lab.
Killeen, P. (1970). Personal communication.
Koestler, A. G. (1965). Personal communication.
Scheckel, C. I. (1965). *J. Comp. Physiol. Psychol.* **59**, 415–418.
Schrier, A. M., and Blough, D. S. (1966). *J. Comp. Physiol. Psychol.* **62**, 457–458.
Smith, J. C. (1970). Conditioned Suppression As An Animal Psychophysical Technique. *In* Stebbins, W. C. (ed.) "Animal Psychophysics." Appleton, New York.
Symmes, O. (1962). *Science* **136**, 714–715.
Taylor, H. L., Smith, J. C., and Hatfield, A. C. (1967). "Immediate Behavioral Detection of X-rays by the Rhesus Monkey," Tech. Rep. No. ARL-TR-67-70. 6571st Aeromed. Res. Lab.
von Békésy, G. (1947). *Acta Oto-Laryngol.* **35**, 411–422.
Yarczower, M., Wolbarsht, M. L., Galloway, W. D., Fligsten, K. E., and Malcolm, R. (1966). *Science* **152**, 1392–1393.

CHAPTER 19

The Importance of Nonhuman Primate Studies of Learning and Related Phenomena For Understanding Human Cognitive Development*

DUANE M. RUMBAUGH

It is now generally recognized that man is a primate, and member of the order *Primates*. The implications of the biological relationship between man and other primate forms are far-reaching, but not necessarily obvious to all. The use of nonhuman primates as animal models for the study of certain disease processes has been the greatest single reason for the justification of primate research. Other reasons, however, have become recognized in very recent years. We now acknowledge that many of man's behavioral propensities can be better understood by including his nonhuman primate relatives into study programs than by studying man alone. Man's inclination toward and control of aggression, his highly complex social behaviors, his growth and development, and even his basic psychological processes of learning, forgetting, perception, and transfer skills also can be better understood through study of their homologs as manifested by the various nonhuman primate forms.

* This paper was supported by an NIH grant (RR-0165).

Le Gros Clark (1959) has argued that extant primate forms can be so ordered as to increasingly *approximate* man. Though such a gradation is not to be taken in any manner as evidence for linear evolution, it is the case that as we review the series of prosimians, New World and Old World monkeys, lesser apes, and great apes, we can see that the *approximation* of man becomes increasingly refined. Few can deny, for example, that the great apes (Pongidae), particularly the chimpanzee (*Pan*) and gorilla (*Gorilla*), are more suggestive of man than are the lemurs (*Lemur*), by comparison relatively primitive primates.

One of the prime correlates of this graduated series of primates is cortical complexity (Connolly, 1950). Among the primates, cortical development is at a maximum in man; next to man the great apes have the most highly developed brains and cortices. Next in this line are the lesser apes, the gibbons and siamangs (Hylobatidae), followed by the Old World and New World monkeys and the prosimians (Prosimiae). This array of *natural* preparations can prove of great value to us as we set about understanding the cognitive development of the normal child and problems of psychological function which characterize certain forms of human retardation. In comparison to man, nonhuman primates provide an array of "normal" retardates. As we learn how to enhance their cognitive processes through radical and, perhaps, high-risk methodologies, we should be better prepared for enhancing similar processes for the human child.

I. Learning-Set Skills

If learning skills for tasks that are relatively complex are provided for by the primate cortex, we should find increased capacity for complex learning as we study the graduated series of nonhuman primates from prosimian to great ape as discussed above. Within the order Primates the evidence is in basic accord with this supposition. Indeed, considering the tree shrew (*Tupaia*) in this regard, though not universally accepted as a true primate, Leonard *et al.* (1966) reported that *T. glis* failed to learn-how-to-learn and to develop learning sets (Harlow, 1949) even after 800 six-trial object-discrimination problems. Though only a relatively small number of primate genera have been studied at all intensively in learning-set training situations, it is the case that, with one or two marked exceptions, learning-set skills of primates increase as cortical complexity increases.

This author (Rumbaugh, 1970) has reported that great apes performed better than other primate forms when tested with a modified discrimination-reversal training method designed to ensure equitable assessment of capacities for complex learning across species. That method, which yields a transfer index (TI) value per animal, brings all subjects to a predetermined level of task mastery, for example, 67 and 84% responses correct, on each of a series of visual discrimination problems prior to cue reversal. Performance on reversal trials 2–10 for a block of 10 problems is used to calculate the TI, percentage choices correct on the reversal trials divided by the prereversal criterional mastery level. Great apes achieved the highest values and were clearly superior to gibbons, one of the lesser ape forms. They were also clearly superior to talapoin (*Miopithecus*) specimens.

Figure 1 portrays a widely cited comparison of the learning-set skills of the rhesus monkey (*Macaca mulatta*) with those of the squirrel monkeys (*Saimiri sciureus*) and marmosets (*Callithrix jacchus*). Figure 2 portrays in summary TI data reported by Rumbaugh (1970). These two figures combined can be taken as qualified support for the generalization that *at least within the order Primates* there appears to be a relationship between cortical complexity and complex learning skills.

Information of this kind underscores the importance of studying learn-

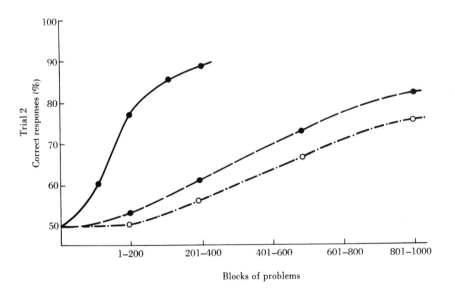

Fig. 1. Learning-set curves for rhesus monkeys, squirrel monkeys, and marmosets. (——) Rhesus monkey; (— —) squirrel monkey; (— ·) marmoset. (After Miles, 1957.)

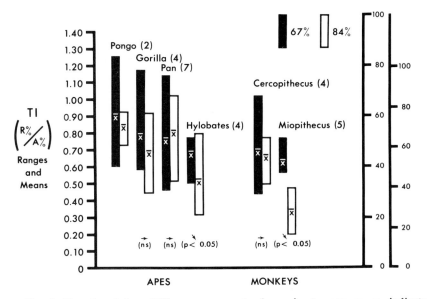

Fig. 2. Transfer index (TI) measurements for primate groups as indicated. Collectively the great apes were superior to the gibbons. Both the gibbons and talapoins (*Hylobates* and *Miopithecus*) manifested significant drops in TI performance as the prereversal criterional level was increased from 67 to 84%. *R*% Correct where *A* = 67% and 84%. (See Rumbaugh, 1970, for a detailed discussion.)

ing in *normal* nonhuman primate genera selected for their characteristic cortical development. As the brains of all nonhuman primates bear close resemblance in terms of basic organization to man's brain, such studies seem particularly warranted. Ontogenetic studies of learning-set skills within and across the genera should help us understand the operation of learning processes at levels which, when compared to man, are primitive yet nonetheless normal. Also, if by intensive study we are able to determine which variables of early experience and training control the development of nonhuman primate complex learning skills, we should be better prepared for understanding the development of learning capacities within the normal and brain-damaged child. There is no doubt that these kinds of skills are fundamental to the cognitive processes of the human child and adult (Levinson and Reese, 1967).

Interestingly, the learning-set paradigm was formulated by Harlow in work first with animals, not humans. It was through the improvement of monkey subjects in discrimination-learning as a function of encounter with a long series of problems, each presented for too few trials to permit mastery, that Harlow revealed how experience provided for one-trial insightful learning of a kind possibly related to that observed by Köhler (1925) with apes.

As reviewed by Robinson and Robinson (1970) there have been a number of studies demonstrating the relationship between rate of learning set formation and mental age (MA). In some instances, retarded children have been even inferior to nonhuman primates in their ability to develop learning sets. This group of studies has also demonstrated that non-institutionalized retardates do better in the establishment of learning sets than do matched institutionalized subjects and that moderately retarded subjects seem particularly sensitive to failure, as it reflects adversely upon learning-set performance. This finding is of interest for it parallels an analogous one reported by Davenport et al. (1969), that chimpanzees reared for the first several months of life in restricted environments are less facile than regular laboratory-reared ones in the initial phases of a variety of learning tasks.

In a comprehensive review of learning-set studies with children, Reese (1962) concludes that, while a simple stimulus–response learning theory might be adequate for infrahuman learning-set data, it is inadequate with data from human subjects. The difficulty is not that children acquire learning sets much more rapidly than monkeys but rather that the within-problem learning is more complex in humans than monkeys. Too, once there is criterional mastery of one problem, immediate and complete formation of learning set is achieved in children but not in monkeys. In other words, children, particularly normal ones, have the capacity to acquire a learning set in a single problem, with subsequent improvement perhaps being the formation of a performance set rather than additional learning set. Finally, there is a considerable difference in the kinds of errors made by monkeys and children. Children are more inclined toward persistence in errors, such as position preference, position alternation, and stimulus alternation, than are monkeys, which, in turn, are relatively more inclined toward errors of stimulus perseveration, i.e., responding to a stimulus regardless of its positive or negative cue value (Kintz et al., 1969; Ellis et al., 1963).

As the capacity for learning set varies as a function of MA and as a function of chronological age and the cortical complexity of nonhuman primates, learning-set performances can be used to great advantage as a dependent measure to discern whether certain extreme manipulations, as in early experience or with various drugs, might produce supranormal cognitive capacities. By reason of their learning-set skills, nonhuman primates should prove to be extraordinarily valuable as process models, if not animal models, for better understanding the factors which affect the development of human cognition.

II. Transfer Skills

It is apparent that in comparison to the human adult, the child and retardate have performance deficits, both in learning and in failure to transfer what has been learned from one situation to another. Failures to discriminate the relevant dimensions of a problem (Zeaman and House, 1963), failure to relinquish responses that are no longer appropriately rewarded (Razran, 1933), and failure to use social behaviors in accordance with the exigencies of a given context can be costly to the interests of adjustment and adaptation.

In the previous section the thesis was supported that various nonhuman primate genera differ naturally in their cognitive capacities that provide for the formation of learning sets and related tasks. In this section we will consider how they also differ with regard to their abilities to appropriately transfer learning from one situation to another and in their readiness to relinquish behaviors that are no longer rewarded.

In my experience the discrimination-reversal task is particularly valuable in research efforts designed to discern species differences in learning. This task usually consists of a visual discrimination problem in which the subject must choose between two objects simultaneously presented on each of a series of trials. To some point in the trial series one stimulus is reliably correct (choice food rewarded) and the other reliably incorrect (choice not rewarded), but thereafter the cue values are reversed for the remaining trials, e.g., correct becomes incorrect and vice versa. The task requires both facile learning of the initial discrimination and a readiness to relinquish choice of the initially correct stimulus once the cues are reversed for high rates of reinforcement to be obtained. Performances on the trials subsequent to cue reversal are of particular interest for they can be taken as evidence of the subject's ability to transfer what has been learned on the initial trials in conjunction with his ability to learn the changed cue values. Appropriate transfer can allow for perfect performance on all reversal trials except the first one where, for the first time, choice of the initially correct stimulus nets no reward. In studies of this type, it is common to give a large number of problems to obtain reliable measurements of interest.

Not all species, let alone individuals within a given species, can execute reversals with equal facility. Further, even though matched for prereversal performance accuracy (percentage responses correct and/or trials to reach criterion) many species differ in performance on the reversal trials. This observation is of particular import for it suggests that certain species differences, such as cortical development, and age influence the capacity of nonhuman primates to transfer learning from one context to another. Accordingly, transfer processes among nonhuman primate taxa can be studied to assist us in understanding transfer-of-training processes and problems as they are manifested in the preverbal and/or mentally retarded human child.

Rumbaugh and Arnold (1971) compared the discrimination-reversal skills of *Cercopithecus* and *Lemur*. In a totally automated test situation, the animals were presented a green circle and a white ×. For 51 trials, choice of one of these stimuli was reliably rewarded, whereas choice of the other was not. At the end of each 51-trial unit, the cue values of the stimuli were reversed so that the stimulus which had been correct became incorrect (not rewarded) and vice versa. On a continuing basis for a total of 200, 51-trial units the cues were similarly reversed.

In accordance with the considerations of the first section of this paper, it was predicted that the *Cercopithecus* monkeys, because of their superior cortical development, would be superior to the *Lemur*. This prediction was borne out in that an analysis of the trials immediately preceding the reversal indicated that the *Cercopithecus* monkeys were, on the average, performing more accurately than were the *Lemur* subjects. More significant was the observation that on the trials immediately following cue reversal the *Cercopithecus* monkeys did better than the *Lemur* subjects. Finally, when matched for performance accuracy immediately prior to cue reversal, for example, 80–85% correct, the *Cercopithecus* monkeys did better (Fig. 3). Though equated for prereversal achievement, they reversed with unequal facility. The *Cercopithecus* monkeys were superior to the *Lemur* in their ability both to learn and to transfer what they had learned.

Nonhuman primates improve in their transfer-of-training skills as a function of experience. Two chimpanzees were trained to achieve criterional mastery of 67% on each of 60 two-choice object-quality discrimination problems of the kind commonly used in learning-set studies (Rumbaugh, 1971a). Upon achievement of the criterion, the cue values of the objects which comprise each problem were reversed and 10 additional trials were given. Transfer-index values were determined for each of six 10-problem blocks as portrayed in Fig. 4. As the average numbers of trials required to reach the criterion held essentially constant across these six blocks (22, 25, 23, 22, 19, and 23) the interblock improvement is believed to be enhancement of transfer-of-training skills. In other words, the chimpanzees became better and better at transferring an operationally defined constant amount of information (67% correct) from the prereversal to the reversal trials of each problem.

Since the human retardate is handicapped in his ability to appropriately

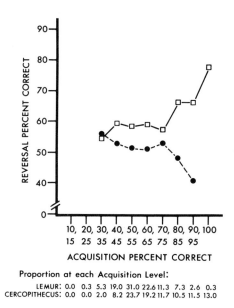

Proportion at each Acquisition Level:
LEMUR: 0.0   0.3   5.3 19.0 31.0 22.6 11.3 7.3 2.6 0.3
CERCOPITHECUS: 0.0   0.0   2.0   8.2 23.7 19.2 11.7 10.5 11.5 13.0

Fig. 3. Reversal performance plotted as a function of prereversal performance levels for four *Cercopithecus* (□) and three *Lemur* (●) subjects. (After Rumbaugh and Arnold, 1971.)

transfer learning from one situation to another and since improvement in transfer as a function of practice can be systematically studied with non-human primates, primates in their great variety might afford us a unique opportunity for better understanding the fundamentals of transfer-of-

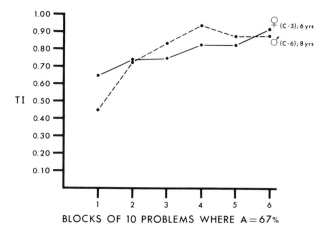

Fig. 4. Improvement in transfer index (TI) performance as a function of practice for two chimpanzees. (After Rumbaugh, 1971a.)

training skills even as they homologously operate at the level of the human retardate and normal human child.

Whereas normal and retarded groups of humans equated for MA performed comparably in tests of simple-stimulus generalization (Barnett, 1958), normal subjects are more likely to use learned cues in the generalization of responses in complex situations (Robinson and Robinson, 1970). Comparable data at the nonhuman primate level are lacking, but it seems likely that the various primate genera might differ in their generalization and transfer processes according to cortical development and the complexity of the test situation. If so, nonhuman primate study should facilitate research in this area.

III. Extinction

A response is said to be extinguished if it is relinquished when no longer reinforced. On the assumption that there might be a cognitive component in the extinction of previously reinforced choices, the aforementioned *Cercopithecus* and *Lemur* subjects were studied comparatively to determine whether the *Cercopithecus* monkeys would extinguish responses more rapidly than would the *Lemur*. If cognitive skills facilitate discrimination and learning, they might also serve to facilitate discrimination of the conditions under which reinforcement is no longer to be obtained.

The animals were first trained on a two-choice discrimination, then given extinction sessions on the following 2 days (Arnold and Rumbaugh, 1971). The observations were in complete accord with the prediction. The *Cercopithecus* monkeys relinquished the choice behaviors more rapidly than did the *Lemur*. Even though the overall pattern of extinction was the same for these two genera, the *Cercopithecus* relinquished the now-unrewarded behaviors much more rapidly, an adjustment aligned with the contingencies of the situation. It was also observed that the *Cercopithecus* monkeys were more inclined to shift from the stimulus which had been initially rewarded than were the *Lemur*, a finding in accord with the previously noted superior discrimination-reversal skills of the *Cercopithecus*. The study supported the contention that differences in cognitive skills that allow for different facilities in learning can also determine the rates with which nonreinforced behaviors are relinquished.

Once again we can see the possibility that the varied forms of nonhuman primates, viewed as natural preparations of varying levels of cortical development, might be uniquely valuable in our attempts to understand the factors which influence behaviors of fundamental importance. To the degree that these factors are understood, we should be better prepared for enhancing the human child's discriminations of reinforcing and nonreinforcing conditions, discriminations vital to both adequate academic achievements and the appropriate use of social behaviors.

IV. Attention and Mediational Deficits

As discussed by Reese (1962) children might well be compromised in discrimination learning because of deficits in some attentional mechanism. Also, the mental retardate is frequently noted for attention deficits and distractability. Their attention can be so deficient as to retard learning and retain performance at chance levels for extraordinary spans of training.

Among human children, differences in the use of mediators, such as verbal labels and other symbolic representations, might well account for differences in learning-set skills, for, as demonstrated by Kendler and Kendler (1967), there are reliable changes in the relative difficulty of reversal and nonreversal shifts as the child's age progresses. Though all nonhuman primates are without the language skills normally manifested by man, they probably have symbolic processes that facilitate and allow for the formation of concepts, strategies, and mediators that do allow for differential mastery levels in complex learning situations. If the limits of learning were diligently sought for the various nonhuman primate forms, we would probably find them to be higher than we now suspect. If so, their learning skills might prove to be particularly valuable for those who wish to determine methods to facilitate cognitive development of the normal child and to alleviate the learning and performance deficits of the retardate.

Individual differences are readily discerned among members of a given primate genus. Reference to the TI range values presented in Fig. 2 reveals considerable variability within groups. Indeed, the poorest chimpanzee was only slightly better than the average talapoin (*Miopithecus*). Quite possibly these individual differences among nonhuman primates are of the same origins as those which produce the marked differences in intelligence and learning skills among humans. It is likely the case that just as the young child and retardates have circumscribed learning skills by reason of mediational deficiencies, so do certain nonhuman primates. If so, the nonhuman primate would once again come to serve as a valuable adjunct for studies undertaken to enhance human learning potential and efficiency.

The performances of five lowland gorillas on a series of 100 fixed-trial discrimination-reversal problems (Rumbaugh and Steinmetz, 1971) is of particular interest. On an alternate problem basis the animals received either 7 or 13 prereversal trials, then 10 reversal trials. In this manner we attempted to determine whether number of prereversal trials would have an effect upon reversal performance apart from the mastery levels which on the average might be produced by these two levels of prereversal trials. We were also interested in the possible relationships between age and prior test experience and both prereversal and reversal performances.

Figures 5 and 6 portray the results. The striking finding was the reversal performance curves of individual animals were almost identical to the prereversal curves, i.e., a high positive correlation existed between prereversal and reversal performances. Good learning was associated with facile reversal performance; poor learning was associated with equally poor reversal performance. Within the limits of this experiment neither age nor prior test history had any appreciable effect upon performance. Individual differences were pronounced.

Reference to the study with *Cercopithecus* and *Lemur* discussed earlier in this chapter (Rumbaugh and Arnold, 1971) suggests that the high correlation attained between prereversal and reversal performances for this group of five gorillas would probably not be obtained with certain

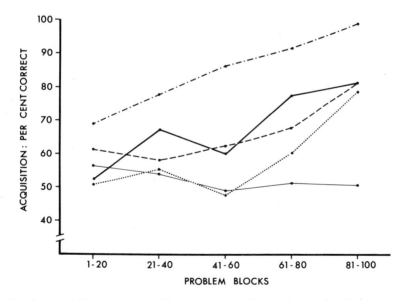

Fig. 5. Acquisition (prereversal) performance levels for each of five gorillas for successive blocks of discrimination-reversal problems. Key: (— ·) Dolly, 36 months; (—) Junior, 50 months; (— —) Mimbo, 60 months; (· · ·) Timbo, 58 months; (—) Yula, 70 months. (After Rumbaugh and Steinmetz, 1971.)

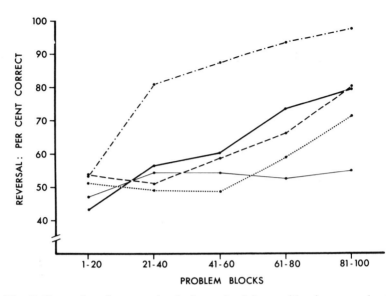

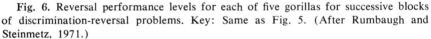

Fig. 6. Reversal performance levels for each of five gorillas for successive blocks of discrimination-reversal problems. Key: Same as Fig. 5. (After Rumbaugh and Steinmetz, 1971.)

other primate forms, notably *Lemur*. As *Lemur* subjects achieved higher and higher prereversal performance levels, their reversal performances deteriorated, i.e., a *negative* correlation was suggested between prereversal and reversal performance levels. It appears that for various primate taxa different correlations exist between prereversal and reversal performance levels. In my estimation it is extremely important that we determine why this is so, for one might intuitively expect that the degree and direction of the correlation would be relatively independent of the species variable. Once again these findings underscore the point that the various nonhuman primate forms can be viewed as *natural* preparations for studies designed to relate a variety of variables, including some of those reliably associated with the species variable, to characteristics which allow for facile reversal and efficient transfer from one situation to another.

V. Changes in Learning Processes

According to Gollin (1965), developmental trends in human learning and cognition reveals shifts from "perceptual" to "conceptual" functioning and from "preverbal" to "verbal" mediation, etc. White (1965) also considers evidence for a hierarchical arrangement of learning processes;

the numerous behavioral changes that occur in the 5–7 year age range for the human child, the emergence of processes reflecting the mediational use of language, and the changing emphasis from "associative" to "cognitive" as a function of development are considered at length.

The work by Kendler and Kendler (1967) clearly attests to the change in learning processes as with development changes are brought about in the inferential and reasoning processes of the child. While they emphasize the dependence of "reasoning" upon the operation of covert mediating mechanisms, such as those involved in human concept formation, they do not rule out the possibility that infrahuman and nonverbal humans are incapable of either representational responses or inferential solutions to problems.

Collectively the work of these researchers supports the point which I press here, namely, that among the varied nonhuman primate genera which vary markedly in central nervous system development and cortical complexity there are probably qualitative differences both in learning and transfer-of-training processes. Though the evidence for this point is admittedly limited at this time, the assertion is not without empirical support.

In a recent paper (Rumbaugh, 1971b) I have presented evidence interpreted to mean that qualitative differences in learning processes exist among nonhuman primate genera as a function of their cortical refinement. Gorillas, gibbons (*Hylobates*), and talapoins were studied in a discrimination-reversal task where prior to cue reversal a performance criterion was achieved. In all problems the first cue-reversal trial was given in accordance with convention ($+$ became $-$; $-$ became $+$). As all the subjects were highly experienced with discrimination-reversal problems, this first trial served to signal cue reversal had been implemented. Subsequent to this first reversal trial three conditions prevailed: (1) conventional reversal trials continued; (2) the stimulus which had been initially $+$ and became $-$ on the first reversal trial was deleted and a new stimulus was introduced in its stead; and (3) the stimulus which had been initially $-$ and became $+$ on the first reversal trial was deleted and a new stimulus served in place of it. As the last two conditions obviated either extinction to the initially correct stimulus or the countermanding of inhibition that might have accrued as a result of nonreinforced selection of the initially incorrect stimulus during criterional training, they should complement reversal performance more than the first reversal condition if the learner's processes were fundamentally of a stimulus–response nature. On the other hand, if the learner's processes were fundamentally of an abstractive nature, no differences among the three reversal conditions would be expected.

On the basis of neurologic considerations it had been predicted that the talapoins might be primarily, if not totally, stimulus–response learners, whereas the gorillas might manifest abstractive learning. The results were in total accord with these predictions. In the case of the gibbon, however, attraction to the new stimulus, introduced on trial 2 and following in two of the reversal conditions, was so marked as to obscure the nature of the prereversal criterional training.

The results of this study provide support for opinions expressed in earlier years by both Köhler (1925) and Yerkes (1916, 1943; Yerkes and Yerkes, 1929) to the effect that of all nonhuman primates the great apes, represented by gorillas in the present study, are particularly capable of abstractive learning. More important still is the reminder that the species variable is very powerful and influential; its influence upon learning processes is now more clearly apparent than ever before. To discern the dynamics of its effects upon psychological processes, including those of learning, transfer, and extinction as illustrated by studies herein discussed, should better equip us for understanding the bases of psychological processes in the developing human child.

In Chapter 20 of this volume Meier aptly states the view that human mental retardation ". . . is a developmental phenomenon in which a plethora of variables interact, concurrently and sequentially . . ." Accordingly, it seems unreasonable to expect that among nonhuman primates we can expect to find one species that provides a totally satisfactory model of human retardation. On the other hand, there is nothing to discourage our hope that among the better than 250 species and subspecies that form the order Primates we will be able to determine selected species whose behaviors in realms of learning, extinction, attention, memory, transfer of training, generalization, etc. will serve as *process* models for research which will contribute directly to alleviating and precluding human retardation in its many forms. In their variety and richness nonhuman primates do, as argued in this paper, provide an array of natural preparations that will surely prove useful as process models, if not animal models, for studies relevant to our achievement of a comprehensive understanding of human cognition and its normal ontogeny.

REFERENCES

Arnold, R. C., and Rumbaugh, D. M. (1971). *Folia Primatol.* 14, 161–170.
Barnett, C. D. (1958). Thesis, George Peabody College for Teachers, Nashville, Tennessee.
Connolly, C. J. (1950). "External Morphology of the Primate Brain." Thomas, Springfield, Illinois.

Davenport, R. K., Rogers, C. M., and Menzel, E. W., Jr. (1969). *Amer. J. Ment. Defic.* **73**, 963–969.

Ellis, N. R., Girardeau, F. L., and Pryer, M. W. (1963). *Abstr. Peabody Stud. Ment. Retard.* **2**, 53.

Gollin, E. S. (1965). *In* "Advances in Child Development and Behavior" (L. P. Lipsitt and C. C. Spiker, eds.), Vol. 2, pp. 159–184. Academic Press, New York.

Harlow, H. F. (1949). *Psychol. Rev.* **56**, 51–56.

Kendler, T. S., and Kendler, H. H. (1967). *In* "Advances in Child Development and Behavior" (L. P. Lipsitt and C. C. Spiker, eds.), Vol. 3, pp. 157–189. Academic Press, New York.

Kintz, B. L., Foster, M. S., Hart, J. O., O'Malley, J. J., Palmer, E. L., and Sullivan, S. H. (1969). *J. Genet. Psychol.* **80**, 189–204.

Köhler, W. (1925). "The Mentality of Apes." Routledge & Kegan Paul, London.

Le Gros Clark, W. E. (1959). "The Antecedents of Man." Edinburgh Univ. Press, Edinburgh.

Leonard, C., Schneider, G. E., and Gross, C. G. (1966). *J. Comp. Physiol. Psychol.* **62**, 501–504.

Levinson, G., and Reese, H. W. (1967). *Monogr. Soc. Res. Child Develop.* **32**, No. 7.

Miles, R. C. (1957). *J. Comp. Physiol. Psychol.* **50**, 356–357.

Razran, G. H. (1933). *Arch. Psychol., N. Y.* No. 148.

Reese, H. W. (1962). *Psychol. Bull.* **59**, 502–509.

Robinson, H. B., and Robinson, N. (1970). *In* "Carmichael's Manual of Child Psychology" (P. H. Mussen, ed.), Vol. II, pp. 615–666. J. Wiley, New York.

Rumbaugh, D. M. (1970). *In* "Primate Behavior: Development in Field and Laboratory Research" (L. A. Rosenblum, ed.), Vol. 1, pp. 1–70. Academic Press, New York.

Rumbaugh, D. M. (1971a). *In* "The Chimpanzee" (G. H. Bourne, ed.), Vol. 4, pp. 19–45. Univ. Park Press, Baltimore, Maryland.

Rumbaugh, D. M. (1971b). *J. Comp. Physiol. Psychol.* **76**, 250–255.

Rumbaugh, D. M., and Arnold, R. C. (1971). *Folia Primatol.* **14**, 154–160.

Rumbaugh, D. M., and Steinmetz, G. T. (1971). *Folia Primatol.* **16**, 144–152.

White, S. H. (1965). *In* "Advances in Child Development and Behavior" (L. P. Lipsitt and C. C. Spiker, eds.), Vol. 2, pp. 187–216. Academic Press, New York.

Yerkes, R. M. (1916). *Behav. Monogr.* **3**, 1–145.

Yerkes, R. M. (1943). "Chimpanzees—A Laboratory Colony." Yale Univ. Press, New Haven, Connecticut.

Yerkes, R. M., and Yerkes, A. W. (1929). "The Great Apes: A Study of Anthropoid Life." Yale Univ. Press, New Haven, Connecticut.

Zeaman, D., and House, B. J. (1963). *In* "Handbook of Mental Deficiency" (N. R. Ellis, ed.), pp. 159–223. McGraw-Hill, New York.

CHAPTER 20

Mental Retardation

GILBERT W. MEIER

I. Introduction and Definition

The intrusion of research into human phenomena with comparative, cross-species analysis brings a necessary reevaluation of the phenomena in question and a redefinition of the investigative problems they pose. This taking-of-stock is especially revealing—and laborious—if the phenomena are behavioral. The description of species-typical behaviors and taxonomic analogs is barely beginning, yet certain behaviors are recognized for the continuation of the species, e.g., reproductive and caretaking behaviors, behaviors which are so obvious, so ubiquitous even, that their roles have been minimized, both theoretically and empirically, in the understanding of behavioral ontogeny and the establishment of individual

431

and species adaptability. Nevertheless we are slowly coming to realize that (1) full understanding of the behavior of the single organism requires an adequate comprehension of the historical antecedents of that animal in conjunction with an appreciation of the prevailing context, and (2) whereas mechanisms of adaptation may be similar or, possibly, identical, the trappings of individuality as related to phylogeny and ontogeny make difficult indeed the realization of such universalities.

This dilemma, this disparity of species-typical behaviors and analogs, is painfully obvious in the comparative study of mental retardation. What behaviors are the behaviors of mental retardation? How do these behaviors differ from those which are necessary, or potentially so, for the adaptation of the individual to his current environment? In what ways are these behaviors contrary to such an adaptation? In other words, what does a mental retardate of a nonhuman species look like?

In an earlier review of the comparative research relevant to mental retardation (Meier, 1970a) I elected to confine myself to a coverage of those studies oriented, explicitly or implicitly, to the analysis of learning phenomena and their variations. However, a broader view of the behaviors of many nonhuman species, certainly all mammals, underscores the essential nature of emotional and social factors in the day-to-day existence and in the overall adaptation of each member. A glance at current efforts in the definition of the problem at the human level alone indicates an evolving perspective with a transition from the cataloguing of "biologic" causes of mental deficiency to the imputing of sociocultural factors in mental retardation (see Robinson and Robinson, 1965, pp. 27–40). This transition must be a salutary process, but one still wonders how a society exists with nonbiologic beings much as he should have wondered earlier how any biologic being, especially of a mammalian species, existed in the absence of social influences. The recognition of sociocultural forces in addition to the biologic factors, that is, the recognition of man as a biologic entity existing in a social and nonsocial environment, has not simplified the task of definition of mental retardation even within the target species, *Homo sapiens.* Edgerton (1970), for example, reviewed the available information on mental retardation in non-Western societies and could only conclude, in despair:

> I must acknowledge that this review of the non-Western world has proceeded in epistemological innocence, for nowhere have I attempted to provide a cross-culturally viable definition of mental retardation. I have not done so, for given our surpassing ignorance about the phenomenology of incompetence in the non-Western world, not to mention in our own, the attempt to do so would be premature. But I am not satisfied with this state of ignorance, nor should anyone be, because the search for such a definition of mental retardation is of the utmost theoretical and practical importance.

> For example, it would be foolish to assume that mental incompetence is entirely relative to the sociocultural system in which it occurs until we have good reason to reject the competing idea that every sociocultural system has a minimal set of demands for competence that is fundamentally similar. We cannot reject this latter assumption cut of hand, for in the wake of anthropology's amour with extreme cultural relativism, we are discovering more and more ways in which men everywhere, and sociocultural systems everywhere, are more alike than they are different (Edgerton, 1970, p. 555).

Some degree of consensus does exist as indicated by the widespread acceptance of the definition advocated by Heber (1961): "Mental retardation refers to subaverage general intellectual functioning which originates during the developmental period and is associated with impairment in adaptive behavior." Undoubtedly, much of the appeal of this definition is attributable to the identification of these facets of the problem: (1) The behavioral phenomena are developmental in character. (2) Any characterization is based on behaviors presently observed (as in childhood) and not on those predicted (as in adulthood). (3) Any characterization is relative to the pattern and variability of behaviors within the population at large. (4) The emphasis must be placed on overall adaptability and not the especial concern for presumed cognitive and/or connative disturbances (see Robinson and Robinson, 1965, pp. 35–37).

Nevertheless, Heber's basic definition (1961, p. 3) raises serious concerns—really procedural and conceptual challenges—for the comparative researcher. In the broad definition, is an allusion to intellectual functioning necessary? What is the operational meaning of "intellectual" which can be applied across cultures (or across species)? Is a culture-free test of intelligence an attainable goal or device? Is a "species-free" measure of learning really a feasibility? Will that measure be free of confounding by connative influences or, if not, will the confounding be evident and, therefore, under the experimenter's control? Yet this definition offers several distinct advantages to this same researcher: (1) It encourages him to observe his species in a "natural" setting and appreciate the breadth of his subject's behavioral repertoire, including its variability and extreme or rarely occurring members. (2) It encourages an analysis of the development process whereby ultimate behavioral deficiencies are realized. (3) It encourages the estimation and evaluation of the contribution of the usual socioecologic context in which the species is found. For reasons apart from those pertinent to researches on mental retardation, each of these emphases receives moral and empirical support from other on-going activities within the broad domain of comparative behavioral research. They are strengthened by the growing revival of terms like "competence," "organismic adaptibility," and "holistic approach," which eschew a process or mechanism orientation focused on specific handicaps and other devices which compartmentalize the organism–environment interaction into rational but unreal segments and fixate on univariate, either/or, relations. In the overall problem of behavioral adaptation, one could do well to acquire an orientation of "developmental retardation" rather than "mental retardation" and appreciate thereby the complexity of the functioning of the total organism, the subtleties of its environmental interactions, and the seemingly limitless population of variants leading to the outcome, adaptation (also see Berkson, 1967a; Mason, 1968; Menzel, 1968; Sackett, 1970).

The goal of this review is to convince the reader of the considerable advantage given us in our current understanding of the etiology of mental retardation and potential management of the individual retardate by the extensive research in nonhuman primate species. To an overwhelming extent, this is research dealing with a single species, the rhesus monkey (*Macaca mulatta*), rather than with a broad or representative survey of the order Primates or other taxonomic subclassification. Still, we must recognize the meaning of this seeming myopia: The rhesus monkey, with its facile adjustment to laboratory environments in temperate climates (Ruch, 1959), and relative ease in care and management in essential isolation from infancy to adulthood (Valerio *et al.*, 1969; Van Wagenen, 1950), is available in comparatively large numbers in the feral state and has shown a range of adaptability to ecologic variety rivaling man (Southwick *et al.*, 1965). Moreover, these same features of adaptation to man and his environment have permitted the ready study of these animals in natural (e.g., Southwick *et al.*, 1965; Lindburg, 1971) and quasi-natural settings (e.g., Koford, 1965). Other species will be studied carefully (Jay, 1968), but each must be considered for its economic as well as its taxonomic significance. Not to be lost sight of in this very general overview are the peculiar advantages all of the higher primates have, to wit: (1) the relatively long life cycle with the long infantile period compatible to the life orientation of the human investigator; and (2) the readily identifiable behavioral–developmental homologs. When observing the daily and social behaviors of a group of macaques, the human observer must periodically jar himself lest he "see" too much in their behaviors, that is, lest he observe in their very humanlike antics and communications, his own behaviors and those of his own conspecifics (cf. Van Hooff, 1967; Vine, 1970). Unequivocally, even the novice can "see" so much more in the behaviors of other primates than of the nonprimates, enabling him to extend definitions of behavioral phenomena, including mental retardation, with fewer definitional doubts and misgivings. For research on very early development, these primate species offer the strategic advantages of a moderately long gestation period with a mother and a conceptus of a size commensurate with the agility, skills, and technology of the investigator.

In spite of these many methodologic advantages of the nonhuman primate, we must realize at the outset the recency of the primate strategy, that is, the relative newness of the orientation of comparative researchers in this area to the utilization of other primate species. Masland *et al.* (1958), for example, in their review of mental subnormality failed to cite *any* research with a primate model consistent with then-current definitions of mental retardation. Although a few studies were described with nonhuman primate subjects, these studies were merely confirmatory of certain physiopharmacologic equivalences in the adult monkey (rhesus) to those observed in man. At the time of their survey, none of the relevant studies carried the necessary developmental–behavioral orientation viewed as necessary for present understanding.

Finally, by way of completing this introduction, let me emphasize that potentially all behavioral research with other primate forms is applicable to our comprehension of the human problem if we insist on focusing on learning as necessary in the adaptation to environmental contingencies. Even a very cursory overview of several of the recent excellent surveys of research on learning in nonhuman primates (e.g., Riopelle and Thomsen, 1968; Rumbaugh, Chapter 19, this volume; Schrier *et al.,* 1965) and those on mental retardates (e.g., Ellis, 1970, and preceding volumes) reveals the considerable overlap of approach and information between these two fields which share the common concern of analysis of the adjustment-modification processes in a primate with limited communication skills with the human observer.

In the earlier review of the comparative research on mental retardation-related problems (Meier, 1970a), I organized the literature around the premise that our fundamental concern as biologic researchers is the understanding of the organism–environment interactions, specifically in this case, of those interactions which are behavioral in character. Further, I divided the analyses along temporal lines, along lines which best represented the origin of the environmental influences in the historical and contemporaneous factors of individual behavior; that is, into genetic, ontogenetic, and ambient concerns. I maintain still that an awareness of the potential and variety of each of the classes of factors is mandatory for the understanding of many of the behavioral data presented in the course of developmental research. Thus, the behaviors observed at any one moment in time are complex, composite functions of three subsets of environmental variables: (1) those active in the phylogenesis of the species and of the individual observed; (2) those active in the ontogenesis of the individual; and (3) those effective in the ambient environment, in the context in which the behavioral analysis is being made (Meier, 1970b). Undue emphasis on any one subset leads to fundamental errors both in research design and in research interpretation; in combination, their contributions provide an empirical basis for estimating the individual differences noted in all species, in all dimensions.

II. Individual Differences

For the psychometrician, the phenomena of mental retardation are conceptually straightforward: They are the behaviors located at one of the extremes—that one judged to be inferior, or substandard, or inadequate— on the normal or Gaussian metric function, which are therefore displayed by only a small segment (1–3%) of the population. For purposes of communication, these behaviors are considered to be typical and differentiating of the individuals who display them such that the individuals may be reliably categorized or "diagnosed." Briefly, this particular problem, mental retardation, is only a part of the larger problem of individual differences, a stable dimension of modern psychology, both in research and application. In establishing a nonhuman model of mental retardation, the comparative researcher can only envy the psychometrician's agility. Only recently has the comparative investigator realized the potency (sometimes, perniciousness) of the differences between his unispecies groups. Slowly, he has come to acknowledge the considerable variability attributable to sex, strain or breed, and, even more recently, maintenance and rearing conditions. Only now is he aware that a true science of these differences appears to be emerging.

Consonant with the notion of behavioral proximity of the nonhuman primate for man, individual specimen differences have been long recognized in special primate settings (e.g., Köhler, 1925; Yerkes and Yerkes, 1929). Current field research reveals the large differences in individual behaviors as related to age, sex, social status, and organization (DeVore, 1965; Jay, 1968; Morris, 1967). The challenge of the primate model for mental retardation, as currently defined, is most clearly drawn in several current programs of field study with the Japanese macaque (*M. fuscata*) and with the rhesus macaque (*M. mulatta*) in central Japan and northern India, respectively. The first program provides for the comparative appraisal of over 30 parks in which the primate populations are maintained independently (coordinated by the Japanese Primate Centre) and offers the possibility of discerning the combined influences of localized genetic and environmental factors. Such interactions have been described for reproductive cycles (Kawai *et al.*, 1967), feeding habits (Frisch, 1968), and social behaviors, including caretaking activities (Miyadi, 1967). Thus, temporal parameters of reproductive activity appear to be locally defined. Depending upon peculiar feeding and provisioning experiences, one troop washes its food, for example, whereas a second seines for it. In other troops adult male involvement in the caretaking of young is commonplace, whereas in most such involvement has not yet been observed. (See, also, Marler and Gordon, 1968.)

Singh (1969), in India, has described appreciable differences in the social behaviors of the rhesus macaque living in the rural, forested areas of Rajasthan as contrasted with those living in the urbanized areas of Jaipur. The latter are much more aggressive toward each other as well as toward their human competitors; the former are generally furtive and elusive of the human population. As would be expected of their distinctive habitats, the two populations of macaques differ in their feeding activities and habits as well. More striking are the differences in social sleep and wakefulness patterns and in those laboratory activities by which the "urban" monkey appears more exploratory in its behaviors, although no more facile in its learning of conventional laboratory tasks, e.g., object discrimination on the Wisconsin General Test Apparatus (Singh, 1968). In many respects, the "rural" monkey is suggestive in its behaviors of the human with some degree of "cultural deprivation" (Zigler, 1967).

III. Genetic Factors

In none of these differences in behaviors reported for particular troops or populations of monkeys can genetic factors be pinpointed, nor can they be wholly excluded. Together with the possibility of tremendous variation in social and other environmental variables, the genetic drift, which occurs with restricted breeding access, could give rise to significant behavioral differences. Data on the first subset of environmental variables, that is, the phylogenetic or genetic factors are virtually nonexistent on the nonhuman primate. Well-established, moderately homogeneous sublines of primates, like the sublines of housemice (*Mus musculus*), have not been established and surveyed, for obvious reasons. Some chromosomal research has been done, however, the most striking of which has been the detection of a "mongoloid" chimpanzee with unique facies and other physical characteristics, retarded growth, and distinctive behavioral retardation (McClure *et al.*, 1969). The unusual karyotype (49 chromosomes rather than the usual 48 with the extra chromosome being in the G group) suggested a possible direct homolog of the trisomy-21 in humans.

Regrettably, this has been a unique observation of an animal born to chromosomally normal parents and now limited, tragically, by the demise of the animal during routine clinical observation when about 2 years old.

IV. Ontogenetic Factors

Epidemiologic correlational research has repeatedly implicated a variety of prenatal and perinatal conditions in complications of early development, including childhood behavioral disturbances (Caldeyro-Barcia, 1969; McNeil *et al.*, 1970; Pasamanick and Knobloch, 1961). Experimental research with nonhuman primates on such conditions has been only promising, at best, a consequence of the relative recency of the initiation of these endeavors. Some significant "starts" have been made: A few offer clear relationships between agent and behavioral retardation and/or defect; most are suggestive of behavioral and other growth deficits given a more extensive and longer range research commitment. Some reveal conceptual and technical advances which extend far beyond the particular research effort to which they were applied.

A. PRENATAL INFLUENCES

1. *Teratologic Study: Baseline Data*

As an interpretative device for subsequent experimental studies, Wilson and Gavan (1967) surveyed the relevant literature and canvassed experienced investigators regarding the incidence of spontaneous congenital malformations in the reproduction of laboratory primates. From these sources, they computed an incidence of 0.44% of malformation which was considerably lower than reported for humans (about 7.5%). They considered only externally detected morphologic anomalies, of course, which, although homologous to the human condition, necessarily limited the completeness and sensitivity of their survey (Lapin and Yakovleva, 1963, report an incidence of 0.48%; see, also, Iwamoto, 1967; Iwamoto and Hirai, 1970; Morris, 1971; Koford *et al.,* 1966; Swindler and Merrill, 1971).

Similarly, Koford *et al.* (1966) reviewed the published data on twinning and those collected from the colonies of the Laboratory of Perinatal Physiology in Puerto Rico. Again, the reviewers were able to report an appreciably lower incidence (1 pair in 1003 deliveries) than in man (1 pair in 88 deliveries) for data which were intrinsically more reliable than those collected on congenital malformations. Combining their data with those received from the Russian laboratory at Sukhumi, these authors report a twinning rate of 1 per 437 births.

2. *Virologic and Biochemical Agents*

Reviews of the experimental teratologic data with agents relevant to human clinical and pharmacologic practices reveal increased incidence of malformations following the prenatal application of certain agents during prescribed periods in the primate gestational period (Courtney and Valerio, 1968; Elizan and Fabiyi, 1970; Wilson, 1966; Wilson and Gavan, 1967). Notable among the agents which produced anomalous development are thalidomide and rubella virus. Both have been reported to produce some, at least, of the human morphologic symptoms of prenatal modification in the nonhuman primate (Barrow *et al.*, 1969; Delahunt and Lassen, 1964; Delahunt and Rieser, 1967). Behavioral changes have not been reported for either, however (e.g., Sever *et al.*, 1966), nor has the incidence of malformations and reproduction complications been entirely consistent (Lucey and Behrman, 1963).

Some of the behavioral phenomena observed in primate teratology are instructive in the understanding of behavioral development and the significance of the thalidomide tragedy, for example, for the individual affected. Lindburg (1969a,b) described remarkable compensation in routine, daily behaviors of infant monkeys with thalidomide-induced limb malformations. Moreover, he reported that these animals developed behaviors, albeit limited, consistent with normal maturational schedules for the species (*M. mulatta*). Most striking was the appearance of approximations of social behaviors (e.g., cage-shaking, mounting) at ages typical of the species. (These animals were raised with their mothers during the period of observation.) Only in the amount of exploratory behavior shown, were these animals "deficient" as compared to their normal controls.

Other studies with prenatal administrations of adenovirus, Rous sarcoma virus (Cotes, 1966), and lysergic acid diethylamide (Kato *et al.*, 1970) have been entirely negative in the production of terata or chromosomal change in the offspring involved.

In a somewhat different vein, in an effort to replicate some of the presumed aspects of biochemical duress of maternal phenylketonuria upon the fetus, Kerr *et al.* (1968) observed the growth, behavior, and other aspects of infants born to female rhesus monkeys fed an excess of L-phenylalanine during pregnancy. The infants born to such females were of low birth weight, although they subsequently grew in length, weight, and head circumference to within normal limits for the Wisconsin laboratory. When examined for learning proficiency after a year of age, these animals showed normal performances on multiple-dimension discrimination and delayed-response (5-sec delay) tasks; their performance on a multiple-problem (600) learning-set test, however, was significantly depressed when compared with that of the normal controls. (All animals were maintained in essential social isolation since birth.)

Preliminary observations in the Wisconsin laboratory indicate that a promising model of cretinism can be effected by the administration of a single dose of ^{131}I between days 68 and 97 of gestation (Davenport et al., 1971). Upon delivery, such infants reveal an absence of thyroid hormone in the blood and, in the cases of the four stillbirths, absence of the thyroid gland. Surviving infants had difficulty maintaining respiration, thermoregulation, and normal food intake. They revealed considerable behavioral retardation in their poor posturing and locomotor deficits.

In still another vein, in one that may relate gene and chromosomal action to behavioral development, Phoenix and Goy (Goy, 1968; Phoenix, et al., 1968), following the lead of W. C. Young (Young et al., 1964) showed that not only the external genitalia of female rhesus fetuses, but also their infantile and adolescent behaviors, could be modified by the protracted regimen of injections of testosterone proprionate prenatally. From a series of studies of behavioral sequelae of infantile castration, of chemical analysis of infantile blood plasma, as well as earlier studies on rats and guinea pigs, these investigators developed a hypothesis on maturational stage at birth and hormonal modifiability. Their observations of pseudohermaphroditic female monkeys substantiated the hypothesis and pointed to other studies on the modifiability of the central nervous system and the establishment of sexual role and behavior. Long-range projections from these studies may afford some understanding of the development of strain- and species-specific behaviors in addition to those which are sex-specific.

Speculations have been offered that prenatal metabolic deficiencies, e.g., some form of fetal hypoxia, may lead to serious behavioral disorders in humans. Clinical data are not definitive on this matter; experimental data (with rodents) are strongly suggestive, although the character of the behavioral change is uncertain (Meier, 1971). Myers (1969a,b) in the San Juan laboratory has shown the technical feasibility of direct intrauterine manipulation of the primate fetus and the reasonableness of acute prenatal asphyxia in the production of the severe brain damage seen in certain human infants in the newborn nursery following an abnormal pregnancy. Myers and his co-workers removed the fetus at stated gestational ages, placed it in a saline bath, and maintained it and the female under physiologic conditions with careful monitoring. The investigators then interrupted this stasis by clamping the umbilical cord for 20–30 min, with greater durations for the younger ages (102, 126, and 142 days of gestation; see Myers, 1967). When the fetuses regained normal physiologic activity, the investigators returned the fetus to the uterus, replaced the amniotic fluid, closed the uterus, and repaired the laparotomy. Later, on the 158–160th gestational day, they delivered the infants surgically. After delivery, these animals revealed varying degrees of spastic diplegia, which were associated with considerable difficulty in maintenance of posture and in limited locomotor activity. Moreover, these animals exhibited difficulty in feeding, required extensive hand care, and vocalized little if at all. Histologic examination of the brains revealed a relatively consistent pattern of brainstem damage, especially prominent in the inferior colliculi and the major sensory pathways. Tegmental damage was also noted, especially in the younger fetuses. None showed any cerebral or cerebellar damage. The overall consistency of brain damage, irrespective of gestational age (including full-term fetuses in other studies), makes this series truly remarkable.

B. Perinatal Influences

Since its inception in 1956, the focus of the Laboratory of Perinatal Physiology in San Juan, Puerto Rico, has been upon the anatomic, physiologic, and behavioral sequelae of metabolic disorders in the around-the-expected-birth period. In an early publication Ranck and Windle (1959) described the neuroanatomic changes which follow severe neonatal asphyxiation. These were, surprisingly for the time, primarily changes in the inferior colliculi, the major sensory pathways, and the ventrolateral thalamic nuclei. Cerebral changes were noted only if respiratory complications developed while the neonate was in the nursery (Windle, 1967). Only in the latter eventuality did a picture like human cerebral palsy appear. [Myers *et al.* (1969) noted extensive cerebral involvement in infants following a prolonged asphyxial insult at birth.]

Subsequently, Saxon (1961a,b) and Saxon and Ponce (1961) reported some behavioral changes in monkeys exposed to neonatal asphyxia when examined during the 6–12-month infantile period. They found the experimental (asphyxiated) monkey less emotional in an open-field situation and more exploratory in its behavior toward the testing environment and toward novel objects in the environment. They could describe no deficiencies in multiple-dimension discrimination and in delayed-response learning or in the development of multiple-problem learning sets in these animals in spite of significant differences in sensorimotor capabilities. Later, Meier (1965a) failed to differentiate these animals from their controls in terms of reproductive behaviors. Both the experimental and the control females showed adequate sexual and caretaking behaviors (contrary to those reported by Harlow and Harlow (1962b) for similarly maintained, isolate-reared macaques; see below). Still later, Sechzer (1969) did show that the neonatally asphyxiated animals, now 8–10 years of age, were unable to solve a delayed-response task, even if the delay was as short as 5 sec. These last changes paralleled the progressive atrophic changes in the central nervous system, primarily in the cerebral cortex (Faro and Windle, 1969). The neurohistologic changes developed, according to Faro and Windle, in spite of what appeared to be progressive *decrease* in the gross clinical symptoms of lethargy, nonreactivity, and depressed emotionality. Over 5 years and more, these animals had come to look, more and more, like isolation-reared, fully intact rhesus monkeys with the recognized stereotypic behaviors, self-directed aggression, and so forth (Berkson, 1967b; Sackett, 1968).

Lucey and others (1964) in the laboratory demonstrated that kernicterus could be established in a primate model. They combined two manipulations neither of which produced demonstrable behavioral or anatomic change when applied singly, neonatal asphyxia for 6 min and hyperbilirubinemia (about 30 mg%), and produced most of the clinical and pathologic symptoms of the human disorder: jaundice, diminished reactivity, opisthotonos, electroencephalographic depression, discoloration of the basal ganglia, and invasion of the cortical neurons by the bilirubin. (Without the asphyxial episode, the bilirubin did discolor the basal ganglia and entered the cortical vessels and the neuroglia, but not the neurons.) Long-term changes, viz., changes beyond the neonatal period, for this multiplicative model of one presumed etiologic factor in mental retardation are not yet known.

Foshee *et al.* (1969) attempted a further analysis of the neonatal insult–behavioral change relation of neonatal asphyxia in a continuing study of monkeys subjected to bilateral inferior collicular lesions (electrolytic) within 12 hr of delivery (either vaginal or surgical). Some of these infants were monitored for sleep–wakefulness cycles (cf. Meier and Berger, 1965); others were observed in an avoidance-conditioning procedure (cf. Meier and Garcia-Rodriguez, 1966a) and, later, in reactivity to an enlarged environment and to novel objects (cf. Saxon, 1961a,b). The patterns of infantile sleep–wakefulness were well within the normal range except for the first 5 days, postnatally, when the infant showed unusual amounts of sleep activity, especially of the high voltage–slow wave variety. Overall, the experimental and control infants responded alike under the conditioning procedures; however, the cesarean-delivered operates, about half of the group, were the most active and the most vocal, whereas the surgically delivered controls were the least active and the least vocal.

Throughout the infantile period, the operates showed considerable motor impairment resembling that encountered in the cerebral palsied child; they showed bilateral asymmetry in palmar and plantar grasp, placing response, and in the balance of extension–flexion of the neck muscles, all of which diminished in severity with subsequent development. All animals showed the idiosyncratic behaviors of isolate-reared monkeys, although they were somewhat more bizarre and exaggerated in the operates during infancy and noticeably less so during subadult periods. The experimental animals demonstrated more exploratory behavior, both in the home cage and in the enlarged cage situation. Moreover, the experimental animals appeared less disturbed by the larger experimental situation. Subsequently, these animals showed the usual difficulties in the maintenance of social structure in a communal situation, although they were effective in their reproductive activities at least to the point of procreation and infant maintenance (Foshee and Meier, 1969). The mothers were unusual in later relations: They permitted a degree of mother–infant separation of remarkable extent in frequency and duration—both high—which resembled that seen more recently in a group of feral-reared monkeys in which the adult females had had the seventh cranial nerves sectioned bilaterally (Meier et al., 1971).

The delivery-control procedures used in most of these Puerto Rican studies raise concerns which have yet to be resolved. Meier (1964) reported differences in reactivity and in frequency of vocalization in the surgically delivered infant during the first 5 days postnatally, which suggested a prolonged postnatal depression. Subsequently, Meier and Garcia-Rodriguez (1966b) reported a failure of these infants to respond effectively to an operant-avoidance situation during this same period, and a cumulative behavioral deficit in those surgically delivered infants first exposed to the behavioral-monitoring situation in the neonatal period. Those first exposed at 30 or 60 days, postnatally, showed no such deficit. Moreover, the conditioned emotional responding was depressed for an even longer period than was the conditioned instrumental responding. These and the data from the animals with the neonatal inferior collicular lesions suggest a complex interaction between mode of delivery, brain damage, and early postnatal manipulations, which may require careful consideration in subsequent neonatal trauma longitudinal investigations.

Additionally, Hyman et al. (1970), in a newer and separate program of studies, were unable to show successful delayed response (5-sec delay) in neonatally asphyxiated monkeys, in clear contrast to their control monkeys. These animals were tested when about 1 year of age under somewhat more stringent conditions than those used by Saxon and Ponce (1961). Possibly, the failure to replicate the earlier study is attributable to the more difficult testing conditions, as the authors aver. Possibly other differences existed in addition to those of the testing situation, differences in the rearing and maintenance conditions, for example, which could have confounded the strict replication of these studies. [Such differences are the suggested basis of other failures to replicate studies of early developmental sequelae, as in Harlow and Harlow (1962b) and Meier (1965a).] Subsequent testing under shock avoidance (Hyman *et al.*, 1971) and visual reinforcement conditions (Waizer *et al.*, 1972) revealed differences consistent with a lower level of emotionality in the experimental animals rather than with a primary deficit in learning ability. Shuttle-box testing indicated that the asphyxiated monkeys had raised auditory thresholds between 0.5 and 8 kHz (Berman *et al.*, 1971). Apparently all of these behavioral differences plus the developmental retardation described by Sechzer *et al.* (1971) are congruent with the anatomical changes first reported by Ranck and Windle (1959).

C. Neonatal Influences: Biochemical Factors

Concern about the long-range, persisting effects of infantile biochemical changes is relatively recent in our appreciation of etiologic factors in mental retardation. Specifically, the research with nonhuman primates has focused on two particular aspects of biochemical modification: (1) errors in metabolic processes which characterize some genetic disorders (Hsia, 1967); (2) deprivation of essential nutrients as in mirasmus and kwashiorkor (Scrimshaw and Gordon, 1968).

Phenylketonuria, or phenylpyruric oligophrenia, is such a disorder, wherein the deficiency in the metabolism of certain substances, specifically of the amino acid phenylalanine, is related to profound mental retardation (Jervis, 1954). Waisman and Harlow (1965) described a series of studies in which they attempted to establish a model of this genetic disorder in a number of ways, most successfully by the feeding of a diet high in phenylalanine throughout infancy. In the principal study, Waisman and Harlow fed infant rhesus monkeys a liquid diet with added amounts of phenylalanine to a maximum intake of 3 gm amino acid/kg body weight-day. This intake began at 5 days of age, first in small amounts then increasing to the maximum at about 19 days of age; and continued to about 2 years of age, for two of the animals, and 3 years of age for the remaining four. During this period, the animals were observed on a battery of learning tasks with the Hebb-Williams maze at 15 days of age and concluding with discrimination-reversal problems at approximately 18 months of age.

Despite relatively normal growth rates the monkeys maintained on a high phenylalanine diet (blood titer $\sim$ 10 mg%) performed unsatisfactorily on all of the behavioral tests. Their long latencies for response, their withdrawal, balking, and inappropriate orientation to the test objects and to the reward foods suggested that these deficiencies were primarily motivational in nature as would be attributed to the direct action of the dietary regimen. Possibly, they were motivational–emotional sequelae to the inability to solve these tasks which were otherwise readily mastered by animals of this age with normal developmental histories. [Confirmation of the latter possibility was given by Chamove and Davenport (1969) in a companion study.] The awkwardness in comprehension and obstacle avoidance, and the epileptiform behavior which occasionally occurred in all of the experimental animals did not present obvious handicaps. Two of these animals were observed for an additional 2 years following the termination of the dietary regimen but did not improve in their test performances over those shown earlier: viz., no recovery of function was shown.

Maintenance on a high phenylalanine diet during the first 12 months postnatal may have profound and lasting effects on social behavior, as well (Chamove *et al.,* 1970). Such animals when tested between 13 and 33 months of age showed increased levels of hostile and withdrawal behaviors with lowered levels of play in their home cages and in the playroom with age peers. Moreover, they failed to maintain a dominance heirarchy among themselves over repeated observational periods during which they were placed together under conditions of water and/or food competition.

The necessary balance of the availability of the essential amino acids, specifically of phenylalanine, was shown in subsequent studies by the University of Wisconsin group. Kerr and associates (1969) maintained one group of monkeys on a diet deficient in phenylalanine from 30 to 135 days of age; they maintained the controls on a standard diet or the deficient diet with amino acid supplementation. The growth rates and the general physical condition of the animals on the deficient diet were severely impaired as compared to the controls, especially those on the standard diet. The learning performances on a multiple-problem discrimination task, a delayed-response task, and a learning-set test when approximately 1 year of age revealed significant group differences. The experimental, nonsupplemented animals were significantly inferior in their learning-set performances, but not in their performance on the discrimination and delayed-response tasks, even though the animals had been returned to a standard diet some 5 months earlier. The controls did not differ in their performances either from each other or from other normative values.

The long-term effects of the animals in the Wisconsin studies is in sharp contrast to others in which the dietary deficiency was instituted long after the neonatal period and sustained for shorter periods of time. In these studies (e.g., Zimmerman, 1969; Zimmerman and Strobel, 1969; Zimmerman *et al.,* 1970), adolescent monkeys did show serious behavioral deficiencies, which were limited in extent by the duration of the dietary regimen. Their behavior became species-typical with the return to full dietary balance. That such an eventuality need not be the case if the dietary insufficiency is instituted shortly after birth is implicit in the report of the incomplete program of studies by Kerr *et al.* (1970). In both the juvenile and the infant, dietary insufficiency, especially protein lack, led to lethargy and general nonreactivity, which may be a secondary, but especially potent developmental complication for the young mammal (cf. Meier, 1968).

Davenport and his associates (1971), attempting to produce a model of cretinism by neonatal manipulation, thyroidectomized four infants by ^{131}I injection during the first week after birth. These animals showed the expected myxedema, biochemical changes in the blood, and depressed growth and activity levels. Even after several weeks on thyroxin replacement, the animals did not appear normal but showed the atypical social behavior of the animals on the high-phenylalanine diet: decreased exploratory behavior, increased withdrawal and hostility, and unstable dominance relations. At 20–30 months of age, these animals were consistently submissive in their interactions with normal animals. Their performances on several learning tasks could only be called defective, with atypical response persistence and perseverance where other patterns were required.

D. INFANTILE INFLUENCES: CENTRAL NERVOUS SYSTEM SURGERY

Of the many etiologic factors implicated in behavioral dysfunction during the last several decades, the one with possibly the greatest professional favor and research appeal is some form of infantile brain damage. In the course of human events this is a complication which is the presumed consequence of a number of clinical disorders, not the least of which is neonatal asphyxia and neonatal and infantile viral infections. Nevertheless, since the pioneering work of Tsang (1937) and Kennard (1938), experimentally induced cortical damage in laboratory animals has been remarkably temperate, if not entirely ineffectual, in producing long-term, significant behavioral defects. More recent research, including much with infant primates, corroborates this evaluation (Isaacson, 1968; Blomquist *et al.,* 1971). Apparently, cerebrocortical lesions to the frontal and parietal cortex, for example, inflicted *during infancy* (prior to 6 months) do not produce significant performance deficits when evaluated during *adolescence*. However, surgical involvement of deeper structures, either separately or additionally, may result in serious impairments in learning performances and, in some instances, aberrant motivational–

emotional behaviors as well in both the infant- and adolescent-operate groups.

Harlow and his associates at the University of Wisconsin (Akert *et al.*, 1960; Harlow *et al.*, 1964) described a stepfunction between age at time of destruction of the frontal association cortex and impairment in delayed-response performance. They operated on the animals at 5 days, 150 days, or at 24 months and compared the performances of the animals on 5- and 40-sec delays at 150 days of age or upon full recovery from surgery, whichever was later. The performances of the 5- and 150-day groups were not different from the nonoperate controls, except for the slow initial rise in accuracy of performance; the performance of the 24-month group reflected virtually no learning and was similar to that of adult operates reported earlier (Harlow *et al.*, 1952). Linear relations were noted between age at time of testing and proficiency of performance on tests of discrimination learning, learning set (discrimination and oddity), and the Hamilton search test (insoluble problem), but none indicated brain-damage effects. Similar observations were reported by Raisler and Harlow (1965) in their study of the performance of several groups (by age) with posterior cortical lesions and learning performances. Only the oldest group (900 days) showed any signs of behavioral impairment, albeit overcome by extended training.

Tucker and Kling (1967) corroborated the results of Harlow and his co-workers while analyzing the effects of more complete frontal lobe destruction and extending the range of behaviors observed. They reported that their infant operates (less than 1 month of age) performed the delayed-response task without evident impairment, whereas their juvenile operates (about 3 years of age) expressed significant difficulty. No significant differences were observed on color-discrimination tasks or on a delayed-alteration task; both age groups were equally impaired. (Tucker and Kling used both *M. mulatta* and *M. speciosa* animals, whereas the Wisconsin group used only *M. mulatta*. These differences, however, appeared to be without significance for this problem. Tucker and Kling used subjects which were either maternally reared or isolate-reared. Although this also was a nonsignificant difference in this study, the broader significance of maternal interactions will be considered below.)

More recently, Harlow *et al.* (1968) compared the extent of behavioral change with extent of surgical injury as a function of age at the time of the operation. They performed frontal lobe topectomies (cortical tissue, only) on monkeys at 5, 12, 18, and 24 months of age and lobectomies (cortical tissue plus underlying fibers) at 2, 5, and 24 months of age. They observed the animals on object-discrimination, delayed-response, and learning-set tasks beginning 2–5 months after surgery. (The longest intervals were with the youngest animals for which testing began at about 200 days of age. Controls were first examined at 200 days as well.) These investigators reported that on the object-discrimination learning task the age at time of testing resulted in the greatest differences in group performances (the older animals performed with fewest errors), although the size of the lesion did account for some of the significant differences (the topectomized animals performed with fewest errors). All animals performed above chance levels despite the lesions. The infants lobectomized at 2 months showed no evidence of defect. Deficits in delayed-response performance shown by all groups except the 2-month-old lobectomized animals were related to severity and age at surgical damage. Differences in learning-set performances were attributable to age at testing (the older animals performing at higher levels) and severity of brain damage. All lobectomized groups showed significant impairment, including the two infant groups. Overall, certain systematic differences in test performances were observed in these animals related to task, age, and severity of injury, yet cage behaviors were indistinguishable between groups or other unoperated controls.

In a subsequent report (Harlow *et al.*, 1970), the Wisconsin researchers indicated that, whereas significant differences existed between the frontal-lobectomized (5- and 24-month operates) and the like-aged control animals on all measures, such deficits were not found for the topectomized animals. Previously reported differences in delayed-response performances may have been due to the absence of comparably aged controls. In course the authors concluded that, "There was little evidence from the present study that surgery at 5 months spared any intellectual functions that were lost after surgery at 24 months."

Goldman and her colleagues (Goldman, 1971; Goldman *et al.,* 1970a, 1970b) at the National Institute of Mental Health have now shown that the prefrontal cortex is not a functionally homogeneous area even during infantile development, that the maturational recovery following prefrontal ablation depends upon the particular sector removed (dorsolateral or orbital). The performance of operates (*M. mulatta*) 10 months and more after surgery (at 48–82 days or at 18–24 months) on spatial delayed response, visual pattern discrimination, spatial delayed alternation, and object discrimination reversal tasks were distinctive and complexly related to the particular task, the size and area of ablation, the age at surgery, and the age at testing. Maximum sparing was seen in the behaviors of animals with infantile removal of the dorsolateral sector, but considerable recovery was seen at retesting in those with infantile removal of the orbital sector when some decay in performance was seen in the dorsolateral animals. [Differences were also noted in the home cage-social situation which corroborated the greater deficit of the infantile orbital lesions (Bowden *et al.,* 1971).] Goldman (1971) concluded:

> The idea which emerges has to do with the concept of "commitment." Only in the infant brain would there be the possibility of intact areas still "uncommitted" at the time of injury and, therefore, capable of modification, while in the adult, all pathways remaining after operation would presumably be "fixed." . . . It may be that the adult brain relies more heavily on strategic modes of recovery while the infant depends more on the structural mode, thus providing an adequate basis for more striking restitutions after equivalent injuries in the young than the old (Goldman, 1971, p. 386).

In an effort to understand the development of the neural mechanisms involved in these phenomena of behavioral-sparing, Kling and Tucker (1967, 1968; Tucker and Kling, 1969) examined the effects of concurrent subcortical and nonfrontal cortical lesions on the apparent resistance of infant frontal operates to behavioral deficits on delayed-response learning. They recorded the behaviors of infant monkeys (*M. speciosa*) with combined lesions (between 2nd and 56th postnatal day) in the frontal cortex and the caudate nucleus finding them to be much more fragile than normal as well as showing somatomotor and growth deficiencies and postoperative seizures. In tests of delayed-response learning, these animals at 7 months of age were incapable of above-chance performance beyond the 0-sec delay condition. Kling and Tucker concluded that this age-performance deficit was a corticosubcortical phenomenon since other groups of animals which had suffered infantile damage (3–17 days postnatal) to the prefrontal and posterior association cortices showed little describable impairment and performed indistinguishably from other animals with only frontal damage; all were capable of above-chance performances even to delays as long as 40 sec.

Current efforts deal with the effects of lesions in subcortical areas on infantile behavioral development. Foshee *et al.* (1969; see above) described a possible early change in sleep–wakefulness patterns, an absence of typical emotional behaviors at the end of 1 year, and a diminished frequency thereafter of stereotypic behaviors typical of isolate-reared macaques. Similar differences in behavioral development have been described by Thompson *et al.* (1969) in rhesus monkeys subjected to bilateral amygdaloidectomies, completed by the age of 2½ months. In their home cages these infants showed none of the serious changes in activity level and appetitive behavior characteristic of the adult following similar surgical manipulation (Weiskrantz, 1956). However, in the test situation, these young animals showed fewer fear behaviors (e.g., crouching), although they did appear to be more readily disturbed by active, normal age peers. In group situations, these operates made more fear responses, especially as the length of the test session increased and as the lapsed time since surgery increased. Thompson (1969) found that these behaviors were complexly related to testing parameters: the operates were hyperactive and had a higher frequency of behavior change upon initial exposure to the test conditions but could not be distinguished on these dimensions at the end of 24 hr of continuous exposure to the test situation. The operates, however, did show a high frequency of nonsocial exploratory activity after 24 hr, significantly more than did the control animals.

Without doubt these studies on brain damage and behavior development represent the highest level of theoretic-methodologic sophistication in this entire field of developmental primatology. They reflect an advanced stage of neurosurgical tradition conjoined with a vast wealth of expertise in the analysis of behavioral sequelae and trends. Thus, we see in these studies the interactive effects of age at surgery, site and extent of brain lesion, nature and complexity of the behavioral analysis, and age at evaluation. Together the results of these investigations reveal considerable plasticity—absence of differentiation—in the infant primate brain which permits subsequent behavioral development toward an end of essentially normal primate behavior. As yet, however, the nature of the path, the peculiar infant–environment interactions, which make this end possible is not known. The significance of rearing conditions, social or isolate, is yet to be evaluated fully and systematically (see below).

E. INFANTILE INFLUENCES: SOCIAL-ENVIRONMENTAL FACTORS

In 1962, Harlow and Harlow (1962a,b) described some truly remarkable reproductive behaviors—really misbehaviors—in young adult rhesus macaques which had been reared in essential social isolation for a significant part of the first postnatal year of life. They reviewed the failure of the males to mount the estrus female effectively (see Mason, 1960, and, more recently, Missakian, 1969), and described in detail the relative difficulty of the females to breed and their almost total lack of caretaking behaviors upon delivery of their infant. The animals surveyed had been studied for their infantile and preadolescent adaptation to isolation-rearing without the primate mother, their behaviors in the surrogate situation, and their establishment (or lack thereof) of affective behaviors (Harlow, 1958). Although preadult uniquenesses have been recognized (e.g., Mason and Green, 1962; see also Candland and Mason, 1968; Hill and McCormack, 1969; Mason, 1963, 1965; Mason and Sponholz, 1963; Pratt and Sackett, 1967), their pervasiveness and persistence are now seen as profound (Suomi et al., 1971). (See Hinde, 1971, for an extensive review.)

Although test observations had not been corroborated in entirety (Meier, 1965a,b), they have been viewed more recently as consistent with

other behavioral differences seen in isolate-reared monkeys when compared with feral-reared age counterparts (Sackett, 1970) and have been
satisfactorily incorporated into a view of affective structure and development (Harlow and Harlow, 1965). Moreover, the inadequacy of caretaking behavior may be overcome with further maturation or with relevant
experience during adulthood. That is, some of the isolate-reared females
who were inadequate with their first infants proved to be satisfactory
with the second (Seay et al., 1964). Clearly, animals such as these show
an awesome array of sterotypic behaviors and unusual emotional responses (Sackett, 1968), some of which persist to adulthood (Cross and
Harlow, 1965; Mitchell, 1968a; Mitchell et al., 1966) but vary in frequency and in severity according to the extent of social deprivation during
infancy (Sackett, 1967). More delimited analyses of these isolate-reared
monkeys showed that these animals may show hyperphagia and polydipsia
under routine conditions (Miller et al., 1969; Miller et al., 1971), an inability to respond to appropriate emotional displays, although able to emit
them (Miller et al., 1967), aberrant social-choice behaviors pertinent to
other adults and to infants (Green and Gordon, 1964; Sackett et al., 1967),
and unusual patterns of tolerance, reactivity, and generalization to painful
electroshock (Lichstein and Sackett, 1971).

Harlow et al. (1969) evaluated the learning performances at about
2 years of age of rhesus monkeys exposed to 6–9 months of social isolation from birth onward. They observed the behaviors on discrimination
learning-set, delayed-response, and oddity learning-set tasks. Although
the groups of animals differed markedly in their adaptation to the test
situations, the overall learning performances were indistinguishable. Only
on the initial phases of the delayed-response task was a significant difference noted: The isolate-reared responded with longer latencies and higher
levels of random choices. From the compilation of data, reported here and
elsewhere, the authors concluded "that social isolation produces emotional
problems in rhesus monkeys which make adjustment to social and learning conditions difficult but leaves their intellectual functioning unimpaired."
(See also Angermeier et al., 1967; Griffin and Harlow, 1966; Mason and
Fitz-Gerald, 1962.)

[The reader should distinguish the researches on early social restriction, the topic of discussion here, from those on early sensory deprivation.
Although both are concerned with the developmental effects of altered
(from usual) sensory input during infancy, the rationale of the first focuses
on the contingent stimulation ("social stimulation," "social reinforcement," "socialization") as part of the infant–environment interaction,
whereas the second focuses on the input effects alone with emphasis on
sense systems, classically defined. Even more distinctive is the level of
behavioral analysis: the first emphasizes emotion–motivation–social vari-

ables without orientation to any given modality; the second emphasizes those behaviors, congenital and otherwise, which are traditionally deemed to be modality-related. Although the results from the second line of research have been impressive, indeed (cf. Riesen, 1961, 1966), the analysis of stimulus and response variables, much less their contingent interaction, is considerably more limited. Some of the data from the second group indicate considerable *a priori* structure with minimal distal-environmental modification (e.g., Berger and Meier, 1968; Ganz and Wilson, 1967); others (e.g., Held and Bauer, 1967) reveal the value of inputs in tandem as a necessary feature for sensorimotor development, a fact which may provide a conceptual basis for contingency analysis of behavioral development. See, also, Levison and Levison (1971) and Walk and Bond (1971).]

Corroboration of these observations with other species has been only partially complete. Abnormal stereotypic behaviors have been described an isolate-reared squirrel monkeys (*Saimiri sciureus;* King and King, 1969), crab-eating macaques (*M. irus;* Berkson, 1968), and chimpanzees (*Pan troglodytes;* Davenport and Menzel, 1963) but not in the marmoset, a prosimian (*Hapale jacchus;* Berkson *et al.*, 1966). Like the macaque, the isolate-reared chimpanzee revealed significant inadequacies in early social behavior (Mason *et al.*, 1968), especially evident in the avoidance of social contact, play, grooming (Turner *et al.*, 1969), and other changes in behavior progressively more evident from the onset of isolation, e.g., differential patterns of reactivity to novelty and other stimulus characteristics (Menzel, 1964). As in the rhesus macaques, severity of behavioral abnormality is directly related to extent of early social restriction (Mason *et al.*, 1968).

Rogers and Davenport (1969) reported essentially adequate sexual behavior in adult chimpanzees socially isolated for the first 3 years of postnatal life. Observations of copulatory behaviors of both sexes of isolate-reared chimpanzees indicated that impregnation was possible by both. Moreover, most of the animals of both sexes behaved in the sexual situation within limits defined by feral-reared animals. Considering the behaviors of other chimpanzees raised in a human, rather than chimpanzee environment, Rogers and Davenport chose to emphasize the caretaker-directed attachments developed in the rearing experience and the conflict thereby generated in sexual situations. They offered an interpretation consistent with the attachment behaviors of squirrel monkeys (King and King, 1969) and with the visual preference behaviors described by Sackett *et al.* (1965) for adult, isolate-reared rhesus macaques. [These animals preferred a human adult (female) stimulus to a conspecific stimulus, a behavior possibly related to the early nursery experiences of the isolate-reared animals.]

Davenport and Rogers (1968) concluded that the tested intellectual behaviors of the isolate-reared chimpanzees were clearly deficient. On a delayed-response task the restricted animals performed poorly in the initial phases of the task but did approach the performances of the feral-reared animals during the last half of the test procedure in terms of correctness of response, although with consistently longer latencies throughout. In a discrimination-learning-set task, the wild-born and early social-reared control animals were consistently superior to the laboratory-born and isolate-reared animals when tested at 7–9 years. The two groups of animals revealed a striking difference in the patterns of responding: The laboratory-(restricted) reared animals made many more perseveration errors than did the wild-born controls. Davenport *et al.* (1969; see, also, Rogers and Davenport, 1971) maintained that the "deficit" is real, but relative rather than absolute, and is most marked at the early stages of learning radically new problems.

> . . . Just where one wishes to draw the line between general adaptation and "real learning" is, of course, a moot point, but the present Restricted Ss have since their first testing at an early age, and in at least a dozen different test situations, had serious difficulties both in pretraining and in later performance, and this is certainly not completely irrelevant to the general question of "learning-to-learn" or the ability to adapt to new situations (i.e., "intelligence"). (Davenport *et al.*, 1969, p. 968.)

One must conclude from the published investigations on the sequelae of early manipulation, a line of research far from complete, that social deprivation during infancy yields adults who are socially inadequate by any definition of primate social order or behavior, the reports of apparently satisfactory reproductive and learning behaviors, notwithstanding. In a typical primate group, as found in the feral state, these animals could not function, even if they could survive. A cautious evaluation of these studies indicates that we must weigh carefully the attributes of the environments during early development and subsequent behavioral analysis, the contingencies which operate in each, and the discrepancies confronting the animal when moved from one environment to another. This we must do before we can hope to devise an empirically sound theory of early development or an adequate rationale for intervention into presumed counterparts in our own species (cf. Mason, 1971; Meier, 1970b).

F. COMPOSITE ACTION: INFANT–MOTHER–ENVIRONMENT INTERACTIONS

The accumulation of primate behavioral investigations in the field and in the laboratory on features of early social interactions and on the consequences of neonatal or infantile experimental manipulation suggests that an unusually productive strategy for research on mental retardation will include features of each. These investigations indicate that only by looking at the infant–environment interactions directly, with programmatic modifications of each, can we achieve a clear appreciation of normal development and the range of alternate routes to species-typical adaptive adult behaviors.

Published reports of research with this social-developmental environment flavor reveal a remarkable congruence of results despite broad diversity in the problems researched and their conceptual–theoretical bases. For some, this methodology has been used to examine the congenital anatomic–physiologic substrate requisite to the behavioral development of the individual infant. Thus we have students of the neural mechanisms in the maternal deprivation syndrome and the forces involved in the sparing of function related to the surgical loss of circumscribed areas of cortical tissue. At the other extreme are those who are interested in the flexibility of the social structure in which the infant is placed and in its tolerance and/or support of a new member with a sensorimotor handicap. In between are those who see maternal-rearing as a convenient, pragmatic solution to the pressing demands of infant care and those who are frankly interested in the interactive process itself, both in the usual and the unusual, and thus divide their energies between the infant and the environment, social and nonsocial.

Taking a group of infant rhesus monkeys, approximately half of which had had cortical lesions during the first postnatal week, Aarons and his associates (Aarons et al., 1962) attempted an analysis of the ameliorative value of intense social stimulation, i.e., optimal human care in the investigators' homes, upon the development of patterns of social–emotional behaviors. Following the 8–44-week rearing period, they returned the animals to the laboratory and observed them over an extended period for aspects of sensorimotor development, for affective behavior toward the human caretakers and monkey peers, and for exploratory-emotional behaviors with novel objects in unfamiliar settings. In all categories, save that of tactile sensitivity and motor dexterity (deficient in the parietal topectomized animals who compensated with increased oral exploratory activity) the controls and the operates were indistinguishable. The authors were duly impressed with the essential normalcy of the behaviors of all the animals (few autisms were shown as would be expected with isolate-rearing), but especially of the affective behaviors of the operates defective in that one form of stimulation, tactile, proposed as necessary for the establishment of early affective bonds in the primate (Harlow, 1958).

Kling and Green (1967) compared the development of fear responses in infant and juvenile macaques (M. mulatta, M. speciosa, and M. radiata) under conditions of isolate- and maternal-rearing. Half of the animals at each age level suffered bilateral amygdaloidectomies (infants, within the first 5 days; juveniles, presumably, after 1 year of life) without serious immediate postoperative complications; all showed prompt recovery. The two infant operates were immediately accepted by the mothers and resumed usual feeding-clinging patterns. Subsequently, all of the animals behaved in a manner which reflected the influence of the rearing conditions without apparent modification due to the surgical treatment, which in the adult may be profound (Weiskrantz, 1956). The isolate-reared isolates showed all of the characteristics of the deprivation syndrome recognized by the authors.

In their studies on the implications of subcortical structures (the caudate nuclei) in the sparing of delayed-response behaviors in the infant frontal-topectomized animals, Kling and Tucker (1967, 1968; Tucker and Kling, 1969) indicated that maternal-rearing was without demonstrable effect where such rearing practices could be established. Thus, they could find no differences in their frontal-topectomized infants as a function of rearing procedure. Since the combined lesions in the frontal cortex and the caudate nuclei led to changes in behavior inimical to survival, they could not assay the effects of rearing procedures. (Note: We must recognize here as we did in the isolation studies that these laboratory-reared infants were not reared devoid entirely of relevant social stimulation. Some seemed to be amply provided by the human caretakers. In extreme conditions, such as those described by Kling and Tucker, that human social involvement may be extreme as well, in order to assure infant survival.)

At the other end of our ideological dimension are those investigators who have analyzed the social process in dealing with the defective infant. That primate groups may deal satisfactorily with the defective member, is evident in the surveys of terata in natural groups (see above, and Furuya, 1966). The limits of the requisite changes in group behaviors remains problematic. Rumbaugh (1965) described adequate but species-atypical behaviors in mother–infant relations in this context. Squirrel monkeys (*Saimiri sciureus*) do not maintain the ventroventral relation between mother and infant typical of the macaque, for example (Hopf, 1967). Usually the infant clings to the mother in a dorsoventral relation. Rumbaugh bound the limbs of an infant squirrel monkey (10 days old) and noted that the mother immediately retrieved and carried the infant, in spite of its handicap, in a ventroventral position. She was seen to carry the infant bipedally and to cradle it—highly improbable behaviors in this species. (Rumbaugh had also observed these maternal behaviors in a mother with a dead infant.) Berkson (1970) observed the subsequent social interactions of two infant crab-eating macaques (*M. irus*), part of a feral, free-ranging troop which had experimentally imposed visual defects (i.e., "travel vision") when approximately 1 month of age. The animals were observed with their mothers and in the larger social context of the natural group until about 7 months of age when they were lost in a particularly arid period. In most respects, these infants showed normal mother–infant relations, emitting age-typical behaviors for their species. They moved away from the mothers with increasing frequency from the second month onward, interacting with age peers and others of the group. Under conditions of stress or of troop movement from other causes the mothers displayed the necessary retrieval and other behaviors mandatory for the blind infants to remain with the group, and thus to survive. These and other behaviors which appeared to limit the intragroup aggression directed toward the infants were unusual but apparently part of the natural mechanisms within the group enhancing the likelihood of survival of its defective members. Comparable behaviors have been noted by this author (Meier, 1972) in a mother rhesus with an infant deafened within 12 hr of birth and maintained together in a communal situation (Meier, 1969). For the first 2 years the mother remained very solicitous of her offspring as shown in the increased frequency of cradling and retrieving behaviors. At 3½ years, the mother and the male offspring still remain close together despite otherwise fully normal group-oriented social activities of each. The young male does show above-normal levels of aggressiveness, especially toward the mature, dominant male, and some peculiar noise-making behaviors but these may be directly related to his primary handicap, deafness.

In between these extremes is the rapidly growing literature, not directly related to mental retardation per se, which underscores with emphasis the contributions of the infant, of the mother, and of the larger environment on early social behaviors, notably those between mother and infant. From these we have come to realize the significance of infant sex (Jensen et al., 1966; Mitchell, 1968b) and sensorimotor capacities (Berkson and Karrer, 1968; Lindburg, 1969a); also, maternal species (Kaufman and Rosenblum, 1969; Rosenblum and Kaufman, 1968), parity (Mitchell and Stevens, 1968), and early social experience (Mitchell et al., 1967; Seay et al., 1964). Apparently, incapacitation of the mother by way of sectioning of the seventh cranial nerve may also modify the frequency and intensity of those behaviors, even in a group situation (Meier et al., 1971). Such a mother seems to be less restrictive of her infant permitting it to move away at an earlier age than would be expected for age and species. Possibly such a mother, by virtue of her facial paralysis, is less able to communicate with her infant, and thereby less able to "control" it. Possibly, such a mother, by virtue of her facial paralysis and that of her age peers, is less able to communicate with them and, therefore, less likely to restrict the "waywardness" of her offspring.

Jensen and his colleagues at the University of Washington have described the considerable influence the total social environment has on mother–infant relations. Jensen *et al.* (1968) reported that the patterns of mother–infant interactions over the first 15 weeks of postnatal development in pigtailed macaque (*M. nemestrina*) pairs were related to the extent of social interaction permitted. Pairs in sound-shielded environments spent more time in physical contact and less time oriented toward the environment as indicated by lower frequency of nonsocial manipulation, locomotion, and climbing. The authors concluded that (1) a stimulus-poor environment intensifies the physical relationships between a mother and her infant, but (2) the environment does not affect the basic nature of the mother's role, and (3) a stimulus-poor environment produces some retardation in infant development. In a subsequent study these same authors (Jensen *et al.*, 1969) focused on one form of response (hitting behavior) presenting data which supported the conclusion about the relatively meager effect of the environment on maternal behavior. More recently, Wolfheim *et al.* (1970) revealed the differences in mother–infant relations in pigtailed macaque (*M. nemestrina*) between pairs maintained within laboratory cages and others within a large group in a communal compound. From the nature, frequency, and duration of separation and locomotive behaviors as well as the nursing and manipulative behaviors, the investigators necessarily concluded that the group-raised infants were more involved with their mothers, rather than less, and that the mothers were more retentive of their offspring, yet less punitive, than were the isolate-caged mothers. Clearly, under these conditions, the overall environmental influence on maternal behavior was considerable.

Not to be overlooked in this train of investigations on individual development in social contact are those studies on separation phenomena. Although acknowledged for their relation to psychiatric concerns (McKinney and Bunney, 1969), these efforts confirm the significant influence of social factors beyond the mother and infant in mother–infant interactions, the variations attributable to species and to sex of offspring, as well as the persisting effects after mother and infant are reunited and returned to the social situation (Abrams, 1969; Kaplan, 1970; Kaufman and Rosenblum, 1967; Rowell, 1968; Rosenblum and Kaufman, 1968). [See also Spencer-Booth (1969) for other aspects of group influence on mother–infant relations.]

V. Summary

In this review I have tried to summarize and evaluate the behavioral research (primarily) on nonhuman primates relevant to the etiology—ontogenesis—of mental retardation. In so doing, I have taken the liberty

of defining mental retardation as a developmental phenomenon conse-
quential to an inadequate infant–environment interaction; that is, the interaction between the infant, especially as he exists following serious anatomic
and/or biochemical aberrations, and an infantile–childhood environment
which is minimally stimulating, viz., does not provide sufficient contingent
stimulation, and therefore does not permit the evolution of socially requisite coping behavior. Later, in a social or scholastic-learning situation,
such a child lacks the learning dexterity recognized as typical for his
chronological age and is classed, "mentally retarded." Consequently, I
have reviewed the published research on those prenatal, perinatal, neonatal, and infantile variables which can so alter the infant that his reactivity to his early environment is markedly altered and which, therefore,
diminish his effectiveness as a stimulus to caretaking behaviors from his
social environment. In sequence, I discussed the long-range behavioral
effects of an early existence in a restricted social environment upon the
ontogeny of a normal newborn primate. Such a primate does develop, but
his social behaviors are clearly abnormal when compared with those of an
animal of similar age reared in a feral situation. They are, however, "typical" behaviors for this species; that is, the stereotypes of the isolate-reared
macaque are still, undeniably, simian. Learning and other cognitive performances are probably within acceptable, normal ranges for the species
as reared under the usual social conditions, but reactivity to novelty and
change are such that serious motivational difficulties complicate the approach to any new task. In looking at the nonhuman primate research in
this fashion, I believe that I have offered corroboration to the belief that
mental retardation is a developmental phenomenon in which a plethora
of variables interact, concurrently and sequentially, and have added substance to the hope that early diagnosis of behavioral abnormality can, with
appropriate social-environment contingent stimulation, alter the ineluctable
destiny of mental deficiency. And finally, I must remind the reader that
these researches on nonhuman primate development span a very short
period in the history of biomedical science—barely 15 years. Nevertheless,
the amount of provocative and innovative information should cause us to
marvel. That we do not may only reflect our jaded sense of newness and
our eagerness to construct a rational basis for social order and social
change.

ACKNOWLEDGMENT

    This chapter was prepared with support from the Joseph P. Kennedy Jr. Founda-
tion and the Program Project Grant No. HD-00973, National Institute of Child

Health and Human Development, United States Department of Health, Education and Welfare to George Peabody College for Teachers, Nashville, Tennessee.

REFERENCES

Aarons, L., Schulman, J., Masserman, J. H., and Zimmar, G. P. (1962). *Recent Advan. Biol. Psychiat.* 4, 347–359.

Abrams, P. S. (1969). *Amer. J. Phys. Anthropol.* 31, 262 (abstr.).

Akert, K., Orth, D. S., Harlow, H. F., and Schiltz, K. A. (1960). *Science* 132, 1944–1945.

Angermeier, W. F., Phelps, J. B., and Reynolds, H. H. (1967). *Psychon. Sci.* 8, 379–380.

Barrow, M. V., Steffek, A. J., and King, C. T. G. (1969). *Folia Primatol.* 10, 195–203.

Berger, R. J., and Meier, G. W. (1968). *Develop. Psychobiol.* 1, 266–275.

Berkson, G. (1967a). *Amer. J. Ment. Defic.* 72, 10–15.

Berkson, G. (1967b). *In* "Comparative Psychopathology" (J. Zubin and H. F. Hunt, eds.), pp. 76–94. Grune & Stratton, New York.

Berkson, G. (1968). *Develop. Psychobiol.* 1, 118–132.

Berkson, G. (1970). *Folia Primatol.* 12, 284–289.

Berkson, G., and Karrer, R. (1968). *Develop. Psychobiol.* 1, 170–174.

Berkson, G., Goodrich, J., and Kraft, I. (1966). *Percept. Motor Skills* 23, 491–498.

Berman, D., Karalitzky, A. R., and Berman, A. J. (1971). *Exp. Neurol.* 31, 140–149.

Blomquist, A. J., Harlow, H. F., Schlitz, K. A., and Mohr, D. (1971). *The Gerontologist* 11, 41 (abstract).

Bowden, D. M., Goldman, P. S., Rosvold, H. E., and Greenstreet, R. L. (1971). *Exp. Brain Res.* 12, 265–274.

Caldeyro-Barcia, R., moderator (1969). "Perinatal Factors Affecting Human Development," p. 253. Pan Amer. Health Organ., Washington, D. C.

Candland, D. K., and Mason, W. A. (1968). *Develop. Psychobiol.* 1, 254–256.

Chamove, A. S., and Davenport, J. W. (1969). *Develop. Psychobiol.* 2, 207–211.

Chamove, A. S., Waisman, H. A., and Harlow, H. F. (1970). *J. Abnorm. Psychol.* 76, 62–68.

Cotes, P. M. (1966). *In* "Some Recent Developments in Comparative Medicine" (R. N. T-W Fiennes, ed.), pp. 309–312. Academic Press, New York.

Courtney, K. D., and Valerio, D. A. (1968). *Teratology* 1, 163–172.

Cross, H. A., and Harlow, H. F. (1965). *J. Exp. Res. Pers.* 1, 39–49.

Davenport, J. W., Kerr, G. R., and Scheffler, G. (1971). *Mainly Monkeys.* 2, 4–6.

Davenport, R. K., Jr., and Menzel, E. W., Jr. (1963). *Arch. Gen. Psychiat.* 8, 99–104.

Davenport, R. K., and Rogers, C. M. (1968). *Amer. J. Ment. Defic.* 72, 674–680.

Davenport, R. K., Rogers, C. M., and Menzel, E. W. (1969). *Amer. J. Ment. Defic.* 73, 963–969.

Delahunt, C. S., and Lassen, L. J. (1964). *Science* 146, 1300–1305.

Delahunt, C. S., and Rieser, N. (1967). *Amer. J. Obstet. Gynecol.* 99, 580–588.

DeVore, I., ed. (1965). "Primate Behavior: Field Studies of Monkeys and Apes," p. 654. Holt, New York.

Edgerton, R. B. (1970). *In* "Social-Cultural Aspects of Mental Retardation" (H. C. Haywood, ed.), pp. 523–559. Appleton, New York.

Ellis, N. R., ed. (1970). "International Review of Research in Mental Retardation," Vol. 4, p. 4. Academic, New York.

Elizan, T. S., and Fabiyi, A. (1970). *Amer. J. Obstet. Gynecol.* 106, 147–165.

Faro, M. D., and Windle, W. F. (1969). *Exp. Neurol.* 24, 38–53.

Foshee, D. P., and Meier, G. W. (1969). Unpublished observations.

Foshee, D. P., Meier, G. W., and Andy, O. J. (1969). *Pap., Southeast. Psychol. Asso., 1969.*

Frisch, J. E. (1968). *In* "Primates: Studies in Adaptation and Variability" (P. C. Jay, ed.), pp. 243–252. Holt, New York.

Furuya, Y. (1966). *Primates* 7, 488–492.

Ganz, L., and Wilson, P. D. (1967). *J. Comp. Physiol. Psychol.* 63, 258–269.

Goldman, P. S. (1971). *Exp. Neurol.* 32, 366–387.

Goldman, P. S., Rosvold, H. E., and Mishkin, M. (1970a). *J. Comp. Physiol. Psychol.* 70, 454–463.

Goldman, P. S., Rosvold, H. E., and Mishkin, M. (1970b). *Exp. Neurol.* 29, 221–226.

Goy, R. W. (1968). *In* "Endocrinology and Human Behaviour" (R. P. Michaels, ed.), pp. 12–31. Oxford Univ. Press, London and New York.

Green, P. C., and Gordon, M. (1964). *Science* 145, 292–294.

Griffin, G. A., and Harlow, H. F. (1966). *Child Develop.* 37, 533–547.

Harlow, H. F. (1958). *Amer. Psychol.* 13, 673–685.

Harlow, H. F., and Harlow, M. K. (1962a). *Sci. Amer.* 207, 136–146.

Harlow, H. F., and Harlow, M. K. (1962b). *Bull. Menninger Clin.* 26, 213–224.

Harlow, H. F., and Harlow, M. K. (1965). *In* "Behavior of Nonhuman Primates" (A. M. Schrier, H. F. Harlow, and F. Stollnitz, eds.), Vol. 2, pp. 287–334. Academic Press, New York.

Harlow, H. F., Davis, R. T., Settlage, P. H., and Meyer, D. R. (1952). *J. Comp. Physiol. Psychol.* 45, 419–429.

Harlow, H. F., Akert, K., and Schiltz, K. A. (1964). *In* "The Frontal Granular Cortex and Behavior" (J. M. Warren and K. Akert, eds.), pp. 126–148. McGraw-Hill, New York.

Harlow, H. F., Blomquist, A. J., Thompson, C. I., Schiltz, K. A., and Harlow, M. K. (1968). *In* "The Neuropsychology of Development: A Symposium" (R. L. Isaacson, ed.), pp. 79–120. Wiley, New York.

Harlow, H. F., Schlitz, K. A., and Harlow, M. K. (1969). *Proc. Int. Congr. Primatol., 2nd, 1968.* Vol. 1, pp. 178–185.

Harlow, H. F., Thompson, C. I., Blomquist, A. J., and Schiltz, K. A. (1970). *Brain Res.* 18, 343–353.

Heber, R. F. (1961). "A Manual on Terminology and Classification in Mental Retardation," 2nd ed. American Association on Mental Retardation, Pineville, Louisiana.

Held, R., and Bauer, J. A., Jr. (1967). *Science* 155, 718–720.

Hill, S. D., and McCormack, S. (1969). *Pap. Psychon. Soc., 1969.*

Hinde, R. A. (1971). *In* "Behavior of Nonhuman Primates" (A. M. Schrier and F. Stollnitz, eds.), pp. 1–68. Academic Press, New York.

Hopf, S. (1967). *In* "Progress in Primatology" (D. Starck, R. Schneider, and H. J. Kuhn, eds.), pp. 255–262. Fischer, Stuttgart.

Hsia, Y. Y. (1967). *In* "Psychopathology of Mental Retardation" (J. Zubin and G. A. Jervis, eds.), pp. 28–44. Grune & Stratton, New York.

Hyman, A., Parker, B., Berman, D., and Berman, A. J. (1970). *Exp. Neurol.* **28**, 420–425.

Hyman, A., Berman, D., and Berman, A. J. (1971). *Exp. Neurol.* **30**, 362–366.

Isaacson, R. L., ed. (1968). "The Neuropsychology of Development: A Symposium," p. 177. Wiley, New York.

Iwamoto, M. (1967). *Primates* **8**, 247–270.

Iwamoto, M., and Hirai, M. (1970). *Primates* **11**, 395–398.

Jay, P. C., ed. (1968). "Primates: Studies in Adaptation and Variability," p. 529. Holt, New York.

Jensen, G. D., Bobbitt, R. A., and Gordon, B. N. (1966). *Recent Advan. Biol. Psychiat.* **9**, 283–293.

Jensen, G. D., Bobbitt, R. A., and Gordon, B. N. (1968). *J. Comp. Physiol. Psychol.* **66**, 259–263.

Jensen, G. D., Bobbitt, R. A., and Gordon, B. N. (1969). *J. Psychiat. Res.* **7**, 55–61.

Jervis, G. W. (1954). *In* "Genetics and the Inheritance of Integrated Neurological and Psychiatric Patterns" (D. Hooker and C. C. Hare, eds.), pp. 259–282. Williams & Wilkins, Baltimore, Maryland.

Kaplan, J. (1970). *Develop. Psychobiol.* **3**, 43–52.

Kato, T., Jarvik, L. F., Roizin, L., and Moralishvili, E. (1970). *Dis. Nerv. Syst.* **31**, 245–250.

Kaufman, I. C., and Rosenblum, L. A. (1967). *Psychosom. Med.* **29**, 648–675.

Kaufman, I. C., and Rosenblum, L. A. (1969). *Ann. N. Y. Acad. Sci.* **159**, 681–695.

Kawai, A., Azuma, S., and Yoshiba, K. (1967). *Primates* **8**, 35–74.

Kennard, M. A. (1938). *J. Neurophysiol.* **1**, 477–496.

Kerr, G. R., Chamove, A. S., Harlow, H. F., and Waisman, H. A. (1968). *Pediatrics* **42**, 27–36.

Kerr, G. R., Chamove, A. S., Harlow, H. F., and Waisman, H. A. (1969). *Pediat. Res.* **3**, 305–312.

Kerr, G. R., Allen, J. R., Scheffler, G., and Waisman, H. A. (1970). *Amer. J. Clin. Nutr.* **23**, 739–748.

King, J. E., and King, P. A. (1969). *Develop. Psychobiol.* **2**, 251–256.

Kling, A., and Green, P. C. (1967). *Nature (London)* **213**, 742–743.

Kling, A., and Tucker, T. J. (1967). *Brain Res.* **6**, 428–439.

Kling, A., and Tucker, T. J. (1968). *In* "The Neuropsychology of Development: A Symposium" (R. L. Isaacson, ed.), pp. 121–146. Wiley, New York.

Koford, C. B. (1965). *In* "Primate Behavior: Field Studies of Monkeys and Apes" (I. DeVore, ed.), pp. 160–174. Holt, New York.

Koford, C. B., Farber, P. A., and Windle, W. F. (1966). *Folia Primatol.* **4**, 221–226.

Köhler, W. (1925). "The Mentality of Apes." Harcourt, New York.

Lapin, B. A., and Yakovleva, L. A. (1963). *In* "Comparative Pathology in Monkeys" (W. F. Windle, ed.), p. 229. Thomas, Springfield, Illinois.

Levison, C. A., and Levison, P. K. (1971). *Psychon. Sci.* **22**, 145–147.

Lichstein, L., and Sackett, G. P. (1971). *Develop. Psychobiol.* **4**, 339–352.

Lindburg, D. G. (1969a). *Psychon. Sci.* **15**, 55–56.

Lindburg, D. G. (1969b). *Develop. Psychobiol.* **2**, 184–190.

Lindburg, D. G. (1971). *In* "Primate Behavior: Developments in Field and Laboratory Research" (L. A. Rosenblum, ed.), Vol. 2, pp. 1–106. Academic Press, New York.

Lucey, J. F., and Behrman, R. E. (1963). *Science* **139**, 1295–1296.

Lucey, J. F., Hibbard, E., Behrman, R. E., Esquivel de Gallardo, R. O., and Windle, W. R. (1964). *Exp. Neurol.* **9**, 43–58.

McClure, H. M., Belden, K. H., Pieper, W. A., and Jacobson, C. B. (1969). *Science* **165**, 1010–1012.

McKinney, W. T., and Bunney, W. E., Jr. (1969). *Arch. Gen. Psychiat.* **21**, 240–248.

McNeil, T. F., Wiegerink, R., and Dozier, J. E. (1970). *J. Nerv. Ment. Dis.* **151**, 24–34.

Marler, P., and Gordon, A. (1968). *In* "Biology and Behavior: Environmental Influences" (D. C. Glass, ed.), pp. 113–129. Rockefeller Univ. Press, New York.

Masland, R. L., Sarason, S. B., and Gladwin, T. (1958). "Mental Subnormality: Biological, Psychological, and Cultural Factors," p. 442. Basic Books, New York.

Mason, W. A. (1960). *J. Comp. Physiol. Psychol.* **53**, 582–589.

Mason, W. A. (1963). *Percept. Motor Skills* **16**, 263–270.

Mason, W. A. (1965). *In* "Primate Behavior: Field Studies of Monkeys and Apes" (I. DeVore, ed.), pp. 514–543. Holt, New York.

Mason, W. A. (1968). *In* "Biology and Behavior: Environmental Influences" (D. C. Glass, ed.), pp. 70–101. Rockefeller Univ. Press, New York.

Mason, W. A. (1971). *In* "Nebraska Symposium on Motivation" (W. J. Arnold and M. M. Page, eds.), pp. 35–67. Univ. of Nebraska Press, Lincoln.

Mason, W. A., and Fitz-Gerald, F. L. (1962). *Percept. Motor Skills* **15**, 594.

Mason, W. A., and Green, P. C. (1962). *J. Comp. Physiol. Psychol.* **55**, 363–368.

Mason, W. A., and Sponholz, R. R. (1963). *J. Psychiat. Res.* **1**, 229–306.

Mason, W. A., Davenport, R. K., Jr., and Menzel, E. W., Jr. (1968). *In* "Early Experience and Behavior" (G. Newton and S. Levine, eds.), pp. 440–480. Thomas, Springfield, Illinois.

Meier, G. W. (1964). *Science* **143**, 968–970.

Meier, G. W. (1965a). *Anim. Behav.* **13**, 228–231.

Meier, G. W. (1965b). *Science* **206**, 492–493.

Meier, G. W. (1968). *In* "Early Experience and Behavior" (G. Newton and S. Levine, eds.), pp. 338–364. Thomas, Springfield, Illinois.

Meier, G. W. (1969). *Proc. Int. Congr. Primatol., 2nd, 1968* **1**, 66–71.

Meier, G. W. (1970a). *In* "International Review of Research in Mental Retardation" (N. R. Ellis, ed.), Vol. 4, pp. 263–310. Academic Press, New York.

Meier, G. W. (1970b). *In* "Miami Symposium on the Prediction of Behavior, 1968: Effects of Early Experience" (M. R. Jones, ed.), pp. 55–60. Univ. of Miami Press, Coral Gables, Florida.

Meier, G. W. (1972). Unpublished data.

Meier, G. W. (1971). *In* "Pharmacological and Biophysical Agents and Behavior" (E. Furchtgott, ed.), pp. 99–142. Academic, New York.

Meier, G. W., and Berger, R. J. (1965). *Exp. Neurol.* **12**, 257–277.

Meier, G. W., and Garcia-Rodriguez, C. (1966a). *Psychol. Rep.* **19**, 1159–1169.

Meier, G. W., and Garcia-Rodriguez, C. (1966b). *Psychol. Rep.* **19**, 1219–1225.

Meier, G. W., Izard, C. E., and Cobb, C. (1971). *Pap. Southeast Psychol. Asso. 1971.*

Menzel, E. W., Jr. (1964). *Psychol. Forsch.* **27**, 337–365.

Menzel, E. W., Jr. (1968). *Develop. Psychobiol.* **1**, 175–184.

Miller, R. E., Caul, W. F., and Mirsky, I. A. (1967). *J. Personal. Soc. Psychol.* **7**, 231–239.

Miller, R. E., Caul, W. F., and Mirsky, I. A. (1971). *Physiol. Behav.* **7**, 127–134.

Miller, R. E., Mirsky, I. A., Caul, W. F., and Sakata, T. (1969). *Science* **165**, 1027–1028.

Missakian, E. A. (1969). *J. Comp. Physiol. Psychol.* 69, 403–407.

Mitchell, G. D. (1968a). *Folia Primatol.* 8, 132–147.

Mitchell, G. D. (1968b). *Child Develop.* 39, 613–620.

Mitchell, G. D., and Stevens, C. W. (1968). *Develop. Psychobiol.* 1, 280–286.

Mitchell, G. D., Raymond, E. J., Ruppenthal, G. C., and Harlow, H. F. (1966). *Psychol. Rep.* 18, 567–580.

Mitchell, G. D., Arling, G. L., and Møller, G. W. (1967). *Psychon. Sci.* 8, 209–210.

Miyadi, D. (1967). *In* "Progress in Primatology" (D. Starck, R. Schneider, and H. J. Kuhn, eds.), pp. 228–231. Fischer, Stuttgart.

Morris, D., ed. (1967). "Primate Ethology," p. 374. Weidenfeld & Nicolson, London.

Morris, L. N. (1971). *Teratology* 4, 335–342.

Myers, R. E. (1967). *In* "Brain Damage in the Fetus and Newborn from Hypoxia or Asphyxia" (L. S. James, R. E. Myers, and G. E. Gaull, eds.), pp. 17–21. Ross Laboratories, Columbus, Ohio.

Myers, R. E. (1969a). *In* "Perinatal Factors Affecting Human Development" (R. Caldeyro-Barcia, moderator), pp. 205–214. Pan Amer. Health Organ., Washington, D. C.

Myers, R. E. (1969b). *Pap., Conf. Exp. Med. Surg. Primates, 2nd, 1969.*

Myers, R. E., Beard, R., and Adamsons, K. (1969). *Neurology* 19, 1012–1018.

Pasamanick, B., and Knobloch, H. (1961). *In* "Prevention of Mental Disorders in Children" (G. Caplan, ed.), pp. 74–94. Basic Books, New York.

Phoenix, C. H., Goy, R. W., and Resko, J. A. (1968). *In* "Reproduction and Sexual Behavior" (M. Diamond, ed.), pp. 33–49. Indiana Univ. Press, Bloomington.

Pratt, C. L., and Sackett, G. P. (1967). *Science* 155, 1133–1135.

Raisler, R. L., and Harlow, H. F. (1965). *J. Comp. Physiol. Psychol.* 60, 167–174.

Ranck, J., and Windle, W. R. (1959). *Exp. Neurol.* 1, 130–154.

Riesen, A. H. (1961). *In* "Functions of Varied Experience" (D. W. Fiske and S. R. Maddi, eds.), pp. 57–80. Dorsey, Homewood, Illinois.

Riesen, A. H. (1966). *In* "Progress in Physiological Psychology" (E. Stellar and J. M. Sparague, eds.), Vol. 1, pp. 117–147. Academic Press, New York.

Riopelle, A. J., and Thomsen, C. E. (1968). *In* "Methods of Animal Experimentation" (W. I. Gay, ed.), Vol. 3, pp. 81–124. Academic Press, New York.

Robinson, H. B., and Robinson, N. M. (1965). "The Mentally Retarded Child: A Psychological Approach." McGraw-Hill, New York.

Rogers, C. M., and Davenport, R. K. (1969). *Develop. Psychol.* 1, 200–204.

Rogers, C. M., and Davenport, R. K. (1971). *Amer. J. Ment. Def.* 75, 526–530.

Rosenblum, L. A., and Kaufman, I. C. (1968). *Amer. J. Orthopsychiat.* 38, 418–426.

Rowell, T. E. (1968). *Folia Primatol.* 9, 114–122.

Ruch, T. C. (1959). "Diseases of Laboratory Primates." Saunders, Philadelphia, Pennsylvania.

Rumbaugh, D. M. (1965). *Psychol. Rep.* 16, 171–176.

Sackett, G. P. (1967). *J. Comp. Physiol. Psychol.* 64, 363–365.

Sackett, G. P. (1968). *In* "Abnormal Behavior in Animals" (M. W. Fox, ed.), pp. 293–331. Saunders, Philadelphia, Pennsylvania.

Sackett, G. P. (1970). *In* "Miami Symposium on the Prediction of Behavior, 1968: Effects of Early Experience" (M. R. Jones, ed.), pp. 11–53. Univ. of Miami Press, Coral Gables, Florida.

Sackett, G. P., Porter, M., and Holmes, H. (1965). *Science* 147, 304–306.

Sackett, G. P., Griffin, G. A., Pratt, C., Joslyn, W. D., and Ruppenthal, G. (1967). *J. Comp. Physiol. Psychol.* 63, 376–381.

Saxon, S. V. (1961a). *J. Genet. Psychol.* 99, 277–282.

Saxon, S. V. (1961b). *J. Genet. Psychol.* **99**, 283–287.

Saxon, S. V., and Ponce, C. G. (1961). *Exp. Neurol.* **4**, 460–469.

Schrier, A. M., Harlow, H. F., and Stollnitz, F., eds. (1965). "Behavior of Nonhuman Primates," Vol. 1, p. 281. Academic Press, New York.

Scrimshaw, N. S., and Gordon, J. E., eds. (1968). "Malnutrition, Learning, and Behavior," p. 566. MIT Press, Cambridge, Massachusetts.

Seay, B., Alexander, B. K., and Harlow, H. F. (1964). *J. Abnorm. Soc. Psychol.* **69**, 345–354.

Sechzer, J. A. (1969). *Exp. Neurol.* **24**, 497–507.

Sechzer, J. A., Faro, M. D., Barker, J. N., Barsky, D., Gutierrez, S., and Windle, W. F. (1971). *Science* **171**, 1173–1175.

Sever, J. L., Meier, G. W., Windle, W. F., Schiff, G. M., Monif, G. R., and Fabiyi, A. (1966). *J. Infec. Dis.* **166**, 21–26.

Singh, S. D. (1968). *Psychon. Sci.* **11**, 83–84.

Singh, S. D. (1969). *Sci. Amer.* **221**, 108–115.

Suomi, S. J., Harlow, H. F., and Kimball, S. D. (1971). *Psych. Rep.* **29**, 1171–1177.

Southwick, C. H., Beg, M. A., and Siddiqui, M. R. (1965). *In* "Primate Behavior: Field Studies of Monkeys and Apes" (I. DeVore, ed.), pp. 111–159. Holt, New York.

Spencer-Booth, Y. (1969). *Extr. Mammalia* **33**, 80–86.

Swindler, D. R., and Merrill, O. M. (1971). *Amer. J. Phys. Anthropol.* **34**, 435–451.

Thompson, C. I. (1969). *Physiol. Behav.* **4**, 1027–1029.

Thompson, C. I., Schwartzbaum, J. S., and Harlow, H. F. (1969). *Physiol. Behav.* **4**, 249–254.

Tsang, Y. C. (1937). *J. Comp. Psychol.* **24**, 221–253.

Tucker, T. J., and Kling, A. (1967). *Brain Res.* **5**, 377–389.

Tucker, T. J., and Kling, A. (1969). *Exp. Neurol.* **23**, 491–502.

Turner, C. H., Davenport, R. K., Jr., and Rogers, C. M. (1969). *Amer. J. Psychiat.* **11**, 1531–1536.

Valerio, D. A., Miller, R. L., Innes, J. R. M., Courtney, K. D., Pallota, A. J., and Guttmacher, R. M. (1969). "Macaca Mulatta (Management of a Laboratory Breeding Colony)," p. 140. Academic Press, New York.

Van Hooff, J. A. R. A. M. (1967). *In* "Primate Ethology" (D. Morris, ed.), pp. 7–68. Weidenfeld & Nicolson, London.

Van Wagenen, G. (1950). *In* "The Care and Breeding of Laboratory Animals" (E. J. Farris, ed.), pp. 1–42. Wiley, New York.

Vine, I. (1970). *In* "Social Behavior in Birds and Mammals" (J. H. Crook, ed.), pp. 279–354. Academic Press, New York.

Waisman, H. A., and Harlow, H. F. (1965). *Science* **147**, 685–695.

Waizer, J., Baumback, D. T., Berman, D., and Berman, A. J. (1972). *J. Comp. Physiol. Psychol.* **78**, 386–390.

Walk, R. D., and Bond, E. K. (1971). *Psychon. Sci.* **23**, 115–116.

Weiskrantz, L. (1956). *J. Comp. Physiol. Psychol.* **49**, 381–391.

Wilson, J. G. (1966). *In* "Proceedings, Conference on Nonhuman Primate Toxicology" (C. O. Miller, ed.), pp. 114–118. U. S. Dept. of Health, Education, and Welfare, Washington, D. C.

Wilson, J. G., and Gavan, J. A. (1967). *Anat. Rec.* **158**, 99–109.

Windle, W. F. (1967). *In* "Psychopathology of Mental Development" (J. Zubin and G. A. Jervis, eds.), pp. 140–147. Grune & Stratton, New York.

Wolfeim, J. H., Jensen, G. D., and Bobbitt, R. A. (1970). *Primates* **11**, 119–124.

Yerkes, R. M., and Yerkes, A. W. (1929). "The Great Apes." Yale Univ. Press, New Haven, Connecticut.

Young, W. C., Goy, R. W., and Phoenix, C. H. (1964). *Science* 143, 212–218.

Zigler, E. (1967). *Science* 155, 292–298.

Zimmerman, R. R. (1969). *Percept. Motor Skills* 28, 867–876.

Zimmerman, R. R., and Strobel, D. A. (1969). *Proc., 77th Annu. Convention, Amer. Psychol. Ass.* 4, 241–242.

Zimmerman, R. R., Strobel, D. A., and Maguire, D. (1970). *Proc., 78th Annu. Convention, Amer. Psychol. Ass.* 5, 197–198.

CHAPTER 21

Primate Studies and Human Evolution*

S. L. WASHBURN

One hundred years ago it seemed clear to Darwin that man's closest relatives were the apes (Pongidae). In the introduction to the *Descent of Man* (1871) he stated, "Nor shall I have occasion to do more than allude to the amount of difference between man and the anthropomorphous apes; for Prof. Huxley, in the opinion of most competent judges, has conclusively shown that in every visible character man differs less from the higher apes, than these do from lower members of the same order of Primates" (p. 390). Huxley (1863, p. 86) had stated, "It is quite certain that the Ape which most nearly approaches man, in the totality of its organization, is either the Chimpanzee or the Gorilla . . ." But this common nineteenth century point of view came progressively under attack. During the first

* This paper is part of a program on primate behavior supported by the United States Public Health Service (Grant No. MH 08623).

half of this century nearly every major group of living primates has been claimed as the one most closely related to man. At the present time responsible scientists regard the separation of the line leading to man as being from 5 to 50 million years ago. In examining sources, such as Flower and Lydekker (1891) or Weber (1928), it appears that, in addition to progress (new data and clarification of issues) scientists have produced confusion, especially in taxonomy. As Simpson has written (1945, p. 181): "The peculiar fascination of the primates and their publicity value have almost taken the order out of the hands of sober and conservative mammalogists and have kept, and do keep, its taxonomy in turmoil. Moreover, even mammalogists who might be entirely conservative in dealing, say, with rats are likely to lose a sense of perspective when they come to the primates, and many studies of this order are covertly or overtly emotional." Or, again, on page 185: "With closer approach to man in the zoological system, the confusion bequeathed to us by swarms of students, of all degrees of competence and shades of judgment, becomes increasingly greater."

Although the taxonomic confusion and the multitude of theories on human origins are undoubtedly due in part to the situation which Simpson describes, there are other causes. The fossil record of monkeys and apes is exceedingly scanty, being composed largely of teeth and fragments of jaws. Romer (1968, p. 161) has complained of this situation: "So great has been this concentration on dentition that I often accuse my "mammalian" colleagues, not without some degree of justice, of conceiving of mammals as consisting solely of molar teeth and of considering that mammalian evolution consisted of parent molar teeth giving birth to filial molar teeth and so on down through the ages." Comparative anatomy provides a vastly greater amount of information than the fossil record; but here there is no agreement on how the information is to be used. With access to the same information, scientists may conclude either that man is particularly closely related to the African apes or that no creatures properly classified as apes (Pongidae) could have been in the ancestry of man. The problem with the interpretation of the fossils is the fragmentary nature of the evidence. The problem with comparative anatomy is lack of rules in the use of a vast amount of descriptive information. These problems are vividly presented in the controversies between LeGros Clark (1967) and Zuckerman (1966) over the interpretation of *Australopithecus*.

It is my belief that study of the contemporary primates can supplement the fossil record, bringing order into the study of comparative anatomy and clarifying the behavioral stages in the origin and evolution of man. Recent developments have settled some of the fundamental issues, opening the way for a reconsideration of comparative anatomy, behavior, and the fossil record. Perhaps we can now see why Huxley was right, where subsequent comparative anatomy went wrong, and why there are major problems in interpreting the fossil record.

I. New Methods

An essential element in traditional paleontology and comparative anatomy was the judgment of the scientist. In the quotations from Simpson one reads that, "the sober and conservative mammalogists" are "likely to lose a sense of perspective." The closer the approach to man in the classification system, the more likely the scientist is to become emotional. In such a situation, methods which are not so dependent on the scientist's personal judgment are needed.

Quantitative methods have been developed that are not dependent on the judgment of the scientist performing the tests. The results of some of these are given in Table I. It can be seen from the table that (whether the comparison is on the basis of DNA, sequence of amino acids, or immunologic difference) man and chimpanzee are far more similar than man and Old World monkey. In sharp contrast to the situation in paleontology or comparative anatomy, the conclusion is the same, regardless of which method of comparison was used or in which laboratory the tests were performed. Wilson and Sarich (1969) have reviewed the evidence (albumin, transferrin, DNA, hemoglobin) and conclude that, of the contemporary primates, man is most closely related to the African apes (chimpanzee and gorilla). In fact the similarity is so great that the methods either fail to distinguish man and chimpanzee, or just make the distinction (Goodman,

TABLE I

EVOLUTIONARY DISTANCE BETWEEN MAN AND CHIMPANZEE AND MAN AND RHESUS MONKEY

Method	Man– chimpanzee	Man– monkey	Reference
DNA	2.5%	10.1%	Kohne (1970)
Hemoglobin	0	15 Mut. dist.	Reviewed by Wilson and Sarich (1969)
Fibrinopeptides	0	7 Mut. dist.	Doolittle and Mross (1970)
Albumin	7	35 ID units	Wilson and Sarich (1969)
Transferrin	3	30 ID units	Wilson and Sarich (1969)
Carbonic anhydrase	4	50 ID units	Nonno et al., (1969)
Albumin	0.0	3.7%	Goodman (1968)
Transferrin	0.0	3.7%	Goodman (1968)
Gamma globulin	0.19	3.4%	Goodman (1968)

1968). The situation is shown in Fig. 1 (Nonno *et al.,* 1969). Man and chimpanzee differ no more than various species of macaques.

The results of the new information may be briefly summarized. Man is particularly closely related to the African apes. The order of similarity among the apes is African apes, orangutan, gibbons. The Old World monkeys form a natural group (Cercopithecidae) which is much less similar to man. Far less similar are New World monkeys and still further removed are various groups of prosimians. In short, the latest quantitative information agrees with the preevolutionary *scala naturae* (Napier and Napier, 1967, pp. 4–5), and with common opinion in the latter half of the nineteenth century. Neither Darwin nor Huxley would find the conclusions surprising, but they are contrary to many later theories that demand a great antiquity for a separate human lineage or which have suggested that the nearest living relative of man is a tarsier or a monkey, at least not an ape.

The importance of the recent studies is not that they suggest radical changes in primate classification. They do not. The phylogenetic tree derived from the DNA hybridization experiments (Kohne, 1970) would have caused no particular comment in 1870. The general arrangement of the primates is what many scientists believed then. The new information

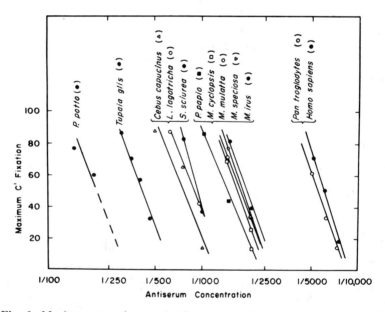

Fig. 1. Maximum complement fixation at equivalence for erythrocyte lysates of various primate species as a function of the concentration of antihuman carbonic anhydrase B in the serologic reaction.

bases the groupings of the primates on techniques which can be replicated, results which are not dependent on the judgment of the person performing the tests. The importance of this may be illustrated in the case of man, where emotions are strongest and where all the difficulties noted by Simpson (1945) are at their greatest. The remarkable similarity of man and the African apes certainly comes as a surprise to those who have been using the traditional comparative methods, whether paleontologic or anatomic. Probably the great majority of theories of human origins (those suggesting an early Miocene or earlier separation of the human line, a separation of more than 20 million years) are not compatible with the molecular data. Probably even the "sober and conservative" scientists are going to be proven wrong. But the essential issue is not who is right or wrong, but that methods have been developed which will settle many of the controversies of the last 100 years.

I regard the problems of the relationship between man, the African apes, and Old World monkeys as now settled. The phylogenetic relationships appear to be of the kind shown in Fig. 2, in which man and the African apes share a long period of common ancestry after the separation of the monkey and ape lineages.

The problems of the construction of a phylogenetic tree from molecular and immunologic data of the contemporary primates have been discussed fully elsewhere (Sarich, 1970), and lie beyond the scope of this paper. However, there are two criticisms which have been raised so often that brief comments are needed here. The first is that the rates of evolution for different parts of animals or for different proteins are different, so that it is

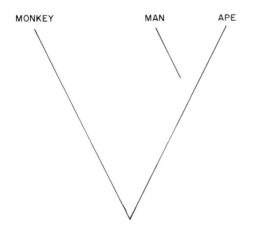

Fig. 2. Relation of Old World monkeys (Cercopithecidae), man (Hominidae), and apes (Pongidae).

impossible to relate degree of morphologic or biochemical difference to time. But the number of differences in the sequence of amino acids in hemoglobin between horse and man is 43; horse and chimpanzee, 43; horse and macaque, 43 (Wilson and Sarich, 1969). There is no evidence of greatly differing rates, as would be expected from what is known of morphologic evolution. It might be expected that the rate of change would correlate with length of generation, rather than with geologic time. But the distance, measured in immunologic distance units (albumin), between carnivores and various primates is: man, 173; chimpanzee, 173; tarsier, 155; *Propithecus,* 150; *Ateles,* 172; Colobinae, 177; and Cercopithecinae, 174 (Sarich, 1972). The amount of change is not identical, but is surprisingly (to a morphologist) comparable. There is no indication that the amount of change is proportional to generation time.

The construction of the phylogenetic tree from the DNA, sequence data, and immunologic information seems fully justified. That man is most closely related to the African apes may now be considered a fact. The determination of an evolutionary time scale from the same data is much more controversial. Theoretically, since there can be only one correct answer to a historical, phylogenetic problem, it should be possible to construct a phylogeny from one regularly evolving molecule and check the historical picture with another molecule. In my opinion this will be possible, and the time of separation of the major groups of primates will be determined from the molecular information. At the moment, albumin is the only protein which has been studied extensively enough so that the internal consistency of the evolutionary picture can be determined. The main outline of the phylogenetic picture derived from albumin agrees with the information from other molecules, but additional information is needed. At present, determination of the rate of evolution of any protein involves major assumptions on the process of evolution, the rate of change, and the estimates of time necessary to calibrate the molecular clocks. In short, the main outlines of the primate phylogenetic tree are determined on the basis of molecular information which is quantitative, countable, and which may be replicated by different scientists. The determination of time scales still involves a much greater degree of personal judgment.

As far as the origin of man is concerned, the problem of man's relation to the apes is the same as that of the interrelations among the macaques—the creatures being compared are so close that the methods are at the limit of their usefulness. As shown in Table I, there are no differences in the sequence of amino acids in hemoglobin or fibrinopeptides, and very small differences in albumin, transferrin, or carbonic anhydrase. An accurate analysis of the differences between man, chimpanzee, and gorilla requires a more rapidly evolving molecular, or a more sensitive, method of analysis.

At the present time, a separation of the lines leading to man and ape of less than 5 million years would be very difficult to reconcile with the fossil record. A separation of more than 10 million years is, probably, not compatible with the molecular and immunologic data.* The rapid discovery of fossils and the greatly increased effort in biochemical taxonomy will undoubtedly narrow these limits in the near future.

II. Field Studies

The field studies support the indications from the molecular data that the behavior of chimpanzees is far more similar to that of man than is the behavior of Old World monkeys. (Chimpanzees are stressed only because there are more data, not because I regard them as necessarily closer to man than gorilla.) Van Lawick-Goodall (1968, 1970) describes extensive object use by chimpanzees; they use sticks in termite "fishing," poking, exploring, agonistic displays, and hitting other animals. They use leaves for cleaning and getting water. They throw rocks in aggressive interactions. Chimpanzees play with objects under natural conditions, and this wide use of objects offers a great contrast to the very limited use of objects by the baboons in the same area. Chimpanzees hunt small mammals and the males may cooperate in this activity over a considerable distance. The female bears her first infant around 8 or 9 years and infants are spaced at more than 3 years. The whole situation is far closer to that of man than that of monkey, for which the comparable figures would be 3–4 years for first infant and subsequent infants spaced at approximately yearly intervals. The ape infant is much more dependent on its mother than is the case in monkeys. The female gorilla carries her infant for the first 6 weeks (Schaller, 1963). The importance of the field studies in reconstructing human evolution is discussed in reconstructing human evolution by Jay (1968).

Since the African apes are knuckle-walkers, the implications of this mode of locomotion for human evolution have been discussed elsewhere (Washburn, 1968).

It was the behavioral similarity between man and ape which led Yerkes to develop the chimpanzee as a laboratory animal (Yerkes and Yerkes, 1929; Yerkes, 1943; Bourne, 1971). Hodos (1970) has shown the importance of considering the actual phylogenetic relations of animals when making neural and behavioral comparisons, and Reynolds (1966) has

---

* It is difficult to relate time in millions of years to the traditional divisions of the age of the mammals. The end of the Miocene is variously estimated from 13 to 5.5 million years ago (Berggren, 1969). More radiometric dates are needed. If the short Pliocene dates are correct, a separation of man and ape lines on the order of 6–8 million years ago might be late Miocene and would make it far easier to reconcile the apparent discrepancies between the different lines of evidence.

stressed the importance of the chimpanzee in the reconstruction of the evolution of human behaviors.

III. Anatomy

Traditional comparative anatomy had two objectives: to describe the differences among animals, and to interpret the information in terms of evolution.* But the role of anatomy is changed if the major phylogenetic problems are settled by the molecular and immunologic information. To take an extreme example, suppose that all the recent information had supported the theory that, of the contemporary primates, man's nearest living relative was the tarsier. Then the similarities of the apes and man would have to be accounted for on the basis of convergent evolution, and one would look at the writings of Jones (1929) and W. E. L. Clark (1934) to see how they had arrived at this conclusion by anatomic means. Clark later changed his mind and his way of using the anatomic information (1955, 1959). A less extreme case would be to suppose that the differences between man, apes, and monkeys (Hominidae, Pongidae, and Cercopithecidae) had proved to be approximately equal, suggesting that the three groups had diverged at more or less the same time, possibly in the Oligocene. This is the belief of many scientists, and this would also involve a very large amount of parallel evolution to account for the similarities of apes and man. The late separation of man and African ape suggests a very different interpretation of the anatomic data. The anatomic similarities are due to long, common ancestry and to recent separation. The point is simply that, if the historical, phylogenetic questions are settled, then all comparative anatomy can contribute is to the understanding of the history. The molecular information tells nothing about the structures and functions of the animals being compared, and the fossils are only bones. The comparative anatomic information is essential in the understanding of what happened in our evolutionary history. It is no longer essential in establishing phylogeny, and it is quite useless in arriving at estimates of time.

The historian of science undoubtedly will be able to find many reasons

---

* This paper is concerned only with the primates and particularly with human evolution. Obviously, a major problem with the use of comparative anatomic information in evolutionary studies is that it is limited to the contemporary primates. Scientists may agree on the anatomic facts but disagree on their interpretation. With access to precisely the same information one may conclude that man's ancestors could not have been brachiators (Straus, 1968) or that this is the most reasonable interpretation of the evidence, both on the basis of the anatomy and on the fact that man still locomotes in that way under appropriate circumstances (Washburn, 1968a). The application of statistical methods does not remove the problem because the evolutionary issue is not the difference between two contemporary forms but the difference of each from a common ancestor some millions of years ago.

why so many scientists departed from the view that man's closest living relatives are the African apes, the view of the *scala naturae,* of Darwin and Huxley, of DNA. Some of the reasons are clear and it is essential to understand them if anatomy is to be used to further the understanding of human evolution. In the earlier studies scientists considered the major anatomic systems of the whole animal. Huxley (1863) was concerned with proportions, bones, muscles, viscera, the brain, etc. This was the common practice of that time, and, if this is the nature of the information fed into the human computer, the arrangement of the primates will be essentially that given by DNA, molecular biology, and immunochemistry.

But in the succeeding era, particularly the first half of the present century, an enormous amount of descriptive anatomic information was collected. The general view of the whole animal was replaced by detailed studies of particular parts. Ruch's *Bibliographia Primatologica* (1941) gives some notion of the vast quantity of information which was accumulated in this way. Since the human mind could not possibly grasp the totality of the information, conclusions were drawn from each specialized investigation, the hope being that ultimately all the different lines of anatomic evidence would yield definitive conclusions. The history of comparative anatomy of the primates shows that this was not the case. There is more disagreement now than in 1890.

The reasons why the traditional anatomic studies did not build toward generally acceptable conclusions, are briefly these. First, the greater the detail with which two animals are compared, the more differences will be found between them. Continued study of the primates led to taxonomic splitting because, as time went on, more and more differences could be enumerated. This has been particularly true in the case of man in which the early studies stress the general similarity with the apes and the later ones emphasize detailed differences. If the phylogenetic position of the animals being compared has been settled by other information, such as fossils and immunology, then the additional anatomic information enriches the understanding of what happened in the history of the primates. But if the anatomic information is being used to measure differences (with no control from other sources), then the greater the differences appear to be, the greater the separation will be. For example, in a general way the arm of man and the arm of the African apes seem very similar, but detailed studies have led to the conclusion that man's ancestors could not have been brachiators (Straus, 1968) or knuckle-walkers (Tuttle, 1969). The basic problem is that the more detailed the study, the more relatively minor structures become the basis for judgment, and, at least at present, there is no method for separating fundamental similarities from minor differences. For example, Napier (1970) stresses the intermembral index, but Schultz (1969) stresses the relation of the limbs to trunk height. According to the intermembral index, man's arms are short, but according to trunk height, they are very long. The choice of the comparative standard determines the result, and, even when many measurements have been taken, the personal judgment of what was worth measuring is the most important factor in determining the result of the comparisons. In summary, rather than clarifying the origin of man, the vast increase of information on the nonhuman primates led to a proliferation of theories of human origins. Competent scientists with access to the same anatomic information select, describe, and evaluate so differently that a wide variety of conclusions may be reached. The increasing quantity of anatomic information has not brought agreement on the major questions of human evolution, *or even on how the information is to be used.*

But, if the phylogenetic questions are regarded as settled by DNA, sequence data, and immunochemistry, then the anatomic information may be evaluated and used effectively. The quantity of anatomic information cannot change the phylogeny, and it can only be used to help in the understanding of the adaptive reasons for the evolutionary history. For example, since the phylogeny shows that man and the apes shared a long time of common ancestry after their lineage had separated from that of the monkeys, much of the similarity of man and the apes in arms, trunk, viscera, teeth, etc. may be accounted for by that long period of common adaptation to arboreal life. Differences may be accounted for by events after the various lineages separated. It can no longer be said that some anatomic difference is so great that there can have been no period of common ancestry after the separation of ape and monkey ancestral lines. If one regards the phylogenetic problems as settled, then the role of comparative anatomy is limited and determined. It can increase the understanding of the evolutionary events, but it cannot change the interpretation of the events.

It should be stressed that there is only one correct answer to a phylogenetic problem. At the present time there may be uncertainties, but ultimately all lines of evidence must accord with one history, with what actually happened. Once the phylogeny is settled, then all the different lines of anatomic evidence (proportions, muscles, viscera, etc.) must be understood in terms of that history. It becomes meaningless to conclude that of the contemporary apes, man is closest to the orangutan in the shoulder and to the African apes in the hand.

IV. Counting the Differences

If the phylogeny is settled, differences may be evaluated and the problems of counting differences reduced. For example, in comparing man and the great apes, similarities in the arm and trunk are the result of the long, common arboreal adaptation while differences in the pelvis and legs are the result of uniquely human adaptations. A list which is composed of items from different adaptive complexes is not useful because the items have very different historical and adaptive meanings. The listed items may be recent or old, correlated or not, of great adaptive importance or not. Even recent lists embody these difficulties (Mayr, 1970, p. 377), and the evaluation of the characters used in the comparisons still presents major problems. For example, to what extent is the shape of the human palate the result of the evolution of the dentition and to what extent may differences in the teeth and palate be evaluated and counted separately? It is my belief that this kind of problem can be settled by experiments and study of the contemporary primates.

V. Adaptive Complexes

The importance of adaptation in determining the course of evolution has been central in the synthetic theory of evolution (Dobzhansky, 1970; Mayr, 1970). How much the classic Darwinian position will have to be modified is now being debated. Even if non-Darwinian evolution has been important at the molecular level, the evolution of adaptive complexes was under the control of selection.* If this is the case, at least at some stage in comparisons, the complexes must be compared. This position is the same as that of LeGros Clark. As I see the matter, when he urged the study and comparison of "total morphological patterns," rather than isolated measurements or anatomic items, he is making the same point (W. E. L. Clark, 1955, p. 15). The morphologic patterns and the adaptive complexes are the same, and the advantage of calling them adaptive is that it clarifies how they are to be identified and helps in interpreting them. For example, the differences in the size of the canine teeth between man and apes have been discussed for more than a century, and have recently been reviewed by Kinzey (1970). But the variations in the canine teeth appear very differently if considered primarily as a dental problem, or as part of an adaptive pattern. In the great apes the large canines are

---

* Even here it is important to distinguish selection as a guiding principle in research from a belief that all structures must be determined in detail by selection. Obviously, the same function (lateral orbital wall in New World and Old World monkeys, for example) may be based on a different arrangement of the bones. It may be that, even with adaptation similar and selection constant, there may be more random variation in structures than has recently been supposed. Perhaps, the Darwinian and non-Darwinian models are not as different as they appear to be at first sight.

present in the males only (Washburn and Avis, 1958). They are part of the anatomy of fighting, and this complex includes: canines, first lower premolars, sex differences in body size, size of muscles of mastication, size of neck muscles, callosities on head (gorilla) or cheek (orangutan), mane which erects, and agonistic displays. The sex differences in aggression are mediated by hormones (Goy, 1970), and practice in patterns of play (reviewed by Dolhinow and Bishop, 1970). In the skull, the related differences appear in the size, shape of palate and nasal aperture, brow ridge, postorbital constriction, sagittal crest, and nuchal crest. Seeing the canine as part of the adaptive complex (total morphologic pattern) of bluffing and fighting leads to an almost entirely different view of the problem than measuring teeth only. Traditionally the kinds of information which are combined in the understanding of the anatomy of aggression were relegated to many separate studies, and field studies are necessary in learning how to identify adaptive complexes.

The importance of considering adaptive complexes may be illustrated by current classifications of primate locomotion. The apes are labeled brachiators, yet the field studies show that the African apes rarely swing below branches (Schaller, 1963; van Lawick-Goodall, 1968). The anatomy must correlate with all the behaviors, not with one part of the total pattern, no matter how dramatic that may be. What is needed is a behavior profile: how the animals sleep, climb up or down, move through the branches or on the ground, feed, etc. From this point of view the African apes are nest-sleepers, reaching-climbers,* knuckle-walkers. The chimpanzees are primarily tree-feeders and the gorillas ground-feeders. Brachiation (swinging under branches) is seen in chimpanzees and young gorillas, and it may be shown experimentally in cages (Avis, 1962), but it is only a small part of the locomotor pattern. The complexity and

---

* I think that the similarities in the arms and trunk of gibbons, great apes, and man are the result of reaching in many directions while climbing and feeding. Mobility in holding and feeding may have been particularly important. The importance of swinging under branches as a method of locomotion has been exaggerated (frequently by me!). The anatomy of climbing-feeding makes brachiation possible and, under appropriate circumstances the great apes may brachiate, but only in the case of gibbons does it account for a large percentage of the locomotor activity. Elsewhere I have reviewed the anatomy of this kind of climbing-reaching-feeding and stressed that many of the detailed actions are common to the apes and man (Washburn, 1968b). Stressing the problems of arboreal feeding, in addition to those of locomotion, Avis (1962), suggests unexpected correlations of anatomic characteristics. For example, the prehensile tail of some New World monkeys is a feeding adaptation, in addition to its being used in some locomotor activities, sleeping, stabilizing, and some social activities. The tail performs the same functions as the hand, or hand and foot, in gibbon and orangutan. From this point of view, there is no adaptation among the New World primates which closely approximates the climbing-feeding adaptation in the Pongidae.

diversity of the behaviors actually seen in the field situation change the interpretation drastically from what seemed reasonable on the basis of anatomy and observations under restricted conditions.

The Old World semibrachiators sleep sitting on callosities, climb and run quadrupedally, and this appears to be a most misleading label both behaviorally and anatomically (Ripley, 1967).

The lorises and pottos, called hangers, are slow-moving quadrupeds, the most extreme in their grasping adaptations of any of the primates (Grand, 1967).

In most mammals feeding is not related to locomotor adaptations, but in the primates the two are closely interrelated. This is because the animals must hold and move while feeding, and this requires adaptations, especially in a large animal. For example, gibbons hang and feed for a large part of their active hours (Carpenter, 1940, 1964; Ellefson, 1968). This unique method of feeding probably accounts for much of the anatomy which has been attributed to brachiation. The long thumb and great toe are adaptations which allow a small animal to grasp a large branch or tree in climbing. The many limitations and consequences of the gibbon feeding-locomotor adaptations have been considered elsewhere (Washburn *et al.*, 1972). They include: size of reproductive group, sexual activity, territorial behavior, vocalizations, temperament, and socialization. Clearly, the kind of adaptive complexes which the fieldwork is revealing are very different and much more complex than the traditional anatomic entities.

If anatomy is organized in terms of adaptive behavioral complexes it is possible to relate these complexes to what the animals actually do, to a behavioral profile. Then this information may be related to ecologic conditions and to the social life of the animals (Campbell, 1963, 1966; Napier and Napier, 1970, p. xiv; Crook, 1970). Then it will be possible to approach problems of comparison and evolution in terms of complex behavioral systems. If, in studying human evolution, the goal is to understand the sequence of adaptive behaviors which led to *Homo sapiens*, then typological classifications and isolated anatomic detail are of limited utility. The adaptive behaviors which lead to the reproductive success of populations are complex, and their bases cut across the traditional methods of analysis. A major goal in the study of the contemporary primates is to learn how to understand adaptation and how to devise methods so that this understanding may be usefully applied to the problems of evolution.

VI. Paleontology

If the main outlines of the phylogeny of the primates are now settled by the molecular and immunologic data, this in no way reduces the importance of paleontology, the direct evidence from times past. For example, the whole controversy over the antiquity of anatomically modern man was based on the Piltdown forgery (Weiner, 1955) and paleontologic mistakes, such as the Galley Hill skull (Oakley, 1964b). Without a fossil record it would not be possible to show that the large size of the human brain was very late in human evolution, long after the manufacture of stone tools (Howell, 1972; Holloway, 1970). The calibration of any molecular clock requires some point of agreement between the fossil record and the molecular information. Potassium–argon and other methods of radiometric dating have revolutionized the whole time perspective on human evolution (Oakley, 1964b; J. D. Clark, 1970). The distribution and ecology of ancestral forms can only be determined from the fossil records.

Knowledge of the structure and behavior of the contemporary primates is necessary both in the interpretation of the fossils (both reconstruction and taxonomy) and in showing the limits of inferences which may be drawn from the fragmentary specimens. For example, there have been numerous attempts to deduce the diet of fossil primates from the dentition. It has been claimed that the large kind of *Australopithecus* (*Australopithecus robustus, A. boisei, Paranthropus,* possibly including *Meganthropus*) was a vegetarian and the small one (*A. africanus*, possibly including *Homo habilis*) was a meat eater. Yet among the contemporary primates, both chimpanzee and gorilla have been dissected. In addition to their dentitions, the anatomy was known, and both species have been kept in captivity and have bred, so a very considerable amount was known about the animals and their dietary needs. Still no one guessed the degree of difference in the diets of the two species. Field studies have shown the chimpanzee is primarily a fruit eater (van Lawick-Goodall, 1968) and the gorilla primarily a ground-feeder (Schaller, 1963). Even on the basis of a very large amount of information, chimpanzee termiting and hunting were not anticipated. One rule for the interpretation of the fossils might be that conclusions should not be drawn from teeth alone which cannot be drawn from extensive knowledge of the whole contemporary animals. The difficulty of drawing dietary conclusions from teeth alone is probably particularly great in the primates which eat a wide range of foods, so that the degree of insect-eating, meat-eating, or various kinds of vegetarian adaptation may vary locally.

The importance of field studies in interpreting primate evolution is not limited to studies of the primates themselves. Sutcliffe (1970) has shown that hyenas do collect bones and do produce fragments which are very similar to those claimed to be the tools of the australopithecines. Schaller (1967) has shown the importance of the leopard and tiger as predators on primates. The van Lawicks (1971) have described the problems of scavenging, and have shown that hunting is an easier way for primates, such as baboons and chimpanzees, to obtain some meat. Ecologic reconstruction depends on knowledge of the behaviors of the contemporary animals as well as the associations of the fossils.

The events of primate evolution cannot be interpreted without the fossil record, but freedom to interpret the record has now been sharply limited by the molecular and immunologic information. Within these limitations a knowledge of the structure and behaviors of the contemporary primates, and at least of some other mammals, is essential to the interpretation of the fossil record.

VII. Parallelism

Many paleontologists have thought that the lineage leading to man separated from that leading to the apes in the Oligocene, something on the order of 25 to 30 million years ago. According to some, the separation between the lineages of the great apes is nearly as ancient (Pilbeam, 1970). There is agreement that the ancestral monkeys and apes of the Oligocene were small quadrupedal forms, and the great separations of the ancestral lines of orangutan, chimpanzee, gorilla, and man would mean that all the anatomic similarities between these forms must be due to parallel evolution. The problem of detailed anatomic parallelism is recognized by Lewis (1969) who, on the basis of a very careful study of the wrist joint, concludes that the joint of man is particularly similar to that of the African apes and that the similarities are too detailed to be due to parallel evolution. The degree of difference between Lewis and Pilbeam clearly shows the necessity for considering the way anatomic and paleontologic facts are interpreted. Competent scientists with access to the same information disagree and, as indicated earlier, the disagreements are greater than they were 100 years ago.

Parallel evolution means that natural selection has been similar so that a common ancestral group, although divided into two or more reproductively isolated groups, evolved in a similar way. If evolution has been parallel, the similarities between two groups will be due to the events which took place after the separation, in addition to those due to the original common ancestry. Parallelism has certainly occurred, and the problem is to decide the importance of the process in particular cases. One method is to examine the detail of the similarity. This is the method used by Lewis in the case of the wrist joint, and the basis is the belief that parallel evolution may cause similarity, *not* identity. For example, the spider monkey may brachiate and, in the length of arms (Erikson, 1963) and some other proportions, *Ateles* parallels the gibbons. The shoulder shows some parallel features, but the hand without a thumb, the wrist, the long, prehensile tail, the teeth, and the immunologic data (Sarich, 1970) show that the locomotor similarities are due to parallel evolution.

The basic principle in separating parallel evolution from similarity due to genetic similarity is that the parallel evolution is due to selection. Therefore, it is unlikely that selection will cause parallel evolution in many different adaptive systems at the same time. In the example cited above, selection has led to some similarities in the arms of spider monkey and gibbon, but the spider monkey has the most prehensile tail of all the primates and the gibbon has the shortest tail (Schultz, 1969). The spider monkey retains three premolar teeth, but the reduction to two in the gibbon's ancestors had taken place by the Oligocene. In the case of man and the African apes the molecular and immunologic information show that the similarities are due to common ancestry, not to parallel evolution.

As shown in Table I, the molecular and immunologic information shows that man and ape are similar in ways that are not related to locomotor adaptations. If the anatomic similarities were due to parallel evolution and man, African ape, and monkey had been separated for approximately an equal period of time, then the immunologic differences between man and ape should be approximately equal to those between man and monkey.

In summary, paleontology and comparative anatomy can help us to reconstruct the events which led to man. They can help in understanding the events which took place during the long period in which man and apes shared common ancestors. They can no longer be used to deny the existence of such a period.

VIII. Conclusions

With the recent advances in molecular biology and immunology, studies of human evolution have swung through a whole circle. Starting with the notion that man was particularly close to the African apes, many other theories were advanced stressing the uniqueness and great antiquity of the human lineage. The new information shows that man shared a long period of common ancestry with the apes, and particularly with the African apes.

Studies of behavior, especially the field studies, strongly support the similarity between man and ape.

Comparative anatomy leads to the same conclusions, provided the anatomy is used to help in the understanding of adaptive complexes. If anatomic topics are investigated separately, almost any conclusions may be drawn from the vast amount of heterogeneous information.

The evidence of the fossils is essential for any understanding of the events of the past, but the conclusions which may be drawn from the fossils are limited by the other kinds of evidence. Particularly, the molecular and immunologic evidence may not be dismissed by unsupported appeals to parallel evolution.

The diversity of the contemporary primates offers many opportunities to enrich the understanding of behavior, adaptation, and the process of primate evolution. The issues are being clarified by new techniques, field studies, behavioral anatomy, and the increasingly rapid discovery of fossils. Some of the old problems are already settled and we may hope for the solution of many more in the near future.

REFERENCES

Avis, V. (1962). *Southwest. J. Anthropol.* **18**, 119.
Berggren, W. A. (1969). *Nature (London)* **224**, 1072.
Bourne, G. H. (1971). "The Ape People." Putnam, New York.
Campbell, B. (1963). *In* "Classification and Human Evolution" (S. L. Washburn, ed.), pp. 50–74. Aldine, Chicago, Illinois.
Campbell, B. (1966). "Human Evolution." Aldine, Chicago, Illinois.
Carpenter, C. R. (1964). "Naturalistic Behavior of Nonhuman Primates." Penn. State Univ. Press, University Park, Pennsylvania.
Clark, J. D. (1970). "The Prehistory of Africa." Thames & Hudson, London.
Clark, W. E. L. (1934). "Early Forerunners of Man." Wm. Wood, Baltimore, Maryland.
Clark, W. E. L. (1955). "The Fossil Evidence for Human Evolution." Univ. of Chicago Press, Chicago, Illinois.
Clark, W. E. L. (1959). "The Antecedents of Man." Edinburgh Univ. Press, Edinburgh.
Clark, W. E. L. (1967). "Man-Apes or Ape-Men?" Holt, New York.
Crook, J. H. (1970). *In* "Social Behaviour in Birds and Mammals" (J. H. Crook, ed.), pp. 103–166. Academic Press, New York.
Darwin, C. (1871). "The Descent of Man." Reprinted: Random House, New York (Modern Library Edition), 1936.
Dobzhansky, T. (1970). "Genetics of the Evolutionary Process." Columbia Univ. Press, New York.
Dolhinow, P. J., and Bishop, N. (1970). *In* "Minnesota Symposia on Child Psychology" (J. P. Hill, ed.), pp. 141–198. Univ. of Minnesota Press, Minneapolis.
Doolittle, R. F., and Mross, G. A. (1970). *Nature (London)* **225**, 643.
Ellefson, J. O. (1968). *In* "Primates: Studies in Adaptation and Variability" (P. C. Jay, ed.), pp. 180–199. Holt, New York.

Erikson, G. E. (1963). *In* "The Primates" (J. Napier and N. A. Barnicot, eds.), pp. 135–164. Zool. Soc. London, London.

Flower, W. H., and Lydekker, R. (1891). "Mammals, Living and Extinct." Adam & Black, London.

Goodman, M. (1968). *In* "Taxonomy and Phylogeny of Old World Primates with References to the Origin of Man" (B. Chiarelli, ed.), pp. 95–107. Rosenberg & Sellier, Torino.

Goy, R. W. (1970). *In* "The Neurosciences, Second Study Program" (F. O. Schmitt, ed.), pp. 196–207. Rockefeller Univ. Press, New York.

Grand, T. I. (1967). *Amer. J. Phys. Anthropol.* **26**, 207.

Hodos, W. (1970). *In* "The Neurosciences, Second Study Program" (F. O. Schmitt, ed.), pp. 26–39. Rockefeller Univ. Press, New York.

Holloway, R. L. (1970). *Nature (London)* **227**, 199.

Howell, F. C. (1972). *In* "Perspectives on Human Evolution" (S. L. Washburn and P. Dolhinow, eds.), Vol. II, pp. 51–128. Holt, New York.

Huxley, T. H. (1863). "Man's Place in Nature." Univ. of Michigan Press, Ann Arbor (Ann Arbor Paperbacks edition, 1959).

Jay, P. C., ed. (1968). "Primates: Studies in Adaptation and Variability." Holt, New York.

Jones, F. W. (1929). "Man's Place among the Mammals." Longmans, Green, New York.

Kinzey, W. (1970). *Pap., Meet. Amer. Anthropol. Ass., 1970.* (Not in print.)

Kohne, D. E. (1970). *Quart. Rev. Biophys.* **3**, 327.

Lewis, O. J. (1969). *Amer. J. Phys. Anthropol.* **30**, 251.

Mayr, E. (1970). "Populations, Species, and Evolution." Harvard Univ. Press, Cambridge, Massachusetts.

Napier, J. R. (1970). "The Roots of Mankind." Random House (Smithsonian Inst. Press), New York.

Napier, J. R., and Napier, P. H. (1967). "A Handbook of Living Primates." Academic Press, New York.

Napier, J. R., and Napier, P. H., eds. (1970). "Old World Monkeys." Academic Press, New York.

Nonno, L., Herschman, H., and Levine, L. (1969). *Arch. Biochem. Biophys.* **136**, 361.

Oakley, K. (1964a). "The Problem of Man's Antiquity." Brit. Mus. (Natur. Hist.), London.

Oakley, K. (1964b). "Frameworks for Dating Fossil Man." Aldine, Chicago, Illinois.

Pilbeam, D. (1970). "The Evolution of Man." Funk & Wagnalls, New York.

Reynolds, V. (1966). *Man* **1**, 441.

Ripley, S. (1967). *Amer. J. Phys. Anthropol.* **26**, 149.

Romer, A. S. (1968). "Notes and Comments on Vertebrate Paleontology." Univ. of Chicago Press, Chicago, Illinois.

Ruch, T. C. (1941). "Bibliographia Primatologica." Thomas, Springfield, Illinois.

Sarich, V. M. (1970). *In* "Old World Monkeys" (J. R. Napier and P. H. Napier, eds.), pp. 175–226. Academic Press, New York.

Sarich, V. M. (1972). Personal communication.

Schaller, G. (1963). "The Mountain Gorilla." Univ. of Chicago Press, Chicago, Illinois.

Schaller, G. B. (1967). "The Deer and the Tiger." Univ. of Chicago Press, Chicago, Illinois.

Schultz, A. H. (1969). "The Life of Primates." Weidenfeld & Nicolson, London.

Simpson, G. G. (1945). "The Principles of Classification and a Classification of Mammals." *Amer. Mus. Natur. Hist. Bull.* **85**, New York.

Straus, W. L., Jr. (1968). *In* "Medicine, Science and Culture" (L. G. Stevenson and R. P. Multhauf, eds.), pp. 161–167. Johns Hopkins Press, Baltimore.

Sutcliffe, A. J. (1970). *Nature* **227**, 1110.

van Lawick-Goodall, J. (1968). "The Behaviour of Free-Living Chimpanzees in the Gombe Stream Reserve." *Animal Behaviour Monographs* **1**, No. 3, 161–311.

van Lawick-Goodall, J. (1970). *In* "Advances in the Study of Behavior." (D. S. Lehrman, R. A. Hinde, and E. Shaw, eds.), Vol. 3, pp. 195–249. Academic Press, New York.

van Lawick, H., and van Lawick-Goodall, J. (1971). "Innocent Killers." Houghton, Boston, Massachusetts.

Washburn, S. L. (1968a). *In* "Changing Perspectives on Man" (B. Rothblatt, ed.), pp. 191–206. Univ. of Chicago Press, Chicago, Illinois.

Washburn, S. L. (1968b). "The Study of Human Evolution." Univ. of Oregon Press, Eugene.

Washburn, S. L., and Avis, V. (1958). *In* "Behavior and Evolution" (A. Roe and G. G. Simpson, eds.), pp. 421–436. Yale Univ. Press, New Haven.

Washburn, S. L., Hamburg, D. A., and Bishop, N. H. (1972). "Social Adaptation in Nonhuman Primates" (in press).

Weber, M. (1928). "Die Säugetiere." Fischer, Jena.

Weiner, J. S. (1955). "The Piltdown Forgery." Oxford Univ. Press, London and New York.

Wilson, A. C., and Sarich, V. M. (1969). *Proc. Nat. Acad. Sci. U. S.* **63**, 1088.

Yerkes, R. M. (1943). "Chimpanzees." Yale Univ. Press, New Haven, Connecticut.

Yerkes, R. M., and Yerkes, A. W. (1929). "The Great Apes." Yale Univ. Press, New Haven, Connecticut.

Zuckerman, S. (1966). *J. Coll. Surg. Edinburgh* **11**, 87.

CHAPTER 22

The Primate Research Center Program of the National Institutes of Health

GEOFFREY H. BOURNE

In 1956 Dr. Karl F. Meyer of California paid a visit to the Soviet Union and visited the Russian Primate Colony at Sukhumi in the Republic of Abkhasia (formerly the Kingdom of Georgia) whose director is Dr. Boris Lapin. In August 1956 Dr. Meyer wrote a letter to the National Heart Institute, giving an account of his visit. That same month the Director of the National Heart Institute, Dr. James Watt, also paid a visit to the Russian Primate Colony. The Sukhumi Center was established in 1927 about the same time as the Yerkes Primate Center was founded.

At the Sukhumi Center the biology, maintenance, and breeding of monkeys have been primary problems for study, and there are also studies on the morphology and biochemistry of the blood of monkeys and apes. The center has had considerable success in acclimatizing monkeys to captivity, including rhesus and green monkeys, and has developed a new nursery for them. Sukhumi now has the seventh successive generation of hamadryad baboons. Further studies have been made on the basal metabolism of monkeys, the regulation of temperature, and the diurnal periodicity of physiologic function. There have been intensive investigations into the higher nervous activity and psychology of monkeys.

After Dr. Meyer and Dr. Watt had visited Sukhumi, they were convinced that the United States should establish one or more primate centers. In September 1956 it was recommended that the United States government support colonies of monkeys for biomedical research. There was some doubt as to whether a number of regional centers should be built or whether there should be one national center. Dr. James Sannon, who was at that time Director of the National Institutes of Health, favored the regional centers rather than the single national center.

Further planning and consultation took place all through 1959, and in 1960 Congress made available funds for the establishment of the first of seven centers and the Council of the National Heart Institute recommended funding of a primate center in Portland, Oregon. In March 1961 the Council recommended that a primate center be established in Seattle, Washington and one in Madison, Wisconsin. In June 1961 it was recommended that an additional center be established in Atlanta, Georgia, with Emory University as the host institution, and based on the Yerkes Laboratories for Primate Biology, which had been located at Orange Park, Florida since 1930 and which had recently been given to Emory University by Harvard and Yale. It also recommended the establishment of a center in New Orleans in association with Tulane and Louisiana State Universities and one in Boston, associated with Harvard University. This meant that six primate centers were established and in March 1962 the seventh was designated to be associated with the University of California at Davis and was called the National Primate Conditioning Center, later called the National Center for Primate Biology. It was intended that this center be concerned mainly with the problem of breeding and conditioning primates; in other words, finding out exactly what their requirements were from every point of view and how these conditions could best be provided.

Since the centers were to be multidisciplinary in their research, it become inappropriate that they should be administered by the National Heart Institute, so in 1962 the Primate Center program was transferred to the Division of Research Facilities and Resources. This division was later renamed the Division of Research Resources.

The functions of the centers were to be as follows:

1. To pursue basic and applied biomedical primate research directed toward a solution of human health and social problems.

2. To establish a resource of scientists in many disciplines who are trained in the use of primates and can help to maintain both continuity and high scientific quality in research.

3. To provide opportunities for research and research training not otherwise available to visiting scientists, postdoctoral fellows, residents, junior faculty members, and to graduate medical, dental, and veterinary students.

4. To determine which problems of medical research are best pursued with nonhuman primates and which species are suitable for particular studies. In other words, the establishment of primate models for certain diseases.

5. To develop improved breeding practices in order to increase the supplies of pedigreed disease-free primates available for research and preserve species in danger of extinction.

6. To study the natural diseases of primates and techniques of importation, conditioning, housing, and management which have an influence on the animal's well-being and suitability for research.

7. To develop new methods and equipment for primate studies.

8. To supply biologic specimens to qualified investigators.

9. To disseminate the findings of studies done at the centers to primate users and others throughout the world.

Collectively the centers are now concerned with more than 500 different projects in human physiology, behavior, and disease, and they maintain over 7000 nonhuman primates of 45 species.

Keeping primates is an expensive business, and primate research which is carried out exclusively by single investigators, each faced with the necessity of keeping up with a colony of primates, would cost far more than maintaining primate centers. This applies especially to the Yerkes center with its collection of great apes. Very few institutions have the space to house such animals properly, and fewer still have the know-how to maintain and handle them. The primate centers, therefore, function as institutions where the development, care or management, diseases, diet, etc. of various primates can be studied in depth and advice given to those laboratories keeping small numbers of research primates on their own premises. Primate centers also carry out the preliminary work necessary to identify a particular type of primate for a particular type of research.

There are other vital reasons why primate centers should exist to support the scientific community. The projects of individual investigators typically are relatively short-lived. Rarely do they extend beyond a few years. Primates that are unique (either by reason of their genotype, phenotype, or treatment) and must be studied over many years should be maintained in a setting that has a long-term commitment to primate research. Whereas it is reasonable for an investigator to obtain and maintain a few small primates for a few years, it is not reasonable for him to do so with rare primate forms.

Eventually this country will have to provide most of the primate material for its research needs. Dwindling primate populations in the field and the increased frequency of embargos on primate importation will require that we develop breeding colonies within the limits of the territorial United States. There is no better place to do this than a primate center. Not only do the seven existing centers hold a large reservoir of animals with potential breeding value but they employ people knowledgeable in the problems of reproduction. Most researchers acknowledge the value of long-term experience with an animal farm for the development of keen insight regarding its capabilities and characteristics.

One of the advantages of having breeding colonies of primates for experimental studies as compared with obtaining them from the wild is that a colony-bred animal has much less pathology than a wild-born animal. For example, some degree of muscle pathology has been found in 100% of wild-born monkeys coming into the Yerkes Primate Center.

The primate centers are associated with various universities, which become the host institutions. The host institution designates the director of the primate center which is associated with it and this appointment is approved by the National Institutes of Health. The director is expected to carry on his own researches as well as direct the center. Each center has a core staff of professional and supporting personnel who have appointments from the host institution. The core staff comes from a variety of disciplines: psychology, neurochemistry, neurology, anatomy, physiology, reproductive biology, virology, veterinary pathology, veterinary medicine, etc. Some members of the core staff hold joint appointments in appropriate academic departments of the university or medical school with which the center is affiliated or even with neighboring or regional institutions. The number of core scientists varies from 10 to about 40 in the various centers and the supporting staff ranges from 80 to 100 or more.

The centers also provide a place where scientists from the host or other institutions can establish research programs. They also house temporarily visiting scientists from other parts of the United States or other parts of the world. The centers also supply blood and other tissues to scientific laboratories all over the world. The centers also have joint research programs with scientists from other universities, even those some distance away. The Yerkes center, for example, has a joint program in immunology with Duke University and two faculty members of the Duke Medical School are also members of the staff of the center. Undergraduate, graduate, and postdoctoral students also contribute to the research program, some of them as summer research assistants.

It has been mentioned that the primate centers provide a reserve pool of animals. One of them has nearly 2000 monkeys and the others have varying numbers. The Yerkes center has the world's greatest collection of apes, about 140, and nearly 1000 monkeys. There are nearly 8000 primates in the seven primate centers.

Some of these animals can be made available to members of the scientific community, and the centers also provide biologic materials. Another function of the centers is the dissemination of scientific information. The Washington Primate Center in Seattle has a magnificent primate information service, unique in the world, which sends up-to-date reference lists of the current publications on primates to anyone who requests them, and will prepare literature lists on special primatologic projects on request. Several of the centers publish their own newsletters and there have been many publications in the form of scientific papers and books. These publications not only include the results of scientific experiments, but include studies of primate housing and husbandry, breeding, nutrition, naturally occurring diseases, and pathology.

The centers have many visitors, including about 1000 scientists in any 1 year. At least another 4000 visitors come to the center mostly in the form of visiting groups, which include college and high school students.

One of the most important functions of the primate centers is of training manpower for research in primates and for handling and maintenance of these animals. Postdoctoral fellows often work at some of the centers and in some cases they have stayed on to become permanent members of the core staff. Numerous workshops, symposia, and seminars serve to disseminate information as well as provide training for various types of personnel.

The centers have also had an international impact and West Germany, Switzerland, France, Holland, and Canada are planning the establishment of primate centers modeled on those in the United States.

Research in the Primate Centers

The spectrum of research carried out in the various primate centers is very wide and covers many disciplines.

The Oregon Primate Center, because of the interest of its director in the biology and pathology of skin, has become internationally known in this area and holds regular international gatherings to exchange scientific information on this subject. The Oregon center is also active in the field of reproductive physiology and behavior. Aspects of this study include hormone effects in reproduction in both prosimians and simians. When male hormones are injected prenatally into female rhesus fetuses sexual and social patterns become masculinized and even pseudohermaphrodites can be produced. The animals, although genetically females, show normal ovaries but masculinized genitalia, a delayed puberty, and a tendency toward masculine behavior. If the hormone is injected into a 6-month-old female she will become aggressive but will still retain her female characteristics.

Other studies at the Oregon center have demonstrated that there are close relationships between hormones and behavior. The center is studying the effect of the hormone progesterone from the adrenal on sexual physiology and behavior, and one of the methods of doing this is by placing adult female pigtail macaques under conditions of social stress. The effects of this stress appears to indicate that the follicular phase of the cycle is markedly lengthened and menstruation tends to be delayed. Other behavioral studies have been carried out on Japanese macaques, rhesus monkeys, and stumptailed macaques and have indicated that in these animals there is a social bond between the mother and her offspring which lasts at least until the offspring are sexually mature, even when there are other young born to the same mother. They have also demonstrated that crowding of groups of animals noticeably increases aggressive behavior, but that the ratio of male and female has little or no effect on aggression. The center at Oregon has also been studying the normal behavior and physiology of prosimians and their reproduction and mating behavior.

Studies at the Oregon center are also associated with the biochemistry of spermatozoa and the enzymes involved in the regulation of fructolysis and also with the effects of fetal adrenal hormones on development and the development of fetal and uterine muscle, including the significance of lipids as an energy source in the female and in the intermediary metabolism of the myometrium. Extensive cardiovascular studies are being carried out in relationship to nutrition. The presence of naturally occurring atherosclerosis in certain South American monkeys in the wild has been demonstrated and it has been shown that cholesterol will tend to deposit in the arteries as a result of certain types of diet, and also smoking and exercise affect the amount of cholesterol deposited. These studies are directed toward an understanding of how the lipid composition of arteries changes with age and with early nutritionally induced atherosclerosis in primates. Nutritional studies are also directed toward the effect of age and diet on the lipid composition of the central and peripheral nervous systems. The degree of nutritional deprivation on the development of the brain and the subsequent learning and behavior of the offspring are also being studied.

Electron microscope studies are being made of the role that various cells play in tissue degeneration and the fine structure of the photoreceptors of the eye is also being investigated. In the area of reproduction studies are being made on the fine structure of the ciliated cell during its life history in the rhesus monkey oviduct. There are also extensive studies in the area of cutaneous biology; these include studies using the electron microscope for the analysis of lysosomes in sebaceous glands of various primates to the growth and behavior of pigment cells from the eyes of rhesus monkeys. They also include the processes of skin wound healing in rhesus monkeys, the testosterone metabolism of bald hair follicles of stumptailed macaques, and the structure and distribution of vibrissae in rhesus monkeys.

The center also supports investigations in the area of immunology, and projects of importance include studies on allergy, serology, and transplantation. Another is on the relation of certain blood factors in rhesus monkeys to adverse transfusion reactions, and their inheritance. The center is interested in the mechanisms of the release of histamine and other vasoactive compounds in rhesus monkeys, and these studies are shedding new light on the problems of hypersensitivity in man.

The importance of the Primate Information Center at the Washington Primate Center has been mentioned earlier. It supplies bibliographic information principally on biomedical research using nonhuman primates and also provides information on the nonhuman primates themselves. This service is available on a worldwide scale and will shortly be completely computerized. Its services will then be made available even more widely and speedily. This is really a very important contribution to the field of primate research. It has a current data base of about 20,000 citations and adds about 4000 additional citations a year. Within 3 months of publication, this center has on its lists about 50% of the world's primate literature, within 3–6 months has another 20% added to it, and within another 6 months has an additional 20%; thus, it covers about 90% of all the literature on nonhuman primates that is produced throughout the world. It will also provide both bibliographies and collections of data. In addition it issues a weekly bibliographic index called *Current Primate References,* which bring to the attention of investigators who receive it the current publications in the area of primatology. About 1350 libraries and investigators receive this bibliography service each week.

The specimen distribution program provides fresh or fixed tissues, blood samples, organs, parts, and cadavers to investigators, both locally and nationally, upon request. During the last year, over 1200 specimens were provided without charge. The goal of the program is to insure the complete and multiple use of all research animals.

The Washington center has specialized in studies of central nervous system control of the cardiovascular system and these studies are considered to be among the best in this area in the world. The areas that are being studied are cardiovascular reactions to stress, anatomic pathways involved in the central regulation of circulation, the effects of cardiovascular reflexes on the circulation, and regulation and cardiovascular responses to social colony behavior. Adult unrestrained pigtail macaques are used in most of these studies and a new series of descriptive and experimental studies on the cardiac nerves of baboon is now proceeding.

Other areas in primatology at the Washington center are those of

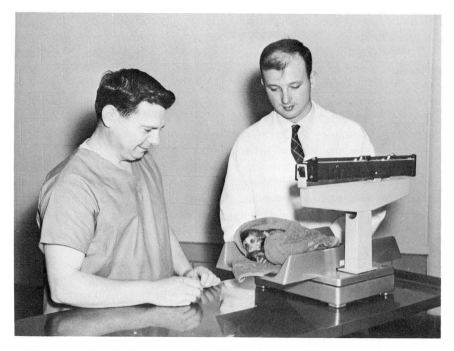

Fig. 1. Dr. Gerald Blakely and Dr. William Morton, Veterinarians at the Primate Field Station, Regional Primate Research Center, University of Washington, Seattle, Washington.

neuroanatomy, social interaction of animals, sensory and motor systems, dental development, neuroendocrinology, and metabolism. The neuro-anatomic investigations involve studies of the interconnections of visual and limbic systems, cortical projections to intralaminar nuclei, commissural connections of the hippocampal formation, corticothalamic projection of the second somatosensory area in the monkey and neuronal ramifications in fetal macaque brains. The social studies among pigtail macaques involve a developmental study of maternal–infant relationships, effects of separation on the mother and the infant, analysis of social organization of captive groups and the leader's role in controlling aggression within the group. These studies have provided important insights into social factors in infant development. They have shown that enriched environments can ameliorate psychopathologies of behavior and development due to maternal and peer deprivation. The studies on the sensory and motor systems include those on the neural basis of reflex and voluntary eye movements, functional relations between motor and cortex cells and jaw muscle in awake monkeys; these studies involve recording from a single cortical cell in awake-behaving monkeys. The center also includes a dental research program and has studies involved in improving endodontic operations on young children and orthodontic procedures such as tooth rotation and remodeling of facial structures by imposed forces. There are also studies on developmental genetics in pigtail macaques, baboon dental development, and cleft palate in the rhesus monkey, which is so far the only living nonhuman primate known to have this condition. The neuroendocrinologic studies include those of thermoregulation in the baboon and on plasma cholesterol metabolism, vitamin D deficiency, and plasma transport of vitamin D. The specific biologic effects of proinsulin, the recently discovered precursor to insulin, are also being explored. The studies have indicated that proinsulin effects include relatively protracted lowering of blood sugar in baboons and these have stimulated the interest in the significance of proinsulin in pathologic states of man.

The New England Primate Center, situated in Southborough, Massachusetts, has studies in the areas of experimental biology, psychiatry, dentistry, pharmacology, psychology, public health, anatomy, nutrition, environmental medicine, physiology, neuropathology, experimental surgery, and reproductive physiology. Recent studies have ranged from in-

Fig. 2. Regional Primate Research Center, University of Washington, Seattle, Washington.

duced ovulation in bonnet monkeys to hypoglycemia in newborn monkeys, protein and ascorbic acid requirements of infant monkeys, the effects of drugs on chimpanzees, liver transplants in rhesus, long-term effects of repeated episodes of strenuous exercise in rhesus monkeys, and the effects of various kinds of visual deprivation in rhesus monkeys. There are also studies of schizophrenia which suggests that it may be the manifestation of an autoimmune disorder. Any demonstration of an autoimmune factor in schizophrenia could shed important light on the etiology of this disease in man.

The New England center actually directs its principal research efforts into the areas of microbiology, comparative pathology, pathobiology, and physiology. The projects tend to cut across the lines dividing these specialties and involve specialists in several different fields. One of the most important areas of the work in microbiology has been with the herpes virus group. Virus isolated from squirrel monkeys has been shown to produce malignant lymphoma in several species, including owl monkeys and marmosets. The discovery of this virus is potentially significant because it provides opportunity for studies of a viral neoplasm of a primate origin. Efforts are also underway to discover suitable low virulance herpes viruses which could be used for immunization purposes. Some studies are also being carried out on the metabolism of vitamin D.

Both the normal requirements of vitamin D and the toxic level are under study. The interrelationship of several factors in metabolic bone disease is also being explored. One of the studies of great interest is that of spontaneous abortion in newly imported wild-caught pregnant monkeys. The fact that these almost universally abort is recorded but is not understood. In the attempt to find out the reason for this abortion, detailed studies of female genital tracts of 14 species are in progress, and an atlas on the subject is planned. Other research in the area of pathobiology is an attempt to establish a model for phenylketonuria and other metabolic defects in nonhuman primates. The phenylalanine tolerance of four species of macaques have been studied and the problems of inheritance of PKU is one of the subjects of investigation. In the neurophysiologic area, four major projects are devoted to the investigation of the role of the "startle" response mechanisms and the organization of motor function, the ability of the infant nervous system to compensate for early loss of movement centers in the cortex, the relationship of postural abnormalities to certain types of brain lesions, and the transmission of sensation in skin supplied by only one nerve root.

The Yerkes Primate Research Center is unique among the other primate centers in that it holds the largest collection of great apes in the world and very likely the largest collection that has ever been got together in the history of the world. Research at the Yerkes center is concentrated into four main areas: (1) studies of the brain and nervous system by anatomic and neurophysiologic techniques; (2) behavior; (3) experimental pathology; and (4) reproductive physiology. The anatomic studies on the brain involve the tracing of the accessory optic tract and the study of the efferent fibers of the visual areas of various primates. For several decades it has been accepted that in primates including man, there were three visual cortical areas arranged in a concentric fashion; that is, area 17 is surrounded by areas 18 and 19. Although the validity of this concept has been questioned now and then, it was not really challenged until recently. With the advent of new methods, it has been shown that in primates there are at least five different visual cortical areas that are not necessarily arranged in a concentric fashion. Two new areas are areas MT and DM. The neuroanatomic laboratory at the Yerkes center is interested in what areas of the brain receive fiber connections from, and thus could be influenced by, visual area MT. It appears that in primates this area is linked to a very large number of cortical sites: for example, nine different areas on the contralateral side of the brain. On the other hand, visual areas 18 and 19 are connected with only three or four cortical areas. This laboratory has also demonstrated that the cortico-cortical connections of area 17 originate in the upper three layers of the

Fig. 3. Yerkes Primate Research Center, Emory University.

cortex. It has also shown that the connective tissue space in the accessory optic system has capillaries which differ from those usually found in the brain in that they possess an extracellular perivascular connective tissue space. The only other areas in the brain that have similar vascular arrangements are the pineal gland, the subcommissural organ, the pituitary, and the area postrema. It is of interest that these structures are all involved in secretion.

The neurophysiologic studies at Yerkes are concerned with (1) neural mechanisms of behavior and (2) neural mechanisms of sleep and wakefulness. A series of experiments have been in progress for the last 6 years in the neurophysiologic laboratory in localizing the neuromechanisms that can control the expression of social and emotional behaviors. Aggressive and sexual behaviors have been produced in monkeys which are equipped with chronically implanted electrodes through which, via telemetry, stimulation can be applied. These electrodes are implanted in the subcortical regions of the brain, and, for instance, the attack responses which are produced by stimulation can be altered social variables as measured by intensity, duration, and object of evoked aggressions. Preliminary estimate of the hierarchy of attack objects suggests that evoked attack is similar to spontaneous aggressive behaviors. Stimulation of other portions of the

Fig. 4. A view from the rear of the Yerkes Primate Center on the Emory University Campus, with a section of the large-primate wing in the foreground. The Center grounds are completely enclosed by an electric chain-link fence for security purposes.

hypothalamus may induce stimulus-bound sexual behavior. This has been found to be sensitive to changes in the hormonal status of the female, in other words, to vary during the different stages of estrus. Similarities between electrically evoked responses and the naturally occurring social acts indicate that this is a useful technique for localizing the areas of the central nervous system that control emotional and social behaviors.

Sleep and wakefulness studies have been conducted on the influence of vestibular stimulation on sleep. This involves rotating animals at constant velocity in a centrifuge. Movements of the animals on the centrifuge while rotation is in progress stimulates the semicircular canals of the inner ear, and this produces vestibuloocular reflexes. Recordings of the eye movements over a 24-hr period, i.e., with EEG's and EMG's, have been taken from animals before, during, and following long periods of continuous rotation. The preliminary findings of these experiments indicate that this form of vestibular stimulation causes significant decrease in a particular stage of sleep, which in the terminology of some investigators has been labeled "deep sleep." The implications of these findings are indicative of some hazards that may be encountered in prolonged manned space missions on rotating space platforms. These studies also provide information on the basic mechanisms underlying sleep and wakefulness.

Other studies on the brain at the Yerkes center include those which are concerned with the effects of low protein intake on RNA, DNA, and enzymes in primate brains, particularly a low protein intake in the ex-

Fig. 5. Indoor dens heated (in winter time) for great apes. Dens communicate with outside run to which arrival has access, except in subzero weather. (Yerkes Primate Center.)

Fig. 6. Large outdoor compound with adjoining caging at the field station. This 9000 square foot compound houses a group of 16 chimpanzees in a seminatural setting in order that their social organization and behavior may be studied.

pectant mother. Observational studies on the human have shown clear evidence of affected mental development after severe periods of malnutrition, particularly during the critical periods of development of the brain. The taxonomic closeness of primates to humans makes them ideal experimental animals compared to widely used rodents to study the impact of protein–calorie malnutrition of the developing nervous system. Like humans, monkeys also have a prenatally developed brain, and work is in progress to show the impact on the nervous system of the fetus and the neonate of experimentally induced maternal malnutrition in pregnancy. Cytochemical observation obtained from young juvenile squirrel monkeys in which protein malnutrition was induced by feeding a diet with 2% protein to a group of experimental animals for a period of 15 weeks, showed the sensitivity of the nervous system to dietary abuse. The motor neurons of the spinal cord and the Purkinje cells of the cerebellum are very sensitive to protein deficiency in the diet. The number of oligodendroglial cells increases sharply. There is also a significant decrease in the amount of RNA in the Purkinje cells and the anterior horn cells of the spinal cords. Some of the larger neurons also show chromatolysis. The different layers of the cerebellar cortex as well as some of the neurons of the spinal cord show decreased activity of succinic dehydrogenase and increased levels of thiamine pyrophosphatase and lactic dehydrogenase in malnourished animals. ATPase activity, though appearing to remain unchanged quantitatively, showed profound disturbance of its intracellular localization, an interesting finding since this type of change would not have been shown by conventional biochemical investigation.

In the behavioral area, the studies on the talapoin monkey and other apes have demonstrated that the historic assertions of both Köhler and Yerkes that great apes are capable of abstractive learning processes not commonly found among monkeys, particularly the more primitive ones, has been confirmed. These tests were carried out on the talapoin monkey and the gorilla. The implications of the research in relation to the evolution of intelligence in the ontogenetic development of intelligence of the human child are important. Intelligence as we see it at the level of man is reflective of his superior cortex. Nonhuman primates have suggestions of this intelligence in accordance with their cortical development. The human child becomes abstractive by virtue of his cortical development, a primate correlate of maturation during the early years of life. These studies will help to throw light on how the learning processes of stimulus–response learners might be developed and cultivated to convert them into abstractive learners. For this purpose, the nonhuman primate presents itself as a promising model for future study. Studies on the chimpanzee have also demonstrated that voluntary smoke inhalation (smoking) is a method of administering psychoactive drugs. A number of animals have reached reliable levels of smoking behavior that permit drug manipulation to be done. The administration of marijuana via smoking was found to effect immediately operant level pulling on a differential reinforcement for low rates of responding (DRL) schedule. The animals, after smoking marijuana, made more incorrect responses and obtained a smaller number of reinforcements than following plain cigarette smoking. This response decrement was caused by general shortening of interresponse time durations. The initial results with methamphetamine suggests that oral doses of 15–30 mg caused large decrements in performance of successive criterional reversal tasks. There is also some evidence that stereotyping behavior increases following the administration of methamphetamine.

In studying the problem of alcohol, young chimpanzees and young rhesus monkeys have been made physically dependent upon ethanol and produced mild to severe symptoms when the substance was abruptly removed from the diet. Animals which have died during these withdrawal symptoms have shown edema of the brain, which suggests that the controversy regarding the forcing of fluids into human patients with delirium tremens would be undesirable. The disappearance of ethanol from blood has been determined for young chimpanzees and rhesus monkeys, both during and following periods of chronic ethanol ingestion. The disappearance rates increased during periods of ethanol administration

Fig. 7. A large male orangutan, Yerkes Primate Center.

and decreased during subsequent abstinence periods, which is a finding consistent with reports using human subjects. Another item of interest is that during periods of chronic ethanol ingestion there is a rise in blood methanol levels, both in young chimpanzees and in rhesus monkeys. Similar changes in blood methanol concentration have recently been reported in man. Although the importance of a chronically elevated blood methanol level has not yet been established, it is possible that methanol may play a significant role in the physical dependence on ethanol.

Studies on cross-modal perceptual integration in apes and monkeys is being carried out at the center. It was found that apes are capable of matching abstract information coming from two different sensory modalities, i.e., vision and touch. Previous attempts to demonstrate this ability had been successful only in man, and some neurologic theories suggested that only man is capable of it. In addition, it has been shown that non-human primates are able to perceive the representational character of photographs.

Work is currently underway to determine whether or not monkeys, who are phylogenetically and morphologically simpler than apes, are able to do the tasks. The results will have evolutionary and neurologic significance. This basic research has suggested a promising approach to the under-standing of some special learning deficits, e.g., dyslexia, which occur in considerable numbers of human children. In addition, the assessment methods developed with apes may be applied almost without change to the assessment of these deficits in humans.

Another area in which new ground is being broken are studies in which the endocrinologic factors associated with status in a group hierarchy are being explored, with special attention being paid to adrenocorticosteroid and testosterone levels. Already it has emerged that animals which are leaders in a monkey society have higher levels of testosterone than others. Aggressive and violent animals show the same thing. It has also been found that male testosterone production in the rhesus monkey shows a seasonal annual cycle.

In the area of experimental pathology an interesting interinstitutional program has developed between Duke University and the Yerkes center. Work with chimpanzees has provided an invaluable experimental model for studies of the role of tissue antigens and tissue transplantation. It appears that direct transplant of organs from chimpanzees to humans is not feasible because of the presence of heterophilic antigens in chimpanzees to which all humans react adversely; however, the chimpanzee has most of the antigens found in the major histocompatibility locus of man, which makes these species extremely useful in studies having a direct clinical relationship to the role of antigens in man. The chimpanzee thus becomes a unique experimental model for the study of problems involved in organ transplantation. Experiments on transplantation immunity which morally and ethically cannot be attempted in man can be tried in this species. Moreover, the experiments can be better controlled in chimpanzees. Chimpanzees are currently being used to evaluate new methods of immunosuppression for human organ transplants. Chimpanzee antisera to chimpanzee transplantation antigens are capable of defining certain human transplantation antigens; thus, it is possible to manufacture "reagents" for human tissue-typing in the body of the chimpanzee without resorting to the hazard of using human volunteers. The histocompatability loci are being investigated in the chimpanzees at the center and the immunologic group is being traced through four generations of chimpanzees in the Yerkes colony and this has already been accomplished for material histocompatability antigens through three of these generations. The colony pedigree over the last 40 years is still being organized and provides a unique opportunity to study the inheritance of these antigens.

The center has also been associated with the Egleston Children's Hospital at Emory University in cross-circulation studies between chimpanzees and humans. Two children in a hepatic coma have been subjected to this procedure, which involves connecting the blood circulation of the child in series with that of the chimpanzee so that the child's blood passes through the liver of the chimpanzee. In the first child, the blood chemistry

Fig. 8. Feeding the apes, Yerkes Primate Research Center.

fell to virtually normal after 4 hr. The cross-circulation was terminated after 6 hr. However, the child did not recover consciousness and subsequently died. The second child remained connected with the chimpanzee for 16 hr, and at the end of that time, she was able to remember her name and address, her telephone number, and even sing a little song. She did not go back into coma and gave every indication of surviving until she died a few weeks later from a sudden hemorrhage which was unrelated to the success of the chimpanzee treatment. There is no doubt that many lives could be saved by this procedure. It is limited by the fact that, although the majority of humans belong to the O blood group, a minority of chimpanzees belong to that group. Selective breeding of chimpanzees for this blood group is possible, but it would take a long time for a significant number of animals to be bred and would be very expensive. If animals used for this purpose could be used more than once, a real breakthrough of the use of chimpanzees for this form of therapy would be possible. Full details of this cross-circulation study are given in Chapter 11.

The veterinary pathology department of the center is involved in studying the effects of feeding milk from leukemic cows to infant rhesus monkeys and infant chimpanzees. None of the animals have developed leukemia so far. It is also holding and monitoring a group of Air Force monkeys which were exposed to atomic radiation during the 1950's. Breeding experiments have been carried out with these animals to look for genetic defects, but none have been found to date, although an increased stillbirth rate has been observed. There also appears to be an increased incidence of cancer in these animals.

The pathology and behavior departments also identified, within the last few years, a chimpanzee born at the center which was a mongoloid with an additional chromosome, retarded mental and physical development, with epicanthal folds in the eyes, and also with heart defects and increased susceptibility to disease, in other words, resembling what is found in human cases of Down's disease. This is the first and only case so far of mongolism in a nonhuman primate.

Recently, chimpanzees have died in the center suffering from pneumonia which turned out to be due to an organism called *Pneumocystic carinii*. This type of pneumonia occurs with some frequency in humans in such conditions as debilitated infants, infants born with some type of immunologic defect, persons with leukemia and other forms of malignancy, and persons receiving immunosuppressive drugs, such as kidney transplant recipients. The disease has not previously been reported in nonhuman primates, although it can be induced in some rodents by prolonged administration of steroids. The occurrence of this disease in nonhuman primate species may serve to provide a model for the study of this poorly understood human disease. Experimental infection with this organism is being studied at the moment.

A young chimpanzee has recently been noted to have a chromosomal abnormality which appears to be a balanced translocation. The parents of this animal are being studied to determine if this anomaly may be present in one of them. Balanced translocations are found with some frequency in the human population but apparently have not previously been recorded in nonhuman primates. The occurrence of this abnormality and that of the Downs-like syndrome indicate that nonhuman primates, especially the great apes, may be used for models in investigating some of the cytogenetic abnormalities that affect man. These studies demonstrate how similar are the gene pools of the great apes and humans.

The center is about to start now on a project to communicate with great apes using a computer system. Great apes have been taught relatively sophisticated communication using American sign language and using colored geometrical shapes, in other laboratories. This new sophisticated computer approach to communication is bound to have important spinoff in teaching human children who have reading difficulties.

The Yerkes laboratories were the first to demonstrate that isolation of

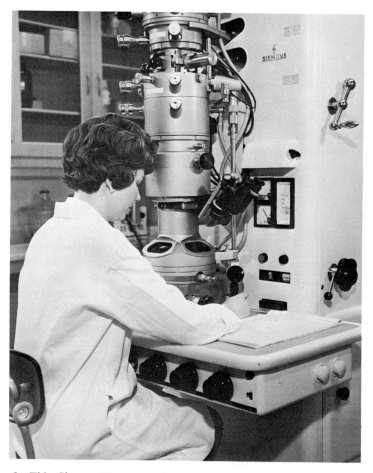

Fig. 9. This Siemen Electron Microscope, with a maximum magnification of 160,000× on the viewing screen, is used to study the ultrastructure of primate tissues. An EM technician is screening thin sections of orangutan leucocytes prior to photographing them.

infant primates, in this case chimpanzees, resulted in abnormal behavior, and in this particular study baby chimpanzees were taken from their mothers at birth and kept in total isolation for 20 months. The behavioral aberrations which developed from this treatment are now well known, especially since the Wisconsin center was subsequently able to show a similar effect in rhesus monkeys treated in the same way. However, the Yerkes center has demonstrated an important difference between aberrent chimpanzees and monkeys. The former when mixed with normal chimpanzees improve behaviorally, and the rhesus monkeys do not improve when mixed with normal specimens of their own kind.

A good example of collaboration with other institutions is shown by the reproductive physiology program which is coordinated at the center and which is funded by the Ford Foundation. Urine collected from apes under study is forwarded to the Emory University Medical School where it is assayed for estrogens and to the Harvard Medical School where it is assayed for gonadotropins. At the center, studies are made of the sexual swelling, temperature changes by an implanted temperature sensor, vaginal smears and uterine histology, and by direct observation of the ovaries by endoscopy. The combined studies will eventually give a valuable coordinated picture of the sexual cycle in great apes which should have important human applications. In fact, these studies have shown a close endocrinologic resemblence between human and chimpanzee in the hormonal control of the menstrual cycle. In fact, it is more closely paralleled by the chimpanzee than by any of the other mammalian species which have been studied. The reproductive biology group in collaboration with the pathology department have found malignant tumors of the uterine corpus in squirrel monkeys continuously treated with large doses of the estrogen-mimetic compound diethylstilbestrol. This is the first time that carcinogenicity of an estrogenic substance in a primate has been demonstrated. It is of considerable significance in view of the long-term estrogen therapy used in humans for ovarian replacement therapy and for contraception and the use of diethylstilbestrol for fattening cattle for human consumption.

There is also a unit in the center which is particularly concerned with histochemistry and with muscle structure and function. One of the interesting findings of this laboratory is that some degree of myopathy is widespread among nonhuman primates, being especially obvious in wild-born or wild-raised animals and is particularly common in the squirrel monkey. This laboratory is also involved in studies of the basic problems involved in manned space flight, using rhesus monkeys. Another finding of interest by this unit is the location of the presence of sarcosporidial infection in the muscle of nonhuman primates. This infection has been found to have a crippling effect on some rhesus monkeys and to be present also in some of the muscle fibers of one of our gorillas.

In the center there are not only collaborative studies of the various research projects but also a cooperative graduate program in primatology with the University of Georgia. This program gives a Ph.D. in primatology with accent on behavior. It is sponsored by the psychology department of The University of Georgia, and it is probable that a similar doctoral program sponsored by the anthropology department will be started at a

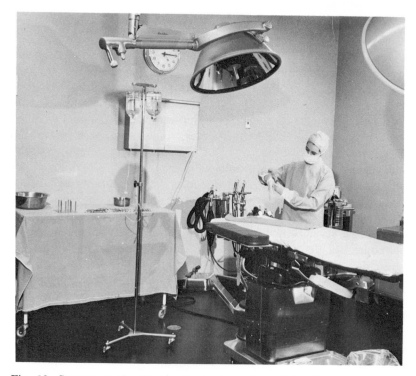

Fig. 10. Surgery at the Yerkes Center is performed in this modern operating room. A nurse is preparing to set up sterile surgical field.

later date. Plans are afoot for a doctoral program with Georgia State University.

Under the auspices of the International Union for Conservation of Natures, the Yerkes center keeps the International Studbook for Orangutans. More than 400 orangutans are known to be in captivity in nearly 25 countries. Approximately two-thirds of these animals are in the United States at nearly 40 zoos and institutions. Thirty-eight of these animals are at the Yerkes center where there have been 21 orangutan births since 1967.

The Wisconsin Primate Center has had a major emphasis on psychological development, and this includes the effects of brain lesions, enriched environment, separation and isolation and studies on the effects of prefrontal lesions on both learned and social behavior. This has produced definitive data which will help to define the extent of resulting deficits and the relation of the deficit to the age at which the lesion occurs. The effect of temporal lobe lesions on learned behaviors are also under study. Basic rearing studies in developmental psychology are also based on the behavior of monkeys reared in various conditions of social contact or isolation, including maternal deprivation. The effects of extrasensory stimulation during the first 2 months of life on the rate of development of perceptual and reflexive behaviors and relationships between the estrus cycle and behavior are also studied. Long-term follow-up studies of the effects of early rearing conditions are also underway. Although learning studies with rhesus monkeys using traditional and automated testing equipment are continued, the center has also begun to give increased emphasis to psychoendocrinology in the maturation of learning. Neonatally thyroidectomized rhesus monkeys show no clear indication of deficits when tested on learning tasks, and the effects of prenatal thyroidectomy appear to be more extensive and to include the appearance of abnormal neurologic signs within 4 hr after birth. In the area of psychochemistry, pituitary adrenal studies, and studies of brain nucleic acid metabolism are particularly important, and in the former, the data suggest a close relationship between fear-provoking situations and plasma cortisol concentrations. In the latter, changes in brain RNA resulting from learning have been demonstrated. The pediatric biochemistry and mental retardation unit and the perinatal biology unit are conducting a broad range of studies of nutrition agents which may have a specific fetal teratogenic effect, and definitions of prematurity and normal fetal growth in rhesus monkeys. These studies have the potential for improving the understanding of the development and management of human infants and the effect of maternal disease states.

Members of the departments of obstetrics and gynecology, anesthesiology, and radiation biology are collaborating in development of techniques for *in vivo* studies of fetal development. In the prematurity studies, infant monkeys are delivered by caesarean section at various known gestational stages and are studied closely during the succeeding days and weeks to determine the effect of different variables on their environment. Their potential value in defining techniques for improving care of prematurely born human infants with respiratory distress is being explored. Work in the area of reproductive physiology is a major emphasis of the physiology and endocrinology unit. Ovarian morphology is being coordinated with progesterone levels in the tissues and blood of rhesus monkeys at various stages of pregnancy. The effect of hypophysectomy during early pregnancy, blood lipid fluctuations, and urinary estrogens are being studied as are gonadotropin excretions and steroid metabolism. Seasonal variations in the occurrence of ovulation in rhesus monkeys are being explored. The experimental pathology unit is emphasizing the study of cardiovascular disorders including hypotension, atherosclerosis, and venooclusive disease, and ultrastructural development of the heart, liver, lungs, and kidneys during fetal life. The toxic effects of chlorinated hydrocarbon, such as pesticides and monocrotaline injections, are being investigated.

The Delta Primate Center, which is located near New Orleans, is involved in the studies of the role of primates in human disease, the effects of adverse environment on development, and the characterization of primates as laboratory models. Some of the aspects of these studies involve communicable diseases, environmental health, reproductive physiology, social and behavioral development, and the study of primate disease and health. The principal studies in the area of microbiology are directed toward the relationship of antibodies, immunologically active cells and their relationships to disease or equilibrium. Efforts are being made to find a suitable animal model for the study of infant diarrhea. In the virologic area, the agent causing infectious hepatitis in chimpanzees and patas monkeys and several other viruses are under study. A vaccine for cholera has been developed and is being tested in chimpanzees. Parasitologic studies are being conducted to determine the extent and range of parasitism among primates at the center, the morphologic and life histories of a variety of parasites, and the suitability of various primates as host for species of filariae found in primates which are related to those in man. The environmental health division is concerned particularly with the effects of radiation of the development of the brains of monkeys and chimpanzee fetuses and on the aging process in the nervous system. The effect of intermittent gamma irradiation of the cerebral cortex of rhesus monkeys are studied through biopsy specimens taken from irradiated and control animals at 6-month intervals. In other studies the acute responses to radiation are compared in baboons housed in the ^{60}Co radiation field, receiving daily radiation, and in animals which have not been in the radiation field. Tissues from chimpanzees, baboons, and rhesus, patas, and cynomolgus monkeys are being studied to determine the response of lymphocytes to irradiation.

In the reproductive physiology division, studies of the mechanisms of fertilization are under way. Talapoin monkeys are being used in studies of the time of ovulation and of further development of hormone-releasing intrauterine contraceptive devices. Aspermatogenesis in chimpanzees induced by injection of testicular homogenates has suggested a possible method for fertility control. These studies will be extended to include male rhesus monkeys, and the effects of injection of testicular antigens on the reproductive activity of rhesus females will also be studied. A comparison of reproductive performance of talapoins under three different conditions of husbandry is designed to identify the main factors in good husbandry and suggest how reproductive efficiency can be improved. The program for the study of primate disease and health attempts to assist the health of all animals brought to the center and to determine standards of primate health both in the wild and captive state. Diseases are induced to determine both contagious and zoonotic factors and to gauge the effectiveness of various measures of therapy.

The primate center at Davis, California, which has been called the National Center for Primate Biology, although its emphasis is primarily in primate medicine and husbandry, has a number of peripheral areas of research which are of particular interest, and the need for strong genetics and nutrition research programs has been recognized as the reproductive biology and breeding programs advanced. The primate medicine, virology, and parasitology programs all center on the biologic problems of primates, not only in essential areas of knowledge in themselves, but also as problems which virtually affect the usefulness of primates as research animals. The behavioral biology program now being given increased emphasis is seen as an essential contribution to the total biologic profile necessary for full understanding of nonhuman, and ultimately human, primates. Distinctive studies have been carried out on the methods of caging primates. Cages of several conventional types have been installed. Commercially produced corn cribs for the storage of unshelled corn in the Midwest, have been found to be useful in this particular situation. Several types of stationary and movable windbreaks have been used, and different types of cage flooring are being tested. Although a well-drained cement base with direct connections for a sanitary sewer is desirable, it is also expensive, and sand, various types of rock and gravel, and redwood chips have been used experimentally. A one-half acre enclosure constructed of chain-link fence has been completed and is being evaluated for the effectiveness of various types of outdoor caging. Animals housed in outdoor cages appear in general to be healthier and more active than those housed indoors. Preliminary observation also suggests that conception rates are more favorable in the random-breeding animals housed in outdoor areas than those kept in indoor cages. Future plans include expansion of the outdoor caging area so that more extensive observations can be made on optimum densities and the effects of different types of outdoor housing on colony health behavior patterns and reproduction.

The reproductive physiology program is being developed in the direction of detecting the time of ovulation. The optimal time for breeding is being identified through time-breeding programs which not only make it possible to identify the time of conception with a high degree of accuracy but also increase the rate of conception. In addition, random outdoor-breeding studies are being conducted to determine the age of female monkeys at the first conception under outdoor caging conditions and to relate this to the season of the year and other environmental factors. Superovulation, and twinning, reproductive failures, spontaneous abortion, and artificial insemination are the other areas under investigation. Both the onset of puberty and sexual maturation and the aging processes of the reproductive organs at menopause are also areas of interest. Extensive studies of comparative embryology will identify different stages of development in terms of size, age, and internal and external characteristics. The studies of normal embryology, combined with experiment teratology, will provide much-needed information on developmental processes in a large number of species and will lead to the development of the most suitable primate model for the study of embryopathies in man.

Primate medicine studies at the Davis center cover diverse problems such as the effects of low-dose fractionated X-ray on the ovary of the developing fetus, on parasites, on the control of intestinal diseases, surgical procedures in support of teratology and fetal physiology. Arthritis studies may eventually prove particularly significant since the disease appears to be quite similar to rhematoid arthritis in man. More than 60 simian viruses have been reported in the primatology literature, and one objective of the Davis center's biology unit is to investigate pathogenesis, diagnosis, prevention, and control of the viral diseases of primates which affect the health of research colony animals and, in some cases, personnel who work with them. Development of safe immunizing vaccines is one aspect of this effort. The recently discovered Yaba-like disease and simian hemorrhagic fever, which has wiped out large segments of primate colonies at research centers in several parts of the world within the past few years, are particularly emphasized in the virology studies. The initial objective is to identify the etiologic agent. Another aspect of the virology program is to find out more about the viral flora typically found in wild-caught and laboratory primates of various species. Much more needs to be known about the microbiologic profile, viral, bacterial, parasitic, and fungal, of Old World monkeys. Such knowledge is essential to successful management of research colonies. Diseases caused by animal parasites can be important to health problems in research colony primates. In addition, nonhuman primates share with human primates approximately 200 species of parasites. Available information on these parasites, however, is fragmentary. Therefore, parasitology is an essential aspect of the work at the Davis center. Life cycles and modes of transmission of various parasites are being traced. Both new parasites and new host records have been discovered in several species of primates. Particular attention is given to the danger of transmission of diseases to monkeys housed in outdoor cages so they are exposed to insects and rodents. Not only the diseases but the types of cages used need to be studied if the health of primate colonies is to be maintained. Behavioral and genetic studies also form part of the Davis center activities.

Conclusion

This chapter gives some conception of the wide range of research activities being carried out in the Primate centers and indicates their importance in the health research area. There is little doubt that the centers have proved to be a remarkable investment by the Federal Government and NIH. They are already spinning off findings of great importance in biomedical research and with significant clinical implications and these benefits may be expected to increase continuously in the future.

Subject Index

A

A13 virus, isolation of, 133

AA153 virus, isolation of, 133

Abbreviata, primate research on, 176

Acanthocephala, simian infections by, 193

Achylia, experimental, in monkeys, 324–325

Acid-fast organisms, cell-culture studies on, 31

ACTH, enhancement of viral infections by, 104

Actinomycin C, as immunosuppressant, 207

Adeno-associated satellite viruses, primate research on, 75

Adenomatosis, infectious, of sheep lungs, 129

Adenovirus 2, primate research on, 76

Adenovirus 4, primate research on, 74

Adenovirus 12, primate research on, 74, 76, 125

Adenoviruses
  cell-culture studies on, 32
  primate research on, 72–76, 133
  tumor studies on, 126

Aëdes aegypti, as yellow fever vector, 1

Aëdes dominicii, in yellow fever research, 4

Aëdes aegypti, in yellow fever transmission, 9, 64

Aëdes africanus, in yellow fever cycle, 79

Aëdes leucocelaenus, in studies on yellow fever, 9

Aëdes scapularis, in studies on yellow fever, 9

African green monkey
  breeding of, 487
  cancer studies on, 214, 220
  in cardiovascular research, 244
  virus research on, 68, 71, 73, 74, 122–125, 129, 131, 133
    adenoviruses, 73, 74, 76
    arboviruses, 82, 83–84
    herpesviruses, 89, 90, 91
    myxoviruses, 94, 97, 100
    papovaviruses, 100–103
    picornaviruses, 103, 104, 110
    poxviruses, 113
    reoviruses, 117

African mustache monkey, virus research on, 104

African sleeping sickness, cell-culture studies of, 33

1211 agent, from poxviruses, 114

Aging, *see also* Degenerative diseases
  in nonhuman primates, 286–287, 303

Air pollution, degenerative disease and, 292

Alastrim virus
  classification of, 111
  primate research on, 66, 111

Aleutian mink disease, 129

Almirante, Panama, yellow fever outbreak in, 11–12

515

QL
737   Bourne, Geoffrey Howard, 1909-
P9   Nonhuman primates and medical research. Ed.
B77n   by Geoffrey H. Bourne.   New York, Academic
         Press, 1973.
              xvi, 537p. illus. 24cm. index.

         Includes bibliographies.

355980

1.Primates as laboratory animals.   2.Primates-Diseases.
I.Title.